# *Theory of Differentiation*

## *A Unified Theory of Differentiation Via New Derivate Theorems and New Derivatives*

**CANADIAN MATHEMATICAL SOCIETY SERIES OF MONOGRAPHS AND ADVANCED TEXTS**

*Monographies et Études de la Société Mathématique du Canada*

---



A complete list of titles in this series appears at the end of this volume.
*Tous les titres de cette collection sont énumérés à la fin du volume.*

# *Theory of Differentiation*

## *A Unified Theory of Differentiation Via New Derivate Theorems and New Derivatives*

KRISHNA M. GARG

*University of Alberta*

A Wiley-Interscience Publication

JOHN WILEY & SONS, INC.

New York • Chichester • Weinheim • Brisbane • Singapore • Toronto

This book is printed on acid-free paper. ♾



***Library of Congress Cataloging in Publication Data:***

Garg, Krishna, M.
Theory of Differentiation / Krishna M. Garg.
p. cm. — (Canadian Mathematical Society series of monographs and advanced texts)
"A Wiley-Interscience publication."
Includes index.
ISBN 0-471-25387-1 (cloth : alk. paper)
1. Differential calculus. I. Title. II. Series.
QA306.G37 1998
515′.33—dc21 97-51534
CIP

10 9 8 7 6 5 4 3 2 1

# *Preface*

By "theory of differentiation" we mean the theory of all aspects of various kinds of derivates and derivatives, as well as the theory of various Perron and Denjoy-Perron integrals which in turn are related to various derivates and derivatives. It is, however, the second title which describes the actual theme and the nature of the present work; the reason behind it will be apparent from what follows. By the term "derivate theorems" in this title we mean, on the other hand, theorems on unilateral (or Dini) derivates.

In the present work we deal with some outstanding basic problems of the theory of differentiation, which in turn leads to a unified theory of various aspects of differentiation theory.

First, there is at present no cohesive theory of unilateral derivates; the numerous results on these derivates have been obtained by various mathematicians over the last 120 years by ad hoc methods, and so they are found to be scattered all over the literature.

Therefore, we present here a unified theory of unilateral derivates. This is achieved with the help of two new fundamental theorems on unilateral derivates. From these two theorems we deduce systematically, step by step, some new theorems along with the various known results on unilateral derivates. Many of the known results are strengthened in this process.

Second, after the discovery of Weierstrass, it became well known that there are continuous functions that are not derivable at any point. The same will be seen here to hold in terms of various generalized derivatives that are known.

Therefore, we present here a theory of some new derivatives in terms of which every continuous function is derivable at a set of points which has cardinality $c$ in every interval. Some of the new derivatives are also defined for real-valued functions on normed vector spaces. This theory plays a central role in the present work and unifies in a sense the various aspects of differentiation theory. Many of the known results in differentiation theory are found to hold in terms of new derivatives without any derivability hypothesis. For example, (a) the mean value theorems, (b) the monotonicity theorems, (c) the commonly known first and second derivative tests for points of extrema, (d) the concavity test, and (e) the l'Hôpital's rule for limits. This fact has indeed some far-reaching consequences as will be seen here. In particular, the unification in the theory of differentiation is achieved through the new derivatives only due to this fact. The various known results on ordinary derivative follow of course from the results on new derivatives because these derivatives generalize the ordinary derivative.

We include here some results on the new derivatives of continuous nonderivable functions—that is, functions that are not derivable in the ordinary sense at any point, which follow from the theory of new derivatives. These results clarify some of the mystery of nonderivable functions, and they reveal some order in the paths of one-dimensional Brownian motion. Also, since many of the results of the theory of new derivatives do not involve any derivability hypothesis, it is clear that this theory is directly related to nonsmooth analysis.

Further, to solve the problem of reconstruction of a primitive function from its new derivatives, we define here new Perron integrals corresponding to each new derivative and obtain the fundamental theorem establishing the relation between the two. Also, we obtain descriptive definitions of the new Perron integrals in terms of new Denjoy-Perron integrals and establish their equivalence with the original integrals.

Quoting the illustrious couple W. H. Young and G. C. Young, A. N. Singh wrote the following in the concluding pages of his well known monograph on nondifferentiable functions:

> It has been suggested by W. H. Young and G. C. Young that efforts should be made to evolve a definition of differentiation, according to which all continuous functions would be differentiable, at least almost everywhere, and correspondingly to find a definition of integration that would always lead back from the derivate to the primitive function.

The theory of new derivatives and the theory of corresponding Perron integrals, presented here, deal exactly with these two problems posed by W. H. Young and G. C. Young, except for the fact that the term "almost everywhere" does not hold in general for the new derivatives considered here.

Next, we present several direct applications of the theory of new derivatives and of the theory of unilateral derivates. Namely, we obtain results on (a) the derivates and derivability of symmetric, smooth and quasismooth functions, (b) the differential structure of a function at its nonmonotonicity points and of nowhere monotone and singular functions, (c) the nature of new derivatives of nonderivable functions and of Brownian paths, and on (d) the properties of strong derivates and strong derivative introduced by Peano. Many of the known results in these directions are strengthened in this process, and also several new results are obtained. For example, the result obtained on the ordinary derivability of quasismooth functions is altogether new.

The applications that are discussed from now on are, on the other hand, more of a theoretical nature since they deal with unified axiomatic theories.

It has been well known for some time that the various known generalized derivatives behave somewhat differently in terms of their results, and it has been an open problem to develop a theory which explains these differences.

Therefore, we present here unified axiomatic theories of three kinds of generalized derivatives which explain these differences. In the first two theories we deal with various known generalized derivatives and generalized symmetric derivatives, and in the third theory we deal with generalized new derivatives which are obtained on combining the notions of new derivatives with the ones of various known generalized derivatives.

The first of these axiomatic theories includes in particular most of the known results on ordinary derivative, and the second theory includes similarly the various known results along with several new results on ordinary symmetric derivative. Further, in the case of various known generalized derivatives and generalized symmetric derivatives, several new results follow from the first two theories whereas the existing results are strengthened.

Most of the results of the three axiomatic theories are deduced from the results of the theory of new derivatives. This indeed is somewhat surprising since the known generalized derivatives are found to be not comparable with new derivatives in general. However, the various results on generalized derivatives are still deducible from the results on new derivatives due to the compatibility of the two derivatives.

The results of each of the three axiomatic theories are obtained in terms of various admissibility and regularity hypotheses on the corresponding generalized derivates, and we determine here the status of admissibility and regularity for several known generalized derivates. This in turn explains the differences in the behavior of various known generalized derivatives. Further, to determine which of the results hold for any new generalized derivative, it is enough to find out its status in terms of admissibility and regularity, which can be determined quickly by the methods used here.

We further present here unified theories of three other aspects of generalized derivates and derivatives.

First, using the axiomatic models of generalized derivates developed for the above purposes, we present a unified theory of generalized derivates of typical continuous functions. Here a proposition is said to hold for a typical continuous function if it holds for a residual set of functions in the Banach space $\mathbb{C}$ of continuous real-valued functions on the interval $[0, 1]$ under the uniform norm. Relative to various generalized derivates we obtain here the nonderivability of typical continuous functions, and we investigate the nature of generalized new derivatives of such functions whose existence is assured due to the continuity of the function.

Next, we present a unified theory of generalized smooth functions to obtain large classes of continuous functions which are derivable in any generalized sense at least at some points.

Finally, to solve the problem of reconstruction of a primitive function from its various kinds of generalized derivatives, we present an axiomatic theory of generalized Perron integrals which applies simultaneously to various generalized derivatives, generalized symmetric derivatives, and generalized new derivatives. Many of the results of the theory of new Perron integrals mentioned earlier are deduced, in fact, from the results of this general theory.

In conclusion, it is hoped that the above theories bring a little more clarity, order, and harmony in the theory of differentiation. Also, the above theories lead to a totally new perspective in the theory of differentiation, and apart from the progress made in the theory of differentiation in the present work, the above theories should lead to much more progress in the future

We wish to add here our thanks to Mrs. Elizabeth Leonard and Mrs. Vivian Spak for their patient and proficient typing of this monograph. We are also grateful to Professor Jonathan Borwein and Mr. Steve Quigley for their interest in this monograph, and thanks are due to Mrs. Jessica Downey and Ms. Lisa Van Horn for their cooperation in its publication.

KRISHNA M. GARG

*Edmonton, Alberta, Canada*
*July 1998*

# Contents

# General Introduction

The main object of the present work is to present a unified theory of unilateral derivates and a theory of some new derivatives. These two theories lead in turn to a unification of various aspects of the theory of differentiation.

What distinguishes the new derivatives from the various known derivatives is the fact that every continuous real-valued function on an interval is derivable in each of the new senses at uncountably many points in every interval. This is not true for any known generalized derivative, as will be seen here. Many of the known results in differentiation theory, like the mean value theorems and monotonicity theorems, are found to hold in terms of new derivatives without any derivability hypothesis.

The present work is divided into six parts, and so we discuss below the essential features of these six parts separately. The results of each of these parts will be discussed later in greater detail in the introduction of that part. We present on the other hand in Chapter 1 several definitions and notations which are used throughout the present work.

First, in Part I, we present a unified theory of unilateral (or Dini) derivates. This is achieved with the help of two new fundamental theorems on unilateral derivates, which we call the First and Second Derivate Theorems. From these two theorems we deduce in this part some basic results of G. C. Young, W. H. Young, A. Denjoy, and A. P. Morse on unilateral derivates, and we obtain the best possible results on the Baire class of unilateral derivates. Also, similar results are obtained on the Baire class of set-valued medians associated with unilateral derivates.

The other known results on unilateral derivates, like the classical Denjoy-Young-Saks theorem and its analogue, are deduced from these two derivate theorems in Parts II and III due to their close connection with the new derivatives considered in those parts. Many of these results are strengthened in this process. As these known results were obtained earlier by ad hoc methods, the present theory brings out their inter-relations and provides a linearization of the results on unilateral derivates.

In the next two parts we present the theory of new derivatives. First, in Part II, we introduce various notions of new derivatives, and we obtain several derivability theorems to establish the existence of new derivatives at some points for every continuous function on an interval. Also, we investigate in this part various rules of Calculus for the new derivatives.

First, suppose $f: X \to \mathbb{R}$, where $X \subset \mathbb{R}$. We define in this case the notions of lower and upper derivatives of $f$ in terms of unilateral derivates. Further, we call $f$ semiderivable at a point $x$ if it is either lower or upper derivable at $x$, and if so, then the lower or upper derivative of $f$ at $x$ is called the semiderivative of $f$ at $x$. Further, we define here a notion of weak derivative which unifies the lower and upper derivatives with right and left derivatives. These new derivatives turn out to be set-valued in general, and in the present case they are found to be single-valued at all but a countable set of points. We further define here "normal" versions of lower, upper and semi- derivatives because these derivatives are found to be so at most of the points. These normal derivatives are single-valued by definition. Also, we call here $f$ differentiable at a point $x$ in any sense if that derivative of $f$ at $x$ includes a finite element. We investigate here the weaker forms of continuity that hold under various new differentiabilities.

We also define the notions of lower, upper and semi- derivatives in the case when $X$ is a normed vector space, and we obtain some of the results in this case. However, the main focus of the present work is on functions defined on a subset of $\mathbb{R}$.

Next, we describe some of the derivability theorems obtained in this part which play a critical role in the present theory. In the First Derivability Theorem we prove the lower differentiability of every $LSC$ function on an interval at a dense set of points. Then, in the Second Derivability Theorem we establish the lower and upper differentiabilities at almost all the points for two classes of functions $GL\text{-}BV_*$ and $GU\text{-}BV_*$ which are similar to $GBV_*$ functions. Such functions need not be derivable however in the ordinary sense at any point. Next, we deduce from the Second Derivability Theorem with the help of the First Derivate Theorem the classical Denjoy-Young-Saks theorem in terms of normal lower and upper differentiabilities. Also, from the Second Derivability Theorem we deduce with the help of the Second Derivate Theorem an analogue of Denjoy-Young-Saks theorem which we call here the Normal Level Structure Theorem. This theorem provides the differential structure of a function $f: X \to \mathbb{R}$ at the points of almost all of its level sets—that is, at the points of the level set $f^{-1}(y)$ for almost every $y$.

Next, we deduce from the Normal Level Structure Theorem the Third Derivability Theorem according to which every continuous function is normally lower and upper derivable at two dense sets of points in almost all of its level sets. Further, we introduce here two properties $(S_1)$ and $(S_2)$ similar to Banach's properties $(T_1)$ and $(T_2)$, and we prove that every (weakly) Darboux function with property $(S_1)$ or $(S_2)$ is derivable or normally semiderivable respectively at all the points of almost all of its level sets. We deduce here from the Normal Level Structure Theorem and the last theorem several known results, including (a) two basic theorems of Saks on unilateral derivates and (b) several known theorems of Banach and Saks on the ordinary derivability of continuous functions under the properties $(T_1)$, $(T_2)$, $(N)$, and $(S)$. Several of these results are strengthened or generalized in this process.

Finally, we investigate in this part the various rules of Calculus for new derivatives. To be specific, the lower and normal lower (or upper) derivatives are found to be additive and positively linear. Also, in terms of a certain notion of compatibility of two semiderivatives, a theorem is obtained on the linearity of semiderivatives. Similar product and quotient rules are obtained for semiderivatives and normal semiderivatives in terms of their compatibility. Further, we obtain various chain rules for new derivatives. In the case of weak derivative, however, only some conditional forms of additivity and product and quotient rules are found to hold. Further, in the case of functions defined on normed vector spaces, we obtain rules of Calculus for semiderivatives similar to above. We include here some comments on the developments in nonsmooth analysis that are related with new derivatives.

Next, in Part III, we deal with the remaining aspects of the theory of new derivatives. First, we obtain here two mean value theorems in terms of lower and semi- derivatives without any derivability hypothesis. Then we deduce from these two theorems a Cauchy's mean value theorem and l'Hôpital's rule in terms of lower and semi- derivatives without any derivability hypothesis on the function in the numerator. Also, Taylor's formulae are obtained for lower and weak derivatives, and results on term-by-term differentiation of sequences and series of functions are obtained for weak and semi- derivatives.

Further, we obtain here three monotonicity theorems in terms of new derivatives without any derivability hypothesis, and we deduce from one of them (a) generalized versions of the commonly known first and second derivative tests for points of extrema and (b) the concavity test, in terms of new derivatives without any derivability hypothesis. Also, we obtain here some median theorems dealing with the properties of various medians of a function $f$ at its nonmonotonicity points (i.e., points in no neighborhood of which $f$ is nondecreasing), and we deduce from them the Goldowski-Tonelli theorem in terms of new derivatives.

Next, we obtain the Borel class, measurability, and the Zahorski property of various new derivability sets. Also, we investigate here the Baire class and measurability of various new derivatives and related medians as multifunctions. Further, we obtain the Darboux property of new derivatives without any derivability hypothesis. Also, in the case of nonangular functions that are lower or weakly derivable nearly everywhere, we obtain the Denjoy property of the new and ordinary derivatives.

Finally, in this part we deal with the problem of reconstruction of a primitive function from its various new derivatives. For this purpose we define here lower, upper, and weak Perron integrals which provide a solution of this problem for lower, upper, and weak derivatives, respectively. We obtain here various results on these new Perron integrals similar to the ones known for ordinary Perron integral. Further, we present descriptive definitions of these new Perron integrals and establish their equivalence with the original integrals.

Next, in Part IV, we deal with several direct applications of the results of the theory of new derivatives obtained in Parts II and III, and of the theory of unilateral derivates of Part I. First, we obtain some results on the derivates and derivability of symmetric, smooth, and quasismooth functions, or their lower and upper versions. For example, we obtain some results on the existence of points of ordinary derivability for lower quasismooth functions which are altogether new.

Further, we obtain some results on the nature of derivates, medians, and various derivatives of a function at its nonmonotonicity points. This in turn leads to various results on the differential structure of nowhere monotone and strongly nowhere monotone functions which generalize some of the known results. Also, we deduce from these results and two derivability theorems of Part II some results on the nature of new derivatives of continuous nonderivable functions (i.e., functions which are not derivable in the ordinary sense at any point). These results seem to clarify some of the mystery of nonderivable functions, and they reveal some order in the paths of one-dimensional Brownian motion.

Further, following a suggestion made by Peano, we investigate here the properties of strong derivates and strong derivative introduced by him. They are found to have stronger properties than those of other derivates and derivatives. Also, the values of strong derivates of a function at any point are found to determine some properties of the function in a neighborhood of that point, which does not hold for any other known derivates. Further, we obtain several characterizations of strong derivability, and we obtain stronger versions of the two classical theorems of Dini on unilateral derivates in terms of strong derivates.

Next, in Part V, we present unified axiomatic theories of three kinds of generalized derivatives. In the first two theories we deal with various known generalized derivatives and generalized symmetric derivatives, and in the third theory we deal with generalized new derivatives which are obtained on combining the notions of new derivatives with the ones of various known generalized derivatives.

It has been an open problem for some time to develop a theory which explains the existing differences in the behavior of various known generalized derivatives. The first two axiomatic theories explain the reasons behind these differences. The first of these axiomatic theories includes most of the known results on ordinary derivative, and the second theory includes the known results along with several new results on ordinary symmetric derivative. Further, in the case of various known generalized derivatives and generalized symmetric derivatives, several new results follow from the first two theories whereas the existing results are strengthened.

Most of the results of these axiomatic theories are deduced from the results of the theory of new derivatives obtained in Parts II and III. This is indeed somewhat surprising since the known generalized derivatives are found to be not comparable with new derivatives in general. However, the various results on generalized derivatives can still be deduced from the results on new derivatives due to the compatibility of the two derivatives. We refer to the Introduction to Part V for some idea about the nature and scope of the results of these three axiomatic theories.

Finally, in Part VI, we present three independent unified theories dealing with some other aspects of generalized differentiation. The axiomatic models of generalized limits and derivates developed in Part V are utilized in these three unified theories.

In the first theory we deal with various properties of generalized derivates of typical continuous functions. Here a proposition is said to hold for a typical continuous function if it holds for a residual set of functions in the Banach space of continuous real valued functions on the interval [0, 1] under the uniform norm. We prove here, in particular, that relative to various known generalized derivates typical continuous functions are nonangular, and not differentiable at any point. Also, we investigate here the nature of generalized new derivatives of such functions whose existence is assured due to the continuity of the function.

In the second unified theory we define and investigate generalized smooth functions to obtain classes of continuous functions which are differentiable in any generalized sense at least at some points. Also, we obtain here the mean value and Darboux properties of generalized derivatives of generalized smooth functions without any derivability hypothesis.

Finally, to solve the problem of reconstruction of a primitive function from its various generalized derivatives, we present in the third theory a unified axiomatic theory of generalized Perron integrals which applies simultaneously to all three kinds of generalized derivatives considered in Part V.

We conclude this general introduction with some general comments that apply to the present entire work.

As for the origins of the present work, it may be asked what led us to the notions of new derivatives. It was indeed our earlier investigations of the properties of nowhere monotone functions in [G2; G3; G4] , the analogue of Denjoy-Young-Saks theorem obtained in [G1], and its applications in [G5; G6; G7] that led us to these notions.*

*The notions of the new derivatives were first announced in 1976 in [G12] with some initial results of Parts II and III. Later, in 1981, more detailed versions of these results were presented at a symposium at Syracuse which appeared in its proceedings in [G13]. The two fundamental theorems of Part I on unilateral derivates were presented on the other hand, in terms of different nomenclature, at the International Congress of Mathematicians held at Warsaw in 1983. However, due to the constantly growing nature of the present work, it took so long to finalize it. Indeed, the works to appear in [G18] and [G19] are also closely connected with the work presented here.

We have used in the present work several weaker forms of continuity, including internal properties, to obtain the results in maximum generality. However, the reader may replace them by continuity at the first reading, and he or she may return to this technical aspect later.

In the present work we present several theorems on some common themes. To eliminate confusion we have used titles for such theorems—for example, as mentioned above, First, Second, etc., Derivate Theorems. We will deal with the following theorems in a similar manner: (i) Derivability Theorems, (ii) Median Theorems, (iii) Mean Value Theorems, (iv) Cauchy's Mean Value Theorems, (v) L'Hôpital's Rules, (vi) Monotonicity Theorems, (vii) Goldowski-Tonelli Theorems, and some other kinds of theorems. This will bring out more clarity and will make it easier to compare these theorems. By Goldowski-Tonelli Theorems we mean new versions of this well-known monotonicity theorem that are obtained here in terms of various derivatives.

Throughout the work, we indicate open problems at the end of various sections which arise naturally from the developments of the section but were either not attempted or remained unresolved.

As the domain of the function varies frequently in the present work, we introduce here some conventions. We will assume throughout the present work, unless stated otherwise, that $X$ is an arbitrary subset of $\mathbb{R}$ and that $I$ is a subinterval of $\mathbb{R}$. Further, $f$ will be assumed, unless stated otherwise, to be a real-valued function defined on $I$. However, $X$ is also used frequently as the domain of $f$, and we state in that case that "$f: X \to \mathbb{R}$," where $X$ will be understood to be an arbitrary subset of $\mathbb{R}$.

Finally, when some assumptions are used throughout some section, we will indicate them immediately after the introduction of that section. Similarly, when some assumptions are used throughout some chapter, they will be indicated at the end of its introduction.

Remark. We include here a few comments on the second title of the present work to explain what we mean by it and to eliminate any misunderstanding about it. As is clear from the above discussions, this title indeed captures the theme and the actual nature of the present work. Since the various results on unilateral derivates are deduced here from the new derivate theorems of Part I, what we need to consider here is only the role of the first three parts I to III in the last three parts IV to VI. There is of course a certain amount of groundwork involved in the last three parts to prepare the base for the application of the results of the first three parts, and this groundwork is independent of the first three parts which is only to be expected.

Apart from the necessary groundwork, all the results of Parts IV to VI are deduced from the results of the first three parts with the following two exceptions in Part VI. This part is of course closely related with earlier parts and is an integral part of the theory of differentiation presented here. Firstly, the results of the first four sections of Chapter 19 are obtained directly, the hard way, from the definitions of the various derivates considered there. However, the results of the last section of this chapter are obtained from the results of the first three parts and from some results that are deduced from them in Chapter 13.

Secondly, since most of the results of Sections 11.2 and 11.3 of Part III are proved only in a more general setting in Chapter 21 for generalized new Perron integrals, this chapter should indeed be considered to be a part of Chapter 11 and thus, to be a part of the theory of new derivatives.

# *PART I*

# A UNIFIED THEORY OF UNILATERAL DERIVATES

**Introduction to Part I.** In this part we present a unified theory of unilateral derivates. This is achieved with the help of two new fundamental theorems on unilateral derivates which we call simply the First and Second Derivate Theorems. From these two theorems we deduce in this part some other new derivate theorems along with some basic known results on unilateral derivates; some of these results are strengthened in this process. Other known results on unilateral derivates like the classical Denjoy-Young-Saks theorem and its analogue are deduced from the First and Second Derivate Theorems in Parts II and III due to their close connection with the new derivatives considered in those parts.

Since the various known results on unilateral derivates were obtained earlier by ad hoc methods, the present treatment provides a synthesis of these results and brings out their interrelations. Thus this leads to a linearization of the results on unilateral derivates, putting them in the form of a unified theory.

Next we deal with the organization of the results of this part.

First, in Chapter 1, we present some definitions, notations, and preliminary results which are used throughout the present work. In particular, we define in this chapter several weaker forms of continuity, and generalized and inherent versions of properties of functions. Also, we define here lower and upper Baire classes of functions and multifunctions, and include some results on these classes.

Next, in Chapter 2, we obtain the First and Second Derivate Theorems. These two theorems are obtained in terms of unilateral monotonicity and Lipschitz properties of functions on a set which also depend on the values of the function outside the set like the properties $VB_*$ and $AC_*$. Also, we deduce from these two theorems the Third and Fourth Derivate Theorems, which deal with properties of functions on some portion of their domain.

Further, we deduce from the First and Second Derivate Theorems extended versions of two theorems of G.C. Young and W. H. Young, and from the Third Derivate Theorem we deduce strengthened versions of some basic theorems of Denjoy and W. H. Young on unilateral derivates.

Finally, in Chapter 3, we first deduce from the First Derivate Theorem a general theorem on unilateral derivates which we call the Fifth Derivate Theorem. Then we deduce from this theorem two theorems on Baire class and measurability of unilateral derivates which are found to be the best possible in their direction. These theorems strengthen several known results in this direction due to Young, Sierpiński, Banach, and Mišík. Also, we obtain similar results on set-valued medians associated with unilateral derivates. Further, we deduce from the First and Third Derivate Theorems a refined version of a theorem of A. P. Morse on unilateral derivates; and in the First Monotonicity Theorem we obtain a similar refined version of his monotonicity theorem in terms of unilateral derivates.

*CHAPTER 1*

# Definitions, Notations, and Preliminary Results

In this chapter we present some definitions, notations, and preliminary results that are used throughout the present work.

First, in §1.1, we present some definitions and notations. Then, in §1.2, several weaker forms of continuity are presented which will be used throughout. In §1.3 we deal with generalized and inherent versions of properties of functions which are also used frequently.

Next, in §1.4, we define lower and upper Baire classes of functions which are used later in Chapter 3 to obtain the best possible results on the Baire class of unilateral derivates. We also include in this section some results on these classes of functions and on the notion of relative measurability. Finally, in §1.5, we define lower and upper Baire classes of multifunctions, or set-valued functions, which are also needed here because the new derivatives are in general set-valued.

**§1.1 Definitions and notations.** In this section we present some nomenclature and preliminary results that are used throughout the present work.

We will employ $\mathbb{R}$ to denote the set of all real numbers, $\mathbb{R}_+$ and $\mathbb{R}_-$ to denote the sets of positive and negative real numbers, respectively, and employ $\overline{\mathbb{R}}$ to denote the set of extended real numbers, namely $[-\infty, +\infty]$. Also, we will use $\mathbb{N}$ to denote the set of all positive integers. Further, for each $a, b \in \overline{\mathbb{R}}$, we will use as usual $a \wedge b$ and $a \vee b$ to denote the minimum and maximum, respectively, of $a$ and $b$.

Further, unless stated otherwise, we will use $X$ to denote an arbitrary subset of $\mathbb{R}$, and we will use $I$ to denote a subinterval of $\mathbb{R}$.

Given $X \subset \mathbb{R}$, we will use $X^d$ to denote the derived set of $X$ in $\mathbb{R}$, and we will use $X^{d-}$, $X^{d+}$, and $X^{db}$ to denote the sets of left, right, and bilateral limit points, respectively, of $X$. Further, we will use $X', X^-, X^+$, and $X^b$ to denote the sets of points in $X^d, X^{d-}, X^{d+}$, and $X^{db}$, respectively, that are in $X$.

For a set $E \subset X$, we will use however $\overline{E}$ to denote the closure of $E$ in $X$, and we will use $E^{d-}, E^{d+}$, and $E^{db}$ to denote the sets of left, right, and bilateral limit points, respectively, of $E$ that are in $X$. The reason for defining these derived sets differently will be clear from the context.

A set $E$ will be called *countable* if it is either finite or countably infinite. Now, suppose $E \subset X \subset \mathbb{R}$. Then $E$ is said to be *meager* in $X$ if it is of the first category in $X$, and when $X \sim E$ is meager in $X$, $E$ will be said to be *residual* in $X$. A nonempty subset of $X$ will be called a *portion* of $X$ if it is of the form $X \cap I$, where $I$ is an open interval in $\mathbb{R}$.

Further, when $X \subset X^{db}$, the set $X$ is called *bilaterally dense in itself*, and when each portion of $X$ includes a set of positive measure, $X$ is called *metrically dense in itself*. Also, the set $E$ will be said to be *c-dense* in $X$ if for every portion $X_0$ of $X$ the set $E \cap X_0$ has cardinality $c$.

Next, if any pointwise proposition $P$ holds at a residual set of points in $X$, then $P$ will be said to hold *residually everywhere* (or r.e.) in $X$; and if $P$ holds at all but a countable set of points in $X$, then following Saks [S6] we will say $P$ holds *nearly everywhere* (or n.e.) in $X$. Also, the abbreviation a.e. will be used for *almost everywhere* in the measure-theoretic sense.

Further, if $Q$ is some other pointwise proposition, and the set of points where $Q$ holds but $P$ does not hold is countable, of measure zero, or meager, then $P$ will be said simply to hold at *nearly*, *almost*, or *residually all the points*, respectively, *where $Q$ holds.*

Next, we will use $\mathbb{C}$ throughout to denote the Banach space of all continuous real-valued functions on the interval $[0, 1]$ under the uniform norm. Now, any proposition will be said to hold for a *typical continuous function* if it holds for a residual set of functions in $\mathbb{C}$.

Next, given $f: X \to \mathbb{R}$ and $c \in \mathbb{R}$, we will use $f_{+c}$ to denote the function

$$f_{+c}(x) = f(x) + cx, \qquad x \in X. \tag{1}$$

The notation $f_{-c}$ will in turn be used to denote the function $f_{+(-c)}$.

Further, when $x \in X^d$, we will use $\underline{f}(x)$ and $\bar{f}(x)$ to denote the bilateral lower and upper limits respectively of $f$ at $x$. Similarly, when $x \in X^{d-}$, we will use $\underline{f}(x - 0)$ and $\bar{f}(x - 0)$ to denote the left lower and upper limits of $f$ at $x$, and when these limits are equal, we will use $f(x - 0)$ to denote their common value which is called the *left limit* of $f$ at $x$. Note that the limit $f(x - 0)$ may be finite or infinite. When $x \in X^{d+}$, the right limits $\underline{f}(x + 0)$, $\bar{f}(x + 0)$ and $f(x + 0)$ of $f$ at $x$ are defined similarly.

We include here a well-known theorem of W. H. Young [Y2] on the symmetry of unilateral limits which will be used repeatedly. Its proof given by Saks [S8, p. 261] for functions on an interval can be extended without difficulty to functions with an arbitrary domain $X \subset \mathbb{R}$.

1.1.1. THEOREM (W. H. YOUNG). *For every function $f: X \to \mathbb{R}$, the following relations hold at all but a countable set of points in $X$:*

$$\underline{f}(x-0) = \underline{f}(x+0) \leq f(x) \leq \bar{f}(x-0) = \bar{f}(x+0).$$

Further, given $f: X \to \mathbb{R}$, $x, y \in X$, $x \neq y$, we will use $Qf(x, y)$ to denote the following difference quotient, or incrementary ratio, of $f$ on $[x, y]$ or $[y, x]$:

$$Qf(x, y) = \{f(y) - f(x)\}/(y - x). \tag{2}$$

Moreover, we will use $\underline{D}f$ and $\overline{D}f$ to denote the bilateral lower and upper derivates respectively of $f$, and similarly $D_-f$, $D^-f$, $D_+f$ and $D^+f$ to denote the left and right lower and upper derivates of $f$ (see [S8, p. 108] for definitions). The last four derivates were introduced by Dini [Di] and are known as Dini or *unilateral derivates*. It is of course when these derivates are equal at a point $x$ that their common value is called the *derivative* of $f$ at $x$ and is denoted by $f'(x)$.

Given $f: X \to \mathbb{R}$, where $X \subset \mathbb{R}$, we associate with $f$ another function $f_-$ on $-X$ which is defined as follows:

$$f_-(x) = f(-x), \qquad x \in -X. \tag{3}$$

Next, we deal in the following lemma with the relations between $D_+f$ and the other three unilateral derivates of $f$. It is thus enough to consider this single unilateral derivate. The lemma is of course well known, and its proof is quite straightforward. It will be used frequently without reference.

1.1.2. LEMMA. *Let $f: X \to \mathbb{R}$. Then*
(a) *if $x \in X^+$, then $D_+(-f)(x) = -D^+f(x)$;*
(b) *if $x \in X^-$, then $D_+(f_-)(-x) = -D^-f(x)$; and*
(c) *if $x \in X^-$, then $D_+(-f_-)(-x) = D_-f(x)$.*

Finally, given $f: X \to \mathbb{R}$, a point $x \in X$ was called in [G11, p. 296} a *knot point* of $f$ if $\underline{D}f(x) = -\infty$ and $\overline{D}f(x) = +\infty$. We will further call $x$ a *left knot point* of $f$ if $D_-f(x) = -\infty$ and $D^-f(x) = +\infty$; a *right knot point* of $f$ if $D_+f(x) = -\infty$ and $D^+f(x) = +\infty$; and a *bilateral knot point* of $f$ if it is simultaneously a left and right knot point of $f$.

Moreover, when $x$ is a knot, left knot, right knot, or bilateral knot point of $f$, $f$ will be said to be *knotted*, *left knotted*, *right knotted*, or *biknotted*, respectively, at $x$. The sets of points where $f$ is knotted, left knotted, right knotted or biknotted will in turn be called the (ordinary) *knot set*, *left knot set*, *right knot set*, or *bilateral knot set*, respectively, of $f$ and will be denoted by $K(f)$, $K_-(f)$, $K_+(f)$, or $K_b(f)$, respectively.

Clearly, a knot point $x$ of $f$ is a point where $f$ lacks (ordinary) derivability in a sense to the extreme; that is, instead of the derivates $\underline{D}f(x)$ and $\overline{D}f(x)$ being equal, they are as far apart as possible. Similarly, a left or right knot point of $f$ is a point where $f$ lacks left or right derivability respectively to the extreme.

**§1.2 Weaker forms of continuity and internal properties.** In this section we present, for the sake of coherence, several weaker forms of continuity (including internal properties) which are used throughout the present work. We will present here only local versions of these properties; a function will be understood to have any of these properties globally when it has that property at each point of its domain.

We assume here throughout that $f: X \to \mathbb{R}$ and $x \in X$.

**A. Quasicontinuities.** Following Kempisty [Ke], we call $f$ *quasicontinuous* at $x$ if for each pair of positive numbers $\varepsilon$ and $\delta$ there is a portion $X_0$ of $X \cap (x - \delta, x + \delta)$ such that

$$|f(y) - f(x)| < \varepsilon \quad \text{whenever } y \in X_0.$$

Further, generalizing the notions of lower and upper semicontinuities (or *LSC* and *USC*), we will call $f$ *quasi-lower semicontinuous* or *quasi-upper semicontinuous* (or *QLSC* or *QUSC*, respectively) at $x$ if for each pair of positive numbers $\varepsilon$ and $\delta$ there is a portion $X_0$ of $X \cap (x - \delta, x + \delta)$ such that

$$f(y) > f(x) - \epsilon \quad \text{or} \quad f(y) < f(x) + \epsilon, \quad \text{respectively, whenever } y \in X_0.$$

Also, $f$ will be called *quasi-bi-semicontinuous* (or *QBSC*) at $x$ if it is simultaneously *QLSC* and *QUSC*, at $x$.

It should be noted here that the property *QBSC* is in fact weaker than quasicontinuity. For example, if $X$ is bilaterally dense in itself (i.e., $X \subset X^{db}$), then every monotone function $f$ on $X$ is *QBSC*, but it is not quasicontinuous at any point $x \in X$ unless it is continuous from at least one side at $x$. It is further interesting to note that if a *QBSC* function $f$ is bounded, nondecreasing, or constant on a dense set of points in its domain $X$, then it is so on the whole of $X$.

**B. Lower and upper continuities.** We now present the notions of lower and upper continuities used earlier in [G16]. We will call $f$ *lower* *continuous* (or *LC*) *from left* or *right* at $x$ if

$$(1) \qquad \bar{f}(x-0) \le f(x) \quad \text{or} \quad f(x) \le \underline{f}(x+0), \quad \text{respectively,}$$

provided $x \in X^-$ or $X^+$, respectively. Similarly, $f$ will be called *upper continuous* (or *UC*) *from left* or *right* at $x$ if

$$(2) \qquad \underline{f}(x-0) \ge f(x) \quad \text{or} \quad f(x) \ge \bar{f}(x+0), \quad \text{respectively,}$$

provided $x \in X^-$ or $X^+$, respectively.

We will use $LC_-$ and $LC_+$ to denote $LC$ from left and right, respectively, and use $UC_-$ and $UC_+$ to denote $UC$ from left and right, respectively. Further, when $f$ is $LC$ or $UC$ from both sides at $x$, it will be called $LC$ or $UC$, respectively, at $x$.

It should be noted here that $f$ is $UC$ from left or right at $x$ iff $-f$ is $LC$ from that side at $x$, and that $f$ is continuous at $x$ iff it is $LC$ and $UC$ at $x$. Further, it is easy to see that the notions of $LC$ and $UC$ are not comparable in general with those of $LSC$ and $USC$, respectively.

Also, when $f$ is nondecreasing or nonincreasing, it is clear that $f$ is automatically $LC$ or $UC$, respectively. See [G16] for uniform notions of $LC$ and $UC$. These notions are indeed closely related with the properties of lower and upper absolute continuities as seen in [G16].

**C. Weak continuities.** We next present some weaker forms of continuity, lower, and upper continuities and semicontinuities which will be used frequently. We need here the property of being regulated (see, e.g., [C3]).

The function $f$ will be said to be *regulated from left* or *right* at $x$ if the limit $f(x-0)$ or $f(x+0)$, respectively, exists provided that $x \in X^{d-}$ or $X^{d+}$, respectively. Further, $f$ will be said to be *regulated* at $x$ if it is so from both sides at $x$. Note that we do not assume here the limits $f(x-0)$ and $f(x+0)$ to be equal, and each of these limits may be finite or infinite.

Now, we will call $f$ *weakly continuous* (or *WC*) *from left* or *right* at $x$ if $f$ is continuous from left or right, respectively, at $x$ whenever it is regulated from that side at $x$. We will use $WC_-$ and $WC_+$ as before to denote $WC$ from left and right, respectively; and when $f$ is $WC_-$ and $WC_+$ at $x$, it will be called $WC$ at $x$. It should be noted here that $f$ is continuous at $x$ iff it is regulated and $WC$ there. The notion of $WC$ is, however, much weaker than that of continuity. For it is clear from the Young's Theorem 1.1.1 that every function $f$ on $X$ is $WC$ n.e.

We will similarly call $f$ *weakly lower continuous* (or *WLC*) *from left* or *right* at $x$ if $f$ is $LC$ from left or right, respectively, at $x$ whenever it is regulated from that side at $x$. We will use $WLC_-$ and $WLC_+$ to denote $WLC$ from left and right, respectively; and when $f$ is $WLC_-$ and $WLC_+$ at $x$, it will be called $WLC$ at $x$. The notions of *weak upper continuities* $WUC_-$, $WUC_+$ and $WUC$ are defined analogously.

Further, $f$ will be called *weakly lower semicontinuous* (or *WLSC*) at $x$ if $f$ is *LSC* at $x$ from the sides from which it is regulated. The notion of *weak upper semicontinuity* (or *WUSC*) is defined analogously.

**D. Partial weak continuities.** Given a subset $E$ of $X$, we will call $f$ *partially weakly continuous from right* (or $PWC_+$) *with respect to E at* $x \in E$ if $f$ is $WC_+$ at $x$ provided $x$ is only a right limit point of $E$. The *partial weak continuity from left* (or $PWC_-$) *with respect to E at x* is defined analogously; and when $f$ is simultaneously $PWC_-$ and $PWC_+$ with respect to $E$ at $x$, $f$ will be called *partially weakly continuous* (or *PWC*) *with respect to E at x*.

Further, when $f$ is $PWC_+$, $PWC_-$, or *PWC* with respect to $E$ at each point of $E$, then $f$ will be said to be $PWC_+$, $PWC_-$, or *PWC*, respectively, *on* $E$.

It should be pointed out here that the term "with respect to $E$" is used here in reference to the term "partially," and it should not be confused with the restriction of $f$ to $E$. Further, it should be noted that $PWC_+$ on $E$ is much weaker than $WC_+$ on $E$.

**E. Internal properties.** We now present some "internal" properties of functions which are considerably weaker than various continuities but are stronger than their weaker versions presented in part C.

We will call $f$ *internal from left* at $x$ if $f(x-0) \leq f(x) \leq \bar{f}(x-0)$ provided $x \in X^-$. The property *internal from right* at $x$ is defined analogously, and when $f$ is internal from both sides at $x$, it will be called *internal* at $x$.

Further, we will call $f$ *lower internal* (or *LI*) *from left* or *right* at $x$ if

$$(3) \qquad \underline{f}(x-0) \leq f(x) \quad \text{or} \quad f(x) \leq \bar{f}(x+0), \quad \text{respectively,}$$

provided that $x \in X^-$ or $X^+$, respectively. Similarly, $f$ will be called *upper internal* (or *UI*) *from left* or *right* at $x$ if

$$(4) \qquad \bar{f}(x-0) \geq f(x) \quad \text{or} \quad f(x) \geq \underline{f}(x+0), \quad \text{respectively,}$$

provided that $x \in X^-$ or $X^+$, respectively.

We will use $LI_-$ and $LI_+$ to denote *LI* from left and right, respectively, and use $UI_-$ and $UI_+$ to denote *UI* from left and right, respectively. Further, when $f$ is *LI* or *UI* from both sides at $x$, it will be called *LI* or *UI*, respectively, at $x$.

It may be noted here that $f$ is *UI* from left or right at $x$ iff $-f$ is *LI* from that side at $x$ and that $f$ is internal at $x$ iff it is *LI* and *UI* at $x$.

Generalizing the notions of *LSC* and *USC*, we will further call $f$ *lower semi-internal* (or *LSI*) at $x$ if $f(x) \leq \bar{f}(x-0)$ and $f(x) \leq \bar{f}(x+0)$, provided that $x \in X^-$ or $X^+$, respectively, and $f$ will be called *upper semi-internal* (or

$USI$) at $x$ if $f(x) \geq \underline{f}(x-0)$ and $f(x) \geq \underline{f}(x+0)$ provided that $x \in X^-$ or $X^+$, respectively.

It should be noted here that each of the following implications holds locally at $x$ from each of the two sides:

$$(5) \qquad LC \Rightarrow LI \Rightarrow WLC; \quad \text{Continuous} \Rightarrow \text{Internal} \Rightarrow WC.$$

Also, in the case when $f$ is regulated at $x$, the first three properties of (5) become equivalent at $x$, and the same holds for the last three properties of (5). Further, the following implications hold bilaterally at each point $x$:

$$(6) \qquad LSC \Rightarrow LSI \Rightarrow WLSC; \quad USC \Rightarrow USI \Rightarrow WUSC.$$

It is clear on the other hand from Theorem 1.1.1 that every function $f$ on $X$ is internal at all but a countable set of points, and that $f$ is continuous at all but a countable set of points where it is $LC$ or $UC$.

**F. Star internal properties.** We finally present some notions of continuity which are obtained on combining the properties $LC$ and $LI$, and similarly $UC$ and $UI$, which are also used frequently.

Given $f: I \to \mathbb{R}$, we will call $f$ *left-star lower internal* (or $L^*LI$) if it is $LC_-$ and $LI_+$, and we will call $f$ *right-star lower internal* (or $R^*LI$) if it is $LI_-$ and $LC_+$. Similarly, $f$ will be called *left-star upper internal* (or $L^*UI$) if it is $UC_-$ and $UI_+$, and $f$ will be called *right-star upper internal* (or $R^*UI$) if it is $UI_-$ and $UC_+$.

Clearly, the properties $L^*LI$ and $R^*LI$ are stronger than $LI$ but weaker than $LC$, and similarly $L^*UI$ and $R^*UI$ are stronger than $UI$ but weaker than $UC$. Thus the term "star" really stands for "strongly."

1.2.1. Remark. The notions of $WC$, $WLC$, and $WUC$ have been used earlier in [G11] under different terminology. The terms "interned from left and right" were used there for $WC_-$ and $WC_+$, respectively, and "bilaterally interned" was used for $WC$. Similarly, the terms "lower interned from left and right" were used there for $WLC_-$ and $WLC_+$, respectively, and "bilaterally lower interned" was used for $WLC$. Similar terms were used for $WUC$.

Further, it should be pointed out here that the term "internal" was used earlier in [G16, p. 22] in the case of functions of bounded variation for a property which is weaker than the one used here, namely, the property $\underline{f}(x) \leq f(x) \leq \bar{f}(x)$; the function $f$ may be called *weakly internal* at $x$ when these two inequalities hold.

## §1.3 Generalized and inherent properties of functions.

In this section we present generalized and inherent versions of any property of functions and obtain their relationship in general.

Let $P$ be any function-theoretic property, and suppose $f: X \to \mathbb{R}$ and $E \subset X$. Then, extending the generalization used in $VBG$ and $ACG$ [S8, pp. 221, 223] to other properties of functions, we will call $f$ *generalized* $P$ (or $GP$) *on* $E$ if $E$ is the union of a sequence of sets $\{E_n\}$ on each of which $f$ has $P$.

Moreover, when the sets $E_n$ can be chosen to be closed or $G_\delta$-sets in $E$, $f$ will be called *F-generalized* $P$ or $G_\delta$*-generalized* $P$, respectively (or *F-GP* or $G_\delta$*-GP*, respectively), *on* $E$.

Further, $f$ will be said to be *nowhere* $P$ *on* $E$ if it does not have $P$ on any portion of $E$. Also, we will call $f$ *inherently* $P$ (or *IP*) *on* $E$ if every subset of $E$ which is closed in $E$ has a portion on which $f$ has $P$.

The last concept does occur sometimes in the literature without any terminology. For example, one of the classical theorems of Baire (see [N2, pp. 143, 146]) can be described briefly in terms of this terminology as follows: A function $f: I \to \mathbb{R}$ is of Baire class 1 iff it is inherently pointwise discontinuous (i.e., continuous at a dense set of points).

In the following theorem we obtain the general relationship between *IP* and *F-GP*. The proof of this theorem is an adaptation of the proof given in [S8, p. 233] for *VBG* and *ACG* for continuous $f$ and $E$ closed in $\mathbb{R}$.

1.3.1. THEOREM. *Let* $f: X \to \mathbb{R}$ *and* $E \subset X$. *If* $f$ *is IP on* $E$, *then it is F-GP on* $E$. *Conversely, if* $E$ *is a* $G_\delta$*-set in* $\mathbb{R}$ *and* $f$ *is F-GP on* $E$, *then* $f$ *is IP on* $E$.

PROOF. First, suppose $f$ is *IP* on $E$. Let $\mathscr{E}$ denote the family of all countable unions of closed subsets of $E$ on which $f$ has $P$. Then $f$ is *F-GP* on each element of $\mathscr{E}$ and $\mathscr{E}$ is clearly stable under countable union. Let, further, $\mathscr{I}$ denote the family of all open intervals $I$ with rational endpoints for which $E \cap I \in \mathscr{E}$. Then since $\mathscr{I}$ is countable, the set $A = \bigcup\{E \cap I : I \in \mathscr{I}\}$ is an element of $\mathscr{E}$. Hence it is enough to show that $B \equiv E \sim A$ is empty.

Suppose $B \neq \varnothing$. Then since $B$ is closed in $E$, there is by hypothesis an open interval $I$, whose endpoints can of course be chosen to be rational, such that $B \cap I \neq \varnothing$ and $f$ has $P$ on $B \cap I$. Then since $B \cap I$ is an $F_\sigma$-set in $E$, $B \cap I \in \mathscr{E}$, and similarly $A \cap I \in \mathscr{E}$. Hence, $E \cap I = (A \cap I) \cup (B \cap I) \in \mathscr{E}$. Thus $I \in \mathscr{I}$, so that $B \cap I \subset A$, which contradicts the definition of $B$.

Next, to prove the converse, suppose $E$ is a $G_\delta$-set in $\mathbb{R}$ and $f$ is *F-GP* on $E$. Then $E$ is the union of a sequence of sets $\{E_n\}$ closed in $E$ such that $f$ has $P$ on each $E_n$. Let $A$ be any nonempty closed subset of $E$. Then $A = \bigcup_n (A \cap E_n)$, where $A \cap E_n$ is closed in $E$ for each $n$. Now, since $A$ is a $G_\delta$-set in $\mathbb{R}$, it is a Baire space [Wi, pp. 185, 186], and so it is nonmeager in itself. Consequently, there is an $n$ such that $E_n$ includes some portion $A_0$ of $A$. Thus $f$ has $P$ on $A_0$, which proves that $f$ is *IP* on $E$. □

**§1.4 Lower and upper Baire classes and relative measurability of functions.** In this section we present (a) a refinement of Baire classification of functions in terms of lower and upper Baire classes, and (b) Lebesgue measurability of functions relative to an arbitrary set in $\mathbb{R}$. These refinements will be used subsequently to obtain the best possible results on Baire class and measurability of unilateral derivates.

First, let $X$ be any *perfect space*—that is, a topological space in which every open set is an $F_\sigma$-set; for example, a metric space. Then for each countable ordinal $\alpha$ we will use $\Sigma_\alpha$ and $\Pi_\alpha$ to denote the classes of sets in $X$ which are of additive or multiplicative class $\alpha$, respectively, and use $\Delta_\alpha$ to denote the class of ambiguous sets of class $\alpha$ (see [Bo1] or [K1, pp. 345–347] for definitions).

Now, a function $f: X \to \overline{\mathbb{R}}$ will be said to be of *lower* or *upper Baire class* $\alpha$, or *B-measurable of lower* or *upper class* $\alpha$, if for every $c \in \mathbb{R}$ the set $\{x \in X : f(x) > c\}$ or $\{x \in X : f(x) < c\}$, respectively, is of $\Sigma_\alpha$.

We will use $B_\alpha$, $LB_\alpha$, and $UB_\alpha$ to denote the classes of all extended real-valued functions of Baire class $\alpha$, lower Baire class $\alpha$, and upper Baire class $\alpha$, respectively, relative to their domain.

Then, since $\overline{\mathbb{R}}$ has a countable base and $X$ is perfect, it is clear that for each $\alpha < \Omega$,

$$B_\alpha = LB_\alpha \cap UB_\alpha \quad \text{and} \quad LB_\alpha \cup UB_\alpha \subset B_{\alpha+1}.$$

Further, $B_0$, $LB_0$, and $UB_0$ are indeed the classes of continuous, *LSC*, and *USC* functions, respectively. Also, for any $\alpha > 0$, the classes $LB_\alpha$ and $UB_\alpha$ share many properties of the classes $LB_0$ and $UB_0$, respectively. For example, the classes $LB_\alpha$ and $UB_\alpha$ are stable under pointwise limits of nondecreasing and nonincreasing sequences of functions, respectively.

Practically all the results on Baire classes of functions extend to the classes $LB_\alpha$ and $UB_\alpha$ with natural modifications including the classical Lebesgue-Hausdorff theorem on analytical representation of such functions. These classes were first investigated by Hausdorff in his celebrated monograph [Hau3] for real-valued functions, but his nomenclature is now somewhat outmoded. An abstract theory of such classes of functions and multifunctions (see §1.5) which is also applicable to nonseparable spaces is to appear in [G19] (see [G14] for an announcement of some of its results). We state here two theorems without proof which seem to have some significance in the present work.

The following version of the classical theorem of Baire on functions of Baire class 1 was first obtained by W. H. Young [Y3, p. 303] for a real valued function on an interval in $\mathbb{R}$.

1.4.1. THEOREM (BAIRE-YOUNG). *If $f: X \to \overline{\mathbb{R}}$ is of $LB_1$ or $UB_1$, then $f$ is LSC or USC, respectively, r.e. in $X$.*

The following interposition theorem was obtained initially by Hahn [Ha1] for $\alpha = 0$, and then by Hausdorff [Hau2] in the general case.

1.4.2. THEOREM (INTERPOSITION). *Suppose $f, g : X \to \overline{\mathbb{R}}$ and $\alpha$ is any countable ordinal. If $f$ is of $UB_\alpha$, $g$ is of $LB_\alpha$ and $f \le g$ everywhere, then there is a function $h : X \to \overline{\mathbb{R}}$ of $B_\alpha$ such that $f \le h \le g$ everywhere.*

Next, we define Lebesgue measurability of a function relative to an arbitrary subset $X$ of $\mathbb{R}$. We will use the usual symbols $m^*$ and $m$ to denote the Lebesgue outer measure and Lebesgue measure on $\mathbb{R}$. Also, following Saks [S8], we will use $|A|$ to denote $m^*(A)$ for any set $A \subset \mathbb{R}$.

Further, given a set $X \subset \mathbb{R}$, we will use $m_X^*$ to denote the outer measure obtained on $X$ by restricting $m^*$ to the class of all subsets of $X$, and $m_X$ to denote the measure induced by $m_X^*$ in the usual manner on $X$. We will call $m_X$ the *extended Lebesgue measure* on $X$, and a subset $E$ of $X$ will be called *(Lebesgue) measurable in* $X$ if it is $m_X^*$-measurable—that is, if for every set $A \subset X$,

$$m^*(A) = m^*(A \cap E) + m^*(A \sim E).$$

It is clear that every subset of $X$ that is measurable in $\mathbb{R}$ is also measurable in $X$, but the converse holds only when $X$ is measurable in $\mathbb{R}$. Further, when $m^*(X) < \infty$, it can be verified by usual argument that a set $E \subset X$ is measurable in $X$ iff $m^*(E) + m^*(X \sim E) = m^*(X)$.

The class of measurable sets in $X$ is thus in general wider than the class of subsets of $X$ that are measurable in $\mathbb{R}$, and the two classes coincide only when $X$ is measurable in $\mathbb{R}$.

In the following theorem we establish the equivalence of this notion of relative measurability with the topological relativization of measurability used earlier by Hausdorff [Hau1].

1.4.3. THEOREM. *Given $X \subset \mathbb{R}$, a subset $E$ of $X$ is measurable in $X$ iff there is a measurable set $M$ in $\mathbb{R}$ such that $E = M \cap X$.*

PROOF. The sufficiency part is obvious. For if there is a measurable set $M$ in $\mathbb{R}$ such that $E = M \cap X$, then for each subset $A$ of $X$,

$$m^*(A) = m^*(A \cap M) + m^*(A \sim M) = m^*(A \cap E) + m^*(A \sim E).$$

To prove the necessity, suppose $E$ is measurable in $X$. Let us first consider the case when $m^*(X) < \infty$. Choose a measurable set $B$ in $\mathbb{R}$ such that $E \subset B$ and $m(B) = m^*(E)$. Then

$$m^*(E) \le m^*(B \cap X) \le m(B) = m^*(E),$$

so that $m^*(B \cap X) = m^*(E)$. Now since $E$ is measurable in $X$, we obtain

$$m^*(E) + m^*(X \sim E) = m^*(X) = m^*(X \cap B) + m^*(X \sim B),$$

and since $m^*(X) < \infty$, we thus obtain $m^*(X \sim E) = m^*(X \sim B)$. Using the measurability of $B$ once again, we have

$$m^*(X \sim E) = m^*\{(X \sim E) \cap B\} + m^*(X \sim B).$$

The set $C = B \cap (X \sim E)$ is therefore of measure zero, and so $M = B \sim C$ is a measurable set in $\mathbb{R}$ for which $M \cap X = B \cap X \sim C = B \cap X \sim (X \sim E) = E$.

In the general case, set for each positive integer $n$, $X_n = X \cap (-n, n)$ and $E_n = E \cap X_n$. Then for each $n$ it is easy to see that $E_n$ is measurable in $X_n$, and so there is by the above a measurable set $M_n$ in $\mathbb{R}$, which may be chosen to be a subset of $(-n, n)$, such that $E_n = M_n \cap X_n$. The set $M = \bigcup_n M_n$ is thus a measurable set in $\mathbb{R}$ for which

$$M \cap X = \bigcup_n (M_n \cap X) = \bigcup_n (M_n \cap X_n) = \bigcup_n E_n = E. \quad \square$$

Finally, we will call a function $f: X \to \overline{\mathbb{R}}$ *Lebesgue measurable*, or *L-measurable*, or *measurable* (in $X$), if it is measurable with respect to the measure $m_X$, and $f$ will be called *Borel measurable*, or *B-measurable*, if it is so with respect to the relative topology of $X$. Thus when $f$ is $B$-measurable, it follows from Theorem 1.4.3 that $f$ is also $L$-measurable in the present sense, although it need not be measurable in the usual sense unless $X$ is measurable in $\mathbb{R}$.

**§1.5 Baire classes and measurability of multifunctions.** Because the new derivatives defined later in Part II, and the medians defined in §3.3, are in general set-valued, to investigate their properties we present here Baire classes and measurability of set-valued functions, or multifunctions. These properties are similar to those of functions defined in the last section.

Suppose $X$ and $Y$ are two topological spaces where $X$ is perfect. We will use $2^Y$ to denote the space of all "nonempty" closed subsets of $Y$ with the Vietoris (or exponential) *topology* [K1, p. 160], namely, the coarsest topology on $2^Y$ for which the sets

$$\{F \in 2^Y : F \cap U \neq \varnothing\} \quad \text{and} \quad \{F \in 2^Y : F \subset U\}$$

are open for every open set $U$ in $Y$.

Now, let $\varphi: X \to 2^Y$ be any multifunction. The continuity and the Baire class of $\varphi$ are defined as usual in terms of the Vietoris topology of $2^Y$. Also, when $X \subset \mathbb{R}$, the $L$-measurability of $\varphi$ is defined similarly with respect to the extended Lebesgue measure on $X$.

Further, given any countable ordinal $\alpha$, $\varphi$ will be said to be of *lower* or *upper Baire class* $\alpha$, or *B-measurable of lower* or *upper class* $\alpha$, if for every open set $U$ in $Y$ the set

$$\{x \in X : \varphi(x) \cap U \neq \varnothing\} \quad \text{or} \quad \{x \in X : \varphi(x) \subset U\},$$

respectively, is of $\Sigma_\alpha$ in $X$.

We will use the notations $B_\alpha$, $LB_\alpha$ and $UB_\alpha$ (see §1.4) to denote the Baire class $\alpha$ and lower and upper Baire classes $\alpha$ of multifunctions as well. Also, when $\varphi$ is of $LB_0$ or $UB_0$, $\varphi$ is called as before *LSC* or *USC*, respectively.

These two semicontinuities of multifunctions are also well known (see [K2, p. 57]); see further [K1, p. 173] for their local versions.

It should be noted here that in the case when $\varphi(x)$ is singleton for each $x \in X$, then for each $\alpha < \Omega$, $\varphi$ is of $B_\alpha$, $LB_\alpha$, or $UB_\alpha$ as a multifunction iff it is of $B_\alpha$ as a function.

The classes $LB_\alpha$ and $UB_\alpha$ have been considered earlier by Kuratowski in [K3] under a different notation, where he extended many of the known results on $LSC$ and $USC$ multifunctions to general classes $LB_\alpha$ and $UB_\alpha$. See also [K4] for further results, and see [G14] and [G19] for an abstract theory of such classes of multifunctions which is also applicable to nonseparable spaces.

In the present work we will deal with multifunctions $\varphi: X \to 2^{\overline{\mathbb{R}}}$ whose values are connected. Let $M_c$ denote the space of all such connected valued multifunctions.

In this case we do have, as before,

$$LB_\alpha \cap UB_\alpha = B_\alpha \quad \text{and} \quad LB_\alpha \cup UB_\alpha \subset B_{\alpha+1}.$$

(See [G19] for more general hypotheses under which these two relations hold.)

With every multifunction $\varphi \in M_c$ we associate two extended real-valued functions $\varphi_*$ and $\varphi^*$ on $X$ which are defined as follows:

$$\varphi_*(x) = \inf\{y : y \in \varphi(x)\}, \quad \varphi^*(x) = \sup\{y : y \in \varphi(x)\}, \quad x \in X.$$

In the following theorem, which will be needed sometimes, we establish the relationship between the Baire class and measurability of $\varphi \in M_c$ with those of $\varphi_*$ and $\varphi^*$.

1.5.1. Theorem. *Suppose* $\varphi: X \to 2^{\overline{\mathbb{R}}}$ *is connected valued and* $\alpha < \Omega$. *Then*

(a) $\varphi$ *is of* $LB_\alpha$ *iff* $\varphi_*$ *is of* $UB_\alpha$ *and* $\varphi^*$ *is of* $LB_\alpha$;
(b) $\varphi$ *is of* $UB_\alpha$ *iff* $\varphi_*$ *is of* $LB_\alpha$ *and* $\varphi^*$ *is of* $UB_\alpha$;
(c) $\varphi$ *is of* $B_\alpha$ *iff* $\varphi_*$ *and* $\varphi^*$ *are so*;
(d) $\varphi$ *is B-measurable iff* $\varphi_*$ *and* $\varphi^*$ *are so; and in case* $X \subset \mathbb{R}$, *then* $\varphi$ *is L-measurable iff* $\varphi_*$ *and* $\varphi^*$ *are so.*

Proof. Set, for each $c \in \mathbb{R}$, $I_c = (c, \infty]$ and $I^c = [-\infty, c)$. Then since $\varphi$ is connected valued, it is easy to see that the following identities hold:

(1) $$E_c \equiv \{x \in X : \varphi(x) \cap I_c \neq \varnothing\} = \{x \in X : \varphi^*(x) > c\},$$

(2) $$E^c \equiv \{x \in X : \varphi(x) \cap I^c \neq \varnothing\} = \{x \in X : \varphi_*(x) < c\},$$

(3) $$F_c \equiv \{x \in X : \varphi(x) \subset I_c\} = \{x \in X : \varphi_*(x) > c\},$$

(4) $$F^c \equiv \{x \in X : \varphi(x) \subset I^c\} = \{x \in X : \varphi^*(x) < c\}.$$

Now, the necessity part of (a) follows clearly from (1) and (2). To prove its sufficiency, suppose $\varphi_*$ is of $UB_\alpha$ and $\varphi^*$ is of $LB_\alpha$. Then $E_c$ and $E^c$ are of $\Sigma_\alpha$ for every $c \in \mathbb{R}$. If $I = (a, b)$ is on the other hand any bounded open interval in $\mathbb{R}$, then $I = I_a \cap I^b$, and thus with $\varphi$ being connected-valued, it follows from (1) and (2) that

$$(5) \qquad \{x \in X : \varphi(x) \cap I \neq \varnothing\} = E_a \cap E^b \in \Sigma_\alpha.$$

Hence, given any open set $U$ in $\mathbb{R}$, let $\{I_n\}$ be the sequence of its connected components, which are of course open in $\overline{\mathbb{R}}$. Then it follows from (1), (2), and (5) that

$$\{x \in X : \varphi(x) \cap U \neq \varnothing\} = \bigcup_n \{x \in X : \varphi(x) \cap I_n \neq \varnothing\} \in \Sigma_\alpha.$$

The necessity part of (b) follows in turn from (3) and (4). To prove its sufficiency, suppose $\varphi_*$ is of $LB_\alpha$ and $\varphi^*$ is of $UB_\alpha$. If $I = (a, b)$ is any bounded open interval in $\mathbb{R}$ (i.e., $I = I_a \cap I^b$), then it follows from (3) and (4) that

$$(6) \qquad \{x \in X : \varphi(x) \subset I\} = F_a \cap F^b \in \Sigma_\alpha.$$

Hence, if $U$ is any open set in $\mathbb{R}$ with connected components $\{I_n\}$, then since $\varphi$ is connected valued, it follows from (3), (4), and (6) that

$$\{x \in X : \varphi \subset U\} = \bigcup_n \{x \in X : \varphi(x) \subset I_n\} \in \Sigma_\alpha.$$

This establishes (a) and (b), and (c) follows directly from (a) and (b).

Finally, regarding (d), when $\varphi$ is $B$-measurable, it is clear from (1) and (2) that $\varphi_*$ and $\varphi^*$ are $B$-measurable. To prove the converse, suppose $\varphi_*$ and $\varphi^*$ are $B$-measurable. Then it follows by arguments used above for (a) and (b), with the class of Borel subsets of $X$ in the place of $\Sigma_\alpha$, that $\varphi$ is $B$-measurable. A similar argument holds for $L$-measurability. □

For $\alpha = 0$ it is easy to see from the above proof that the first two parts of the theorem hold also locally at any point. On combining this result with Theorem 1.4.1 we obtain the following.

1.5.2. Corollary. *If $\varphi \in M_c$ is of $LB_1$ or $UB_1$, then it is LSC or USC, respectively, r.e. in $X$.*

Let a function $f: X \to \overline{\mathbb{R}}$ be called a *selector* of $\varphi \in M_c$ if $f(x) \in \varphi(x)$ for each $x \in X$. On combining Theorem 1.5.1 with the Interposition Theorem 1.4.2, we obtain the following selection theorem (see [G19] for more general hypotheses under which these two corollaries hold).

1.5.3. Corollary (Selection). *Let $\varphi \in M_c$ and $\alpha < \Omega$. If $\varphi$ is of $LB_\alpha$, then it has a selector of $B_\alpha$.*

*CHAPTER 2*

# Some Fundamental Theorems on Unilateral Derivates

In this chapter we obtain the two fundamental theorems on unilateral derivates mentioned earlier, which we call the First and Second Derivate Theorems, and deduce from these two theorems several other basic results on such derivates.

First, in §2.1, we define several unilateral monotonicity and Lipschitz properties of functions on a set which play a central role in the development of the present unified theory. In terms of these properties we obtain in §2.2 the First and Second Derivate Theorems. The second derivate theorem does follow from the first under more restrictive continuity hypothesis, but not in its full generality. Then, in §2.3, we deduce from these two theorems extended versions of two theorems of G. C. Young and W. H. Young (an illustrious couple).

Next, in §2.4, we deduce from the First and Second Derivate Theorems two other theorems on unilateral derivates, called the Third and Fourth Derivate Theorems, which deal with properties of functions on some portion of their domain and are also found to play a significant role in the present development.

Finally, in §2.5, we deduce from the Third Derivate Theorem strengthened versions of some basic theorems of Denjoy and W. H. Young; these results have a long and fascinating history (see Remark 2.5.7).

**§2.1 Unilateral monotonicity and Lipschitz properties on a set.** In this section we define some monotonicity and Lipschitz properties of a function $f: X \to \mathbb{R}$ on a set $E \subset X$ which are used throughout this part. These properties depend on the values of $f$ outside $E$ as well and are somewhat similar to the restricted properties $BV_*$ and $AC_*$ on $E$ (see Saks [S8, pp. 228, 231] for definitions).

Suppose $f: X \to \mathbb{R}$ and $E \subset X$. We will call $f$ *weakly right increasing* (or *WRI*) on $E$ if

(1) $$f(x) \geq f(a) \quad \text{whenever } a, b \in E,\ x \in X, \text{ and } a \leq x \leq b.$$

The property of $f$ being *weakly left increasing* (or *WLI*) on $E$ is defined similarly by replacing the first inequality in (1) by $f(x) \leq f(b)$. Further, when $f$ is simultaneously *WRI* and *WLI* on $E$, then $f$ will be called *weakly bi-increasing* (or *WBI*) on $E$.

Next, if there is some real number $c > 0$ such that the function $f_{-c}$ is *WRI*, *WLI*, or *WBI* on $E$, then $f$ will be called *strongly right*, *left*, or *bi-increasing* (or *SRI*, *SLI*, or *SBI*) respectively on $E$.

Further, when $-f$ is *WRI*, *WLI*, or *WBI* on $E$, then $f$ will be called *weakly right*, *left*, or *bidecreasing* (or *WRD*, *WLD*, or *WBD*), respectively, on $E$. Similarly, when $-f$ is *SRI*, *SLI*, or *SBI* on $E$, then $f$ will be called *strongly right*, *left*, or *bidecreasing* (or *SRD*, *SLD*, or *SBD*), respectively, on $E$.

Let us note here that when $f$ is *WRI* or *WLI* on $E$, then it is indeed weakly increasing (or *WI*) on $E$, or what is usually known as nondecreasing on $E$.

We will further call $f$ *right lower* or *upper Lipschitz* (or *RLL* or *RUL*) on $E$ if there is some real number $c$ such that the function $f_{+c}$ is *WRI* or *WRD*, respectively, on $E$. In case $f$ is simultaneously *RLL* and *RUL* on $E$, then it will be called *right Lipschitz* (or *RL*) on $E$. Also, if there is a real number $c \geq 0$ such that $f_{+c}$ is *WRI* and $f_{-c}$ is *WRD* on $E$, then $f$ will be called *right c-Lipschitz* (or *RL*($c$)) on $E$.

It is now easy to see that $f$ is *RLL* on $E$ iff there is some real number $c$ such that

(2)
$$f(x) \geq f(a) - c(x-a) \quad \text{whenever } a, b \in E,\ x \in X, \text{ and } a \leq x \leq b.$$

Similarly, $f$ is *RL*($c$) on $E$ for some $c \geq 0$ iff

(3)
$$|f(x) - f(a)| \leq c(x-a) \quad \text{whenever } a, b \in E,\ x \in X, \text{ and } a \leq x \leq b.$$

Also, $f$ is *RL* on $E$ iff it is *RL*($c$) on $E$ for some $c \geq 0$.

Further, $f$ will be called *right monotone* on $E$ if it is either *WRI* or *WRD* on $E$. Similarly, $f$ will be called *strongly right monotone* on $E$ if it is either *SRI* or *SRD* on $E$.

Finally, when there is some real number $c$ such that $f_{+c}$ is monotone on $E$, then $f$ is said to be (of) *monotonic type* (or *MT*) on $E$ (see Zygmund [Zy2, p. 64]). If $c$ can be so chosen that $f_{+c}$ is right monotone on $E$, then we will call $f$ *right monotonic type* (or *RMT*) on $E$.

The left and bilateral notions of all the above Lipschitz and monotonicity properties are defined analogously by replacing the term "right" everywhere by "left" and "bi-," respectively.

Let us note here that $f$ is *RMT* on $E$ iff it is either *RLL* or *RUL* on $E$, and also note that $f$ is *BL* on $E$ iff it is right and left Lipschitz on $E$. Also, the properties of right and left monotonicities of $f$ on $E$ are stronger than the ordinary monotonicity of $f$ on $E$ unless $E = X$.

When $E$ coincides on the other hand with $X$, it is clear from the above definitions that the right, left, and bilateral versions of each of the above properties become equivalent, and so in that case these adjectives will not be used in the nomenclature of these properties, and likewise the letters $R$, $L$, and $B$ will be dropped from their notations.

The function $f$ is thus *strongly increasing* (on $X$) iff there is some $c \in \mathbb{R}_+$ such that $f_{-c}$ is nondecreasing, or, equivalently, increasing. Similarly, $f$ is *strongly decreasing* iff there is some $c \in \mathbb{R}_+$ such that $f_{+c}$ is decreasing. The significance of these two properties in differentiation theory was recognized earlier in [G11], where they were called "adequately increasing" and "adequately decreasing," respectively.

**§2.2 Two fundamental theorems on unilateral derivates.** In this section we obtain the two fundamental theorems on unilateral derivates which deal with generalized unilateral monotonicity and Lipschitz property of functions in terms of their derivates. The various known results on unilateral derivates will be deduced from these two theorems later with their extensions. We include here a few elementary consequences of the First Derivate Theorem.

We begin with the following extension lemma on right monotonicity which also plays a central role in the present chapter.

2.2.1. Lemma (First Extension). *Let $f: X \to \mathbb{R}$, $E \subset X$, $A$ be a dense subset of $E$, and suppose $f$ is QLSC at the points of $E \sim A$ relative to $E$. Then if $f$ is WRI or SRI on $A$, then it is so on $E \cap co(A)$.*

Proof. Set $B = E \cap co(A)$, and first suppose $f$ is *WRI* on $A$. To prove that $f$ has this property on $B$, suppose $a, b \in B$, $x \in X$, and $a < x \leq b$. It is then enough to show that $f(x) \geq f(a)$. Clearly, there exist two points $a', b' \in A$ such that $a' \leq a$ and $b \leq b'$. In case $a \in A$, the required inequality follows from the hypothesis. Hence, suppose $a \notin A$, and then $a' < a$.

Now suppose, if possible, that $f(x) < f(a)$. Then since $f$ is *QLSC* relative to $E$ at $a$, there exists a portion $E_0$ of $E \cap (a', x)$ such that $f(t) > f(x)$ for $t \in E_0$. But since $A$ is dense in $E$, $E_0$ contains a point $a_0 \in A$. Then $a_0 < x \leq b'$, but $f(x) < f(a_0)$, which contradicts the fact that $a_0 \in A$.

Next, if $f$ is *SRI* on $A$, then there exists a real number $c > 0$ such that $f_{-c}$ is *WRI* on $A$. Hence it follows from above that $f_{-c}$ is *WRI* on $B$, and $f$ is consequently *SRI* on $B$. □

2.2.2. THEOREM (FIRST DERIVATE). *Let* $f: X \to \mathbb{R}$, $E \subset X$, *and suppose* $D_+f > 0$ *n.e. in* $E$. *Then* $f$ *is* $G_\delta$-*GSRI on* $E$; *and in case* $f$ *is LSC relative to* $E$, *then it is F-GSRI on* $E$.

*More generally, if* $E \subset F \subset X$ *and* $f$ *is LSC relative to* $F$ *n.e. in* $E$, *then there is a sequence of* $G_\delta$-*sets* $\{E_n\}$ *in* $F$ *such that* $E \subset \bigcup_n E_n$ *and* $f$ *is SRI on each* $E_n$. *Further, in case* $f$ *is LSC relative to* $F$, *then* $E_n$*'s can be chosen to be closed in* $F$.

PROOF. We will first prove here the second part which is more general.

Let $A_n$ denote for each positive integer $n$ the set of points $x$ in $E$ for which

$$(1) \qquad f_{-1/n}(t) \geq f_{-1/n}(x) \quad \text{whenever } t \in X \cap (x, x+1/n).$$

Clearly, $A_n$ is the union of a sequence of sets $\{A_{n,i} : i = 1, 2, \ldots\}$ such that each $A_{n,i}$ has a diameter less than $1/n$. Hence it is clear that

$$E_0 \equiv \{x \in E : D_+f(x) > 0\} \subset \bigcup_n A_n = \bigcup_n \bigcup_i A_{n,i}.$$

Now, suppose $E \subset F \subset X$, $F_0$ is the set of points in $F$ where $f$ is *LSC* relative to $F$, and suppose that the set $C \equiv F \sim F_0$ is countable.

Given $n$ and $i$, it is clear from (1) that $f_{-1/n}$ is *WRI* on $A_{n,i}$, so that $f$ is *SRI* on $A_{n,i}$. Hence by the above lemma $f$ is *SRI* on the set

$$(2) \qquad B_{n,i} = F \cap (A_{n,i})^- \cap (\text{co}(A_{n,i})) \sim (C \sim A_{n,i}).$$

It is then clear that $B_{n,i}$ is a $G_\delta$-set in $F$. Since the set $E \sim E_0$ is countable by hypothesis, so is the set $E \sim \bigcup_n \bigcup_i B_{n,i}$, and so it is the union of a sequence of singleton sets $\{S_n\}$. Now since $f$ is trivially *SRI* on each $S_n$, on rearranging the sets $\{B_{n,i} : n = 1, 2, \ldots;\ i = 1, 2, \ldots\}$ and $\{S_n : n = 1, 2, \ldots\}$ in the form of a sequence we obtain a sequence $\{E_n\}$ of $G_\delta$-sets in $F$ such that $E \subset \bigcup_n E_n$ and $f$ is *SRI* on each $E_n$. This proves the first part of the general result.

Next, to prove the other part of the general result, suppose $f$ is *LSC* relative to $F$. Then $C = \emptyset$, and hence for each $n$ and $i$ the above set $B_{n,i}$ is now an $F_\sigma$-set in $F$, and so it is the union of a sequence of closed sets $\{F_{n,i,j} : j = 1, 2, \ldots\}$ in $F$. The required sequence $\{E_n\}$ of closed sets in $F$ is obtained this time by rearranging the sets $\{F_{n,i,j}\}$ and $\{S_n\}$ in the form of a sequence.

The first part follows now from the above general result on choosing $F = E$. For, as $f(x) \leq f(x+0)$ for each $x \in E_0$, and $E \sim E_0$ is countable, it follows from Theorem 1.1.1 that $C$ is countable. □

We include here a few elementary consequences of the First Derivate Theorem which deal with other generalized properties in terms of unilateral derivates.

2.2.3. Corollary. *Let* $f: X \to \mathbb{R}$, $E \subset X$, *and suppose* $D_+ f > -\infty$ *n.e. in* $E$. *Then f is* $G_\delta$-*GRLL on* $E$; *and if f is LSC relative to E, then it is F-GRLL on E.*

Let $A = \{x : D_+ f(x) > -\infty\}$, and set $E_1 = E \sim A$, and for each $n > 1$, $E_n = \{x \in E : D_+ f(x) > -n\} = \{x \in E : D_+ f_{+n}(x) > 0\}$. Then $E = \bigcup_n E_n$, where $E_1$ is countable by the hypothesis. Further, since $f(x) \le \underline{f}(x + 0)$ n.e. in $E$, $f$ is (by Theorem 1.1.1) *LSC* n.e. in $E$. Now, for each $n$, it follows from the above theorem that $E_n \subset \bigcup_i E_{n,i}$, where $E_{n,i}$'s are $G_\delta$-sets in $E$ on which $f_{+n}$ is *SRI*. Consequently, $f$ is *RLL* on each $E_{n,i}$, and hence it is $G_\delta$-*GRLL* on $E$. Further, in the case when $f$ is *LSC* relative to $E$, the sets $E_{n,i}$ can be chosen by the above theorem to be closed in $E$, and so $f$ is then *F-GRLL* on $E$.

Let us recall here that the right knot set $K_+(f)$ denotes the set of points where $D_+ f = -\infty$ and $D^+ f = \infty$ (see §1.1).

2.2.4. Corollary. *Let* $f: X \to \mathbb{R}$, $E \subset X$, *and suppose* $E \cap K_+(f)$ *is countable. Then f is GRMT on E; and if f is continuous relative to E, then it is F-GRMT on E.*

Let $A = \{x \in E : D_+ f(x) > -\infty\}$ and $B = E \sim A$. Then according to the hypothesis, $D^+ f < \infty$ n.e. in $B$. Hence it follows from the above corollary that $f$ is *GRLL* on $A$ and *GRUL* on $B$. Consequently, $f$ is *GRMT* on $E$. Further, if $f$ is continuous relative to $E$, then it is easy to see from the second part of Theorem 2.2.2 and the proof of Corollary 2.2.3 that the sets involved in the representation of $f$ as a *GRMT* function can be chosen to be closed in $E$.

Following is another interesting consequence of Theorem 2.2.2 which deals with the existence of nonempty perfect sets on which the given function is monotone.

2.2.5. Corollary. *Let* $f: X \to \mathbb{R}$, *where X is a Borel set in* $\mathbb{R}$. *Suppose* $D_+ f > 0$ *at a set of points* $E \subset X$ *and f is LSC relative to* $\overline{E}$ *n.e. in E. If E is uncountable, then there exists a nonempty perfect set P in* $\overline{E}$ *on which f is SRI. Moreover, if* $|E| > 0$, *then P can be chosen to be metrically dense in itself.*

Suppose $E$ is uncountable. According to the hypothesis, there is by the second part of Theorem 2.2.2 a sequence of $G_\delta$-sets $\{E_n\}$ in $\overline{E}$ such that $E \subset \bigcup_n E_n$ and $f$ is *SRI* on each $E_n$. Hence there is an $n$ such that $E \cap E_n$ is uncountable. Then $E_n$ is an uncountable Borel set in $\mathbb{R}$, and so it contains some nonempty perfect set $P$ (see, e.g., [K1, p. 44]) on which $f$ is again *SRI*. Further, if $|E| > 0$, then there exists an integer $n$ such that $|E_n| > 0$, and this $E_n$ contains a metrically dense in itself perfect set (see, e.g., [H1, p. 192]).

The next derivate theorem deals with the right Lipschitz property. In the case of continuous functions this theorem follows directly from Corollary 2.2.3, but to obtain it in maximum generality we need here the following extension lemma on this property (see §1.2D for the definition of $PWC_+$).

2.2.6. LEMMA (SECOND EXTENSION). *Let* $f: X \to \mathbb{R}$, $E \subset X$, $c \in \mathbb{R}_+$, *and suppose f is RL*(c) *on a dense subset A of E. If f is* $PWC_+$ *with respect to E at the points of* $E \sim A$, *then it is RL*(c) *on* $E \cap \text{co}(A)$.

PROOF. Suppose the given hypotheses hold. Then since $f$ is $RL$(c) on $A$, we have

(3)
$$|f(x) - f(a)| \leq c(x-a) \quad \text{whenever} \quad a, b \in A,\ x \in X, \text{ and } a \leq x \leq b.$$

To prove that $f$ is $RL(c)$ on $E \cap \text{co}(A)$, suppose $a, b \in E \cap \text{co}(A)$, $x \in X$, and $a \leq x \leq b$. It is then enough to show that

$$|f(x) - f(a)| \leq c(x-a). \tag{4}$$

Suppose $a < x$, because (4) holds trivially otherwise. Clearly, there exist two points $a', b' \in A$ such that $a' \leq a$ and $b \leq b'$. Hence if $c = 0$, it follows from (3) that $f(x) = f(a') = f(a)$, which implies (4). Hence suppose $c > 0$. Also (3) implies (4) when $a \in A$. Hence suppose $a \notin A$. Then $a$ is a limit point of $A$.

First, suppose $a$ is a left limit point of $A$, and let $\varepsilon > 0$. Then we can choose $a' \in A$ such that $a' < a$ and $a - a' < \varepsilon/2c$. Hence, from (3) we obtain

$$\begin{aligned}|f(x) - f(a)| &\leq |f(x) - f(a')| + |f(a) - f(a')| \\ &\leq c(x-a') + c(a-a') \\ &= c(x-a) + 2c(a-a') \\ &< c(x-a) + \varepsilon.\end{aligned}$$

Since this holds for any $\varepsilon > 0$, we thus obtain (4) in the present case.

Next, suppose $a$ is a unilateral right limit point of $A$. Then by the hypothesis, $f$ is $WC_+$ at $a$. Given $\varepsilon > 0$, let $\delta = \varepsilon/2c$ and $s, t \in X \cap (a, a + \delta)$. Then we can choose a point $a' \in A$ such that $a < a' < \min\{s, t\}$. Hence, by (3),

$$\begin{aligned}|f(s) - f(t)| &\leq |f(s) - f(a')| + |f(t) - f(a')| \\ &< c(s-a') + c(t-a') < 2c\delta = \varepsilon.\end{aligned}$$

This proves that the limit $f(a+0)$ exists, and so by the hypothesis, $f$ is right continuous at $a$. Now choose a decreasing sequence of points $\{a_n\}$ in $A \cap (a, x)$ which converges to $a$. Then, by (3),

$$|f(x) - f(a)| = \lim_n |f(x) - f(a_n)| \leq \lim_n c(x - a_n) = c(x-a),$$

so that (4) holds again. □

2.2.7. THEOREM (SECOND DERIVATE). *Let $f: X \to \mathbb{R}$ and $E \subset X$. If $f$ has finite right derivates at each point of $E$, then $f$ is F-GRL on $E$.*

*Moreover, if there is some $c \in \mathbb{R}_+$ such that $D_+f > -c$ and $D^+f < c$ at each point of $E$, then $f$ is F-GRL$(c)$ on $E$.*

PROOF. First, suppose there is some $c \in \mathbb{R}_+$ such that $D_+f > -c$ and $D^+f < c$ at each point of $E$. Then $D_+f_{+c} > 0$ and $D^+f_{-c} < 0$ at the points of $E$. Hence it follows from Theorem 2.2.2 that there are two sequences of sets $\{A_i\}$ and $\{B_i\}$ in $E$ such that $E = \bigcup_i A_i = \bigcup_i B_i$, and for each $i$, $f_{+c}$ is *SRI* on $A_i$ and $f_{-c}$ is *SRD* on $B_i$. Hence, for each $i$ and $j$, $f$ is *RL*$(c)$ on the set $E_{i,j} = A_i \cap B_j$. Now since $f$ is clearly right continuous on $E$, it follows from Lemma 2.2.6 that $f$ is *RL*$(c)$ on $E \cap (E_{i,j})^- \cap \text{co}(E_{i,j})$. Consequently, there is a sequence $\{F_{i,j,k}\}$ of closed sets in $E$ such that $E_{i,j} \subset \bigcup_k F_{i,j,k}$ and $f$ is *RL*$(c)$ on each $F_{i,j,k}$. Now since $E = \bigcup_i \bigcup_j \bigcup_k F_{i,j,k}$, this proves that $f$ is *F-GRL*$(c)$ on $E$.

Next, suppose $D_+f$ and $D^+f$ are finite at the points of $E$. For each positive integer $n$, set

$$E_n = \{x \in E : D_+f(x) > -n, D^+f(x) < n\}.$$

Then $E = \bigcup_n E_n$; and for each $n$ it follows from above that $E_n = \bigcup_i E_{n,i}$, where $f$ is *RL*$(n)$ on each $E_{n,i}$. Now, for each $n$ and $i$, it follows from Lemma 2.2.6 as above that $E_{n,i} \subset \bigcup_j F_{n,i,j}$, where $\{F_{n,i,j}\}$ is a sequence of closed sets in $E$ on which $f$ is *RL*$(n)$. Consequently, $f$ if *F-GRL* on $E$. □

2.2.8. REMARK. It should be pointed out here that on applying the First Derivate Theorem 2.2.2 to $-f$, $f_-$ and $-f_-$, similar results are obtained on generalized *SRD*, *SLD*, and *SLI* properties (see Lemma 1.1.2). Similarly, on applying the Second Derivate Theorem 2.2.7 to $f_-$, a similar result is obtained in terms of generalized left Lipschitz property.

It may be further noted here that on combining the results of this section with the ones obtained on applying them to $-f_-$, similar results are obtained on bilateral properties like bi-monotonicity, *BLL*, *BL*, and *BMT* in terms of bilateral derivates under certain continuity hypotheses. However, we do not include them here since they hold under much weaker continuity hypotheses as will be seen in a subsequent paper [G18] where we present a similar unified theory of bilateral derivates. The same holds for the other results of this chapter.

## §2.3 Extension of two theorems of G. C. Young and W. H. Young.

In this section we obtain from the two fundamental theorems of the last section an extended version of a basic theorem of G. C. Young on unilateral derivates, along with a unilateral version of a theorem of W. H. Young on infinite derivates.

Given $f: X \to \mathbb{R}$, we will call $f$ *angular* at a point $x \in X^b$ if either $D^-f(x) < D_+f(x)$ or $D^+f(x) < D_-f(x)$. Further, if $f$ is not angular at $x \in X^b$, then $f$ will be called *nonangular* at $x$; and $f$ will be called simply nonangular if it is so at each point of $X^b$.

We first deduce from the First Derivate Theorem the following extended version of G. C. Young's theorem [Y.G1], which was obtained by her in the case when $X$ is an interval.

2.3.1. THEOREM (G. C. YOUNG). *Every function $f: X \to \mathbb{R}$ is nonangular at all but a countable set of points.*

PROOF. Let $E_1$ and $E_2$ denote the sets of points $x$ in $X^b$ where $D^-f(x) < D_+f(x)$ or $D^+f(x) < D_-f(x)$ respectively.

First, suppose $E_1$ is uncountable. Then it is clear that there is a rational number $r$ such that the set

$$A = \{x \in E_1 : D^-f(x) < r < D_+f(x)\}$$

is uncountable. Then $D^-f_{-r} < 0 < D_+f_{-r}$ at each point of $A$, and hence it follows from Theorem 2.2.2 that the function $f_{-r}$ is *GSLD* and *GSRI* on $A$. Consequently, there is an uncountable subset $A_0$ of $A$ such that $f_{-r}$ is decreasing as well as increasing on $A_0$, which is of course not possible. This proves that $E_1$ is countable.

Next, on applying the above result to $-f$ it follows that $E_2$ is countable. Consequently, $f$ is nonangular n.e. in $X$. □

Given a function $f: I \to \mathbb{R}$, W. H. Young proved in [Y1] that the set of points where $f$ has at least one derivate infinite is always a $G_\delta$-set; see [G11, Theorem 3.1] for an extension of this theorem to a function defined on any set $X \subset \mathbb{R}$. In the following theorem we deduce a similar result on right derivates from the Second Derivate Theorem.

2.3.2. THEOREM. *Let $f: X \to \mathbb{R}$. The set of points where at least one right derivate of $f$ is infinite is of the form $G_\delta \cup C$ in $X$, where $C$ is a countable set of points where $f$ is not $WC_+$.*

PROOF. Since $X \sim X^+$ is countable, $X^+$ is a $G_\delta$-set in $X$. Hence there is no loss of generality in assuming $X \subset X^+$.

Let $E$ denote the set of points in $X$ where at least one right derivate of $f$ is infinite, and set $A = X \sim E$. Then, by Theorem 2.2.7, there is a sequence of sets $\{A_n\}$ in $A$ such that $A = \bigcup_n A_n$ and $f$ is *RL* on each $A_n$. Now let $C$ denote the countable set of points in $X$ where $f$ is not $WC_+$ (see Theorem 1.1.1). Then, for each $n$, it follows from Lemma 2.2.6 that $f$ is *RL* on $B_n = X \cap \bar{A}_n \cap \operatorname{co}(A_n) \sim C_n$ where $C_n$ is a subset of $C$. It is now clear that $A_n \subset B_n \subset A$, and that $B_n = F_n \sim C_n$, where $F_n$ is an $F_\sigma$-set in $X$. Hence $A = \bigcup_n B_n = \bigcup_n (F_n \sim C_n)$ is of the form $F_\sigma \sim C_0$ in $X$, where $C_0 \subset C$. Consequently, $E$ is of the required form in $X$. □

2.3.3. REMARK. The term "angular" seems to have been used first by Sierpiński [Si1]. He used it for points where a function has two distinct left and right derivatives, and it was proved by him and Levy [Lev] that for a continuous function such points form a countable set. Sierpiński also used the term "cusp" for the graph of the function at the points corresponding to such points. G. C. Young [Y.G1] obtained the result in Theorem 2.3.1 a little later for a function defined on an interval using points of extrema. The above proof of Theorem 2.3.1 is of course totally different.

## §2.4 Properties on some portion in terms of unilateral derivates.

In this section we deduce two other theorems on derivates from the two fundamental theorems of §2.2. These theorems will be called the Third and Fourth Derivate Theorems since they are also found to play a basic role in the development of the unified theory (see, e.g., the next section). In these theorems we obtain unilateral monotonicity and Lipschitz property of a function on some portion of the set under consideration.

2.4.1. THEOREM (THIRD DERIVATE). *Let $f: X \to \mathbb{R}$, $E \subset X$ and suppose $D_+ f > 0$ at a nonmeager set of points in E. Then f is SRI on a dense subset of some portion $E_0$ of E which includes the points of $E_0$ where f is QLSC relative to E. Consequently, if f is QLSC relative to E, then it is SRI on some portion of E.*

PROOF. Let $A$ denote the set of points in $E$ where $D_+ f > 0$. Then, by the First Derivate Theorem 2.2.2, $A = \bigcup_n A_n$, where $f$ is *SRI* on each $A_n$. Since $A$ is by hypothesis nonmeager in $E$, there is an $n$ such that $A_n$ is dense in some portion $E_0$ of $E$. By replacing $E_0$ by $E_0 \cap \operatorname{co}(A_n)$ if necessary, we can assume here that $E_0 \subset \operatorname{co}(A_n)$.

Now let $B$ denote the set of points in $E_0$ where $f$ is *QLSC* relative to $E$. Then it follows from the First Extension Lemma 2.2.1 that $f$ is *SRI* on $(E_0 \cap A_n) \cup B$, which proves the first part. The last part is of course a direct consequence of the first. □

We include here two consequences of the above theorem dealing with the properties *RLL* and *RMT*.

2.4.2. COROLLARY. *Let $f: X \to \mathbb{R}$, $E \subset X$, and suppose f is QLSC relative to E. If $D_+ f > -\infty$ at a nonmeager set of points in E, then f is RLL on some portion of E.*

It is clear from the hypothesis that there is a positive integer $n$ such that the set $A = \{x \in E : D_+ f > -n\}$ is nonmeager in $E$. Then since $D_+ f_{+n}(x) > 0$ for each $x \in A$, $f_{+n}$ is by the above theorem *SRI* on some portion $E_0$ of $E$, and so $f$ is *RLL* on $E_0$.

2.4.3. COROLLARY. *Let $f: X \to \mathbb{R}$, $E \subset X$ and suppose f is QBSC relative to E. If the set $E \sim K_+(f)$ or $E \sim K_b(f)$ is nonmeager in E, then f is RMT or MT, respectively, on some portion of E.*

If $E \sim K_+(f)$ is nonmeager in $E$, then either $D_+f > -\infty$ or $D^+f < \infty$ at a nonmeager set of points in $E$. Hence it follows from the above corollary that $f$ is either *RLL* or *RUL* on some portion of $E$, and so it is *RMT* on that portion of $E$. Further, if $E \sim K_b(f)$ is nonmeager in $E$, then either $E \sim K_+(f)$ or $E \sim K_-(f)$ is nonmeager in $E$, and so the result follows from the one just established.

Next, in the following theorem we deduce a similar result on right Lipschitz property from the Second Derivate Theorem.

2.4.4. THEOREM (FOURTH DERIVATE). *Let $f : X \to \mathbb{R}$, $E \subset X$ and suppose $f$ is $PWC_+$ on $E$. If $D_+f$ and $D^+f$ are finite at a nonmeager set of points in $E$, then $f$ is RL on some portion of $E$.*

*Moreover, if there is some $c \in \mathbb{R}_+$ such that $D_+f > -c$ and $D^+f < c$ at a nonmeager set of points in $E$, then $f$ is $RL(c)$ on some portion of $E$.*

PROOF. It is enough to prove here the second part, for if $D_+f$ and $D^+f$ are finite at a nonmeager set of points in $E$, then there clearly exists a positive integer $c$ for which the hypothesis of the second part holds.

Hence suppose there is some $c \in \mathbb{R}_+$ such that the set

$$A = \{x \in E : D_+f(x) > -c,\ D^+f(x) < c\}$$

is nonmeager in $E$. Then, by Theorem 2.2.7, $A = \bigcup_n A_n$, where $f$ is $RL(c)$ on each $A_n$. Clearly, there is an $n$ such that $A_n$ is dense in some portion $E_0$ of $E$. Hence it follows from the Second Extension Lemma 2.2.6 that $f$ is $RL(c)$ on $E_0 \cap \mathrm{co}(A_n)$, which in turn includes some portion of $E$. □

2.4.5. REMARK. It may be observed here that the Third Derivate Theorem 2.4.1, which is indeed of a topological nature, does not hold in general on replacing the term "nonmeager" by "positive measure," or even "almost everywhere"; see Remark 2.8(b) of [G11, p. 303] for a counterexample.

## §2.5 Extension of some theorems of W. H. Young and A. Denjoy.

In this section we deduce from the Third Derivate Theorem strengthened versions of some basic theorems of W. H. Young and Denjoy.

We begin with Denjoy's first theorem on derivates (see [D1, p. 149] and [D5, p. 187]). He obtained the following theorem for a continuous function $f$ on an interval, and for a perfect set $E$ and $c \in \mathbb{R}$. We consider here only one part of the Denjoy's theorem; his other two parts follow directly from the following strengthened version of the part considered. It may be recalled here that we use $Qf(a, b)$ to denote the difference quotient $\{f(b) - f(a)\}/(b - a)$ when $b \neq a$.

2.5.1. THEOREM (DENJOY). *Let $f : X \to \mathbb{R}$, $c \in \overline{\mathbb{R}}$, and suppose $E \subset X$ is dense in itself. If there is a sequence of points $\{a_n\}$ in $E$ where $f$ is QLSC relative* *to E such that the points $a_n$ are dense in E, and a sequence of points $\{b_n\}$ in X such that $b_n > a_n$ for each n, $b_n - a_n \to 0$ as $n \to \infty$, and $\limsup_{n\to\infty} Qf(a_n, b_n) \le c$, then $D_+f \le c$ r.e. in E.*

PROOF. First, let $c = 0$, and set $A = \{a_n : n = 1, 2, \dots\}$. Suppose the hypothesis holds for $c = 0$ but $D_+f > 0$ at a nonmeager set of points in $E$. Then by Theorem 2.4.1 there is a portion $E_0$ of $E$ such that $f$ is *SRI* on a dense subset $B$ of $E_0$ which includes the set $A_0 = A \cap E_0$. Hence there is a real number $r > 0$ such that $f_{-r}$ is *WRI* on $B$. Now since $E_0$ is dense in itself, the set $A_0$ is clearly infinite, and so according to the hypothesis we can find an $n$ such that $a_n \in A_0$, $b_n \in \text{co}(B)$, and $Qf(a_n, b_n) < r$. But this contradicts the fact that $f_{-r}$ is *WRI* on $B$. Hence the result holds for $c = 0$.

Next, if $c$ is any real number other than zero, the result follows on applying the above to the function $f_{-c}$. Further, if the hypothesis holds for $c = -\infty$, it also holds for $c = -n$ for each positive integer $n$, and so, for each $n$, $D_+f \le -n$ r.e. in $E$ by the result just proved, which in turn implies that $D_+f = -\infty$ r.e. in $E$. Finally, in the case when $c = \infty$, the result is of course obvious. □

As an immediate consequence of the above theorem we obtain the following result which was obtained by Denjoy [D1] for continuous functions.

2.5.2. COROLLARY. *Let $f : X \to \mathbb{R}$, $c \in \overline{\mathbb{R}}$, and suppose $E \subset X$ is dense in itself. If there is a dense set of points in E where f is QLSC relative to E and $D_+f \le c$, then this inequality holds r.e. in E.*

We include here an easy but interesting consequence of the above corollary the last part of which is obtained with the help of Theorem 2.3.1.

2.5.3. COROLLARY. *Let $f : X \to \mathbb{R}$, $c \in \mathbb{R}$, and suppose $E \subset X$ is dense in itself. If either*

(i) $\{x \in E : D_+f(x) = c\}$ *is dense in E, or*

(ii) $\{x \in E \cap E^{db} : \overline{D}f(x) \le c\}$ *is dense in E, or*

(iii) $\{x \in E : D^-f(x) \le c\}$ *is uncountable in each portion of E,*

*then $D_+f \le c$ r.e. in E.*

To deal with the next theorem, we need here another notion of derivates.

Given $f : X \to \mathbb{R}$ and $x \in X^d$, we will call the lower and upper limits of the quotient $Qf(y, z)$ as $y$, $z$ approach $x$ through $X$ the *lower* and *upper strong derivates* of $f$ at $x$, and denote them by $\underline{D}_*f(x)$ and $\overline{D}_*f(x)$, respectively. When these two derivates are equal, $f$ will be said to be *strongly derivable* at $x$ and the common value of the two derivates will be called the *strong derivative* of $f$ at $x$ and denoted by $f'_*(x)$.

The strong derivates and strong derivative were introduced by Peano [P] in 1892 without any nomenclature. These derivates were called later by

Bruckner and Goffman [BGo] "unstraddled" derivates, but we use the term "strong" here since this definition of a finite strong derivative is equivalent in the present setting to another definition of strong derivative employed by Cartan [Car, p. 48] in Banach spaces. We will return to this notion later in Chapter 14.

In the following theorem we deal with a consequence of Denjoy's Theorem 2.5.1 which deals with the identity of unilateral and strong derivates; (see Remark 2.5.7 for further historical background of this result). Another generalized version of this theorem will be obtained later in §3.4.

2.5.4. THEOREM (DENJOY). *Suppose $f : X \to \mathbb{R}$, where $X$ is dense in itself. If $f$ is QLSC, then $D_+ f = \underline{D}_* f$ and $D^- f = \overline{D}_* f$ r.e. in $X$.*
*Similarly, if $f$ is QUSC, then $D_- f = \underline{D}_* f$ and $D^+ f = \overline{D}_* f$ r.e. in $X$.*

PROOF. It is enough to prove here the first part, for the second part follows on applying it to $-f$.

Hence, suppose $f$ is *QLSC*. Since $X \sim X^+$ is clearly meager in $X$, and $\underline{D}_* f \le D_+ f$ at each point of $X^+$, to obtain the first relation it is enough to show that the set $A = \{x \in X^+ : \underline{D}_* f(x) < D_+ f(x)\}$ is meager in $X$. Suppose it is not so. Then there clearly exists a rational number $r$ such that the set

$$B = \{x \in X : \underline{D}_* f(x) < r < D_+ f(x)\}$$

is nonmeager in $X$. Hence $B$ is dense and nonmeager in some portion $X_0$ of $X$, and so it follows clearly from Theorem 2.5.1 that $D_+ f \le r$ r.e. in $X_0$. This contradiction proves the first relation, and the second relation follows from the first on applying it to $f_-$. □

Finally, in the next two theorems we obtain strengthened versions of two theorems of W. H. Young [Y3, pp. 304, 305] who obtained them for continuous functions on an interval.

2.5.5. THEOREM (YOUNG). *Suppose $f : X \to \mathbb{R}$, where $X$ is dense in itself. If $f$ is QLSC, then $D_+ f$ is LSC r.e. in $X$ and $D^- f$ is USC r.e. in $X$.*

PROOF. It is easy to see from the definitions of strong derivates that $\underline{D}_* f$ is *LSC* and $\overline{D}_* f$ is *USC*. Further, $\underline{D}_* f \le D_+ f$ and $D^- f \le \overline{D}_* f$ wherever these derivates are defined. Hence the result follows immediately from Theorem 2.5.4. □

The following theorem, which was deduced by Young from his above theorem, follows directly on combining the two parts of Theorem 2.5.4.

2.5.6. THEOREM (YOUNG). *Suppose $f : X \to \mathbb{R}$, where $X$ is dense in itself. If $f$ is QBSC, then $D_+ f = D_- f$ and $D^+ f = D^- f$ r.e. in $X$.*

2.5.7. Remark. The results of this section have a long and fascinating history. We confine the following discussion to continuous functions on an interval since the results were discovered in the past only in that case.

It was indeed W. H. Young [Y3] who first obtained Theorems 2.5.5 and 2.5.6 in 1908. Shortly after, in 1915, Denjoy [D1] obtained Theorem 2.5.1. We have called Theorem 2.5.4 also Denjoy's theorem for the following two reasons. First, as seen above, it is essentially a direct consequence of his Theorem 2.5.1. Second, Choquet [C1, C2] obtained another consequence of Theorem 2.5.1, in 1947, in terms of the identity of contingent and paratingent of the graph of $f$ at a residual set of points, which he called Denjoy's theorem and which implies Theorem 2.5.4 (see Remark 3.3.7 and Corollary 3.4.2 for some details).

It should be noted here, however, that Denjoy's Theorem 2.5.1 is much deeper than Theorem 2.5.4, since the former holds for any dense in itself subset $E$ of the domain of the function. Indeed, Denjoy immediately recognized the strength of his theorem and deduced several results from the original version of his theorem, some of which played a significant role in the development of the theory of his now well known Denjoy-Perron integral (see [D2, D4]). Due to the generalization of Denjoy's theorem obtained in Theorem 2.5.1, it is clear that his other results can also be generalized in a similar manner.

It is somewhat surprising that Denjoy's Theorem 2.5.1, although discovered in 1915, went unnoticed and was somehow not included in Hobson's treatise [H1, H2] or Saks' monographs [S6, S8]; this may be due to the complicated nature of its enunciation. However, in the absence of any knowledge of Denjoy's theorem, Theorem 2.5.4 was rediscovered by Bruckner and Goffman [BGo] in 1966 using cluster sets. See Manna and Mukhopadhyay [MM] for an extension of a part of their result from continuous to *LSC* functions.

The Young's Theorem 2.5.6 was in turn rediscovered by Neugebauer [Ne1] in 1962 with a totally different proof, and he generalized it subsequently in [Ne2] to measurable symmetric functions. As will be seen later in §12.3, this generalization is also included in the above generalized version of Theorem 2.5.6.

# *CHAPTER 3*

# Baire Class and Measurability of Derivates and Medians

In this chapter we deal with the Baire class and measurability of unilateral derivates and medians, and we obtain a refined version of a theorem of A. P. Morse on unilateral derivates.

First, in §3.1, we deduce from the First Derivate Theorem a general theorem on unilateral derivates which we call the Fifth Derivate Theorem. Then, in §3.2, we deduce from this theorem two theorems on the Baire class and measurability of unilateral derivates which are found to be the best possible in their direction; the results of this section also have an interesting past.

Next, in §3.3, we define various set-valued medians associated with unilateral derivates and obtain results on the Baire class and measurability of these multifunctions. Further, in §3.4, we obtain a generalized version of the Denjoy's theorem on the identity of unilateral and strong derivates along with another related theorem.

Finally, in §3.5, we deduce from the First and Third Derivate Theorems a refined version of a somewhat surprising theorem of A. P. Morse on unilateral derivates. Also, in the First Monotonicity Theorem we obtain a similar refined version of his monotonicity theorem in terms of unilateral derivates.

**§3.1 A general theorem on unilateral derivates.** In this section we deduce from the First Derivate Theorem a general theorem on unilateral derivates which we call the Fifth Derivate Theorem. This theorem will be used in the next section to obtain the Baire class and measurability of these derivates.

Given a function $f: X \to \mathbb{R}$ and $E \subset X$, we will use $\underline{S}_E(f)$ and $\overline{S}_E(f)$ to denote the sets of points in $E$ where $f$ is *LSC* and *USC*, respectively, relative to $E$. Also, when $E = X$, these two sets will be denoted simply by $\underline{S}(f)$ and $\overline{S}(f)$, respectively.

3.1.1. THEOREM (FIFTH DERIVATE). *Let* $f: X \to \mathbb{R}$, $E \subset X^+$ *and* $\underline{S} = \underline{S}_E(f)$. *Then for every* $c \in \overline{\mathbb{R}}$, *the set* $\{x \in E : D_+ f(x) > c\}$ *is of the form* $(F_\sigma \cap \underline{S}) \cup C$ *in* $E$, *where* $C$ *is a countable subset of* $E \sim \underline{S}$.

PROOF. Set, for each $c \in \overline{\mathbb{R}}$, $E_c = \{x \in E : D_+ f(x) > c\}$.

First, let $c = 0$. Then, by the First Derivate Theorem 2.2.2, $E_0 = \bigcup_n A_n$ where $f$ is *SRI* on each $A_n$. Hence, given $n$, it follows from Lemma 2.2.1 that $f$ is *SRI* on the set

$$B_n = \left(\overline{A}_n \cap \underline{S}\right) \cap \operatorname{co}(A_n).$$

Now let $B'_n = B_n \sim \{\sup B_n\}$. Then it is clear that $B'_n \subset E_0$ and that $B'_n$ is an $F_\sigma$-set in $\underline{S}$. Hence $\bigcup_n B'_n$ also is an $F_\sigma$-set in $\underline{S}$, and since $E_0 \cap \underline{S} \sim \bigcup_n B_n$ is clearly countable, it follows that $E_0 \cap \underline{S}$ is an $F_\sigma$-set in $\underline{S}$.

Now set $P = E \sim \underline{S}$, and let $Q$ be the set of points in $P$ where $f$ has equal left and right lower limits relative to $E$. Then for each $x \in Q$, $f(x) > f\!y(x+0)$, so that $D_+ f(x) = -\infty$. Consequently, $E_0 \sim \underline{S} \subset P \sim Q$, which in turn is countable by Theorem 1.1.1. This proves the result for $c = 0$.

Next, for any real number $c$ other than zero, the result follows on applying the above to the function $f_{-c}$. This further yields the result for $c = -\infty$, for $E_{-\infty} = \bigcup_n E_{-n}$. The result for $c = \infty$ is on the other hand obvious since $E_\infty = \varnothing$. □

From the above theorem we obtain the following result on the set complementary to the one considered in the theorem. The last part of this result has been obtained by Zahorski [Z4] in the particular case when $E = X = \mathbb{R}$ and $c \in \mathbb{R}$.

3.1.2. COROLLARY. *Let* $f: X \to \mathbb{R}$, $E \subset X^+$ *and* $\underline{S} = \underline{S}_E(f)$. *Then for every* $c \in \overline{\mathbb{R}}$, *the set* $A = \{x \in E : D_+ f(x) \le c\}$ *is of the form* $G_\delta \cup (G_\delta \sim \underline{S})$ *in* $E$. *Consequently,* $A$ *is of the form* $G_\delta \cup H$ *in* $E$, *where* $H$ *is a set of points where* $f$ *is not continuous relative to* $E$.

The first part of this corollary follows easily from the above theorem, and its second part is obtained from the first since the points in $E$ where $f$ is continuous relative to $E$ form a $G_\delta$-set in $E$.

## §3.2 Baire class and measurability of unilateral derivates.

In this section we investigate the Baire class and measurability of unilateral derivates.

Although the bilateral derivates of a function $f: X \to \mathbb{R}$ are known to be always of Baire class 2 (see, e.g., [G11], p. 313), the unilateral derivates of $f$ are not even measurable in general, and can indeed be of any given Baire class without being of a lower class (see Example 3.2.8(c)). A brief discussion of the historical background of this problem is in order here.

Suppose $f : I \to \mathbb{R}$. The first result obtained in this direction was due to W. H. Young, who proved [Y3] in 1908 that if $f$ is continuous, then $D_+f$ is of $LB_1$. It was indeed from this theorem that he deduced his Theorem 2.5.5.

Next, given any countable ordinal $\alpha$, it was proved by Sierpiński [Si3] in 1922 that if $f$ is of $B_\alpha$, then $D_+f$ is of $B_{\alpha+3}$. In response to a problem posed by Sierpiński, it was proved the same year by Banach [B2] that if $f$ is (Lebesgue) measurable, then so is $D_+f$; and he proved further that when $f$ is bounded and of $B_\alpha$, then $D_+f$ is of $B_{\alpha+2}$.

Strengthening the above results it has been proved by Mišík [Mis1] that if $f$ is of $B_\alpha$, then $D_+f$ is of $LB_{\alpha+1}$. Thus $D_+f$ is of $B_{\alpha+2}$ whenever $f$ is of $B_\alpha$, and in connection with this result Mišík posed the problem ([Mis2], p. 25) as to whether this result is the best possible for $\alpha > 0$.

We deduce here from the Fifth Derivate Theorem results which strengthen all the above-mentioned results. These results provide a negative response to Mišík's problem (see Corollary 3.2.5) and are found to be the best possible in their direction (see Example 3.2.8). Also, we will obtain the results relative to any set in the domain of the function which are somewhat stronger (see Remark 3.2.7).

It is necessary to introduce here some notations and conventions.

Given $f : X \to \mathbb{R}$ and $E \subset X$, we will call $f$ *(Lebesgue) measurable on E*, or *relative* to $E$, if its restriction to $E$ is measurable with respect to the extended Lebesgue measure $m_E$ on $E$ (see §1.4 for definition). Similarly, $f$ will be said to be of some *Baire class on E*, or *relative* to $E$, if its restriction to $E$ is so with respect to the relative topology of $E$. Further, any derivate of $f$ will be said to be measurable or of some Baire class on $E$ if it is so on the set of points in $E$ where it is defined.

Next, a subset of $X$ will be called measurable or a Borel set of some given class if it is so relative to $X$. Finally, for any set $E \subset X$ and $\alpha < \Omega$, we will use $\Sigma_\alpha(E)$, $\Pi_\alpha(E)$, and $\Delta_\alpha(E)$ to denote the classes of sets in $E$ which are of additive, multiplicative, or ambiguous class $\alpha$, respectively, relative to $E$.

We first deduce from the Fifth Derivate Theorem the Baire class and measurability of $D_+f$ in terms of the properties of the set $\underline{S}(f)$. The results of the following lemma are indeed more general than the ones of the theorems deduced from it (see Remark 3.2.7).

3.2.1. Lemma. *Let $f : X \to \mathbb{R}$, $E \subset X$, $\alpha < \Omega$, and $\underline{S} = \underline{S}_E(f)$.*

(a) *If $\alpha > 0$ and $\underline{S}$ is of $\Sigma_\alpha(E)$, then $D_+f$ is of $LB_\alpha(E)$.*

(b) *If $\alpha > 1$ and $\underline{S}$ is of $\Pi_\alpha(E)$, then $D_+f$ is of $UB_\alpha(E)$.*

(c) *If $\alpha > 1$ and $\underline{S}$ is of $\Delta_\alpha(E)$, then $D_+f$ is of $B_\alpha(E)$.*

(d) *If $\underline{S}$ is Lebesgue or Borel measurable in E, then so is $D_+f$ relative to E.*

Proof. To prove (a), suppose $\alpha > 0$ and $\underline{S}$ is of $\Sigma_\alpha(E)$. Then for each $c \in \mathbb{R}$ it follows from Theorem 3.1.1 that the set $\{x \in E : D_+f(x) > c\}$ is of the form $(F_\sigma \cap \Sigma_\alpha) \cup F_\sigma$ in $E$, and thus is of $\Sigma_\alpha(E)$. Consequently, $D_+f$ is of $LB_\alpha(E)$.

Next, to prove (b), suppose $\alpha > 1$ and $\underline{S}$ is of $\Pi_\alpha(E)$. Given $c \in \mathbb{R}$, let $B = \{x \in E : D_+ f(x) < c\}$. Then since

$$B = \bigcup_n \{x \in E : D_+ f(x) \le c - 1/n\},$$

it follows from Corollary 3.1.2 that $B$ is of the form $G_{\delta\sigma} \cup (G_{\delta\sigma} \cap \Sigma_\alpha)$ in $E$, and thus is of $\Sigma_\alpha(E)$. Consequently, $D_+ f$ is of $UB_\alpha(E)$.

Part (c) follows directly from (a) and (b), and (d) is obtained by a similar argument. □

In part (a) of the following theorem we obtain generalized versions of the two theorems of W. H. Young [Y3] mentioned earlier.

3.2.2. THEOREM (YOUNG). *Let* $f : X \to \mathbb{R}$ *and* $E \subset X$.

(a) *If $f$ is* LSC *relative to $E$, then $D_+ f$ is of $LB_1(E)$, and $D_+ f$ is* LSC *relative to $E$ r.e. in $E$.*

(b) *If $f$ is either* LSC *or* USC *relative to $E$ n.e. in $E$, then each unilateral derivate of $f$ is of $B_2(E)$.*

PROOF. Part (a) follows directly from part (a) of the above lemma and the generalized version of the Baire-Young theorem presented earlier in Theorem 1.4.1.

To obtain (b), it is clearly enough to prove the result for $D_+ f$. Set $\underline{S} = \underline{S}_E(f)$ and $\overline{S} = \overline{S}_E(f)$. If the set $E \sim \underline{S}$ is countable, then $\underline{S}$ is a $G_\delta$-set in $E$, and so it is of $\Delta_2(E)$. If on the other hand $E \sim \overline{S}$ is countable, then since $\underline{S} \cap \overline{S}$ is a $G_\delta$-set in $E$, the set $\underline{S} = (\underline{S} \cap \overline{S}) \cup (\underline{S} \sim \overline{S})$ is again of $\Delta_2(E)$. Hence in any case it follows from part (c) of the above lemma that $D_+ f$ is of $B_2(E)$. □

In the following lemma we investigate the properties of $\underline{S}(f)$ in terms of those of $f$.

3.2.3. LEMMA. *Let* $f : X \to \mathbb{R}$ *and* $\alpha < \Omega$.

(a) *If $f$ is of $LB_\alpha$, then $\underline{S}(f)$ is of $\Pi_\alpha(X)$.*

(b) *If $f$ is Lebesgue or Borel measurable, then so is $\underline{S}(f)$ in $X$.*

PROOF. Set $\underline{S} = \underline{S}(f)$, and let $Q$ denote the set of all rational numbers. Then it is easy to see that

$$(1) \qquad X \sim \underline{S} = \bigcup_{\substack{r,s \in Q \\ r > s}} \left[\{x \in X : f(x) > r\} \cap \{x \in X : f(x) < s\}^{-}\right].$$

If $f$ is of $LB_0$, we have indeed $\underline{S} = X$, which of course is closed in $X$. Hence to prove (a), suppose $f$ is of $LB_\alpha$, where $\alpha > 0$. Then it is clear from (1) that

$X \sim \underline{S}$ is of $\Sigma_\alpha(X)$, and thus $\underline{S}$ is of $\Pi_\alpha(X)$. Part (b) also follows from (1) in a similar manner. □

The following theorem follows now directly from Lemmas 3.2.3 and 3.2.1.

3.2.4. THEOREM. *Let $f: X \to \mathbb{R}$ and $E \subset X$.*
(a) *If $f$ is of $LB_1(E)$, then $D_+f$ is of $B_2(E)$.*
(b) *If $f$ is of $LB_\alpha(E)$, where $\alpha > 1$, then $D_+f$ is of $UB_\alpha(E)$.*
(c) *If $f$ is Lebesgue or Borel measurable on $E$, then each of its unilateral derivates is Lebesgue or Borel measurable, respectively, on $E$.*

Judging from Theorems 3.2.2 and 3.2.4 together, it is somewhat surprising to note the reversal in the roles of lower and upper Baire classes as $\alpha$ crosses 1.

In terms of the ordinary Baire class, from the parts (a) and (b) of Theorem 3.2.4 we obtain the following strengthened version of the results of Sierpiński, Banach, and Mišík mentioned earlier; the above-mentioned problem of Mišík is also resolved by the following result.

3.2.5. COROLLARY. *Let $f: X \to \mathbb{R}$ and $E \subset X$. If $f$ is of a Baire class $\alpha > 0$ on $E$, then each unilateral derivate of $f$ is of Baire class $\alpha + 1$ on $E$.*

Finally, as an application of Theorem 3.2.4 we include here some interesting consequences of the results of §2.4 which may be compared with Corollary 2.2.5.

3.2.6. THEOREM. *Let $f: X \to \mathbb{R}$, $c \in \mathbb{R}_+$, and suppose $X$ is a Borel set in $\mathbb{R}$. If $f$ is Borel measurable and the set $A$ of points in $X$ where (i) $D_+f > 0$, (ii) $D_+f > -\infty$, (iii) $f$ is not right knotted, (iv) $D_+f$ and $D^+f$ are finite, or (v) $D_+f > -c$ and $D^+f < c$, is uncountable, then there is a nonempty perfect set $P$ in $X$ on which $f$ is SRI, RLL, RMT, RL, or RL(c), respectively.*

*Moreover, if $f$ is Lebesgue measurable and $|A| > 0$, then in each case $P$ can be chosen to be metrically dense in itself.*

PROOF. We will first consider case (i). Hence, suppose the set $A = \{x \in X: D_+f(x) > 0\}$ is uncountable. First, suppose $f$ is Borel measurable. Then by Theorem 3.2.4, $A$ is a Borel set in $\mathbb{R}$. Let $B$ denote the set of points in $A$ where $f$ is *LSC*. Then since $A \sim B$ is clearly countable (see Theorem 1.1.1), $B$ is an uncountable subset of $A$ which is also a Borel set in $\mathbb{R}$. Hence $B$ includes some nonempty perfect set $E$ (see [K1], p. 44). Now it follows from Theorem 2.4.1 that $f$ is *SRI* on some portion of $E$, which clearly includes a perfect set $P$ on which $f$ is *SRI*.

Next, suppose $f$ is Lebesgue measurable and $|A| > 0$. Then since $A$ is, by Theorem 3.2.4, Lebesgue measurable in $\mathbb{R}$, so is $B$ and $|B| > 0$. Hence by a theorem of Lusin (see, e.g., [H1], p. 192) the perfect subset $E$ of $B$ can be chosen to be metrically dense in itself. Now the portion of $E$ on which $f$ is *SRI* includes a nonempty, metrically dense in itself, perfect set $P$.

In cases (ii), (iv), and (v), the result is obtained by a similar argument from Corollary 2.4.2 and the two parts of Theorem 2.4.4, respectively, and in case (iii) the result follows directly from that of (ii). □

3.2.7. Remark. It is interesting to note here that in the results of this section the hypotheses involve only properties of $f$ relative to $E$, whereas their conclusions deal with properties of derivates of $f$ on $E$ which also depend on the values of $f$ outside $E$. Thus as long as $f$ is, for example, measurable relative to $E$, the measurability of its unilateral derivates on $E$ remains unaltered by any change in the values of $f$ outside $E$.

Further, as mentioned earlier, the results of Lemma 3.2.1 are considerably more general than the ones of the theorems deduced from it. For, let $f$ be, for example, any discontinuous solution of the equation $f(x+y) = f(x) + f(y)$, $x, y \in \mathbb{R}$, and $E = \mathbb{R}$. Then $f$ is nonmeasurable, and since $\underline{S}(f) = \varnothing$, $f$ satisfies the hypothesis of each part of Lemma 3.2.1, but none of the hypotheses of Theorems 3.2.2 and 3.2.4 holds for $f$.

3.2.8. Example. We include here some simple examples to show that the results of this section are the best possible in their direction, and that a unilateral derivate can be of any given Baire class without being of a lower class. We will assume here that $E = X$.

(a) We first show that the conclusion of Theorem 3.2.2(a) cannot be improved even when $f$ is a Lipschitz function. Let $I = [0,1]$, and let $A$ be a measurable subset of $I$ such that $|J \cap A| > 0$ and $|J \sim A| > 0$ for each subinterval $J$ of $I$. Define $f$ to be the indefinite integral of the characteristic function of $A$ on $I$. Then $f$ is a Lipschitz function with derivatives 0 and 1 at two metrically dense sets of points in $I$. The set $\{x \in I : D_+ f(x) \leq 0\}$ is thus a dense $G_\delta$-set in $I$ (see Corollary 3.1.2), and so it is residual in $I$. Consequently, the set $\{x \in I : D_+ f(x) \geq 1\}$ which is also dense in $I$ cannot be a $G_\delta$-set in $I$. Hence $D_+ f$ is not of $UB_1$.

This example shows further that Corollary 3.2.5 does not hold for $\alpha = 0$, and that its conclusion cannot be improved at least in the case when $\alpha = 1$.

(b) We next show that Theorem 3.2.2(a) does not hold if $f$ is *USC* in the place of *LSC*. Let $\{r_n\}$ be an enumeration of the set $Q$ of all rational numbers, and define $f(r_n) = r_n + 1/n$ for each $n$ and $f(x) = x$ for $x \in \mathbb{R} \sim Q$. Then $f$ is clearly USC, but since $D_+ f = -\infty$ or 1 according to whether $x \in Q$ or $\mathbb{R} \sim Q$, and $\mathbb{R} \sim Q$ is not an $F_\sigma$-set, $D_+ f$ is not of $LB_1$.

Also, since $\underline{S}(f) = \mathbb{R} \sim Q$ is a $G_\delta$-set, this example shows further that for $\alpha = 1$, Lemma 3.2.1(a) does not hold if $\underline{S}(f)$ is a $G_\delta$-set in the place of an $F_\sigma$-set.

(c) Let $C_1$ and $C_2$ be any two mutually disjoint countable dense subsets of $\mathbb{R}$. Given an arbitrary set $P \subset \mathbb{R}$, set $A = (P \cup C_1) \sim C_2$, $B = \mathbb{R} \sim A$, and define $f$ to be the characteristic function of $B$ on $\mathbb{R}$. Then $\underline{S}(f) = A$ and $D_+ f(x) = 0$ or $-\infty$ according to whether $x \in A$ or $B$. We will now construct several examples by choosing $P$ suitably.

Given $1 < \alpha < \Omega$, first choose $P$ to be a set of $\Pi_\alpha$ which is not of $\Sigma_\alpha$. Then $f$ is of $LB_\alpha$ and $\underline{S}(f)$ is of $\Pi_\alpha$, but $D_+ f$ is not of $LB_\alpha$. This shows that the hypothesis of Lemma 3.2.1(a) cannot be weakened, and that the conclusions of Lemma 3.2.1(b) and Theorem 3.2.4(b) cannot be strengthened in general.

If $P$ is chosen on the other hand to be a set of $\Sigma_\alpha$ which is not of $\Pi_\alpha$, then $f$ is of $UB_\alpha$ and $\underline{S}(f)$ is of $\Sigma_\alpha$, but $D_+ f$ is not of $UB_\alpha$. This shows that the conclusion of Lemma 3.2.1(a) cannot be strengthened in general, and that the hypotheses of Lemma 3.2.1(b) and Theorem 3.2.4(b) cannot be weakened.

It is further clear from these two examples that for each $\alpha > 1$ there exist Borel measurable functions $f$ on $\mathbb{R}$ for which $D_+ f$ is of $UB_\alpha$ but not of $LB_\alpha$, or of $LB_\alpha$ but not of $UB_\alpha$, respectively, and in each of these two cases $D_+ f$ is of $B_{\alpha+1}$ but not of $B_\alpha$. The same holds for $\alpha = 1$ when $f$ is as defined above in example (a) or (b), and it holds for $\alpha = 0$ when $f$ is an everywhere derivable function whose derivative is not continuous.

Finally, on choosing $P$ to be any nonmeasurable set, or to be a measurable set which is not a Borel set, we obtain a function $f$ for which $D_+ f$ is not $L$-measurable, or is $L$-measurable but not $B$-measurable, respectively.

**§3.3 Baire class and measurability of medians.** In this section we define some set valued medians which are associated with various derivates, and obtain the Baire class and measurability of these medians from the results of the last two sections.

Given $f: X \to \mathbb{R}$ and $x \in X'$, we will call the interval $[\underline{D}f(x), \overline{D}f(x)]$ the (ordinary) *median* of $f$ at $x$. Similarly, when $x \in X^-$ or $X^+$, the interval $[D_- f(x), D^- f(x)]$ or $[D_+ f(x), D^+ f(x)]$ will be called the *left* or *right median*, respectively, of $f$ as $x$. Also, for each $x \in X^d$, the interval $[\underline{D}_* f(x), \overline{D}_* f(x)]$ will be called the *strong median* of $f$ at $x$.

We will use $Mf(x)$, $M_- f(x)$, $M_+ f(x)$ and $M_* f(x)$ to denote the ordinary, left, right, and strong medians, respectively, of $f$ at $x$. Further, each element of these medians will in turn be called a *median derivate, left median derivate, right median derivate,* or *strong median derivate,* respectively, of $f$ at $x$. Also, when $M_* f(x) = \overline{\mathbb{R}}$, $f$ will be said to be *strongly knotted* at $x$, and $x$ will be called a *strong knot point* of $f$.

Further, when an extended real number is simultaneously a left and right median derivate of $f$ at $x$, it will be called a *bilateral median derivate* of $f$ at $x$; the set of all such derivates will in turn be called the *bilateral median* of $f$ at $x$ and denoted by $M_b f(x)$.

Let us note here that $f$ has an ordinary, left, right, or strong derivative at $x$ iff it has only one corresponding median derivate at $x$, and then that median derivate is the desired derivative of $f$ at $x$. Further, $f$ is knotted, or left, right, bi-, or strongly knotted at $x$ iff $Mf(x)$, $M_- f(x)$, $M_+ f(x)$, $M_b f(x)$, or $M_* f(x)$, respectively, is equal to $\overline{\mathbb{R}}$.

Firstly, in the case of ordinary median we obtain the following result directly from Theorem 3.4 of [G11, p. 313] with the help of Theorem 1.5.1.

3.3.1. THEOREM. *Let $f: X \to \mathbb{R}$. Then $Mf$ is always of $B_2$; and in the case when $f$ is WC, $Mf$ is of $UB_1$.*

Next, it is clear from Lemma 1.1.2(c) that $M_- f(x) = M_+(-f_-)(-x)$ for each $x \in X^-$. Hence it is enough to consider $M_+ f$. Therefore in the following theorem we deal with right and bilateral medians.

3.3.2. THEOREM. *Let $f: X \to \mathbb{R}$ and $E \subset X$.*

(a) *If $f$ is of $B_0$, $B_1$, or $B_\alpha$ ($\alpha > 1$) on $E$, then $M_+ f$ and $M_b f$ are of $UB_1$, $B_2$, or $LB_\alpha$, respectively, on $E$.*

(b) *If $f$ is Lebesgue or Borel measurable on $E$, then so are $M_+ f$ and $M_b f$ on $E$.*

PROOF. Define $\varphi(x) = M_b(x)$, $x \in X^b$. Then, for each $x \in X^b$,

$$\varphi_*(x) = \max\{D_-f(x), D_+f(x)\}, \qquad \varphi^*(x) = \min\{D^-f(x), D^+f(x)\}.$$

To prove (a), first suppose $f$ is continuous on $E$. Then it follows from Theorem 3.2.2 and Lemma 1.1.2 that $D_+f$ and $D_-f$ are of $LB_1(E)$ and that $D^+f$ and $D^-f$ are of $UB_1(E)$. Hence $\varphi_*$ is of $LB_1(E)$ and $\varphi^*$ is of $UB_1(E)$. Consequently, it follows from Theorem 1.5.1 that $M_+f$ and $M_bf$ are of $UB_1(E)$.

The other parts are obtained similarly from Theorems 3.2.4 and 1.5.1. □

In the following two corollaries we state some consequences of the above theorem which follow directly with the help of Corollaries 1.5.2 and 1.5.3.

3.3.3. COROLLARY. *Let $f: X \to \mathbb{R}$ and $E \subset X$. If $f$ is continuous relative to $E$, then $M_+f$ and $M_bf$ are USC relative to $E$ r.e. in $E$.*

3.3.4. COROLLARY. *Let $f: X \to \mathbb{R}$. If $f$ is of $B_\alpha$, where $\alpha > 1$, then $M_+f$ and $M_bf$ have selectors of $B_\alpha$; and if $f$ is measurable, then $M_+f$ and $M_bf$ have measurable selectors.*

Further, on taking into account the definition of $UB_1$, we obtain the following result from the above theorem.

3.3.5. COROLLARY. *Let $f: X \to \mathbb{R}$, $E \subset X$ and $c \in \overline{\mathbb{R}}$. If $f$ is continuous relative to $E$, then the set of points where $c$ is a right (or bilateral) median derivate of $f$ is residual in each subset of $E$ in which it is dense.*

In the case when $E$ is dense in itself, we obtain on the other hand the following more general result from Corollary 2.5.2.

3.3.6. THEOREM. *Let $f: X \to \mathbb{R}$, $c \in \overline{\mathbb{R}}$ and suppose $E \subset X$ is dense in itself. If there is a dense set of points in $E$ where $f$ is QBSC relative to $E$ and where* (i) *$c \in M_+f$,* (ii) *$c \in M_bf$,* (iii) *$f$ is right knotted, or* (iv) *$f$ is biknotted, then the latter holds r.e. in $E$.*

3.3.7. REMARK. In the particular case when $f$ is a continuous function on an interval $I$, it is easy to see that for each right (or left) median derivate $c$ of $f$ at a point $x$ there is a sequence of points $\{x_n\}$ in $I$ which converges to $x$ from the right (or left) such that $Qf(x_n, x) \to c$ as $n \to \infty$. The terms right and left median derivates have also been used by Denjoy [D4, p. 176], but he used these terms only for the interior elements of the right and left medians of $f$ at any point.

The various medians defined here are also related in a sense with the contingent and paratingent defined by Bouligand [Bou]. Let $f$ be a continuous function on $X$, $x \in X'$, let $G$ be the graph of $f$, and let $P$ be the point $(x, f(x))$ on $G$ corresponding to $x$. Then $Mf(x)$ is indeed the convex hull of the set of slopes of all half-tangents in the contingent of $G$ at $P$; and when $x \in X^-$ or $X^+$, then $M_-f(x)$ or $M_+f(x)$ is obtained by considering only those half-tangents of $G$ at $P$ which are limits of sequences of half-lines that issue from $P$ and meet $G$ at points on the left or right, respectively, of $x$. The strong median $M_*f(x)$ is on the other hand the convex hull of the set of slopes of all lines in the paratingent of $G$ at $P$; and if $f$ is a Darboux function, this is merely a linearized form of the paratingent of $G$ at $P$. However, the medians are somewhat easier to deal with.

**§3.4 Relations between unilateral and strong derivates and medians.** In this section we deduce from the Third Derivate Theorem a further generalization of Denjoy's Theorem 2.5.4 on the identity of unilateral and strong derivates. Further, we obtain from the Fifth Derivate Theorem the Borel class of the set of points where this identity holds, and also of the set of points where the derivates are not symmetrical.

Given a function $f: X \to \mathbb{R}$, let $\underline{Q}(f)$ and $\overline{Q}(f)$ denote the sets of points in $X$ where $f$ is *QLSC* or *QUSC*, respectively. When there is an $F_\sigma$-set $H$ in $X$ such that $\underline{S}(f) \subset H \subset \underline{Q}(f)$ or $\overline{S}(f) \subset H \subset \overline{Q}(f)$, we will call $f$ *weakly QLSC* or *weakly QUSC*, respectively. Further, when $f$ is simultaneously weakly *QLSC* and weakly *QUSC*, we will call it *weakly QBSC*.

Let us note here that weak *QLSC* is considerably weaker than *QLSC*. For, $f$ is clearly weakly *QLSC* when it is *QLSC*, but when $f$ is on the other hand nowhere *LSC*, it is again weakly *QLSC*. Similar statements hold for weak *QUSC* and weak *QBSC*.

3.4.1. THEOREM. *Suppose $X$ is dense in itself and $f: X \to \mathbb{R}$ is weakly QLSC. Then the following holds r.e. in $X$:*

$$\underline{D}_*f = D_+f \le D_-f \le D^+f \le D^-f = \overline{D}_*f.$$

PROOF. It is enough to prove here that $\underline{D}_*f = D_+f$ r.e. in $X$. For, on applying this result to $f_-$ it follows that $\overline{D}_*f = D^-f$ r.e. in $X$, and the other relations follow easily from these two equations with the help of Theorem 2.3.1.

Now, since $X \sim X^+$ is countable, it is obviously meager in $X$. Hence it is enough to show that the set

$$E = \{x \in X^+ : \underline{D}_*f(x) < D_+f(x)\}$$

is meager in $X$. Suppose $E$ is nonmeager in $X$. Then it is clear that there is a rational number $c$ such that the set

$$E_0 = \{x \in E : \underline{D}_*f(x) < c < D_+f(x)\}$$

is nonmeager in $X$. Hence there is a portion $X_0$ of $X$ such that $E_0$ is nonmeager in each portion of $X_0$. Further, there is by hypothesis an $F_\sigma$-set $H$ in $X$ such that $\underline{S}(f) \subset H \subset \underline{Q}(f)$. Also, it is clear from Theorem 1.1.1 that $f$ is *LSC* n.e. in $E_0$, so that $E_0 \sim H$ is countable. Hence $H \cap X_0$ is a nonmeager $F_\sigma$-set in $X_0$, and so $H$ includes some portion $X_1$ of $X_0$. Thus $X_1 \subset H \subset \underline{Q}(f)$; that is, $f$ is *QLSC* on $X_1$.

Now since $E_0$ is nonmeager in $X_1$, it follows from the Third Derivate Theorem 2.4.1 that $f_{-c}$ is strongly increasing on some portion $X_2$ of $X_1$. But then, for each $x \in X_2$, $\underline{D}_*f_{-c}(x) \geq 0$, or $\underline{D}_*f(x) \geq c$, which is a contradiction. □

On applying the above theorem to $-f$, a similar result is obtained in terms of the derivates $D^+f$ and $D_-f$, and on combining these two results we obtain the following.

3.4.2. COROLLARY. *Suppose X is dense in itself and $f: X \to \mathbb{R}$ is weakly QBSC. Then $M_+f = M_-f = M_*f$ r.e. in X; or, equivalently, $D_+f = D_-f = \underline{D}_*f$ and $D^+f = D^-f = \overline{D}_*f$ r.e. in X.*

It is easy to see from the definition of strong derivates and Theorem 1.5.1 that $M_*f$ is always *USC* (see Theorem 14.1.1). Hence for each $c \in \overline{\mathbb{R}}$, the set $\{x \in X : c \in M_*f(x)\}$ is closed in $X$. The above corollary thus leads further to the following.

3.4.3. COROLLARY. *Suppose $f: X \to \mathbb{R}$ is weakly QBSC, and let $c \in \overline{\mathbb{R}}$. If $c \in M_*f$ or f is strongly knotted at a dense set of points in X, then $c \in M_bf$ or f is biknotted, respectively, r.e. in X. Consequently, each of the sets $\{x \in X : c \in M_bf(x)\}$ and $K_b(f)$ is residual in every portion of X in which it is dense.*

Next, in the following theorem we investigate the Borel class of the set of points where a unilateral derivate coincides with the corresponding strong derivate.

3.4.4. THEOREM. *Given $f: X \to \mathbb{R}$, the sets of points where $D_+f = \underline{D}_*f$ or $D^-f = \overline{D}_*f$ are of the form $G_\delta \cup (G_\delta \sim \underline{S}(f))$ in X. Consequently, if $\underline{S}(f)$ is an $F_\sigma$-set in X, or f is either* LSC *or nowhere* LSC, *then these two sets are $G_\delta$-sets in X.*

PROOF. It is enough to prove the result, as before, for the first set. Set $E = \{x : D_+f(x) = \underline{D}_*f(x)\}$, $S = \underline{S}(f)$, and let $\{c_n\}$ be an enumeration of the set of rational numbers. Then

$$E = X^+ \sim \{x : \underline{D}_*f(x) < D_+f(x)\}$$

$$= X^+ \sim \bigcup_n \{x : \underline{D}_*f(x) \leq c_n < D_+f(x)\}$$

$$= X^+ \cap \bigcap_n [\{x : \underline{D}_*f(x) > c_n\} \cup \{x : D_+f(x) \leq c_n\}].$$

Now since $X \sim X^+$ is countable, $X^+$ is a $G_\delta$-set in $X$. Further, for each $n$, since $\underline{D}_* f$ is $LSC$, the first set within the square brackets is open in $X$, whereas the other set is by Corollary 3.1.2 of the form $G_\delta \cup (G_\delta \sim S)$ in $X$. Hence there are two sequences of $G_\delta$-sets $\{A_n\}$ and $\{B_n\}$ in $X$ such that

$$E = \bigcap_n \{A_n \cup (B_n \sim S)\}.$$

Now, set $H = \{x \in X^+ : f(x) > f(x+0)\}$. Then $H$ and $S$ are clearly disjoint, and the set $C = X \sim (H \cup S)$ is countable by Theorem 1.1.1. Also, $H \subset E$, for $D_+f(x) = -\infty$ for each $x \in H$. Hence,

$$(1) \qquad E = H \cup \bigcap_n [A_n \cup \{B_n \sim (H \cup S)\}] = \left(\bigcap_n A_n\right) \cup H \cup C_0,$$

where $C_0 \subset C$. Now set $C_1 = C \sim C_0$. Then $C = C_0 \cup C_1$, so that $X = H \cup S \cup C_0 \cup C_1$, and hence

$$(2) \qquad H \cup C_0 = X \sim (S \cup C_1) = (X \sim C_1) \sim S.$$

Now since $C_1$ is countable, it follows from (1) and (2) that $E$ is of the form $G_\delta \cup (G_\delta \sim S)$ in $X$. □

On combining the above theorem with the result obtained on applying it to $-f$, we obtain the following.

3.4.5. Corollary. *Let $f: X \to \mathbb{R}$. If $f$ is either continuous, or nowhere LSC or USC, then the sets of points where $M_+ f = M_* f$ or $M_b f = M_* f$ are $G_\delta$-sets in $X$.*

Finally, in the following theorem we investigate the Borel class of the set of points where the unilateral derivates are not symmetrical.

3.4.6. Theorem. *If $f: X \to \mathbb{R}$ is continuous, then the set of points where the derivates of $f$ are not symmetrical is a $G_{\delta\sigma}$-set in $X$.*

Proof. Let $A$ denote the set of points where the derivates of $f$ are not symmetrical, and let $\{c_n\}$ be an enumeration of the set of rational numbers. Then it follows from Corollary 3.1.2 that the set

$$\begin{aligned}\{x : D_+ f < D_- f\} &= \bigcup_n \{x : D_+ f \le c_n < D_- f\} \\ &= \bigcup_n [\{x : D_+ f \le c_n\} \sim \{x : D_- f \le c_n\}]\end{aligned}$$

is a $G_{\delta\sigma}$-set in $X$. The sets of points where $D_-f < D_+f$, $D^+f < D^-f$ or $D^-f < D^+f$ are proved similarly to be $G_{\delta\sigma}$-sets in $X$, and $A$ is clearly the union of these four sets. □

3.4.7. REMARK. Let $A_f$ denote for each continuous function $f$ defined on an interval $I$ the set of points in $I$ where the unilateral derivates of $f$ are not symmetrical—that is, where either $D_-f \neq D_+f$ or $D^-f \neq D^+f$. According to Corollary 3.4.2, $A_f$ is meager in $I$. Young proved in [Y4] that $A_f$ is dense in $I$ in the case when $f$ is nonderivable (i.e., not derivable at any point), and he raised the question in the case of Weierstrass' nonderivable function $f$ whether $A_f$ is countable or a general meager set. As it will be seen later in Theorem 13.5.1, $A_f$ is indeed $c$-dense in $I$ whenever $f$ is nonderivable. Although $A_f$ is of measure zero for the Weierstrass' function (see G. C. Young [Y.G2]), it was shown in [G8] that in the case of the Besicovitch's nonderivable function $f$, $A_f$ covers almost all points of $I$.

3.4.8. PROBLEM. We state here two problems dealing with unilateral derivates which arise naturally in connection with the results of this section.

First, in connection with Theorems 3.4.1 and 3.4.4, it is natural to ask whether for every meager $F_\sigma$-set $E$ in $I$ there is some continuous function $f$ on $I$ for which $E$ is the set of points where $D_+f \neq \underline{D}_*f$.

Next, in connection with Theorem 3.4.6 it is natural to ask whether for every meager $G_{\delta\sigma}$-set $E$ in $I$ there is some continuous function $f$ on $I$ for which $E$ is the set of points where the derivates of $f$ are not symmetrical. See Problem 13.5.9 for another interesting problem on unilateral derivates that has been posed by Zahorski [Z3].

## §3.5 Refinement of a theorem of A. P. Morse.

In this final section of Part I, we deduce from the First and Third Derivate Theorems a refined version of a penetrating theorem of A. P. Morse [Mo2] on unilateral derivates. Also, we obtain here a similar refined version of a monotonicity theorem of Morse from which he deduced his main theorem.

Given a continuous function $f:\mathbb{R} \to \mathbb{R}$ and $r \in \mathbb{R}$, Morse proved in 1954 that if $D^+f \geq r$ at a dense set of points and $D^+f < r$ at some point, then $D^+f = r$ at a set of points whose cardinality is $c$. It is natural to ask here what can be said about the left derivates of $f$ at the points of this uncountable set. In the refined version (Theorem 3.5.5) we investigate this question and obtain a weakening of the continuity hypothesis.

We begin with a reduction lemma regarding continuity hypothesis for monotonicity which will be used frequently. Some nomenclature is needed here.

Let $Q$ be any continuity property for which $WQ$ (or weakly $Q$) has been defined in §1.2C. When a function is simultaneously $IQ$ and $WQ$, we will call it *strongly inherently* $Q$, or $I^*Q$. Further, let $\mathcal{F}$ denote the family of all real valued functions that are defined on an open interval in $\mathbb{R}$.

3.5.1. LEMMA (FIRST REDUCTION). *Let $P$ be any function-theoretic property. If every function in $\mathcal{F}$ which has property $P$ and is $LC$, $LC_+$ or $LC_-$ is nondecreasing, then so is every function in $\mathcal{F}$ which has property* P *and is $I^*LC$, $I^*LC_+$ or $I^*LC_-$, respectively.*

PROOF. We will first prove the result for $LC_+$.

Suppose P is a property such that every function in $\mathcal{F}$ which has property P and is $LC_+$ is nondecreasing. Let $f: I \to \mathbb{R}$ be a function in $\mathcal{F}$ which has the property P but is only $I^*LC_+$. Then it is enough to show that the set $E$ of points in $I$ where $f$ is not $LC_+$ is empty.

Suppose $E \neq \varnothing$. Since $f$ is $ILC_+$, $E$ is clearly nowhere dense in $I$, and so is in turn its closure $\overline{E}$ relative to $I$. Also, due to the same property, there is an open subinterval $J$ of $I$ such that $\overline{E} \cap J \neq \varnothing$ and $f$ is $LC_+$ relative to $\overline{E} \cap J$. Choose $x \in E \cap J$. Now, it is enough to show that $f$ is $LC_+$ at $x$, for this would contradict the definition of $E$.

First, suppose $x \notin E^+$. Then $x$ is the left endpoint of some contiguous interval $(x, y)$ of $\overline{E}$, and so $f$ is by hypothesis nondecreasing on $(x, y)$. But since $f$ is $WLC_+$, it is then nondecreasing on $[x, y)$, and so $f$ is $LC_+$ at $x$.

Next, suppose $x \in E^+$ and let $\varepsilon > 0$. Then since $f$ is $LC_+$ relative to $\overline{E} \cap J$, we can choose a point $z \in \overline{E} \cap J$ such that $x < z$ and

$$(1) \qquad f(t) > f(x) - \varepsilon \quad \text{for} \quad t \in \overline{E} \cap (x, z).$$

When $t \in (x, z) \sim \overline{E}$, $t$ is in some contiguous interval $(a, b)$ of $\overline{E}$ such that $a \in \overline{E} \cap (x, z)$, and since $f$ is nondecreasing on $[a, b)$ as seen above, it follows from (1) that $f(t) \geq f(a) > f(x) - \varepsilon$. Thus the inequality (1) holds whenever $t \in (x, z)$, which proves again that $f$ is $LC_+$ at $x$. The result for $LC_+$ is thus established.

Now, the result for $LC_-$ follows on applying the above result to $-f_-$. Further, on making $LC_-$ a part of P it follows from the above result that $LC$ can be reduced to $LC_-$ and $I^*LC_+$ together, and then on making $I^*LC_+$ a part of P it follows from the result on $LC_-$ that $LC_-$ and $I^*LC_+$ together can be reduced to $I^*LC$. □

We will further need here the following lemma on the preservation of nowhere dense sets by strongly monotone functions. This does not hold in general for ordinary monotone functions, or even for strictly monotone functions, as is easy to see from the restriction of the Cantor's continuous nondecreasing function to $P^b$, where $P$ is the Cantor's nowhere dense perfect set.

3.5.2. LEMMA. *Suppose $E \subset \mathbb{R}$ and $f: E \to \mathbb{R}$ is strongly monotone. Then if $E$ is nowhere dense in $\mathbb{R}$, then so is $f(E)$.*

PROOF. To be definite, suppose $f$ is strongly increasing. Then there is an $r \in \mathbb{R}_+$ such that $f_{-r}$ is increasing on $E$. Let $a, b \in E$, $a < b$. Then $f(a) < f(b)$, and it is enough to show that the interval $J \equiv (f(a), f(b))$ includes an open subinterval which is disjoint with $f(E)$.

Since $\overline{E}$ is also nowhere dense in $\mathbb{R}$, it has a contiguous interval $(c, d)$ such that $a \le c < d \le b$. Define

$$u = \sup\{f(x) : x \in E \cap [a, c]\}, \qquad v = \inf\{f(x) : x \in E \cap [d, b]\}.$$

Then since $f$ is increasing on $E$, $f(a) \le u \le v \le f(b)$. We claim that $u < v$. Set $\varepsilon = r(d - c)/2$. Then we can find $x$ and $y$ in $E \cap [a, c]$ and $E \cap [d, b]$, respectively, such that $f(x) > u - \varepsilon$ and $f(y) < v + \varepsilon$. Now since $f_{-r}$ is increasing on $E$, $f(x) - rx < f(y) - ry$, and hence

$$v - u > f(y) - f(x) - 2\varepsilon > r(y - x) - 2\varepsilon \ge r(d - c) - 2\varepsilon = 0.$$

Thus $(u, v)$ is an open subinterval of $J$ which is disjoint with $f(E)$. □

We will first obtain a refined version of the Morse's monotonicity theorem. He proved part (a) of the following monotonicity theorem for a continuous function $f$ for which $D^+f \ge 0$ at a dense set of points. However, this condition is stronger than $f$ being nowhere strongly decreasing. For, it is easy to see that $f$ is nowhere strongly decreasing iff $\overline{D}_*f \ge 0$ everywhere, or, equivalently, iff $\overline{D}_*f \ge 0$ at a dense set of points (for $\overline{D}_*f$ is always $USC$).

Let us recall here that according to our convention it will be understood in the remaining results of this section that $f : I \to \mathbb{R}$, where $I$ is some subinterval of $\mathbb{R}$.

3.5.3. THEOREM (FIRST MONOTONICITY). *Suppose $f$ is nowhere strongly decreasing. Then $f$ is increasing whenever any of the following two conditions holds:*

(a) *$f$ is $I^*LC_-$ and $LI_+$ and* $\operatorname{card}\{x : D^+f(x) = 0\} < c$;

(b) *$f$ is $I^*LC$ and* $\operatorname{card}\{x : D^+f(x) = 0 \in M_-f(x)\} < c$.

PROOF. Since in each of the two cases $f$ is clearly $WLC$ from both sides, it is enough to prove the result when $I$ is open. Also, since $f$ is nowhere constant in each of the two cases, it is enough to show that $f$ is nondecreasing. Hence, according to Lemma 3.5.1, there is no loss of generality in assuming $f$ to be $LC_-$ in case (a), and to be $LC$ in case (b).

First, to prove the sufficiency of (a), suppose $f$ is $LC_-$ and $LI_+$ but is not nondecreasing. Then there exist $a, b \in I$, $a < b$ such that $f(a) > f(b)$. Set $J = (f(b), f(a))$, and define for each $r \in J$ the following:

$$x_r = \sup\{t \in [a, b] : f(t) \ge r\}.$$

Then it follows from the present continuity hypotheses that $a < x_r < b$ and $f(x_r) = r$. Hence $D^+f(x_r) \le 0$. Now set

$$E = \{x_r : r \in J\}, \qquad A = \{x \in [a, b] : D^+f(x) < 0\}.$$

Then $f(E) = J$, and $D^+f = 0$ at the points of $E \sim A$.

Now, since $f$ is nowhere strongly decreasing, and it is clearly $QUSC$, it follows from the Third Derivate Theorem 2.4.1 that $A$ is meager in $I$. Hence it follows clearly from the First Derivate Theorem 2.2.2 that $A$ is the union of a sequence of nowhere dense sets $\{A_n\}$ on each of which $f$ is $SRD$. Now it follows from Lemma 3.5.2 that $f(A_n)$ is nowhere dense in $\mathbb{R}$ for each $n$. Hence $f(A) = \bigcup_n f(A_n)$ is meager in $\mathbb{R}$, so that $f(E \sim A)$ is residual in $J$. Consequently, $E \sim A$ has cardinality $c$, which proves the sufficiency of (a).

Next, to prove the sufficiency of (b), suppose $f$ is $LC$ but it is not nondecreasing. Let $a$, $b$, $J$, $E$, and $A$ be defined as above, and set further

$$B = \{x \in [a,b] : D^{-}f(x) < 0\}, \qquad C = \{x \in [a,b] : D_{-}f(x) > 0\}.$$

Then on applying the above result to $-f_{-}$ it follows that $f(B)$ is meager in $J$. Further, since $f$ is angular at the points of $E \cap C$, $f(E \cap C)$ is countable by Theorem 2.3.1. Hence, if $D = E \sim (A \cup B \cup C)$, $f(D)$ is residual in $J$, and so $D$ has cardinality $c$. At the points of $D$ it is clear on the other hand that $D^{+}f = 0$ and $D_{-}f \leq 0 \leq D^{-}f$, which proves the sufficiency of (b). □

Let us note here that none of the conditions (a) and (b) of the above theorem is necessary for a function to be increasing; this is clear from any continuous increasing function which is singular.

In the following corollary we deduce from part (a) of the above theorem two sufficient conditions for a function to be nondecreasing which are also necessary including their continuity hypotheses. Similar necessary and sufficient conditions can be deduced from part (b).

A function $f$ was called in [G11, p. 297] *lower singular* if its derivative is nonnegative at almost all the points where it exists, or, equivalently, if $\overline{D}f \geq 0$ a.e. (see Theorem 5.5.2).

3.5.4. Corollary. *Suppose $f$ is $I^{*}LC_{-}$ and $LI_{+}$. Then $f$ is nondecreasing whenever any of the following two conditions holds:*

(a) *$f$ is nowhere strongly decreasing and for every $\varepsilon > 0$ there is an $r \in \mathbb{R}_{+}$ such that $r < \varepsilon$ and* $\operatorname{card}\{x : D^{+}f(x) = -r\} < c$;

(b) *$f$ is lower singular and there is an $r \in \mathbb{R}_{+}$ such that* $\operatorname{card}\{x : D^{+}f(x) = -r\} < c$.

When (a) holds, it follows from the above theorem that for every $\varepsilon > 0$ there is an $r \in \mathbb{R}_{+}$ such that $r < \varepsilon$ and $f_{+r}$ is increasing, and this clearly implies that $f$ is nondecreasing; and when (b) holds, it follows from the above theorem that there is an $r \in \mathbb{R}_{+}$ such that $f_{+r}$ is increasing, and so $f$ is lower Lipschitz, whence $LAC$ (see [G16] for definition), and hence it follows from Proposition 2.24 of [G11] that $f$ is nondecreasing.

Finally, the following refined version of the initially stated theorem of Morse is obtained directly from Theorem 3.5.3 on applying it to $f_{-r}$ and $-(f_{-r})_{-}$ (see Lemma 1.1.2).

3.5.5. THEOREM (MORSE). *Suppose* $\overline{D}_* f \geq r$ ($\in \mathbb{R}$) *at a dense set of points in* $I$ *and* $\underline{D}_* f < r$ *at some point.*

(a) *If* $f$ *is* $I^*LC_-$ *and* $LI_+$, *then* $D^+ f = r$ *at a set of points whose cardinality is* $c$.

(b) *If* $f$ *is* $I^*LC$, *then the sets of points where* $D^+ f = r \in M_- f$ *or* $D^- f = r \in M_+ f$ *have cardinality* $c$.

We conclude this section with some comments on its two theorems.

3.5.6. REMARK. In connection with the Morse's Theorem 3.5.5, it is interesting to note here that it was proved earlier by Sierpiński [Si2] in 1914 that in the case of a continuous nowhere monotone function $f$ each derivate of $f$ assumes the value zero at a dense set of points.

Further, in connection with part (b) of Theorem 3.5.5, it is natural to ask if the two sets considered in it coincide, namely if the set of points where $D^+ f = D^- f = r$ has cardinality $c$. As we see later in Chapter 13 (see Remark 13.4.7), this is not true in general, but it does hold for a special class of functions called lower smooth functions (see Theorem 13.1.6(b)).

Finally, it should be pointed out here that the proofs given by Morse [Mo2] for his initial versions of Theorems 3.5.3 and 3.5.5 have been generalized by Manna [Man2] from continuous functions to $USC$ functions which are $LI_+$. These generalizations are of course included in the above theorems.

# *PART II*

# A THEORY OF SOME NEW DERIVATIVES

**Introduction to Part II.** We present the theory of new derivatives in the present and the next part. In the present part we introduce the various notions of new derivatives, and we obtain some basic results on them along with several derivability theorems to establish the existence of new derivatives at some points for every continuous function on an interval. Further, we deal in this part with the Calculus of new derivatives.

First, in Chapter 4, we present four equivalent definitions of lower and upper derivatives of real-valued functions defined on normed vector spaces: one geometrical and three analytical.

Next, given any function $f : X \to \mathbb{R}$, where $X \subset \mathbb{R}$, we introduce extended versions of lower and upper derivatives of $f$ in terms of unilateral derivates, along with a notion of weak derivative. Also, we call $f$ semiderivable at a point if it is either lower or upper derivable there. As special cases of these new derivatives we define further "normal" lower, upper and semi- derivatives.

Except for the normal new derivatives, these new derivatives turn out to be set-valued in general. We investigate in Chapter 4 the magnitude of the set of points where the new derivatives are not unique (or single-valued); and in the case when $X \subset \mathbb{R}$, this set is found to be countable. Also, we investigate in this chapter the continuity properties that hold under various new derivabilities.

In the next two chapters we obtain several derivability theorems in which we establish the existence of points where a function is derivable in a new or ordinary sense. We discuss here the nature of the results obtained in these theorems in some detail to put these results into some perspective.

First, in Chapter 5, we obtain two derivability theorems. In the First Derivability Theorem we prove the lower differentiability of every *LSC* function on an interval at a dense set of points. Then, in terms of properties *L*-$BV_*$ and *U*-$BV_*$ similar to the property $BV_*$ on a set, we obtain in the

Second Derivability Theorem the lower and upper differentiability of $GL\text{-}BV_*$ and $GU\text{-}BV_*$ functions, respectively, a.e. However, such functions need not be derivable in the ordinary sense at any point. Further, with the help of the First Derivate Theorem of §2.2, we deduce from the Second Derivability Theorem the well-known Denjoy-Young-Saks theorem which establishes the normal lower and upper differentiability of an arbitrary function $f : X \to \mathbb{R}$ at almost all the points where it is not lower or upper knotted, respectively (see §4.3 for definitions).

Next, in Chapter 6, we obtain with the help of the Second Derivate and Second Derivability Theorems a theorem which we call the Normal Level Structure Theorem. This theorem is analogous to the Denjoy-Young-Saks theorem and was obtained earlier in [G1] in a somewhat different form. It provides the differential structure of a function $f : X \to \mathbb{R}$ at the points of almost all of its level sets—that is, at the points of the level set $f^{-1}(y)$ for almost every $y$.

All the other results of this chapter are deduced from this Normal Level Structure Theorem. First we deduce from it two basic theorems of Saks on unilateral derivates. Then we deduce from it the Third Derivability Theorem according to which every continuous function is normally lower and upper derivable at two dense sets of points in almost all of its level sets. Also, we introduce here the notions of regular and singular functions relative to any notion of generalized derivative, and in the case of a regular or singular function relative to lower derivative we obtain in the Fourth Derivability Theorem the existence of points in almost all of its level sets where the normal lower derivative is finite or infinite, respectively.

Further, we deduce from the Normal Level Structure Theorem the Second Level Structure Theorem which relates the derivability and semiderivability of a weakly Darboux function at the points of almost all of its level sets with the local structure of the level set at the point in question. Also, we introduce here two properties of functions $(S_1)$ and $(S_2)$ which are similar to Banach's properties $(T_1)$ and $(T_2)$ but are more directly related to derivability and semiderivability of the function. Then in the Fifth Derivability Theorem we obtain characterizations of these two properties in terms of derivability and normal semiderivability of the function at all the points of almost all of its level sets. From this derivability theorem and the two level structure theorems we obtain in turn strengthened and generalized versions of several known theorems of Banach and Saks on the derivability of continuous functions under the properties $(T_1)$, $(T_2)$, $(N)$, and $(S)$.

Finally, in Chapter 7, we investigate the various rules of Calculus for new derivatives. In particular, the lower and normal lower (or upper) derivatives are found to be additive and positively linear. The weak derivative is found, on the other hand, to be only conditionally additive in the sense that the additivity holds for it only when the sum function is also weakly derivable at the point in question. Further, in terms of a notion of compatibility of two semiderivatives, a theorem is obtained on the linearity of semiderivatives.

Similar product and quotient rules are obtained for semiderivatives and normal semiderivatives in terms of their compatibility. Also, conditional product and quotient rules are obtained for weak derivative. Further, we obtain here various chain rules for new derivatives.

Also, similar rules of Calculus are obtained in this chapter for lower and upper derivatives of functions defined on normed vector spaces.

It is clear from the above-mentioned derivability theorems that the theory of new derivatives is directly related with nonsmooth analysis. For the sake of coherence, we present some comments on the developments in nonsmooth analysis related with new derivatives at the end of Part II in §7.8.

# *CHAPTER 4*

# Definitions of New Derivatives and Preliminary Results

We begin in this chapter with the definitions of new derivatives which are in general set-valued, and we obtain some preliminary results on them.

First, in §4.1, we present the definitions of lower and upper derivatives of real-valued functions defined on normed vector spaces. Indeed, we present here four equivalent definitions, one geometrical and the other three analytical. These two notions of derivability are found to provide a decomposition of ordinary differentiability of such functions. Further, we define a notion of semiderivative in terms of lower and upper derivatives.

Then in the case of real-valued functions defined on the real line, or on any subset of $\mathbb{R}$, we introduce in §4.2 extended versions of lower and upper derivatives in terms of unilateral derivates. Also, a function $f$ is called here semiderivable at a point $x$ if it is either lower or upper derivable at $x$; and if so, then the lower or upper derivative of $f$ at $x$ is called the semiderivative of $f$ at $x$. Further, we define in this case the notion of weak derivative in terms of bilateral weak derivates that are introduced here. This notion unifies in fact the notions of lower and upper derivatives with the ones of right and left derivatives. Next, in §4.3, we define two specialized versions of new derivatives, namely, knotted and normal lower, upper and semi- derivatives. Also, we introduce here notations for various new derivability sets. It will be seen later in Chapter 6 that every continuous function defined on an interval is normally lower and upper derivable at lots of points.

Next, in §4.4, we investigate the magnitude of the set of points where the new derivatives are not unique. In the case of functions defined on a subset of $\mathbb{R}$, this set is found to be countable. Finally, in §4.5, we investigate the continuity properties that hold under various new derivabilities.

**§4.1 Lower, upper, and semi- derivatives on normed vector spaces.** In this section we present definitions of two new derivatives, called lower and upper derivatives, of real-valued functions defined on normed vector spaces, and we establish some preliminary results on these derivatives. In fact, we present here four equivalent definitions of each of these derivatives, one geometrical and three analytical. Also, we define semiderivative in terms of lower and upper derivatives.

We begin with the geometrical definition of lower and upper derivatives.

Let $X$ be a given (real) normed vector space. We first define two convex cones in the space $Z = X \times \mathbb{R}$. Given $x' \in X'$ and $\varepsilon > 0$, we define

$$C_+ \equiv C_{+,x',\varepsilon} = \{(x,t) \in Z : t \geq (x',x) + \varepsilon\|x\|\},$$

$$C_- \equiv C_{-,x',\varepsilon} = \{(x,t) \in Z : t \leq (x',x) - \varepsilon\|x\|\}.$$

Then it is clear that $C_+$ and $C_-$ are two closed convex cones in $Z$, and that $C_- = -C_+$. Further, for each $z \in Z$, we will call $z + C_\pm$ *affine cones* of $Z$ with vertex $z$.

To obtain a hyperplane that separates these two cones, define $z'(x,t) = t - (x',x)$, $(x,t) \in Z$. Then it is clear that $z' \in Z'$—that is, $z'$ is linear and continuous—and it is easy to see that the hyperplane

$$H \equiv \{(x,t) \in Z : z'(x,t) = 0\} = \{(x,t) \in Z : t = (x',x)\}$$

separates the cones $C_+$ and $C_-$.

We now define lower and upper gradients in terms of which lower and upper derivatives are defined.

DEFINITION 1. Given $f: X \to \mathbb{R}$ and $x \in X$, let $z = (x, f(x))$. We will call a linear functional $x' \in X'$ a *lower* (or *upper*) *gradient* of $f$ at $x$ if for every $\varepsilon > 0$ there is a neighborhood $U$ of $x$ such that the graph of $f$ on $U \sim \{x\}$ does not meet the affine cone $z + C_{-,x',\varepsilon}$ (or $z + C_{+,x',\varepsilon}$).

We include here in Figure 1 an illustration of the graph of $f$ in the neighborhood of a point $x$ where $f$ has a lower gradient; here we use $z$ for $(x, f(x))$ and $C_-$ and $H$ as defined above.

It should be noted here that the lower and upper gradients of $f$ at $x$ are not unique in general. For, let $X = \mathbb{R}$ and $f(x) = |x|$, $x \in \mathbb{R}$; then it can be verified that each element of the interval $[-1, 1]$ is a lower gradient of $f$ at 0 (see relation (1) of §4.2 for the definition of lower gradient in this case).

We will call $f$ *lower* (or *upper*) *derivable* at $x$ if it has at least one lower (or upper) gradient at $x$, and when it is so, the set of all lower (or upper) gradients of $f$ at $x$ will be called the *lower* (or *upper*) *derivative* of $f$ at $x$. Further, we will call $f$ *semiderivable* at $x$ if it is either lower or upper derivable at $x$, and if so, then the lower or upper derivative of $f$ at $x$ will be called the *semiderivative* of $f$ at $x$.

We will use $Lf'(x)$ and $Uf'(x)$ to denote the lower and upper derivatives respectively of $f$ at $x$, and will use $Sf'(x)$ to denote the semiderivative of $f$

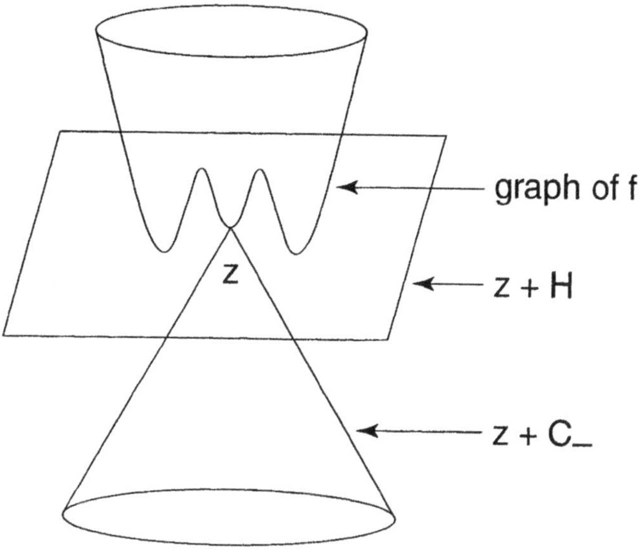


Figure 1

at $x$. In case $Lf'(x)$ is singleton, $f$ will be said to be *uniquely lower derivable* at $x$, and then $Lf'(x)$ will also be used to denote its unique element. The same will apply to $Uf'(x)$ and $Sf'(x)$.

Next, we present an analytical definition of lower and upper gradients.

DEFINITION 2. Given $f: X \to \mathbb{R}$ and $x \in X$, a linear functional $x' \in X'$ is a *lower* or *upper gradient* of $f$ at $x$ iff for every $\varepsilon > 0$ there is a $\delta > 0$ such that for each $y \in X$, if $0 < \|y - x\| < \delta$, then

$$(1) \qquad f(y) > f(x) + (x', y - x) - \varepsilon\|y - x\| \quad \text{or}$$
$$f(y) < f(x) + (x', y - x) + \varepsilon\|y - x\|,$$

respectively.

We claim that the above Definitions 1 and 2 are equivalent. According to Definition 1, $x' \in X'$ is a lower gradient of $f$ at $x$ iff for every $\varepsilon > 0$ there is a $\delta > 0$ such that if $0 < \|y - x\| < \delta$, then

$$(y, f(y)) \notin (x, f(x)) + C_{-, x', \varepsilon},$$

that is,

$$(y - x, f(y) - f(x)) \notin C_{-, x', \varepsilon},$$

that is,

$$f(y) - f(x) > (x', y - x) - \varepsilon\|y - x\|,$$

that is,

$$f(y) > f(x) + (x', y - x) - \varepsilon\|y - x\|.$$

Thus the two definitions of lower gradient are equivalent. The equivalence of the two definitions of upper gradient is seen by a similar argument.

In the following theorem we deal with the relation between lower and upper derivatives.

4.1.1. THEOREM. *Let $f: X \to \mathbb{R}$, where $X$ is a normed vector space. Then $f$ is upper derivable at $x \in X$ iff $-f$ is lower derivable at $x$; and if so, then $Uf'(x) = -L(-f)'(x)$.*

PROOF. According to Definition 2, $x' \in X'$ is an upper gradient of $f$ at $x$ iff for every $\varepsilon > 0$ there is a $\delta > 0$ such that if $0 < \|y - x\| < \delta$, then

$$f(y) < f(x) + (x', y - x) + \varepsilon\|y - x\|,$$

or

$$-f(y) > -f(x) + (-x', y - x) - \varepsilon\|y - x\|.$$

Hence it is clear that $x' \in Uf'(x)$ iff $-x' \in \mathrm{L}(-f)'(x)$. Both parts follow clearly from this fact. □

It is clear from this theorem that it is enough to consider lower derivative. Although the above definitions involve the norm of $X$, they are in fact independent of the norm as we see next.

4.1.2. THEOREM. *Let $f: X \to \mathbb{R}$, where $X$ is a normed vector space, and $x \in X$. Then the lower derivability of $f$ at $x$ and its lower derivative at $x$ are both independent of the choice of the norm on $X$.*

PROOF. Let $\|\ \|_1$ be another norm on $X$ which is equivalent to the given norm $\|\ \|$. Then there exists a constant $M \in \mathbb{R}_+$ such that $\|x\| \leq M\|x\|_1$ for every $x \in X$. First, suppose $x' \in X'$ is a lower gradient of $f$ at $x \in X$ relative to the norm $\|\ \|$. Then, given $\varepsilon > 0$, there exists a $\delta > 0$ such that

$$f(y) > f(x) + (x', y - x) - (\varepsilon/M)\|y - x\| \quad (2)$$

$$\text{whenever} \quad 0 < \|y - x\| < M\delta.$$

Hence when $0 < \|y - x\|_1 < \delta$, we have $\|y - x\| \le M\|y - x\|_1 < M\delta$, and so, by (2), $f(y) - f(x) - (x', y - x) > -(\varepsilon/M)\|y - x\| \ge -\varepsilon\|y - x\|_1$. This proves that $x'$ is also a lower gradient of $f$ at $x$ relative to $\| \; \|_1$. The converse follows on the other hand by symmetry, and hence the result. □

Next, it is clear that the definitions of lower and upper gradients in Definition 2 can be restated in terms of lower and upper limits as follows.

DEFINITION 3. Given $f: X \to \mathbb{R}$ and $x \in X$, a linear functional $x' \in X'$ is a lower gradient of $f$ at $x$ iff

$$\liminf_{y \to x} \frac{f(y) - f(x) - (x', y - x)}{\|y - x\|} \ge 0.$$

Similarly, $x'$ is an upper gradient of $f$ at $x$ iff

$$\limsup_{y \to x} \frac{f(y) - f(x) - (x', y - x)}{\|y - x\|} \le 0.$$

We next show that the lower and upper derivatives provide a decomposition of the notion of derivative which is defined as follows: Given $f: X \to \mathbb{R}$, $x \in X$, a linear functional $x' \in X'$ is called the *derivative* of $f$ at $x$ if

$$|f(y) - f(x) - (x', y - x)| = o(\|y - x\|),$$

or, equivalently, if

$$f(y) = f(x) + (x', y - x) + o(\|y - x\|); \tag{3}$$

and if so, then the derivative $x'$ of $f$ at $x$ is denoted as usual by $f'(x)$ (see, e.g., Cartan [Car, p. 25], where $f$ is further assumed to be continuous at $x$).

In terms of the small zero order, the Definition 2 can be restated in the following form.

DEFINITION 4. Given $f: X \to \mathbb{R}$ and $x \in X$, a linear functional $x' \in X'$ is a lower gradient of $f$ at $x$ iff

$$f(y) \ge f(x) + (x', y - x) + o(\|y - x\|).$$

Similarly, $x'$ is an upper gradient of $f$ at $x$ iff

$$f(y) \le f(x) + (x', y - x) + o(\|y - x\|).$$

It is clear from the above definition and (3) that $x' \in X'$ is the derivative of $f$ at $x \in X$ iff it is simultaneously a lower and upper gradient of $f$ at $x$. The following theorem which compares derivability with lower and upper derivabilities is, however, not so obvious.

4.1.3. THEOREM. *Let $f: X \to \mathbb{R}$, where $X$ is a normed vector space, and $x \in X$. Then $f$ is derivable at $x$ iff it is lower and upper derivable at $x$; and if so, then the lower and upper derivatives of $f$ at $x$ are unique and equal to $f'(x)$.*

PROOF. It is enough to prove here the sufficiency part, because the necessity follows clearly from the definitions. Hence suppose $f$ is lower and upper derivable at $x$. Then $f$ has a lower gradient $x'$ and an upper gradient $y'$ at $x$. It is now enough to show that $x' = y'$, for this would establish the uniqueness and identity of $x'$ and $y'$, which in turn implies as seen above that $x'$ is the derivative of $f$ at $x$.

Hence, suppose $x' \neq y'$. Choose $y \in X$ such that $(x', y) \neq (y', y)$. Since $x'$ and $y'$ are linear, we can clearly assume that $\|y\| = 1$. First, suppose $(x', y) > (y', y)$, and let $0 < \varepsilon < (x' - y', y)/2$. Then we can find a $\delta > 0$ such that whenever $0 < \|y - x\| < 2\delta$,

$$f(y) > f(x) + (x', y - x) - \varepsilon\|y - x\|,$$

$$f(y) < f(x) + (y', y - x) + \varepsilon\|y - x\|.$$

Let $z = x + \delta y$. Then since $0 < \|z - x\| = \delta$, we obtain

$$f(z) > f(x) + (x', z - x) - \varepsilon\|z - x\| = f(x) + \delta(x', y) - \varepsilon\,\delta,$$

$$f(z) < f(x) + (y', z - x) + \varepsilon\|z - x\| = f(x) + \delta(y', y) + \varepsilon\,\delta.$$

Hence, on combining these two inequalities we obtain $(x' - y', y) < 2\varepsilon$, which is a contradiction.

In the case when $(x', y) < (y', y)$, a similar contradiction is obtained by choosing $0 < \varepsilon < (y' - x', y)/2$ and $z = x - \delta\, y$. □

Finally, in the following theorem we deal with the properties of lower and upper derivatives at a point.

4.1.4. THEOREM. *Let $f: X \to \mathbb{R}$, where $X$ is a normed vector space, and $x \in X$. If $f$ is lower (or upper) derivable at $x$, then its lower (or upper) derivative at $x$ is a convex set which is closed in the norm topology of $X'$.*

PROOF. It is enough to prove the result here for lower derivative, for the result on upper derivative follows from it with the help of Theorem 4.1.1.

Suppose $f$ is lower derivable at $x$. To prove the convexity of $Lf'(x)$, suppose $x', y' \in Lf'(x)$ and $z' = tx' + (1-t)y'$, where $0 < t < 1$. Then given $\varepsilon > 0$, we can find a $\delta > 0$ such that whenever $0 < \|y - x\| < \delta$,

$$f(y) > f(x) + (x', y - x) - \varepsilon\|y - x\|,$$
$$f(y) > f(x) + (y', y - x) - \varepsilon\|y - x\|.$$

Hence if $0 < \|y - x\| < \delta$, then

$$\begin{aligned} f(y) &= tf(y) + (1-t)f(y) \\ &> t\{f(x) + (x', y - x) - \varepsilon\|y - x\|\} \\ &\quad + (1-t)\{f(x) + (y', y - x) - \varepsilon\|y - x\|\} \\ &= f(x) + (z', y - x) - \varepsilon\|y - x\|. \end{aligned}$$

This proves that $z' \in Lf'(x)$, and so $Lf'(x)$ is convex.

Next, suppose $x'$ is some limit point of $Lf'(x)$. Then, given $\varepsilon > 0$, there is a $y' \in Lf'(x)$ such that $\|y' - x'\| < \varepsilon/2$. Further, there is a $\delta > 0$ such that whenever $0 < \|y - x\| < \delta$,

$$f(y) > f(x) + (y', y - x) - (\varepsilon/2)\|y - x\|.$$

Hence if $0 < \|y - x\| < \delta$, then since $|(y', y - x) - (x', y - x)| \leq \|y' - x'\| \, \|y - x\| < (\varepsilon/2)\|y - x\|$, we obtain

$$f(y) > f(x) + (x', y - x) - \varepsilon\|y - x\|,$$

which proves that $x' \in Lf'(x)$. Consequently, $Lf'(x)$ is closed. □

4.1.5. REMARK. It should be noted here that the notion of lower gradient considered in Definition 2 is more general than the notion of subgradient used in Convex Analysis for real-valued convex functions on $\mathbb{R}^n$ (see Rockafellar [Ro1, p. 214]).

## §4.2 Lower, upper, semi-, and weak derivatives on the real line.

In this section we present extended versions of the notions of lower and upper derivatives defined in the last section for real-valued functions defined on a subset of $\mathbb{R}$. Also, related notions of lower and upper medians are defined here. We include here another notion of "weak derivative" for such functions which unifies these two derivatives with unilateral derivatives.

Suppose $f: X \to \mathbb{R}$, where $X \subset \mathbb{R}$, and let $x \in X^b$. Then using the Definition 2 of the last section in the present setting, it is clear that $c \in \mathbb{R}$ is a lower gradient of $f$ at $x$ iff for every $\varepsilon > 0$ there is a $\delta > 0$ such that

$$(1) \qquad f(y) > f(x) + c(y-x) - \varepsilon|y-x|$$

whenever $y \in X$ and $0 < |y-x| < \delta$. The definition of a finite upper gradient $c$ of $f$ at $x$ is obtained on the other hand by reversing the inequality in (1) and replacing $-\varepsilon$ by $\varepsilon$.

However, it is not difficult to see that these definitions can be reinterpreted in terms of unilateral derivates of $f$ at $x$ as follows. A real number $c$ is a lower gradient of $f$ at $x$ iff $D^-f(x) \leq c \leq D_+f(x)$, and $c$ is an upper gradient of $f$ at $x$ iff $D^+f(x) \leq c \leq D_-f(x)$.

The latter definitions extend in a natural manner to extended real numbers as well. Hence in the present case we will call any $c \in \overline{\mathbb{R}}$ a *lower* or *upper gradient* of $f$ at $x$ if $D^-f(x) \leq c \leq D_+f(x)$ or $D^+f(x) \leq c \leq D_-f(x)$, respectively.

Thus, in the present case, $f$ will be called *lower derivable* at $x$ iff $D^-f(x) \leq D_+f(x)$; and if so, then the interval $[D^-f(x), D_+f(x)]$ will be called the *lower derivative* of $f$ at $x$ and denoted by $Lf'(x)$.

Similarly, $f$ will be called *upper derivable* at $x$ iff $D^+f(x) \leq D_-f(x)$; and if so, then $[D^+f(x), D_-f(x)]$ will be called the *upper derivative* of $f$ at $x$ and denoted by $Uf'(x)$.

We will further call $f$ *semiderivable* at $x$ if it is either lower or upper derivable at $x$, and then its lower or upper derivative at $x$ will be called the *semiderivative*, or *semi- derivative*, of $f$ at $x$ and denoted by $Sf'(x)$.

In the following two theorems we extend Theorems 4.1.1 and 4.1.3 to the above extended versions of lower and upper derivatives. The following theorem deals with the relation between lower and upper derivatives.

4.2.1. THEOREM. *Let $f : X \to \mathbb{R}$, $X \subset \mathbb{R}$ and $x \in X^b$. Then $f$ is upper derivable at $x$ iff $-f$ is lower derivable at $x$; and if so, then $Uf'(x) = -L(-f)'(x)$.*

PROOF. According to the present definition, $c \in \overline{\mathbb{R}}$ is an upper gradient of $f$ at $x$ iff $D^+f(x) \leq c \leq D_-f(x)$, or, equivalently, iff $-D_+(-f)(x) \leq c \leq -D^-(-f)(x)$, or, equivalently, iff $D^-(-f)(x) \leq -c \leq D_+(-f)(x)$. This proves that $c \in Uf'(x)$ iff $-c \in L(-f)'(x)$. The two results follow clearly from this fact. □

According to the above theorem it is enough as before to consider lower derivative. The following theorem deals with the decomposition of derivative in terms of lower and upper derivatives.

4.2.2. THEOREM. *Let $f : X \to \mathbb{R}$, $X \subset \mathbb{R}$ and $x \in X^b$. Then $f$ is derivable at $x$ iff it is lower and upper derivable at $x$; and if so, then the lower and upper derivatives of $f$ at $x$ are unique and equal to $f'(x)$.*

PROOF. Since the necessity part is obvious in this case, we need to prove only the sufficiency. Hence suppose $f$ is lower and upper derivable at $x$. Then

$$D^-f(x) \le D_+f(x) \le D^+f(x) \le D_-f(x) \le D^-f(x),$$

so that the four derivates of $f$ are equal at $x$. Hence $f$ is derivable at $x$, and it is further clear that the lower and upper derivatives of $f$ at $x$ are unique and equal to $f'(x)$. □

Next, to define the notion of weak derivative, we first define *lower* and *upper weak derivates* $\underline{D}_w f(x)$ and $\overline{D}_w f(x)$ of $f$ at a point $x \in X'$. When $x \in X^b$, we define

$$\underline{D}_w f(x) = \vee D_\pm f(x), \qquad \overline{D}_w f(x) = \wedge D^\pm f(x);$$

and when $x$ is only a left or right limit point of $X$, then these derivates will be assumed to coincide with the left or right derivates, respectively, of $f$ at $x$. In terms of the weak derivates, it is interesting to note here that $f$ is nonangular at a point $x \in X^b$ iff $\underline{D}_w f(x) \le \overline{D}_w f(x)$; and thus the latter inequality holds n.e.

We further define $f$ to be *weakly derivable* at $x$ if $\overline{D}_w f(x) \le \underline{D}_w f(x)$; and if so, then the interval $[\overline{D}_w f(x), \underline{D}_w f(x)]$ will be called the *weak derivative* of $f$ at $x$ and denoted by $Wf'(x)$; also then each element of $Wf'(x)$ will be called a *weak gradient* of $f$ at $x$.

It is now easy to see that $f$ is weakly derivable at $x$ iff it is either semiderivable or unilaterally derivable from some side at $x$. When $f$ is semiderivable at $x$, it is clear that $Wf'(x) = Sf'(x)$, but when $f$ is unilaterally derivable at $x$, it may have weak gradients at $x$ other than its unilateral derivative at $x$; for example, if $f$ is unilaterally derivable from both sides at $x$ and $f'_-(x) < f'_+(x)$, then $Wf'(x) = [f'_-(x), f'_+(x)]$.

Further, when $f$ is unilaterally derivable at $x$, we will use $f'_u(x)$ to denote the unilateral derivative of $f$ at $x$, which may be either $f'_+(x)$ or $f'_-(x)$; when both of these unilateral derivatives exist, $f'_u(x)$ may denote any of them.

When the lower, upper, semi-, or weak derivative of $f$ at $x$ is singleton, $f$ will be said as before to be *uniquely lower, upper, semi-*, or *weakly derivable*, respectively, at $x$, and then $Lf'(x)$, $Uf'(x)$, $Sf'(x)$, or $Wf'(x)$ will also be used to denote its unique element.

Further, when $f$ has a finite lower, upper, semi-, or weak gradient at $x$, we will call $f$ *lower, upper, semi-*, or *weakly differentiable*, respectively, at $x$; and when this gradient is further unique, $f$ will be called *uniquely lower, upper, semi-*, or *weakly differentiable*, respectively, at $x$. Also, when $f$ has a finite ordinary derivative at $x$, $f$ will be called simply *differentiable* at $x$.

Next, when $D^-f(x) < D_+f(x)$, we will call $f$ *lower angular* at $x$, and when $D^+f(x) < D_-f(x)$, $f$ will be called *upper angular* at $x$. When $f$ is not lower or upper angular at $x$, we will call it *lower* or *upper nonangular*, respectively, at $x$.

Thus it is clear that $f$ is upper angular at $x$ iff $-f$ is lower angular at $x$, and that $f$ is angular at $x$ iff it is either lower or upper angular at $x$. Further, it should be noted here that $f$ is lower derivable at each point where it is lower angular; and that if $f$ is lower derivable at $x$, then its lower derivative is unique at $x$ iff it is lower nonangular at $x$. Similar statements hold for upper derivative and upper angularity.

Next, we consider the points where the lower or upper derivative is maximal. Given $f: X \to \mathbb{R}$, we will call a point $x \in X^b$ a *cusp point* of $f$ if $f$ has infinite left and right derivatives at $x$ which are not equal. We will further call $x$ a *lower cusp point* of $f$ if $f'_-(x) = -\infty$ and $f'_+(x) = +\infty$, and we will call $x$ an *upper cusp point* of $f$ if $f'_-(x) = +\infty$ and $f'_+(x) = -\infty$. Also, when $x$ is a lower or upper cusp point of $f$, $f$ will be said to be *lower* or *upper cusped* at $x$.

Thus it is clear that $x$ is a cusp point of $f$ iff it is a lower or upper cusp point of $f$. Further, $f$ is lower cusped at $x$ iff it is lower derivable at $x$ and $Lf'(x) = \overline{\mathbb{R}}$; similarly, $f$ is upper cusped at $x$ iff it is upper derivable at $x$ and $Uf'(x) = \overline{\mathbb{R}}$.

Next, we will define lower and upper medians which relate with lower and upper derivatives the same way as the ordinary median defined in §3.3 relates with the ordinary derivative.

Suppose $x \in X^b$. If $f$ is lower nonangular at $x$, then we will call $[D_+f(x), D^-f(x)]$ the *lower median* of $f$ at $x$; and when $f$ is upper nonangular at $x$, $[D_-f(x), D^+f(x)]$ will be called the *upper median* of $f$ at $x$. We will use $\underline{M}f(x)$ and $\overline{M}f(x)$ to denote the lower and upper medians, respectively, of $f$ at $x$, and each element of these medians will in turn be called a *lower* or *upper median derivate*, respectively, of $f$ at $x$.

It is interesting to note here that $f$ is lower, upper, or weakly derivable at $x$ iff it has at most one lower, upper, or bilateral median derivate, respectively, at $x$. Further, $f$ has a unique lower, upper, or weak derivative at $x$ iff it has one and only one lower, upper, or bilateral median derivate, respectively, at $x$; and if so, then that unique median derivate is the desired unique derivative of $f$ at $x$.

4.2.3. REMARK. The term "cusp" seems to have been used first by Sierpiński [Si1] as mentioned earlier in Remark 2.3.3. See also G. C. Young [Y.G2, p. 167].

Following the geometric definition of lower and upper derivatives presented in the last section, these derivatives can also be defined at the unilateral limit points of $X$ in $X$ in terms of unilateral derivates of $f$. If $x \in X^+ \sim X^-$, it is then easy to see that $Lf'(x) = [-\infty, D_+f(x)]$ and $Uf'(x) = [D^+f(x), \infty]$; and if $x \in X^- \sim X^+$, then similarly $Lf'(x) = [D^-f(x), \infty]$ and $Uf'(x) = [-\infty, D_-f(x)]$. Thus $f$ would then be automatically lower and upper derivable at every unilateral limit point of $X$ in $X$ without being necessarily derivable there. Therefore, to avoid unnecessary complications of this nature, we will not pursue this option here at the unilateral limit points of $X$ which are of course only countable.

## §4.3 New derivatives and knotted and normal new derivatives; new derivability sets.

In this section we first define lower and upper knot points of a real-valued function defined on a subset of $\mathbb{R}$. Using such points we then define knotted and normal lower, upper and semi- derivatives of such a function. Further, we define the terms "new derivatives" and "knotted and normal new derivatives", and then we introduce notations for various new derivability sets and knot sets which will be used frequently in the present work.

In terms of the normal new derivatives it will be seen later that every continuous function has these derivatives at lots of points; indeed there are even continuous functions which are not derivable in the ordinary sense at any point but have these derivatives almost everywhere. Also, in terms of these derivatives the classical Denjoy-Young-Saks theorem and its analogue obtained in [G1] will be seen to assume much simpler and more elegant forms (see Theorems 5.5.2 and 6.1.2).

First, we define lower and upper knot points which relate with lower and upper derivatives the same way as an ordinary knot point defined in §1.1 relates with ordinary derivative.

Suppose $f: X \to \mathbb{R}$, $X \subset \mathbb{R}$, and $x \in X^b$. Then $x$ will be called a *lower knot point* of $f$ if $D_+f(x) = -\infty$ and $D^-f(x) = +\infty$; and $x$ will be called an *upper knot point* of $f$ if $D_-f(x) = -\infty$ and $D^+f(x) = +\infty$. Also, when $x$ is a lower or upper knot point of $f$, $f$ will be said to be *lower* or *upper knotted*, respectively, at $x$. Further, the sets of lower and upper knot points of $f$ will be called the *lower* and *upper knot sets*, respectively, of $f$, and will be denoted by $\underline{K}(f)$ and $\overline{K}(f)$, respectively.

Thus $f$ is clearly upper knotted at $x$ iff $-f$ is lower knotted at $x$, and $f$ is biknotted at $x$ iff it is lower and upper knotted at $x$. Also, $f$ is lower or upper knotted at $x$ iff $\underline{M}f(x)$ or $\overline{M}f(x)$, respectively, is identical to $\overline{\mathbb{R}}$. Further, it is clear that a lower or upper knot point of $f$ is a point where $f$ lacks lower or upper derivability, respectively, to the extreme as defined in §1.1.

Next, we will call $f$ *knottedly lower derivable* at $x$ if it is uniquely lower derivable and upper knotted at $x$, and then the unique lower derivative of $f$ at $x$ will be called the *knotted lower derivative* of $f$ at $x$. Similarly, when $f$ is uniquely upper derivable and lower knotted at $x$, $f$ will be called *knottedly upper derivable* at $x$ and the unique upper derivative of $f$ at $x$ will be called the *knotted upper derivative* of $f$ at $x$.

For an example of knotted lower derivability, it is easy to see that the continuous function $f(x) = \sqrt{|x|}\sin^2(1/x)$, $f(0) = 0$, has a knotted lower derivative 0 at 0. Similarly, it can be seen easily that the function $f(x) = \sqrt{|x|}\sin(1/x)$, $f(0) = 0$, is biknotted at 0.

To provide some insight into the nature of knotted lower and upper derivatives, we include here in Figures 2 and 3 the graph of a continuous function $f$ in the neighborhood of a point $x_0$ where $f$ has a knotted lower or upper derivative $c \in \mathbb{R}$; here we use $z$ to denote the point $(x_0, f(x_0))$ and use $L$ for the line $y = f(x_0) + c(x - x_0)$.

Next, we will call $f$ *normally lower* (or *upper*) *derivable* at $x$ if it is either derivable or knottedly lower (or upper) derivable at $x$; and if so, then its

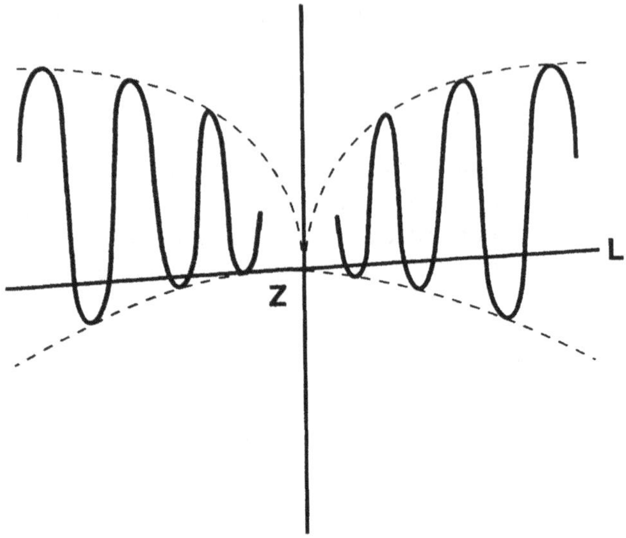


Figure 2

unique lower (or upper) derivative at $x$ will be called the *normal lower* (or *upper*) *derivative* of $f$ at $x$.

It is now easy to see that $f$ has a knotted or normal upper derivative $c$ at $x$ iff $-f$ has a knotted or normal lower derivative, respectively, $-c$ at $x$.

Finally, $f$ will be called, as before, *knottedly* (or *normally*) *semiderivable* at $x$ if it is knottedly (or normally) lower or upper derivable at $x$, and if so, then the unique semiderivative of $f$ at $x$ will be called the *knotted* (or *normal*) *semiderivative* of $f$ at $x$.

We now define what we mean by "new derivatives."

We use the term "new derivatives" in the present work for the derivatives defined in §4.2—namely, for lower, upper, semi- and weak derivatives. Further, we use the term "knotted new derivatives" for knotted lower, upper and semi- derivatives; and we use the term "normal new derivatives" for normal lower, upper and semi- derivatives.

Note that we use the term "new derivatives" only in a generic sense, and no particular derivative is called new derivative. It should be noted further that the general new derivatives are set valued in general, but the knotted and normal new derivatives are unique or single-valued by their definition.

The above new derivatives were discovered in the context of the second outstanding problem of differentiation theory mentioned in the preface—namely, to find a new notion of derivative in terms of which every continuous function is derivable at a $c$-dense set of points, or indeed at enough points so that its properties can be investigated in terms of its new

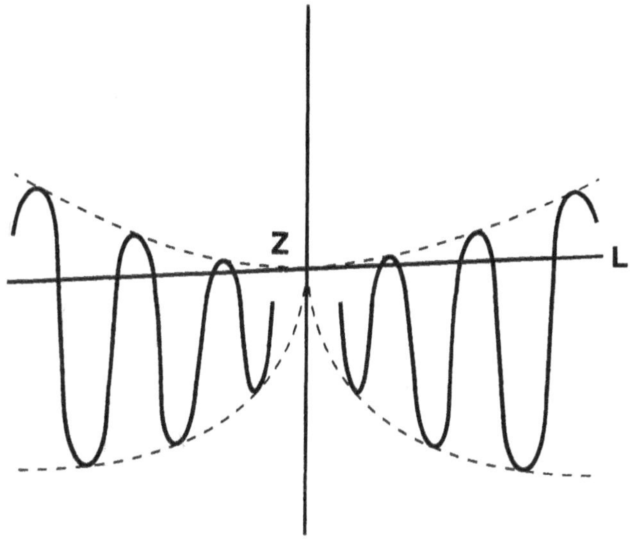


Figure 3

derivative wherever it exists. As it will be seen here, all the general and normal new derivatives meet this criterion, some better for some properties of functions and some for others. In the case of continuous nonderivable functions, the knotted new derivatives meet on the other hand the criterion of existence of new derivative at a *c*-dense set of points.

Next, we introduce some convenient notations for various derivability sets of $f$ which will be used throughout the work.

Suppose $f: X \to \mathbb{R}$, where $X \subset \mathbb{R}$. We will use $\Delta(f)$ to denote the set of points in $X$ where $f$ is derivable in the ordinary sense, and we will use $\Delta^*(f)$ to denote the set of points where $f$ is differentiable; that is, it has a finite derivative. Each of these two derivability sets will be called an *ordinary derivability set* of $f$.

When there is no possibility of confusion, we will drop the reference to $f$ in the above notations. The same applies to all the notations that follow; and for the sake of brevity we will present them without reference to $f$.

We will further use $\Delta_+$ and $\Delta_-$ to denote the sets of points where $f$ is right or left derivable, respectively; the sets $\Delta^*_+$ and $\Delta^*_-$ are defined similarly by replacing the term "derivable" by "differentiable." Among these sets, $\Delta_+$ and $\Delta^*_+$ will be called the *right derivability sets* of $f$, and $\Delta_-$ and $\Delta^*_-$ will be called the *left derivability sets* of $f$. These four derivability sets of $f$ will also sometimes be called the *unilateral derivability sets* of $f$, and $\Delta_u$ and $\Delta^*_u$ will be used to denote the sets $\Delta_+ \cup \Delta_-$ and $\Delta^*_+ \cup \Delta^*_-$, respectively.

Next, we will use $\underline{\Delta}$ to denote the set of points where $f$ is lower derivable, and we will use $\underline{\Delta}_u$, $\underline{\Delta}_k$, and $\underline{\Delta}_n$ to denote the sets of points where $f$ is uniquely, knottedly, or normally lower derivable, respectively. Similarly, $\underline{\Delta}^*$, $\underline{\Delta}^*_u$, $\underline{\Delta}^*_k$, and $\underline{\Delta}^*_n$ will be used to denote the sets of points where $f$ is lower differentiable, or uniquely, knottedly, or normally lower differentiable, respectively. All of these eight sets will be called the *lower derivability sets* of $f$.

It may be noted here that $\underline{\Delta}_k = \underline{\Delta}_u \cap \overline{K}$ and $\underline{\Delta}_n = \underline{\Delta} \cup \underline{\Delta}_k$, and similar relations hold for the lower differentiability sets.

The eight *upper derivability sets* $\overline{\Delta}$, $\overline{\Delta}_u$, $\overline{\Delta}_k$, $\overline{\Delta}_n$, $\overline{\Delta}^*$, $\overline{\Delta}^*_u$, $\overline{\Delta}^*_k$, and $\overline{\Delta}^*_n$ are defined analogously, and so are the *semiderivability sets* $\Delta_s$, $\Delta_{su}$, $\Delta_{sk}$, $\Delta_{sn}$, $\Delta^*_s$, $\Delta^*_{su}$, $\Delta^*_{sk}$, and $\Delta^*_{sn}$. The *weak derivability sets* $\Delta_w$, $\Delta_{wu}$, $\Delta^*_w$, and $\Delta^*_{wu}$ are also defined in an analogous manner.

Further, all the above derivability sets other than the ordinary derivability sets will be called simply the *general weak derivability sets* of $f$. Also, the sets $\underline{\Delta}_k$, $\underline{\Delta}^*_k$, $\overline{\Delta}_k$, $\overline{\Delta}^*_k$, $\Delta_{sk}$, and $\Delta^*_{sk}$ are sometimes called the *knotted derivability sets* of $f$, and the remaining general weak derivability sets will in contrast be called the *nonknotted derivability sets* of $f$.

4.3.1. REMARK. As mentioned earlier, the notions of new derivatives were first presented in 1976 in [G12], and then in 1981 in [G13]. In these presentations, several results of the theory of new derivatives obtained here in Parts II and III were announced without proof. Here the theory of new derivatives is presented with proofs and its various other developments.

## §4.4 Uniqueness of new derivatives.

In this section we investigate the magnitude of the sets of points where the new derivatives are not unique. In the first two theorems we deal with the case when $X \subset \mathbb{R}$, and in the third theorem we deal with the case when $X$ is a normed vector space.

4.4.1. THEOREM. *Let $f : X \to \mathbb{R}$, where $X \subset \mathbb{R}$. Then the set of points where $f$ has a nonunique weak derivative (or, in particular, lower or upper derivative) is countable. Moreover, if $f$ is nonangular, then its weak derivative is unique at every point where it exists.*

PROOF. Let $E$ denote the set of points in $X$ where $f$ has a nonunique weak derivative, and let $x \in E$. Then it is clear that $x \in X^b$ and $\overline{D}_w f(x) < \underline{D}_w f(x)$; that is,

$$\min\{D^- f(x), D^+ f(x)\} < \max\{D_- f(x), D_+ f(x)\}.$$

Consequently, either $D^- f(x) < D_+ f(x)$ or $D^+ f(x) < D_- f(x)$; that is, $f$ is angular at $x$. Hence it follows from Theorem 2.3.1 that $E$ is countable.

The second part also follows clearly from the above argument. □

As we see next, there is indeed a large class of continuous functions for which the weak derivative is unique at every point where it exists.

Let us recall here that any proposition is said to hold for a *typical continuous function* if it holds for a residual set of functions in the Banach space $\mathbb{C}$ of all continuous real-valued functions on the interval $[0, 1]$ under the uniform norm.

We introduce here some terminology regarding nonderivability of functions which will be used frequently in our discussions, and particularly in §13.5 and Chapter 19.

We will call a function $f: I \to \mathbb{R}$ *nonderivable* or *nondifferentiable* if it is not derivable or differentiable, respectively, at any point. It should be pointed out here that in the literature a nonderivable function has sometimes been called nondifferentiable.

Also, in the case when $f$ is not unilaterally derivable at any point, we will call it *unilaterally nonderivable*. The first example of such a continuous function was constructed by Besicovitch [Be]; later other examples were given by Morse [Mo1] and Singh [Sing2]. Further, we will call $f$ *knotted* if it is knotted at each point of its domain—that is, if $\underline{D}f = -\infty$ and $\overline{D}f = +\infty$ everywhere.

It is clear that every knotted function is nonderivable. The Weierstrass' nonderivable function is indeed a knotted function which is further biknotted a.e. (see, e.g., [G8]).

In the following theorem we deal with two basic properties of typical continuous functions. These properties will be obtained later in Chapter 19 in terms of several generalized derivates (see Theorem 19.5.4). For a complete description of the differential structure of typical continuous functions in terms of ordinary derivates see [G10] or Theorem 19.5.5.

4.4.2. Theorem. *A typical continuous function is nonangular and knotted. Consequently, the weak derivative of a typical continuous function is unique at every point where it exists.*

Proof. As for the nonangularity of a typical continuous function, it has been proved by Jarník (see part 1 of [J1, Satz I] and its proof) that for such a function $f$ every extended real number is a derived number of $f$ at each point of $(0, 1)$, which clearly implies that $f$ is nonangular.

The knottedness of a typical continuous function $f$ has also been obtained by Jarník [J1] at the points of $(0, 1)$. However, its knottedness at all the points of $[0, 1]$ follows from Theorem 2.2 of [BG] where a typical continuous function was proved to be not of monotonic type at any point, and this clearly implies that $f$ is knotted.

The second part follows from the first with the help of Theorem 4.4.1. □

When $X$ is on the other hand a general normed vector space, Theorem 4.4.1 does not hold in general for lower and upper derivatives. Let, for example, $X = \mathbb{R}^2$ and $f(x, y) = |x|$ for each $x$ and $y$. Then it is easy to see that $f$ has two distinct lower gradients $(-1, 0)$ and $(1, 0)$ at every point of the $y$-axis, which of course form an uncountable set.

In the following theorem we investigate, for a normed vector space $X$, the category and measure of the set of points where the lower or upper derivative of a function on $X$ is not unique.

4.4.3. THEOREM. *Suppose $f: X \to \mathbb{R}$, where $X$ is a normed vector space with a countable basis. Then the set $A$ of points where $f$ has a nonunique semiderivative is meager in $X$.*

*Moreover, if $X = \mathbb{R}^n$, then $A$ is further of measure zero.*

PROOF. It is clearly enough to prove the two results for the set $E$ of points in $X$ where $f$ has a nonunique lower derivative. Choose a basis $\{x_i\}$ of $X$ within its unit ball, and let $\{r_j\}$ be an enumeration of the set of rational numbers. Now define the following for each pair of positive integers $i$, $k$:

$$(1)\quad A_{i,k} = \{t(x_i + x) : t \in \mathbb{R},\, x \in X,\, \|x\| < k^{-2},\, \|t(x_i + x)\| < k^{-1}\}.$$

Further, for each triple of positive integers $(i, j, k)$, define $E_{i,j,k}$ to be the set of points $y$ in $E$ such that

$$(2)\quad f(z) \ge f(y) + tr_j + |t|k^{-1} \quad \text{whenever} \quad z - y = t(x_i + x) \in A_{i,k}.$$

We will first prove that $E = \bigcup_i \bigcup_j \bigcup_k E_{i,j,k}$. Let $y \in E$. Then $f$ has two distinct lower gradients $x', y' \in X'$ at $y$. Choose $i$ such that $(x', x_i) \neq (y', x_i)$. Suppose, for the sake of definiteness, that $(x', x_i) > (y', x_i)$. Now choose $j$ such that $(y', x_i) < r_j < (x', x_i)$, and let

$$(3)\qquad \varepsilon = (1/4)\min\{(x', x_i) - r_j,\, r_j - (y', x_i),\, 1\}.$$

Then according to the definition of lower gradient, we can find a $\delta > 0$ such that whenever $0 < \|x - y\| < \delta$, we have

$$(4)\qquad f(x) > f(y) + (x', x - y) - \varepsilon\|x - y\| \quad \text{and}$$

$$f(x) > f(y) + (y', x - y) - \varepsilon\|x - y\|.$$

Now choose the smallest integer $k$ which is greater than $\|x'\|$, $\|y'\|$, $1/\varepsilon$ and $1/\delta$. We claim that $y \in E_{i,j,k}$ for this choice of $i$, $j$, and $k$.

Suppose $z - y \in A_{i,k}$. Then $z - y = t(x_i + x)$, where $t \in \mathbb{R}$, $x \in X$, $\|x\| < k^{-2}$ and $\|z - y\| < k^{-1}$. We need to verify here the inequality in (2) for $z$. Let us note first that $\|z - y\| < k^{-1} < \delta$, $\|x'\| < k$, and $k^{-1} < \varepsilon \leq 1$. Hence when $t > 0$, we obtain the following with the help of (4) and (3):

$$\begin{aligned} f(z) &> f(y) + (x', z - y) - \varepsilon\|z - y\| \\ &= f(y) + t(x', x_i + x) - \varepsilon t\|x_i + x\| \\ &\geq f(y) + t\{(x', x_i) - r_j\} + tr_j + t(x', x) - \varepsilon t\{\|x_i\| + \|x\|\} \\ &> f(y) + 4\varepsilon t + tr_j - t\|x'\|\,\|x\| - \varepsilon t(1 + k^{-2}) \\ &> f(y) + tr_j + 3\varepsilon t - tk \cdot k^{-2} - \varepsilon tk^{-2} \\ &> f(y) + tr_j + \varepsilon t + 2tk^{-1} - tk^{-1} - \varepsilon^3 t \\ &= f(y) + tr_j + tk^{-1} + \varepsilon(1 - \varepsilon^2)t \geq f(y) + tr_j + |t|k^{-1}. \end{aligned}$$

Similarly, when $t < 0$ we obtain the following from (4) and (3):

$$\begin{aligned} f(z) &> f(y) + (y', z - y) - \varepsilon\|z - y\| \\ &= f(y) + t(y', x_i + x) + \varepsilon t\|x_i + x\| \\ &\geq f(y) + t\{(y', x_i) - r_j\} + tr_j + t(y', x) + \varepsilon t\{\|x_i\| + \|x\|\} \\ &> f(y) - 4\varepsilon t + tr_j + t\|y'\|\,\|x\| + \varepsilon t(1 + k^{-2}) \\ &> f(y) + tr_j - 3\varepsilon t + \varepsilon tk^{-2} + tk \cdot k^{-2} \\ &> f(y) + tr_j - 2tk^{-1} - \varepsilon t(1 - \varepsilon^2) + tk^{-1} \\ &\geq f(y) + tr_j + |t|k^{-1}. \end{aligned}$$

Because this inequality holds trivially when $t = 0$, this shows that $y \in E_{i,j,k}$. The above-mentioned identity has thus been established.

Now, to prove the first part, it is enough to show that $E_{i,j,k}$ is nowhere dense for each $i$, $j$, $k$. Given $i$, $j$, $k$, let $U$ be any nonempty open set in $X$ and suppose it includes some point $y$ of $E_{i,j,k}$. We claim that $U \cap (y + A_{i,k})$ does not include any point of $E_{i,j,k}$ other than $y$. Suppose it includes some point $z$ of $E_{i,j,k}$ other than $y$. Then $z - y = t(x_i + x)$, where $t \in \mathbb{R}$, $x \in X$, $\|x\| < k^{-2}$, and $\|z - y\| < k^{-1}$. Further, $y - z = -t(x_i + x) \in A_{i,k}$, and so by (2) we obtain

$$f(z) \geq f(y) + tr_j + |t|k^{-1}, \qquad f(y) \geq f(z) - tr_j + |t|k^{-1}.$$

But this implies that $|t|k^{-1} \leq f(z) - f(y) - tr_j \leq -|t|k^{-1}$, that is, $t = 0$, or $z = y$. This contradiction proves the claim.

Now choose $c > 0$ such that $c < 1/(2k)$ and $U$ includes the ball with center $y$ and radius $2c$. Let $V$ denote the open ball with center $y + cx_i$ and radius $ck^{-2}$. Let $x \in V$. Then $x - y = cx_i + z$, where $\|z\| < ck^{-2}$, so that

$$\|x - y\| \leq c\|x_i\| + \|z\| < c(1 + k^{-2}) \leq 2c < k^{-1}.$$

Further, $x - y = c(x_i + c^{-1}z)$, where $\|c^{-1}z\| < k^{-2}$. Hence, by (1), $x - y \in A_{i,k}$, which proves that $V \subset U \cap (y + A_{i,k})$. Because $V$ does not include any point of $E_{i,j,k}$ except possibly $y$, this proves that $E_{i,j,k}$ is nowhere dense in $X$. Consequently, $E$ is meager in $X$.

Next, to prove the second part, suppose $X = \mathbb{R}^n$. For any set $E \subset X$, we will use here $|E|$ to denote the $n$-dimensional Lebesgue outer measure of $E$. Given positive integers $i$, $j$, $k$, let $y \in E_{i,j,k}$. Then as we saw above, for every positive number $c < 1/(2k)$, the open ball $B$ with center $y$ and radius $2c$ includes another ball $V$ with center $y + ca_i$ and radius $ck^{-2}$ that does not include any point of $E_{i,j,k}$ except possibly $y$. Set $V_0 = V \sim \{y\}$. Then $|V_0| = |V|$, so that

$$\frac{|V_0|}{|B|} = \frac{(ck^{-2})^n}{(2c)^n} = 2^{-n}k^{-2n}.$$

Thus it is clear that $y$ is not a point of outer density of $E_{i,j,k}$ (it is indeed not difficult to see that $|B \cap (y + A_{i,k})|/|B| = (2/\pi)\sin^{-1}(k^{-2})$, which is even independent of $n$). Hence it follows from the Lebesgue density theorem [S8, p. 129] that $E_{i,j,k}$ is of measure zero. Consequently, in this case $E$ is of measure zero. □

4.4.4. REMARK. In connection with Theorem 4.4.2 it may be pointed out here that there are many known examples of continuous nonderivable functions which are indeed knotted functions. A general method of construction of such functions has been given by Knopp [Kn] (see also Hobson [H2, pp. 407–412]).

The Weierstrass' nonderivable function is also known to be a knotted function. This function was first presented by Weierstrass in his lectures and was published later in 1875 by du Bois-Reymond [Boi]. Its derivates were studied in detail by G. C. Young [Y.G2], and a complete description of its derivates was obtained later in [G8].

## §4.5 Continuity under new derivabilities.

Since ordinary differentiability implies continuity, it is natural to expect some form of continuity under each of the new derivabilities. In the present section we investigate these continuities.

We begin here with a general $X$.

4.5.1. THEOREM. *Suppose $f: X \to \mathbb{R}$, where $X$ is a normed vector space. If $f$ is lower derivable at $x \in X$, then it is LSC at $x$.*

PROOF. Suppose $f$ is lower derivable at $x$. Then it has a lower gradient $x'$ at $x$. Given $\varepsilon > 0$, we can then choose a $\delta > 0$ such that $\delta < \varepsilon/(\|x'\| + 1)$ and

$$f(y) > f(x) + (x', y - x) - \|y - x\| \quad \text{whenever} \quad 0 < \|y - x\| < \delta.$$

Then whenever $\|y - x\| < \delta$, we have

$$f(y) - f(x) \geq -\|x'\|\,\|y - x\| - \|y - x\| > -\{\|x'\| + 1\}\delta > -\varepsilon,$$

which proves that $f$ is $LSC$ at $x$. □

The next two theorems deal with the case when $X \subset \mathbb{R}$.

4.5.2. THEOREM. *Suppose* $f : X \to \mathbb{R}$, *where* $X \subset \mathbb{R}$, *and let* $x \in X^b$.
(a) *If $f$ is lower derivable at $x$, then it is QLSC at $x$.*
(b) *If $f$ is lower differentiable at $x$, then it is LSC at $x$.*
(c) *If $f$ is uniquely weakly differentiable at $x$, then it is internal at $x$.*

PROOF. To prove (a), suppose $f$ is lower derivable at $x$. Then $D^-f(x) \leq D_+f(x)$, and hence either $D^-f(x) < \infty$ or $D_+f(x) > -\infty$. Consequently, either $f(x) \leq \underline{f}(x - 0)$ or $f(x) \leq \underline{f}(x + 0)$; that is, $f$ is $LSC$ from at least one side at $x$. Hence $f$ is $QLSC$ at $x$.

Next, to prove (b), suppose $f$ is lower differentiable at $x$. Then there is a finite number $c$ such that $D^-f(x) \leq c \leq D_+f(x)$. Hence it is clear that $f(x) \leq \underline{f}(x \pm 0)$, and so $f$ is $LSC$ at $x$.

Finally, to prove (c), suppose $f$ is uniquely weakly differentiable at $x$. Then $f$ has a finite unique weak derivative $c$ at $x$, so that

$$\min\{D^-f(x), D^+f(x)\} = \max\{D_-f(x), D_+f(x)\} = c.$$

It is then easy to see that $c$ is a bilateral median derivate of $f$ at $x$, from which it is clear that $f$ is internal at $x$. □

The usual continuity of $f$ at a point where it is differentiable follows from the part (b) of the above theorem on applying it to $f$ and $-f$. Further, it follows from its part (c) that a regulated function is continuous at every point where it is uniquely weakly differentiable.

Since a $LSC$ function on an interval is known to be Darboux iff it is internal (see Choquet [C1] or Lemma 7.3.1), from the last two parts of the above theorem we further obtain the following.

4.5.3. COROLLARY. *Let* $f : I \to \mathbb{R}$, *where $I$ is open. If $f$ is uniquely lower differentiable everywhere, then it is a LSC, Darboux function.*

Although lower derivability does not imply $LSC$ in general, this implication does hold at nearly all the points as we see in the next theorem. First we present a more general result in the form of a lemma in terms of knot sets which will be used frequently.

4.5.4. LEMMA. *Suppose $f: X \to \mathbb{R}$, where $X \subset \mathbb{R}$. Then $f$ is LSC at nearly all the points where it is not lower knotted. Consequently, $f$ is continuous at nearly all the points where it is not knotted in the ordinary sense.*

PROOF. To prove the first part, let $A = X \sim \underline{K}(f)$ and $B = \{x \in X : \underline{f}(x-0) = \underline{f}(x+0)\}$. Then since $A \sim B$ is countable by Theorem 1.1.1, it is enough to show that $f$ is $LSC$ at the points of $A \cap B$. Suppose $x \in A \cap B$. Then either $D^- f(x) < +\infty$ or $D_+ f(x) > -\infty$. Hence $f(x) \le \underline{f}(x-0) = \underline{f}(x+0)$, which proves that $f$ is $LSC$ at $x$.

The second part follows, on the other hand, on combining the first part with the result obtained on applying it to $-f$. □

4.5.5. THEOREM. *Suppose $f: X \to \mathbb{R}$, where $X \subset \mathbb{R}$. Then $f$ is LSC at nearly all the points where it is lower derivable.*

*Further, $f$ is continuous at nearly all the points where it is derivable, or where it has a finite unilateral derivative.*

PROOF. The first two parts follow directly from the above lemma. To prove the last part, let $A$ denote the set of points in $X$ where $f$ has a finite unilateral derivative, and let $B$ be the set of points where $f$ has symmetrical lower and upper limits. Then $A \sim B$ is countable by Theorem 1.1.1, and at each point of $A \cap B$, it is clear that $f$ is continuous. □

We conclude this section with an example to show the necessity of the term "finite" in the last part of the above theorem.

4.5.6. EXAMPLE. Let $g$ denote the Cantor's continuous nondecreasing singular function on $I \equiv [0,1]$ which is defined in terms of his nowhere dense perfect set $P$. Set $E = \{x \in P : g'(x) = +\infty\}$. Then since $|g(E)| = |g(I)| = 1$ (see [G9, p. 1443]), $E$ has cardinality $c$. Now, define $f(x) = g(x)$ or 1 according to whether $x \in E$ or $I \sim E$. Then $f$ has a right derivative $+\infty$ at each point of $E$, but it is clearly discontinuous at every point of $E$.

## *CHAPTER 5*

# Existence of New Derivatives

In this chapter we obtain two derivability theorems which establish the existence of points of new derivabilities. Further, we deduce from them with the help of the First Derivate Theorem the celebrated Denjoy-Young-Saks theorem. Other derivability theorems of this nature will be obtained in the next chapter.

First, in §5.1, we prove in the First Derivability Theorem the existence of points of lower differentiability for every *LSC* function on an interval. In §5.2 we define properties $L\text{-}BV_*$ and $U\text{-}BV_*$ similar to the well-known property $BV_*$ on a set. Then, in §5.3, we prove in the Second Derivability Theorem the lower and upper differentiability of $GL\text{-}BV_*$ and $GU\text{-}BV_*$ functions, respectively, a.e. Such functions need not be derivable, however, in the ordinary sense at any point.

Next, in §5.4, we obtain from the First Derivate Theorem some variants of the unilateral monotonic type and Lipschitz properties of §2.1 in terms of unilateral derivates. Then, in §5.5, we deduce from the Second Derivability Theorem, with the help of a result obtained in §5.4, the Denjoy-Young-Saks theorem which establishes the normal lower and upper differentiability of an arbitrary function at almost all the points where it is not lower or upper knotted, respectively.

**§5.1 Lower derivability of *LSC* functions.** In this section we prove the existence of points of lower differentiability for every *LSC* function, and when such a function is lower nonangular, it is indeed uniquely lower differentiable at a $c$-dense set of points.

It should be clear from our convention that in the present section $f: I \to \mathbb{R}$ where $I$ is some subinterval of $\mathbb{R}$. We begin with the following lemma which is known in the case of continuous functions (see, e.g., [G9, p. 1442]).

5.1.1. LEMMA. *Suppose $f$ is LSC. Then $f$ is nowhere monotone iff it is nowhere constant and it has relative minima at a dense set of points in $I$.*

PROOF. Suppose $f$ is nowhere monotone. Then $f$ is obviously nowhere constant. Let $J$ be any open subinterval of $I$. Then there exist $a, b \in J$, $a < b$, such that $f(a) > f(b)$, and similarly there exist $c, d \in J \cap (b, \infty)$, $c < d$, such that $f(c) < f(d)$. Then $a < b < c < d$. Now, since $f$ is $LSC$, it attains its infimum on $[a, d]$, say $m$, at some point $x$ of this interval. Then since $m \le f(b) < f(a)$ and $m \le f(c) < f(d)$, $a < x < d$. This proves that $f$ has a relative minimum at $x \in J$.

The converse is obvious, for if $f$ is nowhere constant but monotone on some open subinterval $J$ of $I$, then it is strictly monotone on $J$, and so it does not have a relative minimum at any point of $J$. □

5.1.2. THEOREM (FIRST DERIVABILITY). *Suppose $f$ is LSC. Then $f$ is lower differentiable at a dense set of points in $I$.*

*Moreover, if $f$ is lower nonangular, then it is uniquely lower differentiable at a c-dense set of points in $I$.*

PROOF. In case $f$ is of monotonic type on some subinterval $J$ of $I$, it is clearly of bounded variation on $J$, and so it follows from the Lebesgue's well-known derivability theorem [L] that $f$ is differentiable a.e. in $J$. Hence for the proof of each part we need to consider only subintervals of $I$ on which $f$ is nowhere of monotonic type.

Hence, suppose $f$ is nowhere of monotonic type on some subinterval $J$ of $I$, and let $r \in \mathbb{R}$. Then since $f_{-r}$ is nowhere monotone on $J$, it has by the above lemma a relative minimum at some point $x_r$ in $J$. It is then clear that $f_{-r}$ has a lower gradient zero at $x_r$, and so $f$ has a lower gradient $r$ at $x_r$. This completes the proof of the first part.

Next, when $f$ is lower nonangular, its lower derivative is unique at every point where it exists, and so in this case $x_r \neq x_s$ whenever $r \neq s$. Hence in this case the set $\{x_r : r \in \mathbb{R}\}$ has cardinality $c$. □

A function $f: X \to \mathbb{R}$ is said to have *symmetrical derivates* [G8] if $D_-f = D_+f$ and $D^-f = D^+f$ at each point of $X^b$. Such a function is clearly nonangular, and it is further derivable at every point where it is weakly derivable. Hence from the above theorem we obtain the following result on the existence of points of derivability which was proved by W. H. Young and G. C. Young [YY] in 1909 in the case of continuous functions.

5.1.3. COROLLARY (YOUNG-YOUNG). *Suppose $f$ is either LSC or USC. If $f$ has symmetrical derivates, then it is differentiable at a c-dense set of points in $I$.*

5.1.4. REMARK. Corollary 5.1.3 was also obtained independently by Zalcwasser in the case of continuous smooth functions (see Zygmund [Zy2, p. 48]) which are known to have symmetrical derivates. Another extension of his result will be obtained later in Chapter 12 (see Theorem 12.4.2).

See §7.8 regarding an extension of first part of the First Derivability Theorem 5.1.2 to functions on $\mathbb{R}^n$. The hypothesis of lower nonangularity in its second part is indeed essential. For, in the case of the Weierstrass' continuous nonderivable function $f$, it is easy to see from Theorem 2 of [G8] that $f$ is lower or upper differentiable only at its lower and upper cusp points respectively, which are of course countable, and it is clear that at each of these points the lower or upper derivative of $f$ is $\overline{\mathbb{R}}$. However, $f$ has infinite knotted (and so normal) lower and upper derivatives $\infty$ and $-\infty$ at two $c$-dense sets of points. Thus as far as lower or upper differentiability is concerned, the Weierstrass' function forms an extreme case due to its angularity at a dense set of points.

On the other hand, by Theorem 4.4.2, a typical continuous function is also nonderivable since it is a knotted function, and since it is nonangular, it follows from Theorem 5.1.2 that such a function is knottedly (and so normally) lower and upper differentiable at two $c$-dense sets of points.

It may be noted here, further, that for a nonangular continuous function, or even a function with symmetrical derivates, the points of semiderivability may form a meager set of measure zero. Indeed, there exist continuous smooth functions whose points of derivability, and thus of weak derivability, form a set of measure zero (see Zygmund [Zy2, p. 58]). Further, since such a function is clearly nowhere of monotonic type, it is biknotted r.e. and hence the set in question is also meager (see [G4] or Theorem 13.3.1).

## §5.2 Lower and upper strong bounded variation on a set.

In this section we define two properties of functions on a set, called $L\text{-}BV_*$ and $U\text{-}BV_*$, which are related to the well-known property $BV_*$ (or $VB_*$) used in the theory of Denjoy-Perron integral (see Saks [S8, p. 228]). With the help of these properties we will obtain two classes of functions which will be seen in the next section to be lower and upper differentiable almost everywhere.

Given $f: X \to \mathbb{R}$, $X \subset \mathbb{R}$, let $I = [a, b]$ be any closed interval whose endpoints belong to $X$. Then the *oscillation* of $f$ on $I$ is defined to be

$$O(f; I) = \sup\{f(x) : x \in I \cap X\} - \inf\{f(x) : x \in I \cap X\}.$$

Modifying this oscillation, we define the *lower* and *upper oscillations* of $f$ on $I$, denoted by $\underline{O}(f; I)$ and $\overline{O}(f; I)$, as follows:

$$\underline{O}(f; I) = \max\{f(a), f(b)\} - \inf\{f(x) : x \in I \cap X\},$$

$$\overline{O}(f; I) = \sup\{f(x) : x \in I \cap X\} - \min\{f(a), f(b)\}.$$

Then it is clear that $\overline{O}(f; I) = \underline{O}(-f; I)$ and

$$\text{(1)} \qquad \max\{\underline{O}(f; I), \overline{O}(f; I)\} \le O(f; I) \le \underline{O}(f; I) + \overline{O}(f; I).$$

Next, given $E \subset X$, we define the *lower* and *upper strong variations* of $f$ on $E$, denoted by $L\text{-}V_*(f; E)$ and $U\text{-}V_*(f; E)$ respectively, to be the suprema of $\Sigma_k \underline{O}(f; I_k)$ and $\Sigma_k \overline{O}(f; I_k)$ respectively, where $\{I_k\}$ is any sequence of nonoverlapping closed intervals whose endpoints belong to $E$. The ordinary *strong variation* $V_*(f; I)$ of $f$ on $E$ is defined on the other hand by replacing $\underline{O}$ by $O$ in the definition of $L\text{-}V_*(f; E)$ (see [S8, p. 228]). Thus it follows clearly from (1) that

$$\text{(2)} \qquad \max\{L\text{-}V_*(f;E), U\text{-}V_*(f;E)\} \le V_*(f;E) \le L\text{-}V_*(f;E) + U\text{-}V_*(f;E).$$

Finally, when $L\text{-}V_*(f;E)$ or $U\text{-}V_*(f;E)$ is finite, $f$ will be said to be of *bounded lower* or *upper strong variation* respectively on $E$, or, simply, *lower* or *upper* $BV_*$ respectively on $E$, or, more briefly, $L\text{-}BV_*$ or $U\text{-}BV_*$ respectively on $E$. Also, when $f$ is $GL\text{-}BV_*$ or $GU\text{-}BV_*$ on $E$ as defined in §1.3, $f$ will be said to be of *generalized bounded lower* or *upper strong variation*, respectively, on $E$.

It is easy to see from the above definitions that if $f$ is $L\text{-}BV_*$ or $U\text{-}BV_*$ on $E$, then it is $BV$ on $E$. Further, it is clear from (2) that $f$ is $BV_*$ on $E$ iff it is simultaneously $L\text{-}BV_*$ and $U\text{-}BV_*$ on $E$. Consequently, $f$ is $GBV_*$ on $E$ iff it is $GL\text{-}BV_*$ and $GU\text{-}BV_*$ on $E$, and each of the latter properties implies in turn that $f$ is $GBV$ on $E$.

Further, it is clear that $f$ is $U\text{-}BV_*$ on $E$ iff $-f$ is $L\text{-}BV_*$ on $E$. Hence it is enough to consider $L\text{-}BV_*$. Also, when $f$ and $g$ are $L\text{-}BV_*$ on $E$ and $c > 0$, it is easy to see that $f+g$ and $cf$ are also $L\text{-}BV_*$ on $E$. Moreover, when $f$ is $L\text{-}BV_*$ on $E$, it is easy to see that $f$ is bounded from below on $X \cap \overline{\text{co}}(E)$.

When $E$ is bounded and $f$ is $BV_*$ on $E$, $f$ is known [S8, p. 229] to be $BV_*$ on $\overline{E}$—that is, the closure of $E$ relative to $X$. However, this does not hold for $L\text{-}BV_*$. For, let $I = [0,1]$, $Q$ be the set of rational numbers in $I$ and $E = I \sim Q$. Then the characteristic function of $Q$ on $I$ is clearly $L\text{-}BV_*$ on $E$, but not on $\overline{E} = I$.

In the following theorem we obtain this extension property for $L\text{-}BV_*$ under a mild continuity hypothesis.

5.2.1. THEOREM. *Suppose $f: X \to \mathbb{R}$, and $E \subset X$ is a bounded set such that the extreme points of $\overline{\text{co}}(E)$ are in $X$. If $f$ is $L\text{-}BV_*$ on $E$ and it is LSC at the points of $\overline{E} \sim E$, then $f$ is $L\text{-}BV_*$ on $\overline{E}$.*

*Consequently, if $f$ is LSC and $GL\text{-}BV_*$ on $E$, then it is $F\text{-}GL\text{-}BV_*$ on $E$.*

PROOF. Let $I = [a,b]$ denote the convex hull of $\overline{E}$. Then, by the hypothesis, $a, b \in X$. Let $a = a_0 < a_1 < \dots < a_n = b$ be any finite sequence of points in $\overline{E}$. For each $k = 1, 2, \dots, n$, set

$$I_k = [a_{k-1}, a_k], \qquad c_k = \inf\{f(x) : x \in X \cap I_k\}.$$

Next, we need to choose $b_k$, $k = 0, 1, \dots, n$, in $E$ in a suitable manner. Let $K$ denote the set of integers $k$ for which $a_k \notin E$. Given $k \le n$, if $a_k \in E$, we choose $b_k = a_k$. Otherwise, $f$ is $LSC$ at $a_k$, and so we can find an open interval $U_k$ containing $a_k$ for which the following two relations hold:

(3) $\quad f(x) > f(a_k) - 1/n \quad$ for $\quad x \in X \cap U_k$.

(4) $\quad$ If $\; f(a_k) > c_k \;$ or $\; c_{k+1}, \;$ then $\; f(x) > c_k \;$ or $\; c_{k+1},$
respectively, for $\; x \in X \cap U_k$.

Clearly, the intervals $\{U_k : k \in K\}$ can be chosen to be mutually disjoint and such that no $U_k$ contains $a_{k-1}$ or $a_{k+1}$. Now, for each $k \in K$, we choose $b_k$ to be any point of $E \cap U_k$ which is clearly nonempty. Then it is clear that $a \le b_0 < b_1 < \cdots < b_n \le b$, $a_{k-1} < b_k$ for $0 < k \le n$, and $b_k < a_{k+1}$ for $0 \le k < n$. Now set, for each $k = 1,2,\ldots,n$,

$$J_k = [b_{k-1}, b_k], \qquad d_k = \inf\{f(x) : x \in X \cap J_k\}.$$

First, let $1 < k < n$. Then we claim that

(5) $$\underline{O}(f; I_k) \le \underline{O}(f; J_{k-1}) + \underline{O}(f; J_k) + \underline{O}(f; J_{k+1}) + 1/n.$$

Suppose $\underline{O}(f; I_k) > 0$, for there is nothing to prove otherwise. Then there are three cases which need to be considered separately.

**Case (i):** $c_k < \min\{f(a_{k-1}), f(a_k)\}$. In this case it is easy to see from (4) that $d_k = c_k$, and hence it follows from (3) that

$$\begin{aligned}\underline{O}(f; I_k) &= \max\{f(a_{k-1}), f(a_k)\} - c_k \\ &< \max\{f(b_{k-1}), f(b_k)\} - d_k + 1/n = \underline{O}(f; J_k) + 1/n.\end{aligned}$$

**Case (ii):** $c_k = f(a_{k-1}) \le f(a_k)$. In this case, if $b_{k-1} \le a_{k-1}$, then

$$\underline{O}(f; I_k) = f(a_k) - f(a_{k-1}) < f(b_k) - d_k + 1/n \le \underline{O}(f; J_k) + 1/n,$$

and if $a_{k-1} < b_{k-1}$, then

$$\begin{aligned}\underline{O}(f; I_k) &= f(a_k) - f(a_{k-1}) < f(b_k) - d_{k-1} + 1/n \\ &= f(b_k) - f(b_{k-1}) + f(b_{k-1}) - d_{k-1} + 1/n \\ &\le \underline{O}(f; J_k) + \underline{O}(f; J_{k-1}) + 1/n.\end{aligned}$$

**Case (iii):** $c_k = f(a_k) < f(a_{k-1})$. In this case we obtain the following by an argument similar to the one used for case (ii):

$$\underline{O}(f; I_k) < \underline{O}(f; J_k) + \underline{O}(f; J_{k+1}) + 1/n.$$

The inequality (5) has thus been established for $1 < k < n$.

Next, since $f$ is $BV$ on $E$, it is bounded on $E$ by some finite number $M$. Then, for $k = 1$, we claim that

(6) $$\underline{O}(f; I_1) \le \underline{O}(f; J_1) + \underline{O}(f; J_2) + M - f(a) + 1/n.$$

In cases (i) and (iii) this inequality follows exactly as above, and in case (ii) we now have

$$\begin{aligned}\underline{O}(f; I_1) &= f(a_1) - f(a_0) < f(b_1) - f(b_0) + f(b_0) - f(a_0) + 1/n \\ &\le \underline{O}(f; J_1) + M - f(a) + 1/n.\end{aligned}$$

For $k = n$, it is proved similarly that

(7) $$\underline{O}(f; I_n) \le \underline{O}(f; J_{n-1}) + \underline{O}(f; J_n) + M - f(b) + 1/n.$$

Now, with the help of (5), (6), and (7), we obtain

$$\begin{aligned}\sum_{k=1}^{n} \underline{O}(f; I_k) &\le 3 \sum_{k=1}^{n} \underline{O}(f; J_k) + 2M + 1 - f(a) - f(b) \\ &\le 3L\text{-}V_*(f; E) + 2M + 1 - f(a) - f(b).\end{aligned}$$

This proves that $L\text{-}V_*(f; \overline{E})$ is finite, and hence $f$ is $L\text{-}BV_*$ on $\overline{E}$.

Finally, it is easy to see that the last part follows directly from the first. □

**§5.3 Lower differentiability of $GL\text{-}BV_*$ functions.** In this section we obtain a theorem establishing unique lower differentiability of $GL\text{-}BV_*$ functions at almost all points. A generalization of Lebesgue's differentiation theorem obtained by Lusin, Denjoy, and Saks for $GBV_*$ functions (see [S8, p. 230]), which is used in the theory of Denjoy-Perron integral, follows from the new theorem as a corollary.

Also, we include here an example of a $GL\text{-}BV_*$ function which is not derivable in the ordinary sense at any point (see Remark 5.3.3).

We need here some notations. Given $f: X \to \mathbb{R}$ and $E \subset X$, we will use $Gf(E)$ to denote the *graph of $f$ on $E$*, namely,

$$Gf(E) \equiv \{(x, f(x)) : x \in E\}.$$

It should be pointed out here that Saks [S8, p. 88] used $B(f; E)$ to denote $Gf(E)$.

Further, following Saks [S8], we will use $\rho$ to denote the usual metric on $\mathbb{R}^2$, and for any set $A \subset \mathbb{R}^2$, we will use $\delta(A)$ to denote the diameter of $A$ relative to $\rho$. Also, $\Lambda(A)$ will denote the *length* of $A$ (see Saks [S8, pp. 53,

54]), namely the one-dimensional Hausdorff outer measure of $A$ which is defined as follows:

$$\Lambda(A) = \lim_{\varepsilon \to 0^+} \Lambda^{(\varepsilon)}(A),$$

where for each $\varepsilon > 0$, $\Lambda^{(\varepsilon)}(A)$ is the infimum of the sums $\Sigma_n \delta(A_n)$ for all countable decompositions $\{A_n\}$ of $A$ such that $\delta(A_n) < \varepsilon$ for each $n$.

Let us recall here further that we use $\underline{\Delta}_u(f)$ (see §4.3) to denote the set of points where $f$ has a unique lower derivative (finite or infinite).

5.3.1. THEOREM (SECOND DERIVABILITY). *Let $f : X \to \mathbb{R}$ and $E \subset X$. If $f$ is $GL$-$BV_*$ on $E$, then it is uniquely lower differentiable a.e. in $E$; and further if $N = E \sim \underline{\Delta}_u(f)$, then $|f(N)| = \Lambda\{Gf(N)\} = 0$.*

PROOF. It is enough to prove the result when $E$ is a bounded set such that the extreme points of $\overline{\text{co}}(E)$ are included in $X$, and $f$ is $L$-$BV_*$ on $E$. For, when $f$ is $GL$-$BV_*$ on $E$, $E$ is clearly the union of a sequence of such sets. It will be convenient to use here the notations developed in §4.3 for various derivability sets.

Let $I = [a, b]$ denote the convex hull of $\overline{E}$. We need to define here a function $g$ of bounded variation on $I$ which agrees with $f$ on $E$. First, by modifying the values of $f$ at the points of $\overline{E} \sim E$, we will define another function $h$ on $X$ which is $L$-$BV_*$ on $\overline{E}$.

Let $h(x) = f(x)$ when $x \in E \cup (X \sim \overline{E})$; and when $x \in \overline{E} \sim E$, define

$$h(x) = \sup_{\delta > 0} \inf\{f(t) : t \in X, |t - x| < \delta\}.$$

Then since $f$ is $L$-$BV_*$ on $E$, so is clearly $h$ on $E$; and since $h$ is clearly $LSC$ at the points of $\overline{E} \sim E$, it is $L$-$BV_*$ on $\overline{E}$ by Theorem 5.2.1.

Next, let $\{I_n\}$ denote the contiguous intervals of $\overline{E}$ in $I$. Define $g(x) = h(x)$ when $x \in \overline{E}$; and when $x \in I_n$ for any $n$, define

$$g(x) = \inf\{h(t) : t \in (X \cup \overline{E}) \cap \bar{I}_n\}.$$

Then since $h$ is $BV$ on $\overline{E}$, it is clear that $g$ is $BV$ on $I$. Also, it is clear that $g = h = f$ on $E$, whereas $g \le h \le f$ on $X \cap I$.

Now, let $\Delta = \Delta(g)$ and $\Delta^* = \Delta^*(g)$. Then it follows from a theorem of de la Vallée Poussin [S8, p. 125] that

(1) $$|I \sim \Delta^*| = 0, \quad |g(I \sim \Delta)| = 0 \quad \text{and} \quad \Lambda\{Gg(I \sim \Delta)\} = 0.$$

Next, since $g \le f$ on $X \cap I$, for each $x \in E \cap \Delta$ it is clear that

(2) $$D^- f(x) \le g'(x) \le D_+ f(x),$$

that is, $f$ is lower derivable at $x$. Hence $E \cap \Delta \subset \underline{\Delta}(f)$. Let $C = E \cap \Delta \sim \underline{\Delta}_u(f)$. Then $C$ is countable by Theorem 4.4.1, and for each $x \in E \cap \Delta \sim C$, $f$ has by (2) a unique lower derivative at $x$ equal to $g'(x)$.

Now, let $N = E \sim \underline{\Delta}_u(f)$ and $N^* = E \sim \underline{\Delta}^*(f)$. Then it is clear from above that $N \subset (E \sim \Delta) \cup C$, and similarly $N^* \subset (E \sim \Delta^*) \cup C$. Now, since $f = g$ on $E$, we obtain the following from (1):

$$|N^*| \le |E \sim \Delta^*| + |C| \le |I \sim \Delta^*| = 0,$$
$$|f(N)| \le |f(E \sim \Delta)| + |f(C)| = |g(E \sim \Delta)| \le |g(I \sim \Delta)| = 0,$$
$$\Lambda\{Gf(N)\} \le \Lambda\{Gg(I \sim \Delta)\} + \Lambda\{Gg(C)\} = 0.$$

This establishes the two parts of the theorem. □

On applying the above theorem to $f$ and $-f$, we obtain in the following corollary the Lusin-Denjoy-Saks theorem mentioned earlier.

5.3.2. COROLLARY (LUSIN-DENJOY-SAKS). *Let $f: X \to \mathbb{R}$ and $E \subset X$. If $f$ is $GBV_*$ on $E$, then it is differentiable a.e. in $E$; and further if $N = E \sim \Delta(f)$, then $|f(N)| = \Lambda\{Gf(N)\} = 0$.*

5.3.3. REMARK. The Lusin-Denjoy-Saks theorem obtained in Corollary 5.3.2 was obtained initially by Lusin [Lu2, p. 69] and Denjoy [D4] independently in the case of continuous functions in a less complete form, and it was obtained later by Saks [S4, p. 133] in the above form. This theorem plays a basic role in the theory of the Denjoy-Perron integral.

In connection with the Second Derivability Theorem 5.3.1, we first observe here that a $GL\text{-}BV_*$ function need not be derivable in the ordinary sense at any point. For, let $f$ be the characteristic function of the set of rational numbers on $I = [0, 1]$. It is easy to see that $f$ is $GL\text{-}BV_*$ and it is clearly not derivable at any point. Secondly, because this function $f$ is not $GBV_*$ by Corollary 5.3.2, this shows further that the class of $GL\text{-}BV_*$ functions is much wider than that of $GBV_*$ functions.

**§5.4 Lower and upper strongly monotonic type and Lipschitz properties on a set.** In this section we define some variants of monotonic type and Lipschitz properties of a function on a set which are related with $L\text{-}BV_*$ and $U\text{-}BV_*$, and in a sense with lower and upper derivatives. Also, we deduce here from the First Derivate Theorem some results on these properties in terms of unilateral derivates.

Given $f: X \to \mathbb{R}$ and $E \subset X$, we will call $f$ *lower strongly monotonic type* (or $L\text{-}MT_*$) on $E$ if it is either $RLL$ or $LUL$ on $E$. Similarly, $f$ will be called *upper strongly monotonic type* (or $U\text{-}MT_*$) on $E$ if it is either $RUL$ or $LLL$ on $E$.

It is then clear that if $f$ is $L\text{-}MT_*$ or $U\text{-}MT_*$ on $E$, then in each of these two cases $f$ is either $LMT$ or $RMT$ on $E$.

Further, we will call $f$ *lower strongly Lipschitz* (or $L\text{-}L_*$) on $E$ if it is $RLL$ and $LUL$ on $E$. Also, if there is a real number $c \ge 0$ such that $f_{+c}$ is $WRI$

and $f_{-c}$ is *WLD* on $E$, then $f$ will be called *lower strongly c-Lipschitz* (or $L\text{-}L_*(c)$) on $E$.

Similarly, $f$ will be called *upper strongly Lipschitz* (or $U\text{-}L_*$) on $E$ if it is *RUL* and *LLL* on $E$. And, in case there is a real number $c \geq 0$ such that $f_{+c}$ is *WLI* and $f_{-c}$ is *WRD* on $E$, then $f$ will be called *upper strongly c-Lipschitz* (or $U\text{-}L_*(c)$) on $E$.

Further, we will call $f$ *strongly Lipschitz* (or $L_*$) on $E$ if it is simultaneously $L\text{-}L_*$ and $U\text{-}L_*$ on $E$, or, equivalently, if $f$ is bilaterally Lipschitz on $E$. The *strong c-Lipschitz property* $L_*(c)$ on $E$ is defined similarly.

Let us note here that $f$ is $L\text{-}L_*$ on $E$ iff there is some real number $c \geq 0$ such that $f$ is $L\text{-}L_*(c)$ on $E$, or, equivalently, such that

$$f(x) \geq \max\{f(a) - c(x-a), f(b) - c(b-x)\}$$

wherever $a, b \in E$, $x \in X$ and $a \leq x \leq b$.

Further, it is easy to see that $f$ is $U\text{-}MT_*$ or $U\text{-}L_*$ on $E$ iff $-f$ is $L\text{-}MT_*$ or $L\text{-}L_*$, respectively, on $E$. Consequently, it is enough to consider the properties $L\text{-}MT_*$ and $L\text{-}L_*$ on $E$.

Next, we obtain some results on the above properties of $f$ on $E$ in terms of unilateral derivates of $f$.

5.4.1. THEOREM. *Let $f: X \to \mathbb{R}$, $E \subset X$, and suppose $E \cap \underline{K}(f)$ is countable. Then $f$ is $G_\delta$-$GL\text{-}MT_*$ on $E$. Moreover, if $f$ is LSC relative to $E$, then it is $F$-$GL\text{-}MT_*$ on $E$.*

PROOF. First, suppose $f$ is *LSC* relative to $E$. Let $A$ denote the set of points in $E$ where $D_+f > -\infty$, and set $B = E \sim A$. Then it follows from Corollary 2.2.3 that $A = \bigcup_n A_n$, where $f$ is *RLL* on each $A_n$. Now, for each $n$, it follows easily from Lemma 2.2.1 that there is a sequence of closed sets $\{A_{n,i}\}$ in $E$ such that $A_n \subset \bigcup_i A_{n,i}$ and $f$ is *RLL* on each $A_{n,i}$. Further, since $D^-f < \infty$ n.e. in $B$, it follows similarly on applying Corollary 2.2.3 and Lemma 2.2.1 to $f_-$ that $B \subset \bigcup_n \bigcup_i B_{n,i}$, where each $B_{n,i}$ is a closed set in $E$ and $f$ is *LUL* on $B_{n,i}$. Consequently, $f$ is $F$-$GL\text{-}MT_*$ on $E$.

Next, to prove the general result, let $C$ denote the set of points in $E$ where $f$ is not *LSC* relative to $E$. Then since $f$ is clearly *LSC* from at least one side at each point of $E \sim \underline{K}(f)$, it follows from Theorem 1.1.1 that $C$ is countable. Hence $E_0 = E \sim C$ is a $G_\delta$-set in $E$. Now since $f$ is by above $F$-$GL\text{-}MT_*$ on $E_0$, it is clear that $f$ is $G_\delta$-$GL\text{-}MT_*$ on $E$. □

5.4.2. THEOREM. *Let $f: X \to \mathbb{R}$ and $E \subset X$. If $D_+f > -\infty$ and $D^-f < \infty$ at each point of $E$, then $f$ is $F$-$GL\text{-}L_*$ on $E$.*

*Moreover, if there is some $c \in \mathbb{R}_+$ such that $D_+f > -c$ and $D^-f < c$ at each point of $E$, then $f$ is $F$-$GL\text{-}L_*(c)$ on $E$.*

PROOF. In each of the two cases $f$ is clearly *LSC* at the points of $E$. First, suppose $D_+f > -\infty$ and $D^-f < \infty$ at the points of $E$. Then it follows from Corollary 2.2.3 that $f$ is *F-GRLL* on $E$, and on applying this corollary to $f_-$ it follows that $f$ is *F-GLUL* on $E$. It is now easy to see that $f$ is $F\text{-}GL\text{-}L_*$ on $E$.

The second part is obtained similarly from Theorem 2.2.2 on applying it to $f_{+c}$ and $(f_{-c})_-$. □

5.4.3. THEOREM. *Let $f: X \to \mathbb{R}$, $E \subset X$, and suppose $f$ is QLSC relative to $E$. If the set $E \sim \underline{K}(f)$ is nonmeager in $E$, then $f$ is $L\text{-}MT_*$ on some portion of $E$.*

PROOF. Suppose the hypothesis holds. Then either $D_+f > -\infty$ or $D^-f < \infty$ at a nonmeager set of points in $E$. Hence on applying Corollary 2.4.2 to $f$ or $f_-$ it follows that $f$ is either *RLL* or *LUL* on some portion of $E$. Consequently, $f$ is $L\text{-}MT_*$ on some portion of $E$. □

5.4.4. THEOREM. *Let $f: X \to \mathbb{R}$, $E \subset X$, and suppose $f$ is QLSC relative to $E$. If there is a nonmeager set of points in $E$ where $D_+f > -\infty$ and $D^-f < \infty$, then $f$ is $L\text{-}L_*$ on some portion of $E$.*

*Moreover, if there is some $c \in \mathbb{R}_+$ such that $D_+f > -c$ and $D^-f < c$ at a nonmeager set of points in $E$, then $f$ is $L\text{-}L_*(c)$ on some portion of $E$.*

PROOF. It is enough to prove here the second part, for the first part follows from the second as seen earlier in the proof of Theorem 2.4.4.

Hence suppose there is some $c \in \mathbb{R}_+$ such that the set

$$A = \{x \in E : D_+f(x) > -c, D^-f(x) < c\}$$

is nonmeager in $E$. Then it follows from Theorem 5.4.2 that $A = \bigcup_n A_n$, where $f$ is $L\text{-}L_*(c)$ on each $A_n$. Hence there is an $n$ such that $A_n$ is dense in some portion $E_0$ of $E$. Now since $f_{+c}$ is *WRI* and $f_{-c}$ is *WLD* on $A_n$, it follows from Lemma 2.2.1 that the same holds on the set $E_0 \cap \operatorname{co}(A_n)$, which in turn contains some portion $E_1$ of $E$. Consequently, $f$ is $L\text{-}L_*(c)$ on $E_1$. □

Finally, from the last two theorems we obtain the following theorem with the help of Corollary 2.2.4 by arguments similar to the ones used for Corollary 2.2.5.

5.4.5. THEOREM. *Let $f: X \to \mathbb{R}$, $c \in \mathbb{R}_+$, and suppose $X$ is a Borel set in $\mathbb{R}$. If $f$ is Borel measurable and the set $A$ of points in $X$ where* (i) *$f$ is not lower knotted,* (ii) *$D_+f > -\infty$ and $D^-f < \infty$, or* (iii) *$D_+f > -c$ and $D^-f < c$, is uncountable, then there is a nonempty perfect set $P$ in $X$ on which $f$ is $L\text{-}MT_*$, $L\text{-}L_*$, or $L\text{-}L_*(c)$, respectively.*

*Moreover, if $f$ is Lebesgue measurable and $|A| > 0$, then in each case $P$ can be chosen to be metrically dense in itself.*

**§5.5 Normal lower derivability outside lower knot set; Denjoy-Young-Saks theorem.** In this section we deduce from the First Derivate and Second Derivability Theorems the celebrated theorem of Denjoy, Young, and Saks [S8, p. 271] on unilateral derivates indicating its relationship with lower and upper differentiability. Other related theorems will be dealt with in the next chapter.

We begin with an elementary lemma dealing with the relation between $L\text{-}MT_*$ and $L\text{-}BV_*$.

5.5.1. LEMMA. *Let $f: X \to \mathbb{R}$ and $E \subset X$. If $f$ is $L\text{-}MT_*$ on $E$, then it is $L\text{-}BV_*$ on each portion of $E$ which includes its bounds.*

*Consequently, if $f$ is $GL\text{-}MT_*$ on $E$, then it is $GL\text{-}BV_*$ on $E$.*

PROOF. Suppose $f$ is $L\text{-}MT_*$ on $E$, and let $E_0$ be any portion of $E$ which includes its bounds. Then $f$ is either $RLL$ or $LUL$ on $E$. First, suppose $f$ is $RLL$ on $E$. Then there is a $c \in \mathbb{R}_+$ such that $f_{+c}$ is right increasing on $E$. Define $g(x) = -cx$, $x \in X$. Then $f = f_{+c} + g$, and since $E_0$ includes its bounds, it is clear that $f_{+c}$ and $g$ are $L\text{-}BV_*$ on $E_0$. Consequently, $f$ is $L\text{-}BV_*$ on $E_0$. A similar argument holds when $f$ is $LUL$ on $E$.

The second part is of course a direct consequence of the first. □

The following theorem was proved originally by Denjoy [D1] for continuous functions, then by G. C. Young [Y.G3] for measurable functions, and finally by Saks [S1] for arbitrary functions.

5.5.2. THEOREM (DENJOY-YOUNG-SAKS). *Every function $f: X \to \mathbb{R}$ is normally lower differentiable at almost all the points where it is not lower knotted. Consequently, $f$ is normally semidifferentiable at almost all the points where it is not biknotted.*

PROOF. Let $\underline{K}$, $\overline{K}$, $\underline{\Delta}^*$, $\overline{\Delta}^*$, and so on, be the lower and upper knot sets and derivability sets of $f$ as defined in §4.3, and set

$$A = X \sim (\underline{K} \cup \underline{\Delta}^*), \qquad B = X \sim (\overline{K} \cup \overline{\Delta}^*).$$

Then according to Theorem 5.4.1 which was deduced from the First Derivate Theorem, $f$ is $GL\text{-}MT_*$ on $A$. Hence, according to the above lemma, $f$ is $GL\text{-}BV_*$ on $A$, and so it follows from the Second Derivability Theorem 5.3.1 that $|A| = 0$. Also, on applying this result to $-f$, we obtain $|B| = 0$.

It is further clear that

$$X \sim (A \cup B) = (\underline{K} \cup \underline{\Delta}^*) \cap (\overline{K} \cup \overline{\Delta}^*)$$
$$= K_b \cup \underline{\Delta}^*_k \cup \overline{\Delta}^*_k \cup \Delta^* .$$

Each of the two parts follows now from this equation since $|A \cup B| = 0$. □

We include here a consequence of the above theorem dealing with the set of points where a function has an infinite weak derivative or gradient. The following result was obtained originally by Lusin [Lu1] for the ordinary derivative of a continuous function, and then by Banach [B1] for unilateral derivatives of arbitrary functions. Following is an extended version of their theorem which follows directly from the second part of the above theorem.

5.5.3. COROLLARY. *Given $f : X \to \mathbb{R}$, if $A$ is the set of points where $f$ has an infinite weak gradient, then $|A| = 0$.*

# CHAPTER 6

# Normal Level Structure and Other Derivability Theorems

In this chapter we obtain the two level structure theorems which deal with the differential structure of a function at the points of almost all of its level sets. Further, we deduce from them several derivability theorems.

First, in §6.1, we obtain the Normal Level Structure Theorem, which is analogous to the Denjoy-Young-Saks theorem. This theorem turns out to be fundamental in its nature since all the results of this chapter are deduced from this theorem. In this section we deduce from this theorem two basic theorems of Saks on unilateral derivates.

Next, in §6.2, we deduce from the Normal Level Structure Theorem the Third Derivability Theorem, which establishes the normal lower derivability of a continuous function at a dense set of points in almost all of its level sets. Further, in the Fourth Derivability Theorem of §6.3, this derivative is found to be finite or infinite when the function is regular or singular, respectively, relative to lower derivative. Also, in §6.4, we deduce the Second Level Structure Theorem from the Normal Level Structure Theorem.

Next, in §6.5, we investigate derivability in terms of properties of level sets. We define here for this purpose properties $(S_1)$ and $(S_2)$ similar to Banach's properties $(T_1)$ and $(T_2)$. Then, in the Fifth Derivability Theorem, we obtain characterizations of these two properties in terms of derivability and semiderivability of the function at all the points of almost all of its level sets. Further, from this derivability theorem and the two level structure theorems we deduce in this section and §6.6 strengthened and generalized versions of several known theorems of Banach and Saks on the derivability of continuous functions under properties $(T_1)$, $(T_2)$, $(N)$, and $(S)$; see, in particular, the Sixth and Seventh Derivability Theorems.

Finally, in §6.7, we obtain some inequalities and other results on the length of the graph of a function, including a theorem of Kolmogoroff and Verčenko.

**§6.1 Normal level structure theorem; two theorems of Saks.** In this section we deduce from some of the consequences of the first two derivate theorems and the Second Derivability Theorem an analogue of the Denjoy-Young-Saks theorem obtained earlier in [G1] which is also related with lower and upper derivabilities. Also, we obtain from this theorem two basic theorems of Saks on unilateral derivates as corollaries, and we show that this analogue is the best possible in its direction (see Remark 6.1.5). Numerous other applications of this theorem will be given in the remaining sections of this chapter.

This analogue deals essentially with the differential structure of a function at the points of most of its level sets, and so it is called here the Normal Level Structure Theorem. We will need here some terminology to state the theorem.

Given $f: X \to \mathbb{R}$, for each $c \in f(X)$ the set

$$f^{-1}(c) = \{x \in X : f(x) = c\}$$

is called a *level set* of $f$.

We will call a level set $S$ of $f$ *differentially normal* if $f$ has a "nonzero" normal semiderivative (finite or infinite) at each point of $S$ where it is not biknotted.

Further, we will call $S$ sometimes *differentially lower* (or *upper*) *normal* if $f$ has a nonzero unique lower (or upper) derivative at each point of $S$ where it is not lower (or upper) knotted. Now it is easy to see that $S$ is differentially normal iff it is differentially lower and upper normal.

Further, if there is a set $A \subset f(X)$ of measure zero such that the level set $f^{-1}(c)$ has some given property $P$ for every $c \in f(X) \sim A$, then we will say that *almost every level set* of $f$ has $P$. Similarly, when the same holds for a meager set $A$, it will be said that *residually every level set* of $f$ has $P$.

First, we deduce here from the Second Derivate Theorem, and its consequence obtained in Theorem 5.4.2, the following lemma on functions that are generalized Lipschitz on a set. We need here only one part of this lemma, the other part will be used later in §6.7.

6.1.1. LEMMA. *Let $f: X \to \mathbb{R}$, $E \subset X$ and $c \geq 0$. If $f$ is $G_\delta$-$GL(c)$ on $E$, then*

$$(1) \qquad |f(E)| \leq c|E| \quad \textit{and} \quad \Lambda\{Gf(E)\} \leq \sqrt{c^2+1} \cdot |E|.$$

*Consequently, the same holds when some lower derivate of $f$ is $\geq -c$ n.e. in $E$ and some upper derivate of $f$ is $\leq c$ n.e. in $E$.*

PROOF. First, suppose $f$ is $G_\delta$-$GL(c)$ on $E$. Then $E = \bigcup_n E_n$, where $\{E_n\}$ is a sequence of $G_\delta$-sets in $E$ on each of which $f$ is $c$-Lipschitz. Set $A_1 = E_1$, and for each $n > 1$, set $A_n = E_n \sim \bigcup_{i<n} E_i$. Then $\{A_n\}$ is a disjoint sequence of

sets such that $E = \bigcup_n A_n$, and as it follows from Theorem 1.4.3, each $A_n$ is measurable with respect to the extended Lebesgue measure $m_E$ on $E$.

Now let $\varepsilon > 0$. Given $n$, choose a sequence of open intervals $\{I_{n,k} : k = 1, 2, \ldots\}$ such that $A_n \subset \bigcup_k I_{n,k}$ and

$$(2) \qquad \sum_k |I_{n,k}| < |A_n| + \varepsilon(c+1)^{-1} 2^{-n}.$$

Then for each $x, y \in A_n \cap I_{n,k}$, since $f$ is $c$-Lipschitz on $A_n$, we obtain

$$|f(y) - f(x)| \le c|y - x| \le c|I_{n,k}|,$$

and

$$\rho((y, f(y)), (x, f(x))) = \sqrt{(f(y) - f(x))^2 + (y - x)^2} \le \sqrt{c^2 + 1} \cdot |I_{n,k}|.$$

Hence, we obtain

$$(3) \quad |f(A_n \cap I_{n,k})| \le c|I_{n,k}|, \qquad \Lambda\{Gf(A_n \cap I_{n,k})\} \le \sqrt{c^2 + 1} \cdot |I_{n,k}|.$$

Now, from the first inequality of (3) we obtain with the help of (2) the following:

$$|f(A_n)| \le \sum_k |f(A_n \cap I_{n,k})| \le c \sum_k |I_{n,k}| < c|A_n| + \varepsilon \cdot 2^{-n}.$$

Hence, using the countable additivity of the measure $m_E$ (see §1.4), we obtain

$$|f(E)| \le \sum_n |f(A_n)| < c \cdot \sum_n m_E(A_n) + \varepsilon = c|E| + \varepsilon.$$

Because this holds for an arbitrary $\varepsilon$, this proves the first inequality of (1). Its second inequality is obtained similarly from the other inequality of (3).

Next, to prove the last part, suppose $D_+ f \ge -c$ n.e. in $E$, and let $d > c$. Then if $D^+ f \le c$ n.e. in $E$, it follows from Theorem 2.2.7 that $f$ is $G_\delta$-$GL(d)$ on $E$, and if $D^- f \le c$ n.e. in $E$, the same is obtained from Theorem 5.4.2. Hence in each of the two cases it follows from the first part that the two inequalities of (1) hold with $d$ in place of $c$. But since this holds for every $d > c$, (1) indeed holds for $c$ as well.

Finally, in the case when $D_- f \ge -c$ n.e. in $E$, the result is obtained on applying the result just established to $-f_-$ (see Lemma 1.1.2). □

The following theorem was obtained earlier in [G1] for a connected domain $X$ and was called there the "Denjoy analogue." The theorem is stated here in a somewhat different form which is found more useful in its application, and its new title is more descriptive of the nature of the theorem.

6.1.2. THEOREM (NORMAL LEVEL STRUCTURE). *For every function $f: X \to \mathbb{R}$, almost every level set of $f$ is differentially normal.*

PROOF. It is enough to prove here that almost every level set of $f$ is differentially lower normal. For, on applying this result to $-f$ it follows that almost every level set of $f$ is differentially upper normal, and the result follows clearly on combining these two results.

Let $\underline{K} = \underline{K}(f)$ and $\underline{\Delta}_u = \underline{\Delta}_u(f)$ (see §4.3). Set

$$A = X \sim (\underline{K} \cup \underline{\Delta}_u), \qquad B = \{x \in \underline{\Delta}_u : Lf'(x) = 0\},$$

and let $E = A \cup B$. Then according to Theorem 5.4.1, which was deduced from the First Derivate Theorem, $f$ is $GL\text{-}MT_*$ on $A$. Hence according to Lemma 5.5.1, $f$ is $GL\text{-}BV_*$ on $A$, and so it follows from the Second Derivability Theorem 5.3.1 that $|f(A)| = 0$. Further, on decomposing $B$ into a sequence of sets of finite measure if necessary, it follows from the first part of the above lemma, by choosing $c = 0$, that $|f(B)| = 0$. Consequently, $|f(E)| = 0$.

Now, for each $c \in f(X) \sim f(E)$, it is clear that the level set $f^{-1}(c)$ is differentially lower normal. □

Next, to deduce the two theorems of Saks from the above theorem, let us observe first that in any differentially normal level set of $f$, if $f$ has a finite unilateral derivate at any point, that derivate has to be normal semiderivative of $f$ at that point.

As an immediate consequence of the above theorem we thus obtain the following result of Lusin [Lu2, p. 105] and Saks [S2], which was obtained initially by Lusin for the set of points where $f$ has a zero derivative, and it was extended later to the present form by Saks.

6.1.3. COROLLARY (LUSIN-SAKS). *Let $f: X \to \mathbb{R}$ and $E \subset X$. If at each point of $E$ at least one unilateral derivate of $f$ is zero, then $|f(E)| = 0$.*

Following Lusin [Lu2], a function $f: X \to \mathbb{R}$ will be said to have (*Lusin's*) *property* $(N)$ on a set $E \subset X$ if for every subset $A$ of $E$, $|f(A)| = 0$ whenever $|A| = 0$.

To obtain the other theorem of Saks [S2] (or see [S8, p. 271]) on this property $(N)$, suppose that at each point of $A$ at least one unilateral derivate of $f$ is finite and $|A| = 0$. Let $A_n$ denote for each $n$ the set of points in $A$ where the derivate in question is in magnitude at most $n$. Then it follows from the above theorem with the help of Lemma 6.1.1 that $|f(A_n)| = 0$ for each $n$, and hence $|f(A)| = 0$. Thus we obtain the following.

6.1.4. COROLLARY (SAKS). *Let $f: X \to \mathbb{R}$ and $E \subset X$. If at each point of $E$ at least one unilateral derivate of $f$ is finite, then $f$ has property $(N)$ on $E$.*

6.1.5. Remark. In the case of the Denjoy-Young-Saks Theorem 5.5.2, it has been shown by Denjoy [D1] that its result is the best possible in its direction even for continuous functions. With the help of the results deduced later from the Normal Level Structure Theorem, we show here that this theorem is also the best possible in its direction even in the case of continuous functions.

For this purpose it is enough to show in terms of each of the following properties of level sets that there exist continuous functions which have this property: namely, there exist continuous functions $f$ for which almost every level set of $f$ include points where (i) $f$ is biknotted; (ii) $f$ has finite, nonzero, knotted lower and upper derivatives; or (iii) $f$ has infinite knotted lower and upper derivatives.

Suppose $f$ is a continuous function on $[0, 1]$. First, when $f$ is nonderivable, almost every level set of $f$ includes a dense set of points where $f$ is biknotted (see Corollary 6.4.4). Next, when $f$ is unilaterally nonderivable (e.g., the Besicovitch's function [Be]), almost every level set of $f$ includes two dense sets of points where $f$ has finite knotted lower and upper derivatives (see Corollary 13.5.2), which in turn may be assumed to be nonzero due to Corollary 6.1.3. Finally, when $f$ is biknotted a.e. (e.g., the Weierstrass' function), almost every level set of $f$ includes two dense sets of points where $f$ has infinite knotted lower and upper derivatives (see Corollary 13.5.3).

## §6.2 Normal lower derivability of continuous functions.

In this section we deduce from the Normal Level Structure Theorem established in the last section the existence of points of normal lower and upper derivability for every continuous function, on an interval, in almost all of its level sets.

We will need here a decomposition of the well-known Darboux property of functions [Da] into lower and upper Darboux properties. We define these properties for extended real-valued functions defined on any subset of $\mathbb{R}$, because they will be needed later while investigating the properties of various derivatives (see, e.g., [G11, p. 325]).

Given $f: X \to \overline{\mathbb{R}}$, we will call $f$ *lower* or *upper Darboux* if for each $a, b \in X$, $a < b$, whenever $f(a) > c > f(b)$ or $f(a) < c < f(b)$ respectively, there exists some point $x$ in $X \cap (a, b)$ where $f(x) = c$. Also, $f$ will be called *weakly Darboux* if it is either lower or upper Darboux.

When $f$ is simultaneously lower and upper Darboux, it will be said to be a *Darboux function* or to have the *Darboux property*. Every continuous function on an interval is known to have this property.

Clearly, $f$ is upper Darboux iff $-f$ is lower Darboux; and when $f$ is nondecreasing, it is automatically lower Darboux. Thus every monotone function on $X$ is weakly Darboux. As for the ordinary Darboux property, it is easy to see that a monotone function on an interval does not have this property unless it is continuous.

Analogous to the Darboux property of continuous functions, when a function $f: I \to \mathbb{R}$ is $LC$, then it is lower Darboux. In the following lemma we obtain a more general version of this result.

6.2.1. Lemma. *Suppose $f: I \to \overline{\mathbb{R}}$. If $f$ is $L^*LI$ or $R^*LI$, then it is lower Darboux; and if $f$ is $L^*UI$ or $R^*UI$, then it is upper Darboux.*

PROOF. Consider first the case when $f$ is $R^*LI$, namely, $LI_-$ and $LC_+$. Suppose $a, b \in I$, $a < b$ and $f(a) > c > f(b)$. Then the set

$$E = \{t \in [a, b] : f(t) \leq c\}$$

is obviously nonempty. Let $x$ be the infimum of $E$. Then it is clear from the hypothesis that $a < x < b$ and that $c \leq \bar{f}(x - 0) \leq f(x) \leq \underline{f}(x + 0) \leq c$. Hence $f(x) = c$, which proves that $f$ is lower Darboux.

When $f$ is $L^*LI$, the result is obtained on applying the above result to $-f_-$. The two results on upper Darboux property are obtained on the other hand on applying the proved results to $-f$. □

6.2.2. THEOREM (THIRD DERIVABILITY). *Suppose $f : I \to \mathbb{R}$ is either $R^*LI$ or $L^*UI$. Then almost every level set of $f$ includes a dense set of points where $f$ is normally lower derivable. Consequently, $f$ is normally lower derivable at a c-dense set of points in $I$.*

PROOF. It is enough to prove the result when $f$ is $R^*LI$, for on applying this result to $f_-$, the result is obtained in the other case.

Hence, suppose $f$ is $R^*LI$. On taking into account Theorem 6.1.2, it is enough to prove the result in the first part for a differentially normal level set $S$ of $f$. Let $J$ be any open subinterval of $I$ for which $S_0 \equiv S \cap J \neq \varnothing$. It is now enough to show that $S_0$ includes some point where $f$ is normally lower derivable.

Let $x \in S_0$, and suppose $f$ is not normally lower derivable at $x$. Then $f$ is either biknotted or it has a nonzero knotted upper derivative at $x$. In both of these cases it is clear that either $D_-f(x) < 0 < D^-f(x)$ or $D_+f(x) < 0 < D^+f(x)$. Hence, in any case we can find two points $a$, $b \in J$ on one side of $x$ such that $a < b$ and $f(a) > f(x) > f(b)$. Let $y$ be the infimum of the set of points $t$ in $[a, b]$ for which $f(t) \leq f(x)$. Then as seen in the proof of the above lemma, $a < y < b$ and $f(y) = f(x)$. Hence $y \in S_0$ and $D^-f(y) \leq 0$. Since $f$ is thus not lower knotted at $y$, it must be normally lower derivable at $y$. This completes the proof of the first part.

Next, to prove the second part, let $A$ denote the set of points in $I$ where $f$ is normally lower derivable, and let $J$ be an arbitrary open subinterval of $I$. In case $f$ is nondecreasing on $J$, it is clear that $A \cap J$ has cardinality $c$. Otherwise, there exist $c, d \in J$, $c < d$, such that $f(c) > f(d)$. Now since $f$ is lower Darboux by the above lemma, $[f(d), f(c)] \subset f(J)$. Hence $|f(J)| > 0$. Also, according to the first part, $|f(J) \sim f(A \cap J)| = 0$. Hence $|f(A \cap J)| > 0$, and thus $A \cap J$ has cardinality $c$. □

In the case of a function with symmetrical derivates (e.g., a smooth function), we obtain the following result on the existence of points of ordinary derivability on applying the above theorem to $f$ and $-f$.

6.2.3. COROLLARY. *Suppose $f: I \to \mathbb{R}$ is left or right strongly LI or UI, and it has symmetrical derivates n.e. Then $f$ is derivable at a dense set of points in almost all of its level sets, as well as at a c-dense set of points in $I$.*

**§6.3 Derivability of generalized regular and singular functions.** Generalizing the notions of absolute continuity and singularity of functions of bounded variation, we define in this section the notions of lower and upper regularity, and lower and upper singularity, of a function relative to any generalized derivative.

First, we deal in the first four theorems with some basic results on the above properties. Then in the Fourth Derivability Theorem we obtain for a function that is regular or singular relative to lower derivative the existence of points in almost all of its level sets where the function has a finite or infinite normal lower derivative, respectively.

Subsequently, in Chapter 13, we deduce from the Third and Fourth Derivability Theorems some results on the nature of lower and upper derivatives of continuous nonderivable functions (see §13.5).

Let $G$-derivative be any *generalized derivative*—that is, a notion of derivative such that every function $f: X \to \mathbb{R}$ is $G$-derivable at each point $x \in X$ where it is derivable in the ordinary sense, and the two derivatives of $f$ coincide at such a point $x$.

Then we will call a function $f: X \to \mathbb{R}$ *lower* or *upper $G$-regular* if it maps the points where $Gf'$ exists and is $-\infty$ or $+\infty$, respectively, into a set of measure zero; and $f$ will be called *$G$-regular* if it is simultaneously lower and upper $G$-regular.

Next, we will call $f$ *lower* or *upper $G$-singular* if the set of points in $X$ where $Gf'$ exists and is negative or positive, respectively, is of measure zero. Further, $f$ will be called *$G$-singular* if it is simultaneously lower and upper $G$-singular, or, equivalently, if $Gf'$ is zero at almost all the points where it exists.

In the case when $G$ is the ordinary derivative, the prefix "$G$-" will be dropped from the above terminology. Further, when $f$ is $G$-regular or $G$-singular, it will also be said sometimes to be regular or singular *relative to* the $G$-derivative; the same will apply to their lower and upper versions.

In the cases of lower, upper, semi-, and weak derivatives, we will use the prefixes $L$-, $U$-, $S$-, and $W$-, respectively. Then it is clear that $f$ is $S$-regular iff it is simultaneously $L$-regular and $U$-regular; and $f$ is similarly $S$-singular iff it is simultaneously $L$-singular and $U$-singular; the same holds for the lower and upper versions of each of these two properties.

As for the background of the above notions, the ordinary lower and upper regularities of a function have been considered earlier by Saks in [S4], where, in analogy with Lusin's condition $(N)$, a lower or upper regular function was said to fulfill *condition* $(N^{-\infty})$ or $(N^{+\infty})$ respectively, and a regular function was said to fulfill *condition* $(N^{\infty})$. The ordinary lower and upper singularities were considered on the other hand in [G11]. It should be pointed out here that the term "singular" is usually used in the literature only for functions of bounded variation (the term "bisingular" was used in [G11] for "singular" to avoid this misunderstanding).

It should be observed here that the class of singular functions as defined above is much wider than what is usually considered as singular functions. For, every nonderivable function is automatically singular as well as regular. On the other hand, there exist nonderivable functions which are $L$-regular and $U$-regular but not $L$-singular or $U$-singular, and vice versa (see Corollaries 13.5.2 and 13.5.3).

Because the above concepts are essentially new, we first obtain some basic results on them that follow from their definitions and from the results obtained in previous sections.

We begin with a theorem on the convertibility of the above upper $G$-properties into lower $G$-properties, and of the $U$-properties into $L$-properties. Some nomenclature is needed here.

Given any generalized $G$-derivative, we will call it *admissible* if for every function $f: X \to \mathbb{R}$, if $f$ is $G$-derivable at $x \in X$, then so is $-f$ and $G(-f)'(x) = -Gf'(x)$. Further, the $G$-derivative will be called *laterally admissible* if for every function $f: X \to \mathbb{R}$, if $f$ is $G$-derivable at $x \in X$, then $f_-$ is so at $-x$ and $G(f_-)'(-x) = -Gf'(x)$.

Most of the known generalized derivatives will be seen later in Chapter 15 to be admissible. On the other hand, the lower and upper derivatives are easily seen to be laterally admissible but not admissible, whereas the semiderivative is admissible.

6.3.1. THEOREM. *Let $f: X \to \mathbb{R}$.* (a) *If a generalized $G$-derivative is admissible (or laterally admissible), then $f$ is upper $G$-regular or upper $G$-singular iff $-f$ (or $f_-$) is lower $G$-regular or lower $G$-singular, respectively.*

(b) *The function $f$ is lower (or upper) $U$-regular or $U$-singular iff $-f$ is upper (or lower) $L$-regular or $L$-singular, respectively. Consequently, $f$ is $U$-regular or $U$-singular iff $-f$ is $L$-regular or $L$-singular, respectively.*

PROOF. Part (a) follows directly from the definitions of admissible and laterally admissible generalized derivatives, and part (b) is obtained similarly from the fact that $Uf'(x) = -L(-f)'(x)$ at each point $x$ where $f$ is upper derivable. □

It is now clear from the above theorem that for any admissible or laterally admissible generalized $G$-derivative, it is enough to consider the lower $G$-regularity and lower $G$-singularity; and among the various $U$- and $L$-regularities and $U$- and $L$-singularities, it is enough to consider $L$-regularity and $L$-singularity.

Next, it should be observed here, as is clear from the above definitions, that each of the regularity and singularity properties with respect to any generalized derivative is stronger than its ordinary version. In the following theorem we obtain the converse in some special cases. We need here some nomenclature.

Any generalized $G$-derivative is called here *weakly compatible with lower derivative* if for every function $f : X \to \mathbb{R}$, at each point $x \in X$ where $f$ is simultaneously lower and $G$-derivable, $Gf'(x) \in Lf'(x)$. Weak compatibility with upper derivative is defined analogously, and as we will see later in Chapters 16 and 17, most of the known generalized derivatives are indeed weakly compatible with lower and upper derivatives.

6.3.2. THEOREM. *Let $f : X \to \mathbb{R}$. (a) If $f$ is $GBV_*$, then it is lower regular or lower singular iff it is so relative to any generalized derivative.*

(b) *If $f$ is $GL$-$BV_*$, and it is lower $L$-regular or lower $L$-singular, then it is lower regular or lower singular respectively relative to every generalized derivative that is weakly compatible with lower derivative.*

PROOF. Part (a) follows directly from the Lusin-Denjoy-Saks theorem (see Corollary 5.3.2), and part (b) is obtained similarly from the Second Derivability Theorem 5.3.1 with the help of Theorem 4.4.1. □

Next, in the following theorem we obtain characterizations of various ordinary and $L$-singularities in terms of derivates and medians.

6.3.3. THEOREM. *Suppose $f : X \to \mathbb{R}$. Then $f$ is lower singular, upper singular, or singular iff $\overline{D}f > 0$, $\underline{D}f \leq 0$, or $0 \in Mf$, respectively, a.e.*

*Similarly, $f$ is lower $L$-singular, upper $L$-singular, or $L$-singular iff $D^-f \geq 0$, $D_+f \leq 0$, or $0 \in \underline{M}f$, respectively, a.e.*

PROOF. Each part is easily seen to follow directly from the Denjoy-Young-Saks Theorem 5.5.2. □

The ordinary and $L$-regularities also generalize Lusin's property $(N)$. In the following theorem we obtain the equivalence of these properties in some special cases.

6.3.4. THEOREM. *Let $f : X \to \mathbb{R}$. If $f$ has property $(N)$, then it is $S$-regular. Moreover,*

(a) *if $f$ is $GBV_*$, then it is regular iff it has property $(N)$; and if $f$ is continuous and $X$ is closed, then $f$ is regular iff it is $GAC_*$; and*
(b) *if $f$ is $GL$-$BV_*$, then it is $L$-regular iff it has property $(N)$.*

PROOF. The first part follows directly from Corollary 5.5.3. Further, in part (a), the first part is obtained easily from Corollaries 5.3.2 and 6.1.4, and its second part has been obtained by Saks [S4]. Part (b) is obtained on the other hand from Theorem 5.3.1 and Corollary 6.1.4. □

Next, in the case of $L$-regular and $L$-singular functions, we deduce in the following theorem from the Third Derivability Theorem the existence of points in various level sets where the normal lower derivative is finite or infinite, respectively. Part (a) of this theorem is somewhat similar to the well-known theorem of Banach [B4] on the differentiability of continuous functions with property $(N)$ (see Theorem 6.6.5 for a strengthened version of Banach's theorem).

6.3.5. THEOREM (FOURTH DERIVABILITY). *Suppose* $f : I \to \mathbb{R}$ *is either* $R^*LI$ *or* $L^*UI$. (a) *If f is L-regular, then it is normally lower differentiable at a dense set of points in almost all of its level sets, as well as at a metrically dense set of points in I.*

(b) *If f is L-singular, then it has an infinite normal lower derivative at a dense set of points in almost all of its level sets, as well as at a c-dense set of points in I provided that f is a nowhere constant Darboux function.*

PROOF. The first part of part (a) follows directly from the Third Derivability Theorem 6.2.2. To obtain its second part, let $A$ denote the set of points where $f$ is normally lower differentiable, and suppose there is a subinterval $J$ of $I$ for which $|A \cap J| = 0$. Then by Lebesgue's theorem, $f$ is nowhere monotone on $J$, and so it is clear from Lemma 6.2.1 that $f(J)$ has a nonempty interior. Thus $|f(J)| > 0$, and hence by the first part, $|f(A \cap J)| > 0$. But since $f$ has, by Corollary 6.1.4, property $(N)$ on $A$, we have $|f(A \cap J)| = 0$, which is a contradiction. Also, under each of the two continuity hypotheses, $f$ is, by Theorem 1.1.1, $LSC$ n.e., and so it follows easily from Theorem 3.2.4 that $A$ is measurable (or see Theorem 10.1.6).

The first part of part (b) is obtained from Theorem 6.2.2 with the help of Corollaries 6.1.3 and 6.1.4, and its second part is a direct consequence of the first. □

At each point where $f$ has an infinite knotted lower derivative, it is easy to see that $f$ has an infinite unilateral derivative from one of the two sides. Hence from part (b) of the above theorem we obtain the following result on the existence of infinite unilateral derivative.

6.3.6. COROLLARY. *Suppose* $f : I \to \mathbb{R}$ *is L-singular and that it is either* $R^*LI$ *or* $L^*UI$. *Then f has an infinite unilateral derivative at a dense set of points in almost all of its level sets, as well as at a c-dense set of points in I provided that f is a nowhere constant Darboux function.*

When $f$ is either $GBV_*$ or has symmetrical derivates n.e., we obtain the following result on ordinary derivative on applying the above theorem to $f$ and $-f$, and on taking into account Theorem 6.3.2 and Corollary 5.3.2 in the former case.

6.3.7. COROLLARY. *Suppose* $f : I \to \mathbb{R}$ *is left or right strongly LI or UI, and that it is either* $GBV_*$ *or has symmetrical derivates n.e. If f is regular or singular,* *then it has a finite or infinite derivative, respectively, at a dense set of points in almost all of its level sets.*

Finally, in the following theorem we obtain a generalized version of a well-known theorem of Lebesgue; namely, every absolutely continuous singular function is constant.

6.3.8. THEOREM. *Let $f: I \to \mathbb{R}$ be a Darboux function which is either $R^*LI$ or $L^*UI$. If $f$ is L-regular and L-singular, then it is constant.*

PROOF. Suppose $f$ is $L$-regular and $L$-singular, and let $A$ denote the set of points in $I$ where $f$ has an infinite lower derivative. Then $|f(A)| = 0$ by hypothesis, and so it follows from part (b) of Theorem 6.3.5 that $|f(I)| = 0$. But since $f$ is a Darboux function, this implies that $f$ is constant. □

**§6.4 Second level structure theorem.** In this section we deduce from the Normal Level Structure Theorem another theorem on the level structure of weakly Darboux functions which we call the "Second Level Structure Theorem." This theorem will be found to play a central role in obtaining the results of the next two sections.

We need here some nomenclature. Given a subset $S$ of $\mathbb{R}$, let us first recall that we use $S'$ and $S^b$ to denote the sets of points in $S$ which are limit points or bilateral limit points, respectively, of $S$. A point $x \in S$ will be called as usual an *isolated point* of $S$ if $x \notin S'$, and we will call $x$ an *unilaterally isolated point* of $S$ if $x$ is an isolated point of $S$ at least from one side. Also, $x \in S$ will be called a *unilateral limit point* of $S$ if it is a limit point of $S$ from one and only one side.

The set $S$ will in turn be called an *isolated* or *unilaterally isolated set* if each point of $S$ is an isolated or unilaterally isolated point, respectively, of $S$.

The first two parts of the following theorem were first obtained in part by Stoilow [St2, St3] for a continuous function on a compact interval (see Iosifescu [Io1, Io2] for extension of some of his results to Darboux functions). The complete theorem was obtained in [G5] in the case of a Darboux function on an interval. The term "nonzero" in the first two parts of the following theorem is placed within brackets to indicate that the proposition holds with this term as well as without this term.

6.4.1. THEOREM (SECOND LEVEL STRUCTURE). *Suppose $f: X \to \mathbb{R}$ is weakly Darboux. Then for any differentially normal level set $S$ of $f$, or for almost every level set $S$ of $f$, a point $x \in S$ is*

(a) *an isolated point of $S$ iff $f$ has a (nonzero) derivative at $x$,*
(b) *a unilateral limit point of $S$ iff $f$ has a (nonzero) knotted semiderivative at $x$, and*
(c) *a bilateral limit point of $S$ iff $f$ is biknotted at $x$.*

PROOF. On account of the Normal Level Structure Theorem 6.1.2, we may assume here as before that $S$ is a differentially normal level set of $f$. Let $x$ be any point of $S$.

Then one of the following three possibilities hold at $x$: (i) $f$ has a nonzero derivative at $x$, (ii) $f$ has a nonzero knotted semiderivative at $x$, or (iii) $f$ is biknotted at $x$. Hence, in any case, each of the four unilateral derivates of $f$ is nonzero at $x$.

We claim that $x \in S^+$ iff $D_+f(x) < 0 < D^+f(x)$. The necessity part of the claim is obvious due to the fact just observed. To prove the sufficiency, suppose the condition holds, and let $\delta > 0$. First, suppose $f$ is lower Darboux. Then, first choose a point $y \in X \cap (x, x + \delta)$ such that $f(y) < f(x)$, and then choose a point $z \in X \cap (x, y)$ such that $f(z) > f(x)$. It is then clear that there is a point $x'$ in $X \cap (z, y)$ such that $f(x') = f(x)$. When $f$ is upper Darboux, $x'$ is obtained by reversing the order of choice of $y$ and $z$. This proves that $x \in S^+$, which establishes the claim. It is proved similarly that $x \in S^-$ iff $D_-f(x) < 0 < D^-f(x)$.

Now, using these two facts, it is clear that $x$ is an isolated point of $S$ iff (i) holds, a unilateral limit point of $S$ iff (ii) holds, and a bilateral limit point of $S$ iff (iii) holds. Moreover, as these three cases remain mutually exclusive on dropping the term "nonzero" in them, the theorem holds also without this term. □

We include here two consequences of the above theorem. Firstly, we obtain from it the following result on the properties of level sets.

6.4.2. COROLLARY. *Suppose $f: X \to \mathbb{R}$ is weakly Darboux. Then for almost every level set $S$ of $f$,*

(a) *$S$ is isolated (or unilaterally isolated) iff $f$ is derivable (or normally semiderivable) at each point of $S$, and*
(b) *$S$ is dense in itself (or bilaterally dense in itself) iff $f$ is not derivable (or semiderivable) at any point of $S$.*

Next, from part (b) of the above corollary we obtain the following result. The necessity part of the last part of this result was obtained by Minakshisundaram [Mi] for a continuous nonderivable function on a compact interval (see also Marcus [M2] and [G6]).

6.4.3. COROLLARY. *Suppose $f: X \to \mathbb{R}$ is weakly Darboux. Then $f$ maps the points where it is derivable or semiderivable into a set of measure zero iff almost every level set of $f$ is dense in itself or bilaterally dense in itself, respectively. Consequently, if $f$ is continuous and $X$ is closed in $\mathbb{R}$, then $|f(\Delta(f))| = 0$ iff almost every level set of $f$ is perfect.*

Now since every perfect set includes a dense set of it bilateral limit points, on combining the last part of the above corollary with part (c) of the Second Level Structure Theorem 6.4.1 we obtain the following.

6.4.4. Corollary. *If $f$ is a continuous nonderivable function on a closed interval $[a, b]$, then it is biknotted at a dense set of points in almost all of its level sets.*

## §6.5 Derivability in terms of properties of level sets.

In this section we deduce from the two level structure theorems some results on the derivability and semiderivability of a function in terms of the properties of its level sets. We introduce here for this purpose two new properties of functions, called properties $(S_1)$ and $(S_2)$, which are similar to Banach's properties $(T_1)$ and $(T_2)$ but play a more direct role in the derivability and semiderivability of the function.

We obtain, in fact, characterizations of these two new properties in terms of derivability and normal semiderivability of the function at all the points of almost all of its level sets (see Theorem 6.5.2). Further, we deduce from this theorem the existence of points of differentiability and semidifferentiability, along with an extended version of a theorem of Saks on the derivability of functions with property $(T_1)$ (see Theorem 6.5.6). Finally, in Theorem 6.5.8 we obtain properties in terms of level sets under which a function is differentiable or semidifferentiable a.e.

After the appearance of Stoilow's paper [St1], Banach [B4] and Saks [S4] recognized immediately the close connection between the derivability of a function and the structure of its level sets. Properties $(T_1)$ and $(T_2)$ were introduced by Banach in the same connection.

Given $f: X \to \mathbb{R}$, where $X \subset \mathbb{R}$, $f$ will be said to have *property* $(T_1)$ or $(T_2)$ if almost every level set of $f$ is finite or countable, respectively. These properties were introduced by Banach in [B4] for continuous functions on a compact interval, and they were referred by him as conditions $(T_1)$ and $(T_2)$.

We define, similarly, $f$ to have *property* $(S_1)$ or $(S_2)$ if almost every level set of $f$ is isolated or unilaterally isolated, respectively (see §6.4 for definition).

Let us note here that the following implications among the above four properties hold in general:

$$(T_1) \Rightarrow (S_1) \Rightarrow (S_2) \Rightarrow (T_2).$$

Further, in the case when $f$ is continuous and $X$ is compact, it is clear that $f$ has property $(S_1)$ iff it has $(T_1)$; (see Lemma 6.5.5 for its extension).

We begin with a general lemma which was obtained earlier in [G5] in the case of a connected domain $X$. Some applications of this lemma were also obtained there.

6.5.1. Lemma. *Let $f: X \to \mathbb{R}$. Then for almost every level set $S$ of $f$,* (a) *$f$ is lower or upper knotted at each limit point of $S$ in the set, and* (b) *$f$ is biknotted at each bilateral limit of $S$ in the set.*

Proof. On account of the Normal Level Structure Theorem 6.1.2, it is enough to consider here a differentially normal level set $S$ of $f$. Then, given $x \in S$, one of the following three possibilities holds at $x$: (i) $f$ has a nonzero derivative at $x$, (ii) $f$ has a nonzero knotted semiderivative at $x$, or (iii) $f$ is biknotted at $x$. In the case (i), $x$ must be an isolated point of $S$, and in case (ii), $x$ can be a limit point of $S$ at most from one side. Each of the two results follows clearly from these facts. □

We now obtain the derivability theorem in terms of properties $(S_1)$ and $(S_2)$. By the term "in general" in the following theorem we mean without the weak Darboux property.

6.5.2. Theorem (Fifth Derivability). *Let $f : X \to \mathbb{R}$. If $f$ is weakly Darboux and it has property $(S_1)$ (or $(S_2)$), then it is derivable (or normally semiderivable) at all the points of almost all of its level sets; the converse holds in general.*

Proof. The first part follows directly from Corollary 6.4.2(a), and the converse is obtained clearly from the above lemma. □

We include here two consequences of this derivability theorem. First, under properties $(S_1)$ and $(S_2)$ we obtain the following result on property $(N)$ similar to Theorem 6.3.4.

6.5.3. Corollary. *Let $f : X \to \mathbb{R}$. If $f$ is weakly Darboux and has property $(S_1)$ (or $(S_2)$), then it has property $(N)$ iff it is regular (or S-regular).*

The necessity part follows directly from Theorem 6.3.4, and the sufficiency part follows in turn from the above theorem with the help of Corollary 6.1.4.

In the following consequence we deal with the existence of points of differentiability and semidifferentiability.

6.5.4. Corollary. *Let $f : X \to \mathbb{R}$. If $f$ is weakly Darboux and has properties $(S_1)$ (or $(S_2)$) and $(N)$, then it is everywhere differentiable (or normally semidifferentiable) in almost all of its level sets; the converse holds in general.*

*Moreover, if the hypothesis of the first part holds and $|X| > 0$, then the set $E$ of points where $f$ is differentiable (or normally semidifferentiable) is of positive outer measure.*

The necessity and sufficiency parts of the first part follow on combining the above theorem with Corollaries 6.5.3 and 6.1.4, respectively. Further, suppose that the hypothesis of the last part holds and that $f$ is not monotone, for the result holds trivially otherwise. Then $f(X)$ includes a nondegenerate interval, and so it follows from the first part that $|f(E)| > 0$, and hence $|E| > 0$.

**A. Derivability under property $(T_1)$.** We first obtain a general result on the equivalence of properties $(S_1)$ and $(T_1)$. Here we will call $X$ *nearly closed* when $\overline{X} \sim X$ is countable. It is interesting to note here that the set of irrational numbers is nearly closed, but the set of rational numbers is not so.

Further, we will call a regulated function $f: X \to \mathbb{R}$ *strongly regulated* when $f$ also has unilateral limits at the limit points of $X$ in $\overline{\mathbb{R}}$, which are not in $X$, from the sides from which they are limit points of $X$.

6.5.5. LEMMA. *If $f: X \to \mathbb{R}$ is strongly regulated and $X$ is nearly closed, then $f$ has property $(S_1)$ iff it has $(T_1)$.*

PROOF. We need to prove here only the necessity part. Hence suppose the hypotheses hold and $f$ has property $(S_1)$. Then there is a subset $E_1$ of $f(X)$ of measure zero such that $f^{-1}(c)$ is isolated for each $c \in f(X) \sim E_1$. Let $A$ denote the set of points in $X$ where $f$ is discontinuous, and let $B$ be the set of limit points of $X$ in $\overline{\mathbb{R}}$ which are not in $X$. Then since $X$ is nearly closed, the set $C = A \cup B$ is countable. Now, let $E_2$ denote the set of left and right limits of $f$ at the points of $C$. Then since $E_2$ is countable, the set $E = E_1 \cup E_2$ is of measure zero.

Now, given $c \in f(X) \sim E$, it is enough to show that the level set $S \equiv f^{-1}(c)$ is finite. Suppose it is not so. Then $S$ has some limit point $x$ in $\overline{\mathbb{R}}$. Since $S$ is an isolated set, $x \notin S$. Further, since $f$ is strongly regulated, $c$ is clearly the left or right limit of $f$ at $x$. Hence, if $x \in X$, then since $x \notin S$, $f$ is discontinuous at $x$, so that $x \in A$; otherwise $x \in B$. Thus $x \in C$, and so $c \in E_2$. But since $E_2 \subset E$, this contradicts the choice of $c$. □

With the help of the above lemma, we obtain from the Fifth Derivability Theorem 6.5.2 the following extended version of a theorem of Saks [S4] (or see [S8, p. 278]) on the derivability of functions with property $(T_1)$, which was obtained by him for a continuous function on a compact interval.

6.5.6. THEOREM (SAKS). *Given $f: X \to \mathbb{R}$, let $N$ denote the set of points where $f$ is not derivable. If $f$ is weakly Darboux and has property $(T_1)$, then $|f(N)| = 0$. Conversely, if $X$ is nearly closed, $f$ is strongly regulated, and $|f(N)| = 0$, then $f$ has property $(T_1)$.*

We include here a result on the relations between various $BV$ properties and properties $(T_1)$, $(T_2)$, $(S_1)$, and $(S_2)$. For a continuous function $f$ on a compact interval $I$, it was proved by Banach [B3] in 1925 (see also Vitali [Vi]) that if $f$ is $BV$, then it has property $(T_1)$. Later Saks [S4] (or see [S8, p. 279]) extended this result to a continuous function which is $GBV_*$; and when $f$ is continuous and $GBV$, he proved that it has property $(T_2)$.

In the following theorem we obtain extended versions of these results along with a similar result on $GL\text{-}BV_*$.

6.5.7. THEOREM. *Let $f: X \to \mathbb{R}$. If $f$ is BV, GBV, $GBV_*$, or GL-$BV_*$, then it has property $(T_1)$, $(T_2)$, $(S_1)$, or $(S_2)$, respectively.*

*Moreover, if $f$ is strongly regulated and $GBV_*$, and $X$ is nearly closed, then $f$ has property $(T_1)$.*

PROOF. When $f$ is $GBV_*$ or GL-$BV_*$, it follows directly from Theorem 6.5.2 with the help of Theorem 5.3.1 and Corollary 5.3.2 that $f$ has property $(S_1)$ or $(S_2)$, respectively. The last part of the theorem follows now with the help of Lemma 6.5.5..

Next, suppose $f$ is $BV$. Then $f$ has an extension $g$ to $\mathbb{R}$ which is also $BV$ (see, e.g., [S8, p. 221]). Hence it follows from the last part of the theorem just obtained that $g$ has property $(T_1)$. Consequently, its restriction $f$ also has property $(T_1)$. Finally, when $f$ is $GBV$, it follows clearly from the result just proved that $f$ has property $(T_2)$. □

**B. Differentiability under properties $(S_1^*)$ and $(S_2^*)$.** In connection with the above results it is natural to ask whether there is any property in terms of level sets under which a function is derivable or semiderivable a.e. Marchaud [Ma] proved in this direction, in 1933, that if all the level sets of a continuous function $f$ on an interval are finite, then $f$ is derivable a.e.; (however, this result is implicit in a lemma obtained by Saks [S4] in 1931; see also Marcus [M3] and Iosifescu [Io2] for an extension of this result to Darboux functions).

Strengthening properties $(S_1)$ and $(S_2)$, let a function $f: X \to \mathbb{R}$ be said to have *property* $(S_1^*)$ or $(S_2^*)$ if *all* the level sets of $f$ are isolated or unilaterally isolated, respectively. Further, for each $x \in X$, by the *level set of $f$ corresponding to $x$* we mean the level set $f^{-1}(f(x))$.

The last part of the following theorem is indeed a derivability theorem which may be compared with the Fifth Derivability Theorem. Clearly this part includes a strengthened version of Marchaud's theorem.

6.5.8. THEOREM. *Suppose $f: X \to \mathbb{R}$ is weakly Darboux. Then $f$ is differentiable (or normally semidifferentiable) at almost all the points of $X$ which are isolated (or unilaterally isolated) points of the corresponding level sets of $f$. Consequently, if $f$ has property $(S_1^*)$ (or $(S_2^*)$), then it is differentiable (or normally semidifferentiable) a.e.*

PROOF. Let $A$ denote the set of points where $f$ is either normally semidifferentiable or biknotted, $B$ be the set of points in $X$ where $f$ has a strict relative maximum or minimum, and set $C = A \sim B$. Then since $B$ is countable, it follows from the Denjoy-Young-Saks Theorem 5.5.2 that $|X \sim C| = 0$. Let $x \in C$ and $S = f^{-1}(f(x))$.

First, suppose $x$ is an isolated point of $S$ but $f$ is not differentiable at $x$. Then $f$ is clearly lower or upper knotted at $x$, say lower knotted. Then since

$f$ does not have a strict relative maximum at $x$, $x$ is clearly a limit point of the set

$$E = \{t \in X : f(t) \geq f(x)\}.$$

Now, for the sake of definiteness, suppose $f$ is lower Darboux and $x$ is a right limit point of $E$. Then, given $\delta > 0$, since $D_+f(x) = -\infty$, we can find a point $y \in X \cap (x, x + \delta)$ such that $f(y) < f(x)$, and then a point $z \in E$ such that $x < z < y$. Now since $f(z) \geq f(x) > f(y)$, there exists a point $x'$ in $X \cap [z, y)$ such that $f(x') = f(x)$. This proves that $x$ is a limit point of $S$, which is a contradiction. A similar argument holds in all the other cases. This proves the unbracketed part of the first result.

Next, when $x \in C$ is an unilaterally isolated point of $S$, it is proved by a similar argument that $f$ is not biknotted at $x$, and hence $f$ is normally semidifferentiable at $x$. This in turn proves the other part of the first result.

The last part is of course a direct consequence of the first. □

6.5.9. REMARK. First, we show here that the hypothesis on $X$ in Lemma 6.5.5 is essential. Let $f$ be any continuous nonderivable function on $[0, 1]$. Then according to the last part of Corollary 6.4.3, almost every level set of $f$ is perfect. Choose an infinite isolated subset from each of these perfect level sets of $f$, and let $X$ be the union of these sets. Then the restriction of $f$ to $X$ is clearly a strongly regulated function which has property $(S_1)$ but not $(T_1)$.

Further, it may be pointed out here that with the help of Lemma 6.5.1 and Corollary 6.4.2, it is possible to determine the measure of the sets of values $c$ in the range of $f$ for which the level set $f^{-1}(c)$ is isolated, unilaterally isolated, dense in itself, bilaterally dense in itself, perfect, finite, or infinite, in terms of the differential structure of $f$. See [G6] for some results in this direction. Using the generalizations obtained in the present section, these results also can be extended to weakly Darboux functions defined on an arbitrary subset of $\mathbb{R}$.

**§6.6 Ordinary derivability under properties $(T_2)$, $(N)$, and $(S)$; some theorems of Banach and Saks.** In this section we deduce from the Second Level Structure Theorem and the Fifth Derivability Theorem extended versions of some theorems of Banach [B4] and Saks [S4] on the existence of points of ordinary derivability of continuous functions under properties $(T_2)$, $(N)$, and $(S)$. Also, two theorems of Rademacher [R] and Banach [B4] on property $(N)$, and another theorem of Banach on property $(S)$, are extended here from continuous functions to measurable functions on any subset of $\mathbb{R}$.

Given a function $f: X \to \mathbb{R}$, we will call $f$ *nearly* of $B_1$ if there is a countable set $C$ in $X$ such that $f$ is of $B_1$ on $X \sim C$. Such a function is not always of $B_1$; for example, the characteristic function of the set of rational numbers is nearly of $B_1$ but not of $B_1$.

It should be noted, further, that when $f$ is left or right strongly lower or upper internal, namely, $L^*LI$, $R^*LI$, $L^*UI$, or $R^*UI$, then as we saw earlier in Lemma 6.2.1, $f$ is weakly Darboux, and it is easy to see from Theorem 1.1.1 that $f$ is nearly of $B_1$.

We begin with the existence of points of derivability for functions with property $(T_2)$. In the particular case of a continuous function on an interval, the last part of the following theorem is implicit in a theorem of Saks [S4] (or see [S8, p. 280]).

6.6.1. THEOREM (SIXTH DERIVABILITY). *Let $X$ be a $G_\delta$-set in $\mathbb{R}$. Suppose $f: X \to \mathbb{R}$ is weakly Darboux and nearly of $B_1$, and suppose it has property $(T_2)$. Then $f$ is derivable at a dense set of points in almost all of its level sets. Consequently, if $|X| > 0$, then $\Delta(f)$ has cardinality c.*

PROOF. There is by hypothesis a countable set $C$ such that $f$ is of $B_1$ on $X \sim C$, and a set $E_1$ of measure zero such that $f^{-1}(c)$ is countable for each $c \in f(X) \sim E_1$. Further, by the Second Level Structure Theorem 6.4.1, there exists another set $E_2$ of measure zero such that for every $c \in f(X) \sim E_2$, $f$ is derivable at each isolated point of the level set $f^{-1}(c)$. Set $E = E_1 \cup E_2 \cup f(C)$. Then $|E| = 0$.

Now, given $c \in f(X) \sim E$, it is enough to show that the level set $S \equiv f^{-1}(c)$ includes a dense set of isolated points. Suppose it is not so. Then there is an open interval $I$ such that $S \cap I$ is nonempty and dense in itself. Now, note that $S$ is a subset of $X \sim C$ which is a $G_\delta$-set in $\mathbb{R}$. Since $f$ is of $B_1$ on $X \sim C$, it thus follows that $S$ is a $G_\delta$-set in $\mathbb{R}$, and so is in turn the set $S \cap I$ which is dense in itself. Consequently, $S \cap I$ is uncountable, which contradicts the fact that $c \notin E_1$. The first part is thus established.

Next, to prove the last part, suppose $|X| > 0$. Then, when $f$ is monotone, the result follows from Lebesgue's theorem. Otherwise, $f(X)$ includes a nondegenerate interval, and so the result follows from the first part. □

**A. Differentiability under property $(N)$.** Before coming to the derivability of functions with property $(N)$, we first extend in the next three propositions some of the known results on this property to measurable functions.

The following characterization of property $(N)$ was obtained originally by Rademacher [R] for continuous functions on an interval, and was later deduced from it by Ellis [E] for a measurable function on an interval.

6.6.2. THEOREM (RADEMACHER-ELLIS). *Suppose $f: X \to \mathbb{R}$ is measurable relative to $X$. Then $f$ has property $(N)$ iff $f(A)$ is measurable whenever $A \subset X$ is measurable in $\mathbb{R}$.*

PROOF. First, suppose $f$ has property $(N)$, and that $A \subset X$ is measurable in $\mathbb{R}$. Then since $f$ is measurable on $A$ in the usual sense, we can find, by Lusin's theorem [S8, p. 72], a sequence of compact sets $\{K_n\}$ in $A$ such that $f$ is continuous on each $K_n$ and $B \equiv A \sim \bigcup_n K_n$ is of measure zero. Then $f(K_n)$ is compact for each $n$, and $|f(B)| = 0$. Consequently, $f(A) = f(B) \cup \bigcup_n f(K_n)$ is measurable.

The converse holds in general. Suppose $f$ does not have property $(N)$. Then there is a set $A \subset X$ of measure zero such that $|f(A)| > 0$. Hence $f(A)$ contains a nonmeasurable set $E$. Now, since $H \equiv A \cap f^{-1}(E)$ is of measure zero, it is measurable in $\mathbb{R}$, but $f(H) = E$ is not measurable. This proves the converse. □

Next, we obtain the following version of a theorem of Banach [B4, Theorem 2] whose proof unfortunately contained a slight oversight. The following lemma strengthens Lemma 7.2 of Saks [S8, p. 283] and is closer to Banach's version. When a function $f: X \to \mathbb{R}$ is Lebesgue measurable in the usual sense (i.e., $X$ also is measurable), we will call it *L-measurable* to distinguish it from the measurability relative to $X$.

6.6.3. LEMMA. *If $f: X \to \mathbb{R}$ is L-measurable and has property $(N)$, then there is a measurable set $A \subset X$ such that $f$ is one-to-one on $A$ and $|f(X) \sim f(A)| = 0$.*

PROOF. Since $f$ is $L$-measurable, we can find, by Lusin's theorem [S8, p. 72], a nondecreasing sequence of compact sets $\{K_n\}$ in $X$ such that $f$ is continuous on each $K_n$ and $|X \sim \bigcup_n K_n| = 0$. Set $K = \bigcup_n K_n$. Then $\{f(K_n)\}$ is a nondecreasing sequence of compact sets with limit $f(K)$.

Next, we will define by induction a disjoint sequence of closed sets $\{F_n\}$ in $f(K)$ such that, for each $n$,

$$(1) \qquad F_n \subset f(K_n), \quad \left| f(K_n) \sim \bigcup_{i \le n} F_i \right| < \frac{1}{n}.$$

Choose $F_1 = f(K_1)$, and suppose, for some $n > 1$, $F_i$ has been defined for each $i < n$. Set

$$E_n = f(K_n) \sim \bigcup_{i<n} F_i.$$

Then since $E_n$ is measurable, it contains a closed set $F_n$ such that $|E_n \sim F_n| < 1/n$. Then $F_n \cap F_i = \varnothing$ for $i < n$, $F_n \subset E_n \subset f(K_n)$, and

$$\left| f(K_n) \sim \bigcup_{i \le n} F_i \right| = |E_n \sim F_n| < \frac{1}{n}.$$

This defines the sequence $\{F_n\}$. It is now clear from (1) that

$$(2) \qquad \left| f(K) \sim \bigcup_n F_n \right| \le \lim_n \left| f(K_n) \sim \bigcup_{i \le n} F_i \right| = 0.$$

Now, set $C_n = K_n \cap f^{-1}(F_n)$ for each $n$. Let $n$ be any fixed positive integer. Then since $f$ is continuous on $K_n$, $C_n$ is compact and $f$ is also continuous on $C_n$. Hence, by a lemma of Mazurkiewicz (see [S8, p. 282]), there exists a measurable set $A_n \subset C_n$ such that $f$ is one-to-one on $A_n$ and $f(A_n) = f(C_n)$. Now, since $F_n \subset f(K_n)$, indeed $f(A_n) = f(C_n) = F_n$, and since the sets in $\{F_n\}$ are mutually disjoint, $f$ is also one-to-one on the measurable set $A = \bigcup_n A_n$. Finally, since $f$ has property $(N)$ and $|X \sim K| = 0$, we have $|f(X) \sim f(K)| \leq |f(X \sim K)| = 0$. Consequently, by (2),

$$|f(X) \sim f(A)| = |f(K) \sim f(A)| = \left| f(K) \sim \bigcup_n F_n \right| = 0. \quad \square$$

Following is an extended version of another theorem of Banach [B4] (or see [S8, p. 284]) which was obtained originally for a continuous function on an interval, and then was deduced later by Ellis [E] for a measurable function on an interval.

6.6.4. THEOREM (BANACH-ELLIS). *If an L-measurable function* $f: X \to \mathbb{R}$ *has property* $(N)$, *then it also has property* $(T_2)$.

We omit the proof here since the original proof of Banach [B4, Theorem 3] becomes valid, even under the present hypothesis, on using the above Lemma 6.6.3 in the place of his Theorem 2 with some obvious modifications. (The proof given in Saks [S8, p. 284] may be used on the other hand for a bounded $X$, replacing its Lemma 7.2 [S8, p. 283] by the above lemma, and then the general result follows on using the $\sigma$-compactness of $\mathbb{R}$.)

Now we are in a position to obtain the following strengthened version of the theorem of Banach [B4] mentioned earlier on the existence of points of differentiability for continuous functions with property $(N)$.

6.6.5. THEOREM (SEVENTH DERIVABILITY). *Let $X$ be a $G_\delta$-set in $\mathbb{R}$. Suppose that $f: X \to \mathbb{R}$ is weakly Darboux and nearly of $B_1$ and that it has property $(N)$. Then $f$ is differentiable at a dense set of points in almost all of its level sets. Consequently, if $|X| > 0$, then $|\Delta^*(f)| > 0$.*

PROOF. Let $A$ denote the set of points where $f$ has an infinite derivative. Then since $|A| = 0$ (see Corollary 5.5.3), $|f(A)| = 0$. Now the first part follows directly from Theorem 6.6.1 since $f$ has property $(T_2)$ by the last theorem.

Next, suppose $|X| > 0$. Set $B = \Delta^*(f)$, and suppose, if possible, $|B| = 0$. Then $f$ is clearly not monotone, and so its range includes a nondegenerate interval. Hence, by the first part, $|f(B)| > 0$. However, due to property $(N)$ of $f$, $|f(B)| = 0$. This contradiction proves that $|B| > 0$. $\square$

**B. Differentiability under property $(S)$.** We next obtain extended versions of two theorems on Banach's property $(S)$ from the Fifth Derivability Theorem (or its Corollary 6.5.4) with the help of Lemma 6.6.3. Extending

Banach's definition, we will say that a function $f: X \to \mathbb{R}$ has *property* $(S)$ if for every $\varepsilon > 0$ there is a $\delta > 0$ such that for each measurable set $E \subset X$, $|f(E)| < \varepsilon$ whenever $|E| < \delta$.

Given a continuous function $f$ on a compact interval $I$, Banach proved in [B4] that $f$ has property $(S)$ iff it has properties $(N)$ and $(T_1)$. Subsequently, in 1928, it was proved by Banach and Saks [BS] that $f$ has property $(S)$ iff it maps the points where it is not differentiable into a set of measure zero; in the same year Bary and Menchoff [BM] proved the latter property to be equivalent to $f$ being expressible as a composition of two absolutely continuous functions.

In the following theorem we obtain from Lemma 6.6.3 an extension of Banach's theorem in [B4] to measurable functions. Its proof is an adaptation of Banach's proof (or see [S8, p. 286]).

6.6.6. THEOREM (BANACH). *Suppose $f: X \to \mathbb{R}$ is L-measurable. If $f$ has property $(S)$ and $|X| < \infty$, then $f$ has properties $(N)$ and $(T_1)$.*

*Conversely, if $f$ has properties $(N)$ and $(T_1)$, and $|f(X)| < \infty$, then $f$ has property $(S)$.*

PROOF. Let $Y$ denote the set of values $y \in f(X)$ for which $f^{-1}(y)$ is infinite. First, suppose $f$ has property $(S)$ and $|X| < \infty$. Then it is clear that $f$ has property $(N)$. Suppose $f$ does not have property $(T_1)$, i.e., $|Y| > 0$. We will define here by induction a disjoint sequence of measurable sets $\{A_n\}$ such that the following holds for each $n$: (i) $f$ is 1-1 on $A_n$, and (ii) $|f(A_n)| \geq |Y|$.

First, choose $A_1$ for $f$ as determined by Lemma 6.6.3. Now, given $n > 1$, suppose $A_i$ has been defined for each $i < n$. Set $E_n = X \sim \bigcup_{i<n} A_i$. Then it is clear from (i) that $Y \subset f(E_n)$. Hence, by Lemma 6.6.3, there exists a measurable subset $A_n$ of $E_n$ such that $f$ is 1-1 on $A_n$ and $|f(A_n)| = |f(E_n)| \geq |Y|$. This defines the sequence $\{A_n\}$.

Now, since $\{A_n\}$ is disjoint and $|X| < \infty$, it is clear that $|A_n| \to 0$ as $n \to \infty$. Hence it follows from property $(S)$ of $f$ that $|f(A_n)| \to 0$ as $n \to \infty$, and this contradicts (ii).

Next, to prove the converse, suppose $|f(X)| < \infty$ and $f$ has properties $(T_1)$ and $(N)$ but not $(S)$. Then there exists an $\varepsilon > 0$ and a sequence of sets $\{E_n\}$ such that for each $n$,

$$|E_n| < 1/2^n \quad \text{and} \quad |f(E_n)| \geq \varepsilon. \tag{3}$$

Set $E = \limsup_n E_n$ and $F = \limsup_n f(E_n)$. Then we claim that

$$F \sim f(E) \subset Y. \tag{4}$$

Given $y \in F \sim f(E)$, there is clearly an increasing sequence of positive integers $\{n_i\}$ and a sequence of points $\{x_i\}$ such that $x_i \in E_{n_i}$ and $f(x_i) = y$ for each $i$. Further, since $y \notin f(E)$, on replacing the sequence $\{n_i\}$ by a subsequence if necessary we can assume that $x_i \neq x_j$ whenever $i \neq j$. Hence $y \in Y$, which proves (4).

Now, it is clear from (3) that $|E| = 0$, and hence $|f(E)| = 0$. Further, let $F_n = \bigcup_{i \geq n} f(E_i)$ for each $n$. Then $\{F_n\}$ is a nonincreasing sequence of sets with limit $F$. Since $|f(X)| < \infty$, it thus follows from (3) that $|F| = \lim_n |F_n| \geq \varepsilon$. Hence $|F \sim f(E)| \geq \varepsilon$, and so it follows from (4) that $|Y| > 0$, which contradicts property $(T_1)$ of $f$. □

It should be noted here that on dropping the hypothesis $|X| < \infty$ in the first part of the above theorem, it still remains valid on replacing property $(T_1)$ by $(S_1)$. In the following theorem we deduce from Corollary 6.5.4 with the help of the above theorem the following extended version of the theorem of Banach and Saks [BS].

6.6.7. THEOREM (BANACH-SAKS). *Given $f : X \to \mathbb{R}$, let $N$ denote the set of points where $f$ is not differentiable. If $f$ is L-measurable and weakly Darboux and has property $(S)$, then $|f(N)| = 0$.*

*Conversely, if $X$ is nearly closed, $f$ is strongly regulated, $|f(X)| < \infty$, and $|f(N)| = 0$, then $f$ has property $(S)$.*

PROOF. First, suppose the hypothesis of the first part holds. Then it follows from the first part of Theorem 6.6.6 as mentioned above that $f$ has properties $(N)$ and $(S_1)$, and hence it follows from Corollary 6.5.4 that $|f(N)| = 0$. Next, suppose the hypothesis of the converse holds. Then it follows from Corollary 6.5.4 that $f$ has properties $(N)$ and $(S_1)$, and hence from Lemma 6.5.5 that it has property $(T_1)$. Consequently, $f$ has property $(S)$ by Theorem 6.6.6. □

In the case of a continuous function we obtain from Theorems 6.6.6 and 6.6.7 the following direct extension of the above-mentioned two characterizations of property $(S)$.

6.6.8. COROLLARY. *If $f$ is a continuous function on a compact set $X \subset \mathbb{R}$, then $f$ has property $(S)$ iff it has properties $(T_1)$ and $(N)$, or, equivalently, iff it is everywhere differentiable in almost all of its level sets.*

6.6.9. REMARK. It should be pointed out here that in Theorem 6.6.6 the hypotheses on the measures of $X$ and $f(X)$ are essential. First, let $f(x) = \sin x$, $x \in \mathbb{R}$. Then since $f$ has a bounded derivative, it is clear that $f$ has the property $(S)$. But $f$ has only property $(S_1)$, not $(T_1)$.

Next, define $f(x) = 1/x$ for $x \neq 0$ and $f(0) = 0$. Then it is clear that $f$ has property $(T_1)$, and hence it follows from Corollary 6.5.3 that $f$ also has property $(N)$. However, it is easy to see that $f$ does not have property $(S)$.

## §6.7 Length of graph; Kolmogoroff-Verčenko theorem.

In this section we obtain some results on the length of the graph of a function, including a theorem of Kolmogoroff and Verčenko which was obtained in connection with the Denjoy-Young-Saks theorem. Further, we obtain integral versions of two inequalities obtained in §6.1, which in turn have been obtained by Saks [S2, S7] in some particular cases.

It may be recalled here that a set $A \subset \mathbb{R}^2$ is said to be of *σ-finite length* if it is a countable union of sets which are of finite length. The general result of Kolmogoroff and Verčenko on the contingent of any subset of $\mathbb{R}^2$ can be deduced easily from the following theorem (see [KV1, KV2], or [S8, p. 266]).

6.7.1. THEOREM (KOLMOGOROFF-VERČENKO). *Given $f: X \to \mathbb{R}$, let $E = X \sim K_b(f)$ and $E_0 = E \sim \Delta_{sn}(f)$. Then $Gf(E)$ is of σ-finite length and* $\Lambda\{Gf(E_0)\} = 0$.

PROOF. It follows from Theorem 5.4.1 that $f$ is $GMT$ on $E$. Hence, $f$ is $GBV$ on $E$. Now, since the graph of a function $BV$ on a bounded subset of $\mathbb{R}$ is known to be of finite length, it is thus clear that $Gf(E)$ is of σ-finite length.

Next, let $A = X \sim \underline{K}(f)$ and $B = X \sim \overline{K}(f)$. Then $E = A \cup B$. Further, set $A_0 = A \sim \underline{\Delta}_u(f)$ and $B_0 = B \sim \overline{\Delta}_u(f)$. Then it is easy to see that

$$\begin{aligned} E &\sim (A_0 \cup B_0) \\ &= \{(A \sim A_0) \cap (B \sim B_0)\} \cup \{(A \sim A_0) \sim B\} \cup \{(B \sim B_0) \sim A\} \\ &= \Delta(f) \cup \underline{\Delta}_k(f) \cup \overline{\Delta}_k(f) = \Delta_{sn}(f). \end{aligned}$$

Consequently, $E_0 = A_0 \cup B_0$.

Now, it follows from Theorem 5.4.1 that $f$ is $GL\text{-}MT_*$ on $A_0$, and hence from Lemma 5.5.1 that $f$ is $GL\text{-}BV_*$ on $A_0$. Consequently, by Theorem 5.3.1, $\Lambda\{Gf(A_0)\} = 0$. Further, on applying this result to $-f$ we obtain $\Lambda\{Gf(B_0)\} = 0$. Hence, $\Lambda\{Gf(E_0)\} = 0$. □

With the help of the above theorem and Theorems 5.5.2 and 6.1.2, it is easy to see that Theorem 5.3.1 on $GL\text{-}BV_*$ functions can be strengthened to the following form.

6.7.2. THEOREM. *Let $f: X \to \mathbb{R}$ and $E \subset X$. If $f$ is $GL\text{-}BV_*$ on $E$, then it is normally lower differentiable a.e. in $E$, and further if $N = E \sim \underline{\Delta}_n(f)$, then* $|f(N)| = \Lambda\{Gf(N)\} = 0$.

Next, in the following theorem we obtain integral versions of the two inequalities obtained earlier in Lemma 6.1.1. The first of the following two inequalities is known for the ordinary derivative in the case of a measurable set $E$ in $\mathbb{R}$ (see Saks [S8, p. 227]).

6.7.3. THEOREM. *Let $f: X \to \mathbb{R}$ and $E \subset X$. If for each $x \in E$, $g(x)$ is a finite lower gradient of $f$ at $x$, then $g$ is measurable on $E$,*

$$(1) \qquad |f(E)| \le \int_E |g| \, dm_E \quad \text{and} \quad \Lambda\{Gf(E)\} \le \int_E \sqrt{g^2 + 1} \, dm_E.$$

*Moreover, if $f$ is measurable on $E$, the same holds when $g(x)$ is for each $x \in E$ a finite weak gradient of $f$ at $x$.*

PROOF. On account of Theorem 4.4.1, we can clearly assume that $E \subset \Delta_{wu}(f)$. First, suppose $g(x) = Lf'(x)$ for each $x \in E$. Then $f$ is $LSC$ on $E$ by Theorem 4.5.2(b), and so it follows from Theorem 3.2.2(b) that $g$ is measurable on $E$. Further, since $m_E$ is $\sigma$-finite, we can clearly assume without loss of generality that $|E| < \infty$.

Let $0 < \varepsilon < 1$, and set the following for each positive integer $n$:

$$E_n = \{x \in E : (n-1)\varepsilon \le |g(x)| < n\varepsilon\}.$$

Then $E = \bigcup_n E_n$, where $\{E_n\}$ is a disjoint sequence of measurable sets in $E$. Now, given $n$, for each $x \in E_n$, $D^- f(x) = g(x) < n\varepsilon$ and $D_+ f(x) = g(x) > -n\varepsilon$. Hence, by Lemma 6.1.1, we have

$$|f(E_n)| \le n\varepsilon |E_n| \le \int_{E_n} (|g| + \varepsilon)\, dm_E,$$

$$\Lambda\{Gf(E_n)\} \le \sqrt{n^2\varepsilon^2 + 1}\, |E_n| \le \int_{E_n} \sqrt{(|g| + \varepsilon)^2 + 1}\, dm_E.$$

Consequently,

$$(2) \qquad |f(E)| \le \sum_n |f(E_n)| \le \int_E |g|\, dm_E + \varepsilon |E|,$$

$$(3) \qquad \Lambda\{Gf(E)\} \le \sum_n \Lambda\{Gf(E_n)\} \le \int_E \sqrt{(|g| + \varepsilon)^2 + 1}\, dm_E.$$

Now, the first inequality of (1) follows from (2) since it holds for arbitrary $\varepsilon$. Its second inequality holds on the other hand trivially when $g$ is not integrable on $E$ relative to $m_E$. Otherwise, since $\sqrt{(|g| + \varepsilon)^2 + 1} \le |g| + \varepsilon + 1$, this inequality follows from (3) with the help of Lebesgue's dominated convergence theorem on making $\varepsilon \to 0$.

Next, suppose $f$ is measurable on $E$, and that $g(x) = Wf'(x)$ for each $x \in E$. Set

$$(4) \quad A = E \sim \Delta_{sn}(f), \qquad B = E \cap \underline{\Delta}_n(f), \qquad C = E \sim (A \cup B).$$

Then $E = A \cup B \cup C$. Now, by Theorem 5.5.2, $|A| = 0$, and by Theorem 6.1.2, $|f(A)| = 0$. Also by Theorem 6.7.1, $\Lambda\{Gf(A)\} = 0$. Further, it follows easily from Theorem 3.2.4(c) (or see Corollary 10.1.5(b)) that $B$ is measurable in $E$, and so is in turn $C$. Also, it is clear that $g$ is the lower or upper derivative of $f$ at the points of $B$ and $C$, respectively. Hence it follows from the first part that $g$ is measurable on $E$ and that

$$|f(E)| \leq |f(B)| + |f(C)| \leq \int_B |g|\, dm_E + \int_C |g|\, dm_E = \int_E |g|\, dm_E.$$

The other inequality of (1) is obtained similarly. □

From the last theorem we obtain the following result on unilateral derivates, whose first part has been obtained by Saks [S2] when $E$ is measurable in $\mathbb{R}$.

6.7.4. Corollary. *Let $f : X \to \mathbb{R}$ and $E \subset X$. If $g$ is some fixed unilateral derivate of $f$ which is finite n.e. in $E$, then $g$ is measurable on $E$,*

$$|f(E)| \leq \int_E |g|\, dm_E \quad \text{and} \quad \Lambda\{Gf(E)\} \leq \int_E \sqrt{g^2 + 1}\, dm_E.$$

*Moreover, when $f$ is measurable on $E$, the same holds if for nearly every $x$ in $E$, $g(x)$ is any finite unilateral derivate of $f$ at $x$.*

Here $g$ can be clearly assumed to be finite on $E$, and if $A = E \cup \Delta_{sn}(f)$, then $|A| = |f(A)| = \Lambda\{Gf(A)\} = 0$ as before, and since $g(x)$ is then the unique semiderivative of $f$ at $x$ for each $x \in E \sim A$, both parts follow from the last theorem.

Further, from the above corollary we obtain the following strengthened version of the second part of Lemma 6.1.1.

6.7.5. Corollary. *Let $f : X \to \mathbb{R}$, $E \subset X$ and $c \geq 0$. Suppose at each point of $E$ there is some unilateral derivate of $f$ which is in magnitude not greater than $c$. Then*

$$|f(E)| \leq 2c|E| \quad \text{and} \quad \Lambda\{Gf(E)\} \leq 2\sqrt{c^2 + 1}\,|E|. \tag{5}$$

*Moreover, if either $f$ is measurable on $E$, or the unilateral derivate in question is fixed, then*

$$|f(E)| \leq c|E| \quad \text{and} \quad \Lambda\{Gf(E)\} \leq \sqrt{c^2 + 1}\,|E|. \tag{6}$$

The second part follows directly from the above corollary. To obtain the first, let $A$, $B$, $C$ be as defined in (4). Then $|A| = |f(A)| = \Lambda\{Gf(A)\} = 0$ as seen above, and since $D_+f$ and $D^+f$ are bounded in magnitude by $c$ at the points of $B$ and $C$ respectively, it follows from the second part that

$$|f(E)| \leq |f(B)| + |f(C)| \leq c|B| + c|C| \leq 2c|E|.$$

A similar argument holds for the other inequality of (5).

The first inequality of (6) has been obtained in the second case by Saks [S7, p. 239] (or see [S8, p. 271]).

6.7.6. REMARK. It may be observed here that the coefficient 2 on the right sides of the two inequalities of (5) is the best possible in general. Let $E = X = (0, 1)$, and let $A$ be any nonmeasurable subset of $E$ such that $|A| = |E \sim A| = 1$. Now, define $f(x) = x$ or $-x$ according to whether $x \in A$ or $E \sim A$. Then at each point of $E$ there is a unilateral derivate of $f$ which is in magnitude 1, and since $E$ is measurable in $\mathbb{R}$, we obtain

$$|f(E)| = |f(E) \cap E| + |f(E) \sim E| = |f(A)| + |f(E \sim A)|$$
$$= |A| + |E \sim A| = 2 = 2|E|.$$

The other inequality of (5) can also be seen similarly to become an equality in this case.

# CHAPTER 7

# Calculus of New Derivatives

In this chapter we investigate various rules of Calculus for new derivatives.

First, in §7.1, we introduce some terminology regarding (a) the similarity and dissimilarity of two semiderivatives and (b) the compatibility of two weak or semi- derivatives. Then, we compare in §7.2 the new derivatives of a restriction of a function with those of the original function.

Next, in §7.3, we investigate the additivity and linearity of new derivatives. In particular, the lower derivative and knotted and normal lower derivatives are found to be additive and positively linear. The weak derivative is found, on the other hand, to be only conditionally additive in the sense that the additivity holds only when the sum function is also weakly derivable at the point in question. Further, in terms of the compatibility of semiderivatives, a theorem on the linearity of semiderivative is established.

Next, in §7.4, similar product and quotient rules are obtained for semiderivatives and for knotted and normal semiderivatives, in terms of their compatibility. Also, conditional product and quotient rules are obtained for weak derivative.

In the next section, chain rules are obtained for new derivatives in the particular case when the inner or outer function of the composite function is differentiable in the ordinary sense at the point in question. Further, in §7.6, a general chain rule is obtained for semiderivative along with a conditional general chain rule for weak derivative.

Next, in §7.7, similar rules of Calculus are obtained for semiderivatives of functions defined on normed vector spaces with independent arguments. The results become somewhat simpler and more elegant in this case. Finally, in §7.8, we present some comments on the developments in nonsmooth analysis that are related to new derivatives.

In this chapter, instead of following our original convention, we assume throughout that $f: X \to \mathbb{R}$, where $X$ is assumed in §§7.1 to 7.6 to be a subset of $\mathbb{R}$, and in §§7.7 and 7.8 to be a real normed vector space. More details of assumptions will be given in each section after its introduction.

**§7.1 Similar, dissimilar, and compatible semi- and weak derivatives.** To begin with, we introduce in this section some terminology which will be needed throughout this chapter. Also, we include here some commonly used operations of addition and multiplication in $\overline{\mathbb{R}}$ along with some related notations.

First, to be able to compare semi- and weak derivatives of two functions, we define their similarity, dissimilarity, and compatibility.

Suppose $f: X \to \mathbb{R}$, $g: Y \to \mathbb{R}$, where $X, Y \subset \mathbb{R}$, and let $x \in X^b$ and $y \in Y^b$. First, suppose $f$ and $g$ are semiderivable at $x$ and $y$, respectively. Then we will call $Sf'(x)$ and $Sg'(y)$ *similar* if they are simultaneously lower or upper derivatives. If one of them is a lower derivative and the other an upper derivative, the two semiderivatives will be called *dissimilar*. Also, when $Sf'(x)$ and $Sg'(y)$ are similar or dissimilar, we will say sometimes that $Sf'(x)$ is similar or dissimilar, respectively, to $Sg'(y)$.

Further, $Sf'(x)$ will be said to be *totally similar* or *dissimilar* to $Sg'(y)$ if the following two conditions hold:

(i) If $Sg'(y)$ is a lower derivative, then $Sf'(x)$ is a lower or upper derivative, respectively, and if $Sg'(y)$ is an upper derivative, then $Sf'(x)$ is an upper or lower derivative, respectively.

(ii) If $Sg'(y)$ is unique or normal, then so is $Sf'(x)$.

Next, let $c \in \mathbb{R}$. Then a semiderivative will sometimes be called a *c-semiderivative* if it is a lower or upper derivative according to whether $c \geq 0$ or $\leq 0$. Further, the semiderivatives $Sf'(x)$ and $Sg'(y)$ will be called *c-compatible* if they are similar or dissimilar according to whether $c \geq 0$ or $\leq 0$; and $Sf'(x)$ will be said to be *c-compatible with* $Sg'(y)$ if it is similar or dissimilar to $Sg'(y)$ according to whether $c \geq 0$ or $\leq 0$.

It is interesting to note here that $Sf'(x)$ is a 0-semiderivative iff it is a derivative, and, similarly, $Sf'(x)$ is 0-compatible with $Sg'(y)$ iff the former is a derivative.

The above notions of similar, dissimilar, and compatible semiderivatives will also apply to functions defined on normed vector spaces.

Next, two unilateral derivatives $f'_u(x)$ and $g'_u(y)$ will be called *symmetric* or *asymmetric* if they are from the same or opposite sides, respectively. Also, given $c \in \mathbb{R}$, $c \neq 0$, these unilateral derivatives will be called *c-compatible* if they are symmetric or asymmetric according to whether $c > 0$ or $< 0$.

Finally, suppose $f$ and $g$ are weakly derivable at $x$ and $y$, respectively. Then $Wf'(x)$ will be said to be *totally similar* or *dissimilar* to $Wg'(y)$ if the following three conditions hold:

(i) If $Wg'(y)$ is a semiderivative, then so is $Wf'(x)$ and the latter is totally similar or dissimilar, respectively, to the former semiderivative.

(ii) If $Wg'(y)$ is a unilateral derivative, then so is $Wf'(x)$ and the two are symmetric.

(iii) If $Wg'(y)$ is unique, then so is $Wf'(x)$.

Also, the weak derivatives $Wf'(x)$ and $Wg'(y)$ will be said to be *totally similar* or *dissimilar* if each of them is so relative to the other. The same will apply to totally similar and dissimilar semiderivatives.

Further, given $c \in \mathbb{R}$, the weak derivatives $Wf'(x)$ and $Wg'(y)$ will be called *totally c-compatible* if they are totally similar or dissimilar according to whether $c \geq 0$ or $\leq 0$; and $Wf'(x)$ will be called *totally c-compatible with* $Wg'(y)$ if $Wf'(x)$ is totally similar or dissimilar to $Wg'(y)$ according to whether $c \geq 0$ or $\leq 0$.

Next, since the values of new derivatives are in general subsets of $\overline{\mathbb{R}}$, we define here the operations of addition and multiplication in $\overline{\mathbb{R}}$ and present some notations which will be used frequently in this chapter.

The operations of addition and multiplication in $\overline{\mathbb{R}}$ are obtained by extending these operations from $\mathbb{R}$ to $\overline{\mathbb{R}}$ (see, e.g., [HS, pp. 54, 55]) as follows: Given $c \in \mathbb{R}$, we define

$$c + (\pm\infty) = (\pm\infty) + c = (\pm\infty) + (\pm\infty) = \pm\infty,$$

$$c(\pm\infty) = (\pm\infty)c = \pm\infty \quad \text{or} \quad \mp\infty \quad \text{according as} \quad c > 0 \quad \text{or} \quad < 0,$$

$$(\pm\infty)(\pm\infty) = +\infty \quad \text{and} \quad (\pm\infty)(\mp\infty) = -\infty.$$

Further, if $c, d \in \overline{\mathbb{R}}$, $c - d$ is defined to be $c + (-1)d$ provided that this sum is defined. The operations $(\pm\infty) + (\mp\infty)$ and $(\pm\infty) - (\pm\infty)$ remain undefined as usual. We do define, however, as in measure theory (see, e.g., [HS, p. 55]), the following:

$$0(\pm\infty) = (\pm\infty)0 = 0,$$

and these products will be distinguished sometimes from the others by calling them *indeterminate products*. The use of indeterminate products will simplify considerably the enunciation of results in this chapter. Further, when $A$, $B$ are two nonempty subsets of $\overline{\mathbb{R}}$, and $c \in \overline{\mathbb{R}}$, we define

$$cA = \{cd : d \in A\},$$

$$A + B = \{c + d : c \in A,\, d \in B,\, c + d \text{ is defined}\}.$$

Clearly, $cA$ is always a nonempty subset of $\overline{\mathbb{R}}$, but $A + B$ becomes empty when one of $A$ and $B$ consists of only $+\infty$ and the other of only $-\infty$. Thus we will call the sum $cA + dB$ to be *defined* if it is nonempty—that is, if there exist $a \in A$, $b \in B$ such that $ca + db$ is defined.

Finally, when each element of $A$ is positive, we will call $A$ also *positive* and write $A > 0$. The inequalities $A < 0$, $A \geq 0$, and $A \leq 0$ are defined similarly.

**§7.2 New derivatives of restrictions of functions.** We begin in this section with a simple result comparing the new derivatives of restrictions of a function with those of the original function. The result is included here basically due to a significant variation in this regard in the case of new derivatives as compared with the ordinary derivative.

Given $f: X \to \mathbb{R}$, $E \subset X$, we will use $f_E$ to denote the *restriction* of $f$ to $E$, which is also called sometimes a "partial function" of $f$ (see Kuratowski [K1, p. 105]).

7.2.1. THEOREM. *Suppose $f: X \to \mathbb{R}$, $E \subset X \subset \mathbb{R}$, and $x \in E^b$. If $f$ is weakly derivable at $x$, then so is $f_E$ and*

$$Wf'(x) \subset W(f_E)'(x). \tag{1}$$

*Further, if $f$ is lower, upper, left, or right derivable at $x$, then so is $f_E$.*

PROOF. It is clear from the definitions of unilateral derivates that

$$D_-f(x) \le D_-f_E(x), \qquad D^-f_E(x) \le D^-f(x), \tag{2}$$

and, similarly,

$$D_+f(x) \le D_+f_E(x), \qquad D^+f_E(x) \le D^+f(x). \tag{3}$$

Each of the results follows clearly from these four relations. □

7.2.2. REMARK. In the case of the ordinary derivative, as well as unilateral derivatives, identity holds in the place of (1). However, the reverse inclusion of (1) does not hold in general for weak and semi- derivatives. Let $A$ be the set of irrational numbers, and set $E = A \cup \{0\}$. Then the characteristic function $f$ of $A$ has a knotted lower derivative zero at 0, but the lower derivative of $f_E$ is $\overline{\mathbb{R}}$ at 0.

When $f_E$ is on the other hand nonangular at $x$, then since its weak derivative is unique at $x$ (see Theorem 4.4.1), (1) automatically becomes an identity. Hence (1) can fail to be an identity only at a countable set of points (see Theorem 2.3.1).

The converse of the above theorem, of course, does not hold in general. It is further interesting to note here, as it follows easily from (2) and (3), that when $f_E$ has a knotted semiderivative at $x$, the same is true of $f$ provided that it is weakly derivable at $x$.

**§7.3 Additivity and linearity of new derivatives.** In this section we investigate the additivity and linearity of new derivatives. First, in Theorem 7.3.1, the additivity is obtained in the case when one of the functions is differentiable in the ordinary sense at the point in question. Then, in Theorems 7.3.2, we obtain the additivity of lower derivative, and Theorem 7.3.5 deals with the additivity of knotted and normal lower derivatives.

Next, in Theorem 7.3.7, a conditional form of additivity is obtained for weak derivative. Finally, in Theorems 7.3.10 and 7.3.11, the linearity of weak and semi- derivatives is investigated. Also, some examples are presented in Remark 7.3.12 to indicate the various problems that arise in this connection with new derivatives.

We will assume throughout this section that $f, g : X \to \mathbb{R}$, where $X \subset \mathbb{R}$, and $x \in X^b$.

We first obtain in the following theorem the additivity of weak derivative when one of the functions is differentiable at the point in question.

7.3.1. THEOREM. *Suppose $f$ is differentiable at $x$. Then $f + g$ is weakly derivable at $x$ iff $g$ is so; and if so, then the two weak derivatives are totally similar and*

$$W(f+g)'(x) = f'(x) + Wg'(x). \tag{1}$$

PROOF. Given $y \in X$, $y \neq x$, we clearly have

$$Q(f+g)(x,y) = Qf(x,y) + Qg(x,y), \tag{2}$$

where $Qf(x, y)$ converges to $f'(x)$ as $y \to x$ through $X$. Hence we obtain

$$D_+(f+g)(x) = f'(x) + D_+g(x), \qquad D^+(f+g)(x) = f'(x) + D^+g(x),$$

and two similar relations hold with $-$ in the place of $+$. Now since $f'(x)$ is finite, each part of the result follows clearly from these four relations. □

In the following theorem we deal with the additivity of lower derivative. A similar theorem on the additivity of upper derivative is obtained on applying this theorem to $-f$ and $-g$.

7.3.2. THEOREM. *Suppose $f$ and $g$ are lower derivable at $x$.*

(a) *If the following sum is defined, then $f + g$ is lower derivable at $x$ and*

$$Lf'(x) + Lg'(x) \subset L(f+g)'(x). \tag{3}$$

(b) *If from each side either $f$ or $g$ is unilaterally differentiable at $x$, then $f + g$ is lower derivable at $x$ and*

$$L(f+g)'(x) = Lf'(x) + Lg'(x); \tag{4}$$

*and if $Lf'(x)$ and $Lg'(x)$ are simultaneously unique or normal, then so is $L(f+g)'(x)$.*

PROOF. To prove (a), suppose the sum in (3) is defined. Let $c \in Lf'(x)$, $d \in Lg'(x)$ such that $c + d$ is defined. Then it is enough to show that

$$D^-(f+g)(x) \le c + d \le D_+(f+g)(x). \tag{5}$$

It is clear from (2) that

$$D^-(f+g)(x) \le D^-f(x) + D^-g(x) \le c + d,$$

provided that the sum of $D^-f(x)$ and $D^-g(x)$ is defined. Otherwise, one of them is $+\infty$, and so either $c$ or $d$ is $+\infty$, and hence $c + d = +\infty \ge D^-(f+g)(x)$. This proves the first inequality of (5); the other inequality is proved similarly. Now the two conclusions of (a) follow clearly from (5).

Next, to prove (b), suppose for the sake of definiteness that $f$ is left differentiable at $x$ and $g$ is right differentiable at $x$. Then it follows from (2) that

$$(6)\quad D^-(f+g)(x)$$
$$= f'_-(x) + D^-g(x) \le D_+f(x) + g'_+(x) = D_+(f+g)(x).$$

Now, it is clear from (6) that $f + g$ is lower derivable at $x$ and that (4) holds. Further, if $Lf'(x)$ and $Lg'(x)$ are unique, it is clear from (6) that $L(f+g)'(x)$ is also unique. Similar arguments hold in the other three cases.

Also, when $Lf'(x)$ and $Lg'(x)$ are normal, it is clear from the present hypothesis that one of $f$ and $g$ is differentiable at $x$, and so it follows from Theorem 7.3.1 that $L(f+g)'(x)$ is also normal. □

Let us note here that the sum in the inclusion (3) of Theorem 7.3.2 is always defined when $f$ or $g$ is lower differentiable at $x$. Further, as we will see later in Remark 7.3.12(a), the reverse inclusion of (3) does not hold in general.

We include here two interesting consequences of the above theorem. From part (a) of Theorem 7.3.2 we obtain the following result on applying it to $-(f+g)$ and $g$.

7.3.3. COROLLARY. *If $f$ and $g$ are lower derivable at $x$ and $f+g$ is upper differentiable at $x$, then $f$ and $g$ are derivable at $x$.*

Further, when $f$ is a convex function on an interval $I$, it is well known that $f$ is subdifferentiable (and so lower derivable) at every interior point of $I$, and that it is unilaterally differentiable from both sides at such a point (see, e.g., Rockafellar [Ro1]). Hence from part (b) of Theorem 7.3.2 we obtain the following result when one of the functions is convex.

7.3.4. COROLLARY. *Suppose $f, g : I \to \mathbb{R}$, $f$ is convex and $x \in I^\circ$. Then if $g$ is lower derivable at $x$, then so is $f+g$ and* (4) *holds.*

Next, in the following theorem we deduce from Theorem 7.3.2 the additivity of knotted and normal lower derivatives.

7.3.5. THEOREM. *Suppose the hypothesis of Theorem 7.3.2(a) holds and $f + g$ is lower nonangular at $x$. Then $f + g$ has a unique lower derivative at $x$ and (4) holds. Further,*

(a) *if $f$ is lower differentiable at $x$ and $Lg'(x)$ is a knotted lower derivative, then so is $L(f + g)'(x)$; and*

(b) *if $f$ and $g$ are normally lower differentiable at $x$, then so is $f + g$.*

PROOF. Since $f + g$ is lower nonangular at $x$, $L(f + g)'(x)$ is unique, and so (3) reduces to the identity (4).

To prove (a), suppose its hypothesis holds. Then there is a $c \in \mathbb{R}$ such that $D^-f(x) \le c \le D_+f(x)$, and $g$ is upper knotted at $x$. Hence it is clear that

$$D_-(f+g)(x) \le D^-f(x) + D_-g(x) = -\infty,$$

$$D^+(f+g)(x) \ge D_+f(x) + D^+g(x) = +\infty.$$

Thus $f + g$ is also upper knotted at $x$, and so $L(f + g)'(x)$ is a knotted lower derivative.

Next, to prove (b), suppose its hypothesis holds. Then $L(f + g)'(x)$ is finite by (4). Hence if $Lf'(x)$ or $Lg'(x)$ is a knotted lower derivative, then so is $L(f + g)'(x)$ by (a). Otherwise $f$ and $g$ are both differentiable at $x$, and so is in turn $f + g$. □

Next, to deal with the additivity of weak and semi- derivatives, we first obtain in the following lemma the relations between these derivatives of $-f$ and $f_-$ (see §1.1 for notation) and those of $f$. This lemma will be found helpful in reducing several proofs to fewer cases.

Let us recall here that we use $f_u'(x)$ to denote a unilateral derivative of $f$ at $x$.

7.3.6. LEMMA. *The function $f$ is weakly derivable at $x$ iff $-f$ is so, or, equivalently, iff $f_-$ is weakly derivable at $-x$, and if so, then*

$$Wf'(x) = -W(-f)'(x) = -W(f_-)'(-x). \tag{7}$$

*The same holds for unilateral and semiderivabilities of $f$.*

*Further, if $f$ is semiderivable at $x$, then $S(-f)'(x)$ and $S(f_-)'(-x)$ are totally dissimilar and similar, respectively, to $Sf'(x)$; and if $f$ is unilaterally derivable at $x$, then $(-f)_u'(x)$ and $(f_-)_u'(-x)$ are symmetric and asymmetric, respectively, to $f_u'(x)$.*

PROOF. For each of the functions $-f$ and $f_-$, each result follows easily from the relations between its four unilateral derivates and those of $f$ (see Lemma 1.1.2). □

As we will see later in Remark 7.3.12(c), (d), the weak and semi- derivatives lack additivity in general. However, when $f + g$ is also weakly derivable at the point in question, the following "conditional additivity" remains valid for weak and semi- derivatives.

7.3.7. THEOREM. *Suppose that $f$ and $g$ have unique weak derivatives at $x$ of which the sum is defined, and also suppose that $f + g$ is weakly derivable at $x$. Then*

$$Wf'(x) + Wg'(x) \in W(f+g)'(x). \tag{8}$$

*Further, if $f$ and $g$ have either (i) dissimilar semiderivatives at $x$, or (ii) asymmetric unilateral derivatives at $x$, then*

$$W(f+g)'(x) = Wf'(x) + Wg'(x). \tag{9}$$

PROOF. Let $c = Wf'(x)$ and $d = Wg'(x)$. Then $c + d$ is defined by hypothesis. Since $f$ and $g$ are by the hypothesis nonangular at $x$, each of these weak derivatives is either a semiderivative or a unilateral derivative. Further, since the roles of $f$ and $g$ are interchangeable here, there are indeed only three basically different cases that need to be considered separately.

**Case (i):** $c = Sf'(x)$ and $d = Sg'(x)$. When these semiderivatives are similar, the result follows from Theorem 7.3.2(a) and the above lemma.

Hence, suppose they are dissimilar, say $c = Lf'(x)$ and $d = Ug'(x)$. Then $D^-f(x) = D_+f(x) = c$ and $D_-g(x) = D^+g(x) = d$. Now, since

$$D_-(f+g)(x) \le D^-f(x) + D_-g(x) \le D^-(f+g)(x),$$

it follows that $c + d \in M_-(f+g)(x)$. It is proved similarly that $c + d \in M_+(f+g)(x)$. Hence, as was observed in §4.2 just before Remark 4.2.3, $W(f+g)'(x)$ is unique and equal to $c + d$. Consequently, (9) holds.

**Case (ii):** $c = f'_u(x)$ and $d = g'_u(x)$. First, suppose these unilateral derivatives are symmetric, say right derivatives. Then $c = f'_+(x)$ and $d = g'_+(x)$; and since $c + d$ is defined, it is clear that $f + g$ is right derivable at $x$ and $c + d = (f+g)'_+(x) \in W(f+g)'(x)$.

Next, suppose $c$ and $d$ are asymmetric unilateral derivatives, say $c = f'_-(x)$ and $d = g'_+(x)$. Then since $f$ and $g$ are nonangular at $x$, it is clear that $c \in M_+f(x)$ and $d \in M_-g(x)$. Hence,

$$\begin{aligned} &D_-(f+g)(x) \\ &\quad = f'_-(x) + D_-g(x) \le c + d \le f'_-(x) + D^-g(x) = D^-(f+g)(x), \end{aligned}$$

so that $c + d \in M_-(f+g)(x)$. It is proved similarly that $c + d \in M_+(f+g)(x)$. Hence it follows that $c + d \in W(f+g)'(x)$ and that (9) holds.

**Case (iii):** $c = Sf'(x)$ and $d = g'_u(x)$. Suppose, to be definite, that $c = Lf'(x)$ and $d = g'_+(x)$. Then $D^-f(x) = D_+f(x) = c$ and, as seen before, $d \in M_-g(x)$. Hence,

$$D_-(f+g)(x)$$
$$\leq D^-f(x) + D_-g(x) \leq c + d = D_+f(x) + g'_+(x) = D_+(f+g)(x).$$

Thus $c + d = \max\{D_-(f+g)(x), D_+(f+g)(x)\} \in W(f+g)'(x)$, and so (8) holds. When $d = g'_-(x)$, the result follows on applying the result just proved to $f_-$ and $g_-$. This proves the result whenever $c = Lf'(x)$, and on applying this result to $-f$ and $-g$, the result is obtained when $c = Uf'(x)$. □

When $f + g$ is nonangular at $x$, the inclusion (8) in the above theorem clearly becomes an equality.

On combining the second part of the above theorem, case (i), with Theorem 7.3.2(a), we obtain the following interesting version of Theorems 7.3.1 and 7.3.2(b).

7.3.8. Corollary. *If $f$ is derivable at $x$, $g$ has a unique lower derivative at $x$, and the following sum is defined, then $f + g$ has a unique lower derivative at $x$ and*

$$L(f+g)'(x) = f'(x) + Lg'(x).$$

Next, to deal with linearity of weak and semi- derivatives, we first obtain the following lemma on scalar products.

7.3.9. Lemma. *Let $c \in \mathbb{R}$. If $f$ is weakly derivable at $x$, then so is $cf$; and if so, then*

$$W(cf)'(x) = cWf'(x), \tag{10}$$

*where $W(cf)'(x)$ is totally $c$-compatible with $Wf'(x)$.*

Proof. The result holds trivially when $c = 0$, for then the product on the right side of (10) is by definition zero even when it is indeterminate.

When $c > 0$, it is easy to see that

$$D_-(cf)(x) = cD_-f(x), \qquad D^-(cf)(x) = cD^-f(x),$$

and when $c < 0$, we have similarly,

$$D_-(cf)(x) = cD^-f(x), \qquad D^-(cf)(x) = cD_-f(x).$$

Similar relations hold for right derivates, which are obtained by replacing $-$ in these relations by $+$.

Each of the results follows easily from these relations on unilateral derivates. □

Now, the following theorem on the linearity of semiderivative follows directly on combining the above lemma with Theorem 7.3.2.

7.3.10. THEOREM. *Let $h = af + bg$, where $a, b \in \mathbb{R}$ and $a \neq 0$. Suppose $f$ and $g$ have ab-compatible semiderivatives at $x$.*

(a) *If the following sum is defined, then $h$ has a semiderivative at $x$, a-compatible with $Sf'(x)$, and*

$$Sh'(x) \supset aSf'(x) + bSg'(x). \tag{11}$$

(b) *If from each side either $f$ or $g$ is unilaterally differentiable at $x$, then $h$ has a semiderivative at $x$ as in* (a) *and* (11) *becomes an identity; and if $Sf'(x)$ and $Sg'(x)$ are simultaneously unique or normal, then so is $Sh'(x)$.*

The following theorem on the linearity of weak derivative is obtained, on the other hand, on combining Theorem 7.3.7 with Lemma 7.3.9.

7.3.11. THEOREM. *Let $h = af + bg$, where $a, b \in \mathbb{R}$. Suppose $f$ and $g$ have unique weak derivatives at $x$, the following sum is defined and that $h$ is weakly derivable at $x$. Then*

$$aWf'(x) + bWg'(x) \in Wh'(x). \tag{12}$$

*Further, if $f$ and $g$ have either* (i) *-ab-compatible semiderivatives at $x$, or* (ii) *asymmetric unilateral derivatives at $x$, then* (12) *becomes an equality.*

7.3.12. REMARK. We include here some examples and comments to indicate the problems that arise in connection with the additivity of weak and semi- derivatives.

(a) Firstly, the reverse inclusion of (3) in Theorem 7.3.2 does not hold in general even when $Lf'(x)$ and $Lg'(x)$ are unique and finite. For example, let

$$f(x) = \sqrt{|x|}\cos^2(1/x), \qquad g(x) = \sqrt{|x|}\sin^2(1/x), \qquad x \neq 0,$$

and $f(0) = g(0) = 0$. Then $f$ and $g$ are continuous functions with knotted lower derivatives equal to zero at $x = 0$. But since $f(x) + g(x) = \sqrt{|x|}$, $x \in \mathbb{R}$, it is clear that $f + g$ is lower cusped at 0, and so its lower derivative at 0 is $\overline{\mathbb{R}}$. However, (3) can fail to be an identity only at a countable set of points as it follows from Theorem 4.4.1.

(b) It is further clear from the above example that $L(f + g)'(x)$ does not have to be unique when $Lf'(x)$ and $Lg'(x)$ are unique, and even finite. Conversely, $L(f + g)'(x)$ can be unique without $Lf'(x)$ and $Lg'(x)$ being unique. Let, for example, $f(x) = |x|$ and $g(x) = x^{1/3}$, $x \in \mathbb{R}$. Then $g$ and $f + g$ have an infinite derivative $+\infty$ at zero, whereas $Lf'(0) = [-1, 1]$. When $L(f + g)'(x)$ is on the other hand unique and finite in Theorem 7.3.2(a), then so are $Lf'(x)$ and $Lg'(x)$, for then (3) becomes an identity.

(c) Theorem 7.3.2 holds clearly as long as $f$ and $g$ have similar semiderivatives at $x$. But when these semiderivatives are dissimilar, even if they are unique and finite, $f + g$ is not always semiderivable, or even weakly derivable, at $x$. If $f$ and $g$ are defined as in (a), then $f$ and $-g$ are continuous functions with dissimilar knotted semiderivatives equal to zero at $x = 0$, but if $h = f + (-g)$, then clearly $h(x) = \sqrt{|x|}\cos(2/x)$, $h(0) = 0$, and $h$ is easily seen to be biknotted at 0, and so it is not even weakly derivable at 0.

(d) In the case of measurable functions, this lack of additivity of dissimilar semiderivatives, and thus also of weak derivatives, can indeed occur nearly everywhere. Let $A$ and $B$ be two disjoint countable dense sets in $\mathbb{R}$, and set $C = \mathbb{R} \sim B$ and $E = \mathbb{R} \sim (A \cup B)$. Let $f$ and $g$ be the characteristic functions of $A$ and $C$, respectively. Then it is easy to see that $f$ and $g$ have dissimilar knotted semiderivatives equal to zero at the points of $E$, but since $f(x) + g(x) = 0$, 1, or 2 according to whether $x \in B$, $E$, or $A$, $f + g$ is biknotted at each point of $E$, and so it is not weakly derivable at any point of $E$.

**§7.4 Product and quotient rules for new derivatives.** In this section we investigate the product and quotient rules for new derivatives. First, in Theorem 7.4.1, we obtain the product rule in the case when one of the functions is differentiable in the ordinary sense at the point in question. Then, in Theorem 7.4.3, we obtain the product rule for semiderivatives, and Theorem 7.4.5 deals with a similar product rule for knotted and normal semiderivatives.

Next, in Theorem 7.4.6, we obtain a conditional product rule for weak derivative. Finally, in Theorems 7.4.9 and 7.4.10, we obtain quotient rule for weak and semi- derivatives. Also, in Remark 7.4.11 we present some examples to indicate the problems that arise in this connection with new derivatives.

We continue to assume in this section that $f, g : X \to \mathbb{R}$, where $X \subset \mathbb{R}$, and $x \in X^b$. We introduce here a convenient nomenclature which will be used sometimes in the rest of the work. Given a function $f: X \to \mathbb{R}$ and $x \in X'$, we will call $f$ *D-finite* (or *derivatewise finite*) *at* $x$ if its four derivates are finite at $x$. Clearly, if $f$ is differentiable at $x$, then it is $D$-finite at $x$; and whenever $f$ is $D$-finite at $x$, then it is continuous at $x$. It is further clear that every Lipschitz function is $D$-finite everywhere, and the same holds for every convex function on an open interval.

7.4.1. THEOREM. (a) *If $f$ is differentiable at $x$ and $f(x) \neq 0$, then $fg$ is weakly derivable at $x$ iff $g$ is so; and if so, then $W(fg)'(x)$ is totally $f(x)$-compatible with $Wg'(x)$ and*

(1) $$W(fg)'(x) = f(x)Wg'(x) + g(x)f'(x).$$

(b) *Suppose $f$ is continuous at $x$, $g$ is weakly derivable at $x$, and $g(x) = 0$. If either* (i) *$f(x) \neq 0$, or* (ii) *$g$ is $D$-finite at $x$, then $fg$ has a weak derivative at $x$, totally $f(x)$-compatible with $Wg'(x)$, and*

(2) $$W(fg)'(x) = f(x)Wg'(x).$$

PROOF. First, to prove (a), suppose $f$ is differentiable at $x$ and $f(x) \neq 0$. It is enough to consider here the case when $f(x) > 0$; for when $f(x) < 0$, the result follows with the help of Lemma 7.3.6 on applying this result to $-f$ and $g$.

Given $y \in X$, $y \neq x$, it is easy to see here that

$$Q(fg)(x,y) = f(y)Qg(x,y) + g(x)Qf(x,y). \tag{3}$$

Now, since $f$ is clearly continuous at $x$, $f(y) > 0$ when $y$ is sufficiently close to $x$. Further, as $y \to x$ through $X$, $f(y) \to f(x)$ and the second term on the right side of (3) converges to $g(x)f'(x)$. Hence, we obtain

$$D_-(fg)(x) = f(x)D_-g(x) + g(x)f'(x),$$

$$D^-(fg)(x) = f(x)D^-g(x) + g(x)f'(x),$$

and two similar relations hold for right derivates which are obtained on replacing $-$ everywhere by $+$. Now it is easy to see that the two assertions of (a) follow directly from these four relations.

Next, to prove (b), suppose $f$ is continuous at $x$, $g$ is weakly derivable at $x$, and $g(x) = 0$. Then the second term on the right side of (3) becomes zero. Now, when $f(x) \neq 0$, the result follows from the modified version of (3) by arguments similar to the ones used for (a).

Hence, suppose $f(x) = 0$ and $g$ is $D$-finite at $x$. Then there exist $M$, $\delta \in \mathbb{R}_+$ such that

$$|Qg(x,y)| < M \quad \text{when} \quad y \in X, 0 < |y - x| < \delta.$$

Further, given $\varepsilon > 0$, there is a $\delta_0 > 0$ such that $\delta_0 < \delta$ and $|f(y)| < \varepsilon/M$ when $y \in X$, $|y - x| < \delta_0$. Hence for such $y$ we obtain from (3),

$$|Q(fg)(x,y)| = |f(y)|\,|Qg(x,y)| < (\varepsilon/M)M = \varepsilon.$$

This proves that $fg$ has a derivative equal to zero at $x$. Hence (2) holds, for the product on its right side is zero by definition even when it is indeterminate, and the remaining assertion of (b) is obvious in this case. □

Following are two interesting consequences of the above theorem.

7.4.2. COROLLARY. (a) *If $f$ is continuous at $x$, $g$ is differentiable at $x$, and $g(x) = 0$, then $fg$ is derivable at $x$ and $(fg)'(x) = f(x)\, g'(x)$.*

(b) *If $f$ is differentiable at $x$ and $g$ is continuous and weakly derivable at $x$, then $fg$ has a weak derivative at $x$, totally $f(x)$-compatible with $Wg'(x)$, and (1) holds.*

Part (a) follows clearly from Theorem 7.4.1(b), and (b) follows from Theorem 7.4.1(a) when $f(x) \neq 0$; otherwise it follows from part (a) on interchanging $f$ and $g$.

Next, we obtain the following product rule for semiderivative.

7.4.3. THEOREM. *Suppose $f$ is continuous at $x$, $f(x) \neq 0$, and $f$ and $g$ have $f(x)g(x)$-compatible semiderivatives at $x$.*

(a) *If the following sum is defined, then $fg$ has a semiderivative at $x$, $f(x)$-compatible with $Sg'(x)$, and*

$$f(x)Sg'(x) + g(x)Sf'(x) \subset S(fg)'(x). \tag{4}$$

(b) *If from each side either $f$ or $g$ is unilaterally differentiable at $x$, then $fg$ has a semiderivative at $x$ as in* (a) *and*

$$S(fg)'(x) = f(x)Sg'(x) + g(x)Sf'(x); \tag{5}$$

*and if $Sf'(x)$ and $Sg'(x)$ are simultaneously unique or normal, then so is $S(fg)'(x)$.*

*Moreover, the above holds also when $f(x) = 0$ provided $f$ is differentiable at $x$ and $g$ is continuous at $x$.*

PROOF. The last part of the result for $f(x) = 0$ follows directly from Corollary 7.4.2(b), and when $f(x) \neq 0$ but $g(x) = 0$, the result follows from Theorem 7.4.1(b). Thus it is enough to prove the result when $f(x) > 0$ and $g(x) > 0$; for on applying this result first to $f$ and $-g$ and then applying the two results to $-f$ and $g$, the result will be obtained whenever $f(x)$ and $g(x)$ are nonzero.

Hence, suppose $f(x) > 0$ and $g(x) > 0$. Then $Sf'(x)$ and $Sg'(x)$ are similar by the hypothesis. We give the proof when these semiderivatives are lower derivatives; a similar argument holds when they are upper derivatives.

Now, to prove (a), suppose the sum on the left side of (4) is defined. Then there exist $c \in Lg'(x)$ and $d \in Lf'(x)$ for which $cf(x) + dg(x)$ is defined. We claim that

$$D^-(fg)(x) \leq cf(x) + dg(x) \leq D_+(fg)(x). \tag{6}$$

Let $y \in X$, $y \neq x$. Then since $f$ is continuous at $x$, $f(y) > 0$ when $y$ is sufficiently close to $x$, and $f(y) \to f(x)$ as $y \to x$ through $X$. Now since $g(x) > 0$, it clearly follows from (3) that

$$D^-(fg)(x) \leq f(x)D^-g(x) + g(x)D^-f(x) \leq cf(x) + dg(x), \tag{7}$$

provided that the sum in the middle is defined. Otherwise, one of the derivates $D^-g(x)$ and $D^-f(x)$ is $+\infty$, and so either $c$ or $d$ is $+\infty$. Hence,

$cf(x) + dg(x) = +\infty \geq D^-(fg)(x)$. This establishes the first inequality of (6); the other inequality is proved similarly. Now the two assertions of (a) follow clearly from (6).

Next, to prove (b), suppose for the sake of definiteness that $f$ is left differentiable at $x$ and $g$ is right differentiable at $x$; similar arguments hold in the other three cases. In this case we obtain from (3), as before,

$$(8) \qquad \begin{aligned} D^-(fg)(x) &= f(x)D^-g(x) + g(x)f'_-(x) \\ &\leq f(x)g'_+(x) + g(x)D_+f(x) = D_+(fg)(x). \end{aligned}$$

It is now clear from these relations that $fg$ is lower derivable at $x$ and that the identity (5) holds. Further, if $Lf'(x)$ and $Lg'(x)$ are unique, it follows clearly from (8) that $L(fg)'(x)$ is also unique. Also, when $Lf'(x)$ and $Lg'(x)$ are normal, under the present hypothesis at least one of $f$ and $g$ is differentiable at $x$, and hence it follows from Theorem 7.4.1(a) that $L(fg)'(x)$ is normal. □

Let us note here that the sum in the inclusion (4) of Theorem 7.4.3 is always defined when $f$ or $g$ is semidifferentiable at $x$. Further, as we see later in Remark 7.4.11, the reverse inclusion of (4) does not hold in general, and also this product rule for semiderivative does not hold in general in the absence of the compatibility of semiderivatives defined in the above theorem.

However, when $fg$ is nonangular at $x$, (4) reduces to the identity (5) as seen earlier in connection with additivity of semiderivative.

When one of the functions $f$ and $g$ is convex, we obtain the following result from the above theorem as in the last section.

7.4.4. Corollary. *Suppose $f, g : I \to \mathbb{R}$, $f$ is convex, and $x \in I^\circ$.*

(a) *If $f(x) \neq 0$ and $g$ is $f(x)g(x)$-semiderivable at $x$, then $fg$ has a semiderivative at $x$, $f(x)$-compatible with $Sg'(x)$, and (5) holds.*

(b) *If $f(x) = 0$ and $g$ is continuous at $x$, then $fg$ is $g(x)$-semiderivable at $x$ and $S(fg)'(x) = g(x)\ Lf'(x)$.*

For since $f$ is lower derivable and unilaterally differentiable from both sides at $x$ (see [Ro1]), part (a) follows from Theorem 7.4.3(b), and (b) follows from Theorem 7.4.1(b) on interchanging $f$ and $g$.

Next, in the following theorem we deduce from the above theorem product rule for knotted and normal semiderivatives.

7.4.5. Theorem. *Suppose the hypothesis of Theorem 7.4.3(a) holds and $fg$ is nonangular at $x$. Then $fg$ has a semiderivative at $x$ as in that theorem and (5) holds. Further,*

(a) *if $f$ is semidifferentiable at $x$ and $Sg'(x)$ is a knotted semiderivative, then $S(fg)'(x)$ is also a knotted semiderivative, and*

(b) *if $f$ and $g$ are normally semidifferentiable at $x$, then so is $fg$.*

PROOF. Since $fg$ is nonangular at $x$, $S(fg)'(x)$ is unique, and so (4) reduces to the identity (5). As we saw in the proof of Theorem 7.4.3, it is enough to consider the case when $f(x) > 0$ and $g(x) > 0$. Then the semiderivatives $Sf'(x)$ and $Sg'(x)$ are similar by the hypothesis, say lower derivatives.

Now, to prove (a), suppose its hypothesis holds. Then there exists $c \in \mathbb{R}$ such that $D^-f(x) \le c \le D_+f(x)$ and $g$ is upper knotted at $x$. Hence, it follows from (3) that

$$D_-(fg)(x) \le f(x)D_-g(x) + g(x)Df^-(x) \le -\infty + cg(x) = -\infty,$$

$$D^+(fg)(x) \ge f(x)D^+g(x) + g(x)D_+f(x) \ge +\infty + cg(x) = +\infty.$$

Thus $fg$ is upper knotted at $x$, and so the result follows from the fact that it has a unique lower derivative at $x$.

Next, to prove (b), suppose its hypothesis holds. Then $L(fg)'(x)$ is finite by (5). Hence, if $Lf'(x)$ or $Lg'(x)$ is a knotted lower derivative, then so is $L(fg)'(x)$ by (a). Otherwise, $f$ and $g$ are both differentiable at $x$, and so is in turn $fg$. □

As we will see later in Remark 7.4.11(b), the product rule does not hold in general for weak derivative. But when $fg$ is weakly derivable at the point in question, the following "conditional product rule" remains valid for weak derivative.

7.4.6. THEOREM. *Suppose $f$ is continuous at $x$, $f(x) \neq 0$, $f$ and $g$ have unique weak derivatives at $x$ such that the following sum is defined, and $fg$ is weakly derivable at $x$. Then*

$$f(x)Wg'(x) + g(x)Wf'(x) \in W(fg)'(x). \tag{9}$$

*Further, if $f$ and $g$ have either* (i) *$-f(x)g(x)$-compatible semiderivatives at $x$, or* (ii) *asymmetric unilateral derivatives at $x$, then* (9) *becomes an equality.*

*Moreover, the same holds when $f(x) = 0$ provided that $f$ is D-finite at $x$ and $g$ is continuous at $x$.*

PROOF. The last part of the result, for $f(x) = 0$, follows directly from Theorem 7.4.1(b), and when $f(x) \neq 0$ but $g(x) = 0$, the result follows again from Theorem 7.4.1(b). Hence, as in the case of Theorem 7.4.3, there is no loss of generality in assuming that $f(x) > 0$ and $g(x) > 0$.

Now, let $c = Wf'(x)$ and $d = Wg'(x)$. Then the sum $e \equiv cg(x) + df(x)$ is defined by the hypothesis. As we saw in the proof of Theorem 7.3.7, it is enough to prove the result here in the following three particular cases; in all the other cases the result follows from these three cases.

**Case (i)**: $c = Sf'(x)$ and $d = Sg'(x)$. When these semiderivatives are similar, the result follows directly from Theorem 7.4.3(a).

Hence, suppose they are dissimilar, say $c = Lf'(x)$ and $d = Ug'(x)$. Then $D^-f(x) = D_+f(x) = c$ and $D_-g(x) = D^+g(x) = d$. Now, since $f(x)$ and $g(x)$ are positive and $f$ is continuous at $x$, it follows clearly from (3) that

$$D_-(fg)(x) \leq f(x)D_-g(x) + g(x)D^-f(x) \leq D^-(fg)(x).$$

Hence $e \in M_-(fg)(x)$, and similarly $e \in M_+(fg)(x)$. This proves that $W(fg)'(x)$ is unique and equal to $e$ (see §4.2). Consequently, the equality version of (9) holds.

**Case (ii):** $c = f'_u(x)$ and $d = g'_u(x)$. When these unilateral derivatives are symmetric, say right derivatives, we have $c = f'_+(x)$ and $d = g'_+(x)$, and so it follows from (3) that $e = (fg)'_+(x) \in W(fg)'(x)$.

Next, suppose $c$ and $d$ are asymmetric, say $c = f'_-(x)$ and $d = g'_+(x)$. Then since $f$ and $g$ are by hypothesis nonangular at $x$, $c \in M_+f(x)$ and $d \in M_-g(x)$. Hence, we obtain the following with the help of (3):

$$\begin{aligned} D_-(fg)(x) &= f(x)D_-g(x) + g(x)f'_-(x) \leq df(x) + cg(x) \\ &\leq f(x)D^-g(x) + g(x)f'_-(x) = D^-(fg)(x). \end{aligned}$$

Thus $e \in M_-(fg)(x)$. It is proved similarly that $e \in M_+(fg)(x)$. Hence it follows that $e \in W(fg)'(x)$ and that the equality version of (9) holds.

**Case (iii):** $c = Sf'(x)$ and $d = g'_u(x)$. To be definite, suppose $c = Lf'(x)$ and $d = g'_+(x)$. Then $D^-f(x) = D_+f(x) = c$, and, as before, $d \in M_-g(x)$. Hence from (3) we obtain

$$\begin{aligned} D_-(fg)(x) &\leq f(x)D_-g(x) + g(x)D^-f(x) \leq df(x) + cg(x) \\ &= f(x)g'_+(x) + g(x)D_+f(x) = D_+(fg)(x). \end{aligned}$$

Consequently, $e = \max\{D_-(fg)(x), D_+(fg)(x)\} \in W(fg)'(x)$, and so (9) holds. □

The above inclusion (9) becomes an equality as before when $fg$ is nonangular at $x$.

On combining the second part of the above theorem, case (i), with Theorem 7.4.3(a), the following result is obtained in the direction of Theorem 7.4.1(a).

7.4.7. Corollary. *Suppose $f$ is derivable at $x$ and $g$ has a unique lower derivative at $x$. If one of $f$ and $g$ is continuous and nonzero at $x$ and the following sum is defined, then $fg$ has a unique $f(x)$-semiderivative at $x$ and*

$$S(fg)'(x) = f(x)Lg'(x) + g(x)f'(x). \tag{10}$$

*Moreover, the same holds when $f(x) = g(x) = 0$, provided that one of $f$ and $g$ is continuous at $x$ and the other is D-finite at $x$.*

Next, to obtain the quotient rule, we first prove the following lemma dealing with the reciprocal of a function.

7.4.8. LEMMA. *Suppose $f$ is continuous at $x$ and $f(x) \neq 0$. Then $1/f$ is weakly derivable at $x$ iff $f$ is so; and if so, then the two weak derivatives at $x$ are totally dissimilar and*

$$W(1/f)'(x) = -\{1/f(x)\}^2 Wf'(x).$$

PROOF. Let $y \in X$, $y \neq x$. Since $f$ is continuous at $x$ and $f(x) \neq 0$, $f(y)f(x) > 0$ when $y$ is sufficiently close to $x$. Hence for such $y$ it is easy to see that

$$Q(1/f)(x,y) = -\{f(y)\,f(x)\}^{-1} Qf(x,y).$$

From this equation we obtain, using the continuity of $f$ at $x$ again,

$$D_-(1/f)(x) = -\{1/f(x)\}^2 D^- f(x),$$

$$D^-(1/f)(x) = -\{1/f(x)\}^2 D_- f(x),$$

and two similar relations hold for right derivates. Each part of the result follows easily from these four relations. □

Now, the following quotient rule on semiderivative follows directly on combining the above lemma with Theorem 7.4.3.

7.4.9. THEOREM. *Suppose $g$ is continuous at $x$, $g(x) \neq 0$, and $f$ and $g$ have $-f(x)g(x)$-compatible semiderivatives at $x$.*

(a) *If the following difference is defined, then $f/g$ has a semiderivative at $x$, $g(x)$-compatible with $Sf'(x)$, and*

$$(11) \qquad S(f/g)'(x) \supset \{1/g(x)\}^2\{g(x)Sf'(x) - f(x)Sg'(x)\}.$$

(b) *If from each side either $f$ or $g$ is unilaterally differentiable at $x$, then $f/g$ has a semiderivative at $x$ as in* (a), *and* (11) *becomes an identity; and if $Sf'(x)$ and $Sg'(x)$ are simultaneously unique or normal, then so is $S(f/g)'(x)$.*

The following conditional quotient rule on weak derivative is obtained on combining Theorem 7.4.6 with Lemma 7.4.8.

7.4.10. THEOREM. *Suppose $g$ is continuous at $x$, $g(x) \neq 0$, $f$ and $g$ have unique weak derivatives at $x$ such that the following difference is defined, and $f/g$ is weakly derivable at $x$. Then*

$$(12) \qquad \{1/g(x)\}^2\{g(x)Wf'(x) - f(x)Wg'(x)\} \in W(f/g)'(x).$$

*Further, if $f$ and $g$ have either* (i) *$f(x)g(x)$-compatible semiderivatives at $x$, or* (ii) *asymmetric unilateral derivatives at $x$, then* (12) *becomes an equality.*

7.4.11. REMARK. We present here modified versions of the examples used earlier in Remark 7.3.12 to show that the comments made in that remark on additivity of weak and semi- derivatives also apply to product rule for these derivatives.

(a) Firstly, the reverse inclusion of (4) in Theorem 7.4.3 does not hold in general even when $Sf'(x)$ and $Sg'(x)$ are unique and finite. For example, let $f$ and $g$ be as defined in Remark 7.3.12(a). Then $f + 1$ and $g + 1$ are also continuous functions with knotted lower derivatives equal to zero at $x = 0$. Now, if $h = (f + 1)(g + 1)$, it is clear that

$$h(x) = 1 + \sqrt{|x|} + (|x|/4)\sin^2(2/x), \qquad h(0) = 1.$$

Hence,

$$(1/x)\{h(x) - h(0)\} = \pm\{1/\sqrt{|x|} + (1/4)\sin^2(2/x)\}$$

according to whether $x > 0$ or $< 0$. Thus it is clear that $h$ is lower cusped at 0, and so its lower derivative at 0 is $\overline{\mathbb{R}}$.

This shows further that $S(fg)'(x)$ need not be unique when $Sf'(x)$ and $Sg'(x)$ are unique, and even finite.

(b) Next, when the semiderivatives $Sf'(x)$ and $Sg'(x)$ lack the compatibility as defined in Theorem 7.4.3, $fg$ is not always semiderivable at $x$, or even weakly derivable at $x$. Let $f$ and $g$ be as above. Then $1 + f$ and $1 - g$ have knotted dissimilar semiderivatives equal to zero at $x = 0$, and since $(1 + f)(0) \cdot (1 - g)(0) = 1 > 0$, they lack the compatibility. Again, if $h = (1 + f)(1 - g)$, we have this time

$$h(x) = 1 + \sqrt{|x|}\cos(2/x) - (1/4)|x|\sin^2(2/x), \qquad h(0) = 1.$$

Hence,

$$(1/x)\{h(x) - h(0)\} = \pm\left(1/\sqrt{|x|}\right)\cos(2/x) - (1/4)\sin^2(2/x)\}$$

according to whether $x > 0$ or $< 0$. Now it is easy to see that $h$ is biknotted at 0, and thus it is not even weakly derivable at 0.

Further, in the case of measurable functions, this phenomenon can occur nearly everywhere. If $f$ and $g$ are as defined in Remark 7.3.12(d), it can be easily seen that this assertion holds for $f + 1$ and $g + 1$.

## §7.5 Chain rules in terms of ordinary and new derivatives.

In this section we obtain chain rules for new derivatives in the particular case when the inner or outer function of a composite function is differentiable in the ordinary sense at the point in question (see Theorems 7.5.3 and 7.5.8). These chain rules are deduced here from two fundamental lemmas (see Lemmas 7.5.1 and 7.5.5) in which we compare the unilateral derivates of some restrictions of the composite function with those of its inner and outer functions.

The general chain rules for new derivatives will be obtained in the next section. We include here some examples in Remark 7.5.9 to show the necessity of the hypotheses in the two theorems.

We assume throughout this section, as well as in the next section, that $f: X \to \mathbb{R}$, where $X \subset \mathbb{R}$, $g: f(X) \to \mathbb{R}$, and $x \in X^b$. Also, in these sections we will use $y$ to denote $f(x)$ and use $h$ to denote the composite function $g \circ f$.

To be able to compare the unilateral derivates of $h$ with those of $f$ and $g$, we associate three sets with $f$ and $x$ as follows. We will use $P$, $Z$, and $N$ to denote the sets obtained by adjoining $x$ to the sets of points $t \in X$, $t \neq x$, for which $Qf(x,t) \equiv \{f(t) - f(x)\}/(t - x) > 0$, $= 0$, or $< 0$, respectively.

Further, we will use $\leq'$ to denote a conditional form of the inequality $\leq$ which holds provided that none of the products involved in it is indeterminate.

We begin with a basic lemma comparing the derivates of the restrictions of $h$ and $f$ to $P$, $Z$, and $N$. This lemma will be found to play a fundamental role in the present section as well as in the following section.

7.5.1. Lemma (Fundamental). *Suppose $f$ is continuous at $x$.*

(a) *If $x \in Z'$, then $f_Z$ and $h_Z$ have zero derivatives at $x$.*

(b) *If $x \in P^+$, then*

$$D_+g(y)D_+f_P(x) \wedge D_+g(y)D^+f_P(x) \leq' D_+h_P(x)$$
$$\leq' D^+g(y)D_+f_P(x) \wedge D^+g(y)D^+f_P(x),$$
$$D_+g(y)D_+f_P(x) \vee D_+g(y)D^+f_P(x) \leq' D^+h_P(x)$$
$$\leq' D^+g(y)D_+f_P(x) \vee D^+g(y)D^+f_P(x).$$

(c) *If $x \in N^+$, then*

$$D^-g(y)D_+f_N(x) \wedge D^-g(y)D^+f_N(x) \leq' D_+h_N(x)$$
$$\leq' D_-g(y)D_+f_N(x) \wedge D_-g(y)D^+f_N(x),$$
$$D^-g(y)D_+f_N(x) \vee D^-g(y)D^+f_N(x) \leq' D^+h_N(x)$$
$$\leq' D_-g(y)D_+f_N(x) \vee D_-g(y)D^+f_N(x).$$

*Moreover, when $x \in P^-$,* (b) *holds on replacing $+$ everywhere by $-$, and when $x \in N^-$,* (c) *holds on interchanging $+$ and $-$ everywhere.*

Proof. The part (a) is obvious, for the functions $f_Z$ and $h_Z$ are clearly constant.

To prove (b), it is enough to prove its two left inequalities, for its two right inequalities follow clearly on applying the left inequalities to $f$ and $-g$ (see Lemma 1.1.2).

Hence, suppose that $x \in P^+$ and that the products $D_+g(y)D_+f_P(x)$ and $D_+g(y)D^+f_P(x)$ are not indeterminate. Let $a$ and $b$ denote the infimum and supremum of these two products. Then to prove the two left inequalities of (b), we need to show that

$$(1) \qquad a \leq D_+ h_P(x) \quad \text{and} \quad b \leq D^+ h_P(x).$$

We will consider here two cases separately.

**Case (i):** $D_+ g(y) = 0$. Then it is clear from the definition of $P$ and the hypothesis that $0 \leq D_+ f_P(x) \leq D^+ f_P(x) < +\infty$. Consequently, $a = b = 0$, and there exist $M, \delta \in \mathbb{R}_+$ such that

$$(2) \qquad 0 < Qf(x,t) < M \quad \text{when} \quad t \in P, 0 < t - x < \delta.$$

Also, given $\varepsilon > 0$, since $D_+ g(y) = 0$, there is an $\eta > 0$ such that

$$(3) \qquad Qg(y,z) > -\varepsilon/M \quad \text{when} \quad 0 < z - y < \eta.$$

Now, since $f$ is continuous at $x$, there is a $\delta' > 0$ such that $\delta' < \delta$ and $|f(t) - f(x)| < \eta$ when $t \in X$ and $|t - x| < \delta'$. Hence when $t \in P$ and $0 < t - x < \delta'$, it follows clearly from (2) and (3) that

$$Qh(x,t) = Qg(y, f(t)) \cdot Qf(x,t) > -(\varepsilon/M)M = -\varepsilon.$$

This proves that $D_+ h_P(x) \geq 0 = a$, and hence also $D^+ h_P(x) \geq 0 = b$.

**Case (ii):** $D_+ g(y) \neq 0$. Choose a sequence of points $\{x_n\}$ in $P$ which converges to $x$ from the right such that

$$(4) \qquad D_+ h_P(x) = \lim_n Qh(x, x_n) = \lim_n \{Qg(y, f(x_n)) \cdot Qf(x, x_n)\}.$$

Further, replacing $\{x_n\}$ by a subsequence if necessary, we can assume that the sequence $\{Qg(y, f(x_n))\}$ converges to some $c \in \overline{\mathbb{R}}$. Now, since $f$ is continuous at $x$ and $x_n \in P$ for each $n$, it is clear that $\{f(x_n)\}$ converges to $y = f(x)$ from the right. Hence $D_+ g(y) \leq c$.

Now, if $c > 0$, then since $D_+ f_P(x) \geq 0$ and the indeterminate products are zero by definition, from (4) we obtain

$$D_+ h_P(x) \geq c \liminf_{n \to \infty} Qf(x, x_n) \geq D_+ g(y) D_+ f_P(x) \geq a;$$

and when $c < 0$, from (4) we obtain

$$D_+ h_P(x) \geq c \limsup_{n \to \infty} Qf(x, x_n) \geq c D^+ f_P(x) \geq D_+ g(y) D^+ f_P(x) \geq a.$$

Further, when $c = 0$, it is clear in the present case that $D_+ g(y) < 0$. Hence if $D^+ f_P(x) = +\infty$, we have $a = -\infty \leq D_+ h_P(x)$; otherwise, since $c = 0$ and $D_+ f_P(x) \geq 0$, it follows from (4) that $D_+ h_P(x) = 0 \geq a$.

It remains to prove the second inequality of (1) in the case (ii). First, suppose $D_+g(y) > 0$. Choose this time the sequence $\{x_n\}$ of points in $P$ which converges to $x$ from the right such that

$$D^+f_P(x) = \lim_n Qf(x, x_n).$$

Then since $\{f(x_n)\}$ converges as before to $y$ from the right, it follows that

$$\begin{aligned} D^+h_P(x) \ge \limsup_{n\to\infty} Qh(x, x_n) &= \limsup_{n\to\infty} \{Qg(y, f(x_n)) \cdot Qf(x, x_n)\} \\ &\ge D_+g(y)D^+f_P(x) = b. \end{aligned}$$

When $D_+g(y) < 0$, this inequality is obtained by a similar argument by replacing $D^+f_P(x)$ by $D_+f_P(x)$. This completes the proof of (b).

Next, let ($b_-$) and ($c_-$) denote the versions of (b) and (c), respectively, that are described in the concluding part. Then it is easy to see from Lemma 1.1.2 that ($c_-$) follows from (b) on applying it to $f$ and $g_-$; and then (c) and ($b_-$) are obtained on applying ($c_-$) and (b), respectively, to $f_-$ and $-g_-$. □

In the following lemma we state a direct consequence of the parts (b) and (c) of the above lemma which also will be used repeatedly in the sequel.

7.5.2. Lemma. *Suppose $f$ is continuous at $x$. In case $x \in P^+$ and $g$ is right differentiable at $y$, then if $g'_+(y) > 0$, then*

$$D_+h_P(x) = g'_+(y)D_+f_P(x) \quad \textit{and} \quad D^+h_P(x) = g'_+(y)D^+f_P(x);$$

*and if $g'_+(y) < 0$, then*

$$D_+h_P(x) = g'_+(y)D^+f_P(x) \quad \textit{and} \quad D^+h_P(x) = g'_+(y)D_+f_P(x).$$

*Moreover, when $x \in N^+$ and $g$ is left differentiable at $y$, the same holds on replacing $P$ by $N$ and $g'_+(y)$ by $g'_-(y)$.*

We are now in a position to obtain the chain rule when the outer function $g$ is differentiable at the point in question.

7.5.3. Theorem. *Suppose $f$ is continuous at $x$, $g$ is differentiable at $y = f(x)$ and $g'(y) \neq 0$. Then $h = g \circ f$ is weakly derivable at $x$ iff $f$ is so; and if so, then their weak derivatives are totally $g'(y)$-compatible and*

$$Wh'(x) = g'(y)Wf'(x). \tag{5}$$

Proof. First, suppose $g'(y) > 0$. Then we claim that

$$D_+h(x) = g'(y)D_+f(x), \qquad D^+h(x) = g'(y)D^+f(x). \tag{6}$$

When $z \in Z^+$, it follows from Lemma 7.5.1(a) that $D_+h_Z(x) = 0 = D_+f_Z(x)$, and when $x \in P^+$ or $N^+$, it follows from Lemma 7.5.2 that $D_+h_P(x) = g'(y)D_+f_P(x)$ or $D_+h_N(x) = g'(y)D_+f_N(x)$, respectively. Clearly, $D_+h(x)$ is the infimum of $D_+h_P(x)$, $D_+h_Z(x)$, and $D_+h_N(x)$ that are defined, and the same is true of $f$. The first equation of (6) follows now easily since $D_+f_N(x) \leq 0 = D_+f_Z(x) \leq D_+f_P(x)$. A similar argument holds for the other equation of (6). Further, on applying (6) to $f_-$ and $g$, we obtain the following with the help of Lemma 1.1.2:

$$D_-h(x) = g'(y)D_-f(x), \qquad D^-h(x) = g'(y)D^-f(x).$$

Now, from the above four equations the result follows easily in the present case.

When $g'(y) < 0$, the result is obtained on applying the above result to $f$ and $-g$ with the help of Lemma 7.3.6. □

As we see later in Remark 7.5.9, the above theorem does not hold in general when $g'(f(x))$ is zero or infinite. However, when it is zero and $f$ is $D$-finite at $x$, it follows from Lemma 7.5.1 that $h$ has a zero derivative at $x$. On combining this fact with the above theorem we obtain the following.

7.5.4. Corollary. *Suppose $f$ is weakly derivable and D-finite at $x$, and $g$ is differentiable at $y = f(x)$. Then $h = g \circ f$ has a weak derivative at $x$, totally $g'(y)$-compatible with $Wf'(x)$, and* (5) *holds.*

Next, to obtain the other chain rule we need some more lemmas. Some nomenclature is needed here.

Given $f: X \to \mathbb{R}$, $x \in X'$ and $c \in \overline{\mathbb{R}}$, we will call $c$ a *proper right* or *left median derivate* of $f$ at $x$ if there is a sequence of points $\{x_n\}$ in $X$ which converges to $x$ from the right or left, respectively, such that the ratio $Qf(x, x_n)$ converges to $c$ as $n \to \infty$.

The sets of proper right and left median derivates of $f$ at $x$ will be called simply the *proper right* and *left medians*, respectively, of $f$ at $x$ and will be denoted by $M_+^*f(x)$ and $M_-^*f(x)$, respectively.

Thus it is clear that $M_+^*f(x) \subset M_+f(x)$ in general, and that $M_+^*f(x)$ includes $D_+f(x)$ and $D^+f(x)$ as elements. Further, when $f$ has the Darboux property, it is clear that these two right medians become identical. The same holds for the proper left median of $f$ at $x$.

The following lemma will also play a fundamental role in the rest of this section and the next section.

7.5.5. Lemma (Fundamental). *Suppose $X = I$ and $f$ is continuous. If $x \in P^+$ or $N^+$ and $c \in M_+^*g(y)$ or $M_-^*g(y)$, respectively, then*

(7)
$$D_+h(x) \leq' cD_+f(x) \vee cD^+f(x) \quad \text{and} \quad cD_+f(x) \wedge cD^+f(x) \leq' D^+h(x).$$

*Moreover, when $x \in P^-$ or $N^-$ and $c \in M^*_-g(y)$ or $M^*_+g(y)$, respectively, the same holds on replacing + everywhere by −.*

PROOF. First, suppose that $x \in P^+$ and $c \in M^*_+g(y)$ and that none of the products in (7) is indeterminate. Then there exists $x' \in X$, $x' > x$, such that $f(x') > f(x) = y$, and there is a decreasing sequence of points $\{y_n\}$ in the interval $(y, f(x'))$ with limit $y$ such that

$$c = \lim_n Qg(y, y_n). \tag{8}$$

Now, let $J = [x, x']$. Given $n$, since $f$ is continuous on $J$, $A_n \equiv J \cap f^{-1}(y_n)$ is a nonempty closed set. Let $x_n$ denote for each $n$ the infimum of $A_n$. Then it is easy to see that $\{x_n\}$ is a decreasing sequence of points in $J$, and so it converges to some point $a_0 \in J$. Further,

$$f(a_0) = \lim_n f(x_n) = \lim_n y_n = y = f(x).$$

We claim that $a_0 = x$. If $a_0 > x$, then there is by hypothesis a point $a_1$ in $(x, a_0)$ such that $f(a_1) > f(x)$. Hence, on choosing $n$ large enough, we obtain a point $a_2$ in $(x, a_1)$ such that $f(a_2) = y_n$, which contradicts the fact that $x_n = \inf A_n$.

Thus $\{x_n\}$ converges to $x$ from the right. Hence,

$$D_+h(x) \le \liminf_{n\to\infty} Qh(x, x_n) = \liminf_{n\to\infty} \{Qg(y, y_n)\cdot Qf(x, x_n)\} = a \text{ (say)}.$$

Let $b = cD_+f(x) \vee cD^+f(x)$. Then if $c > 0$ or $< 0$, it is easy to see from (8) that $a \le b$.

Hence, suppose $c = 0$. Then $D_+f(x)$ and $D^+f(x)$ are finite by hypothesis, and so there exist positive integers $M$ and $n_1$ such that $|Qf(x, x_n)| < M$ for $n > n_1$. Now, given $\varepsilon > 0$, there is by (8) another positive integer $n_2$ such that $|Qg(y, y_n)| < \varepsilon/M$ for $n > n_2$. Hence if $n > n_1 \vee n_2$, then $|Qg(y, y_n)\cdot Qf(x, x_n)| < (\varepsilon/M)M = \varepsilon$. This proves that $a = 0 = b$.

Thus, in the present case we have $D_+h(x) \le a \le b$, which establishes the first inequality of (7). The second inequality is obtained on applying the first to $f$ and $-g$ (see Lemma 1.1.2).

When $x \in N^+$ and $c \in M^*_-g(y)$, the result is obtained on applying the above result to $-f$ and $g_-$. The last part follows on applying the established results to $f_-$ and $g$. □

Next, from the above lemma we deduce two other lemmas which also will be used repeatedly in the present and the next section.

7.5.6. LEMMA. *Suppose $X = I$ and $f$ is continuous. If $f$ is right differentiable at $x$ and $f'_+(x) > 0$, then*

$$D_+h(x) = f'_+(x)D_+g(y) \quad \text{and} \quad D^+h(x) = f'_+(x)\, D^+g(y);$$

*and if* $f'_+(x) < 0$, *then*

$$D_+h(x) = f'_+(x)D^-g(y) \quad \text{and} \quad D^+h(x) = f'_+(x)D_-g(y).$$

*Moreover, when f is left differentiable at x, the same holds on interchanging + and − everywhere.*

PROOF. First, suppose $f$ is right differentiable at $x$ and $f'_+(x) > 0$. Then on choosing $c$ to be $D_+g(y)$ or $D^+g(y)$ in Lemma 7.5.5, we obtain

$$D_+h(x) \le f'_+(x)D_+g(y) \quad \text{and} \quad D^+h(x) \ge f'_+(x)D^+g(y)$$

respectively. Further, since there clearly exists a $\delta > 0$ such that $X \cap (x, x+\delta) \subset P$, the reverse inequalities follow from Lemma 7.5.1(b). This proves the result in the present case.

Similar arguments hold in the other three cases. □

7.5.7. LEMMA. *Suppose* $X = I$, *f is continuous and that it has a zero right (or left) derivative at x.*

(a) *If* $D^-g(y)$ *and* $D_+g(y)$ *are finite, then* $D_+h(x)$ (*or* $D^-h(x)$) $= 0$.

(b) *If* $D_-g(y)$ *and* $D^+g(y)$ *are finite, then* $D^+h(x)$ (*or* $D_-h(x)$) $= 0$.

PROOF. First, suppose $f$ has a zero right derivative at $x$ and $D^-g(y)$ and $D_+g(y)$ are finite. Then if $x \in Z^+$, $P^+$ or $N^+$, it follows from Lemma 7.5.1 that

$$D_+h_Z(x) = 0, \quad D_+h_P(x) \ge f'_+(x)D_+g(y) = 0 \quad \text{or}$$

$$D_+h_N(x) \ge f'_+(x)D^-g(y) = 0,$$

respectively. Consequently, $D_+h(x) \ge 0$. As for the reverse inequality, when $x \in Z^+$, we already have $D_+h(x) \le D_+h_Z(x) = 0$; otherwise, on choosing $c = D_+g(y)$ or $D^-g(y)$ according to whether $x \in P^+$ or $N^+$, it follows from Lemma 7.5.5 that $D_+h(x) \le cf'_+(x) = 0$. This proves that $D_+h(x) = 0$.

Similar arguments hold in the other three cases. □

Finally, we obtain the following chain rule in the case when the inner function is differentiable at the point in question.

7.5.8. THEOREM. *Suppose f is differentiable at x,* $f'(x) \ne 0$, *and g is weakly derivable at* $y = f(x)$. *Then* $h = g \circ f$ *is weakly derivable at x and*

$$Wh'(x) \supset f'(x)Wg'(y). \tag{9}$$

*Further, if g is semiderivable at y, then so is h at x, and the two semiderivatives are similar; and if g is unilaterally derivable at y, then so is h at x, and the two unilateral derivatives are* $f'(x)$*-compatible.*

*Moreover, if $X = I$ and $f$ is continuous in a neighborhood of $x$, then (9) becomes an identity; and (i) if $g$ is semiderivable at $y$, then $Sh'(x)$ and $Sg'(y)$ are totally similar, and (ii) if $f'(x) = 0$ and $g$ is D-finite at $y$, then $h$ has a zero derivative at $x$.*

PROOF. The result (ii) in the last part follows directly from Lemma 7.5.7. Hence it is enough to prove the result when $f'(x) > 0$, for when $f'(x) < 0$ the result is obtained on applying this result to $-f$ and $g_-$ with the help of Lemma 7.3.6.

Hence, suppose $f'(x) > 0$. Then there is a $\delta > 0$ such that $X \cap (x, x + \delta) \subset P$. Hence it follows from Lemma 7.5.1(b) that

$$D_+h(x) \geq f'(x)D_+g(y), \qquad D^+h(x) \leq f'(x)D^+g(y);$$

and according to the concluding statement of that lemma we further obtain

$$D_-h(x) \geq f'(x)D_-g(y), \qquad D^-h(x) \leq f'(x)\, D^-g(y).$$

It is now easy to see that the first part of the result follows from these four relations.

Next, suppose $X = I$ and $f$ is continuous in a neighborhood of $x$. Then the above four inequalities become equalities due to Lemma 7.5.6, and the stronger version of the result is obtained in this case from these equalities. □

7.5.9. REMARK. We include here two examples to show that Theorems 7.5.3 and 7.5.8 do not hold in general when the derivative $g'(y)$ or $f'(x)$, respectively, is zero or infinite.

(a) Let $f(x) = x^{1/3}$, $x \in \mathbb{R}$; $g(x) = x^2\sin(1/x)$, $g(0) = 0$; $h = g \circ f$ and $x = 0$. Then $f$ and $g$ are continuous functions on $\mathbb{R}$, $y = f(0) = 0$, and it is clear that $f'(0) = +\infty$ and $g'(0) = 0$. Further, since

$$h(x) = x^{2/3}\sin(x^{-1/3}), \qquad h(0) = 0,$$

it is easy to see that $h$ is biknotted at 0. This shows that Theorem 7.5.3 does not hold when $g'(y) = 0$, and Theorem 7.5.8 does not hold when $f'(x)$ is infinite.

(b) Reversing the roles of $f$ and $g$ in (a), let $f(x) = x^2\sin(1/x)$, $f(0) = 0$; $g(x) = x^{1/3}$, $x \in \mathbb{R}$; $h = g \circ f$ and $x = 0$. Then again $y = f(0) = 0$, and this time $f'(0) = 0$ and $g'(0) = +\infty$. Further,

$$h(x) = x^{2/3}\sin^{1/3}(1/x), \qquad h(0) = 0,$$

which is again biknotted at 0. This shows that Theorem 7.5.3 does not hold when $g'(y)$ is infinite, and Theorem 7.5.8 does not hold when $f'(x) = 0$.

**§7.6 General chain rules for new derivatives.** In this section we obtain general chain rules for weak and semi- derivatives. First, in Theorem 7.6.1, we obtain the chain rule for semiderivative, and Theorem 7.6.3 deals with the exceptional case when zero is a semigradient of the outer function at the point in question.

Next, in Theorem 7.6.6, we obtain a conditional chain rule for weak derivative in the case when the composite function is weakly derivable at the point in question. Also, we include in Remark 7.6.8 some examples to show that the results of this section are the best possible in their direction.

The proofs of the general chain rules become more involved and are indeed based on all the results of the last section.

We continue to assume in this section that $f : X \to \mathbb{R}$, where $X \subset \mathbb{R}$, $g : f(X) \to \mathbb{R}$, $x \in X^b$, $y = f(x)$, and $h = g \circ f$.

7.6.1. THEOREM. *Suppose $f$ is continuous and semiderivable at $x$, $g$ is semiderivable at $y = f(x)$, $0 \notin Sf'(x) \cup Sg'(y)$, and $Sf'(x)$ and $Sg'(y)$ are $Sg'(y)$-compatible. Then $h = g \circ f$ has a semiderivative at $x$ similar to $Sg'(y)$ and*

$$Sg'(y)Sf'(x) \subset Sh'(x). \tag{1}$$

*Further, if $X = I$, $f$ is continuous, and either* (i) *$f$ is unilaterally differentiable from both sides at $x$,* (ii) *$g$ is so at $y$, or* (iii) *$f$ and $g$ have finite $-Sf'(x)$-compatible unilateral derivatives at $x$ and $y$, respectively, then*

$$Sh'(x) = Sg'(y)Sf'(x); \tag{2}$$

*and if $Sf'(x)$ and $Sg'(y)$ are simultaneously unique or normal, then so is $Sh'(x)$.*

*Moreover, the same holds when $0 \in Sf'(x)$ provided that $Sg'(y) \subset \mathbb{R}$.*

PROOF. Since $0 \notin Sg'(y)$, it is enough to prove here all the results in the case when $g$ is lower derivable at $y$ and $Lg'(y) > 0$. On applying these results to $-f$ and $g_-$, the results would follow with the help of Lemma 7.3.6 when $Lg'(y) < 0$; and then on applying these two results to $f$ and $-g$, the results would follow when $g$ is upper derivable at $y$.

Hence, suppose $g$ is lower derivable at $y$ and $Lg'(y) > 0$. Then $f$ is by the hypothesis lower derivable at $x$.

Now, to prove the first two parts, it is enough to prove the results when $Lf'(x) > 0$, for on applying these results to $f_-$ and $g$ the results would follow when $Lf'(x) < 0$.

Hence, suppose $Lf'(x) > 0$. Then in the present case we have

$$0 < D^-f(x) \le D_+f(x), \qquad 0 < D^-g(y) \le D_+g(y). \tag{3}$$

We will first prove that

$$D^-h(x) \le D^-g(y)D^-f(x) \le D_+g(y)D_+f(x) \le D_+h(x). \tag{4}$$

The middle inequality of (4) follows directly from (3). Further, when $x \in Z^-$, $P^-$, or $N^-$, then due to (3) it follows from Lemma 7.5.1 that $D^-h_Z(x) = 0$, $D^-h_P(x) \leq D^-g(y)D^-f_P(x) \leq D^-g(y)D^-f(x)$, or $D^-h_N(x) \leq D_+g(y)D^-f_N(x) \leq 0$, respectively. Now, since $D^-h(x)$ is the supremum of the derivates $D^-h_Z(x)$, $D^-h_P(x)$, and $D^-h_N(x)$ that are defined, and $D^-g(y)D^-f(x) > 0$, we thus obtain the first inequality of (4). Further, since $D_+f(x) > 0$, $x \notin N^+$ or $Z^+$; and since $D_+g(y) > 0$, we obtain the following from Lemma 7.5.1(b):

$$D_+h(x) = D_+h_P(x) \geq D_+g(y)D_+f_P(x) = D_+g(y)D_+f(x).$$

This establishes (4).

It is now clear from (3) and (4) that $h$ is lower derivable at $x$, and

$$Lg'(y)Lf'(x) = [D^-g(y)D^-f(x), D_+g(y)D_+f(x)] \subset Lh'(x).$$

The first part is thus established.

Next, to prove the second part, suppose that $X = I$, that $f$ is continuous, and that one of its hypotheses (i), (ii), and (iii) holds. Then, in the present case, from each side at least one of $f$ and $g$ has a finite positive unilateral derivative at $x$ or $y$, respectively. When $f$ is left or right differentiable at $x$, it follows from Lemma 7.5.6 that the first or the last inequality of (4), respectively, becomes an equality. When $g$ is, on the other hand, left differentiable at $y$, it follows clearly from the last part of Lemma 7.5.1 that

$$D^-h(x) \geq D^-h_P(x) = g'_-(y)D^-f_P(x) = D^-g(y)\,D^-f(x),$$

so that the first inequality of (4) becomes an equality; and when $g$ is right differentiable at $y$, since $x \notin N^+$ or $Z^+$, we obtain similarly the following from Lemma 7.5.2:

$$D_+h(x) = D_+h_P(x) = g'_+(y)D_+f_P(x) = D_+g(y)D_+f(x),$$

so that the last inequality of (4) becomes an equality. Hence, in any case we presently have, in the place of (4),

$$(5) \qquad D^-h(x) = D^-g(y)D^-f(x) \leq D_+g(y)D_+f(x) = D_+h(x).$$

Now, it is clear from (5) that in the present case the inclusion (1) becomes the identity (2). Further, when $Lf'(x)$ and $Lg'(y)$ are unique, the middle inequality of (5) clearly becomes an equality, and hence $Lh'(x)$ also is unique. If, on the other hand, $Lf'(x)$ and $Lg'(y)$ are normal, then under the present hypothesis either $f$ or $g$ has a finite positive derivative at $x$ or $y$, respectively, and hence it follows from Theorem 7.5.3 or 7.5.8 that $Lh'(x)$ is normal.

Finally, to prove the last part, suppose $0 \in Lf'(x)$ and $Lg'(y) \subset \mathbb{R}$. Then in the present case we have

$$(6) \quad D^-f(x) \le 0 \le D_+f(x), \qquad 0 < D^-g(y) \le D_+g(y) < +\infty.$$

We will first prove in this case that

$$(7) \quad D^-h(x) \le D_+g(y)D^-f(x) \le 0 \le D_+g(y)D_+f(x) \le D_+h(x).$$

Because the two middle inequalities of (7) follow directly from (6), it is enough to prove here the last inequality of (7); its first inequality follows from the last on applying it to $f_-$ and $g$ (see Lemma 1.1.2).

In case $D_+f(x) > 0$, it follows from Lemma 7.5.1(b) due to (6) that $D_+h(x) = D_+h_P(x) \ge D_+g(y)D_+f(x)$. Hence, suppose $D_+f(x) = 0$. Now, if $x \in P^+$, we have $D_+h_P(x) \ge 0$ as just seen; and if $x \in Z^+$, we have by that lemma, $D_+h_Z(x) = 0$. When $x \in N^+$, we have on the other hand $D_+f_N(x) = 0$ (for $D_+f(x) = 0$), and so it follows from Lemma 7.5.1 with the help of (6) that $D_+h_N(x) \ge D^-g(y)D_+f_N(x) = 0$. Hence, $D_+h(x) \ge 0 = D_+g(y)D_+f(x)$ in this case as well. This establishes (7).

Now, it follows from (6) and (7) that $h$ is lower derivable at $x$ and

$$Lg'(y)\ Lf'(x) = [D_+g(y)D^-f(x), D_+g(y)D_+f(x)] \subset Lh'(x).$$

This proves the first part in the present case.

Next, to prove the second part, suppose $X = I$ and $f$ is continuous. The hypotheses (i) and (ii) need to be considered here separately; the hypothesis (iii) is not applicable in this case since $0 \in Lf'(x)$.

First, suppose (i) holds. Then $-\infty < f'_-(x) \le 0 \le f'_+(x) < +\infty$, and we claim in this case that

$$(8) \quad D^-h(x) = f'_-(x)D_+g(y) \le 0 \le f'_+(x)D_+g(y) = D_+h(x).$$

It is enough to prove here, as before, the last equality of (8). However, this equality follows directly from Lemma 7.5.6 when $f'_+(x) > 0$, and from Lemma 7.5.7 when $f'_+(x) = 0$. The assertions of the second part follow now from (8) exactly as before.

Next, suppose (ii) holds. Then $0 < g'_-(y) \le g'_+(y) < +\infty$, and we claim this time that

$$(9) \quad D^-h(x) = g'_+(y)D^-f(x) \le 0 \le g'_+(y)D_+f(x) = D_+h(x).$$

It is again enough to prove the last equality of (9). In the case when $D_+f(x) > 0$, this equality follows directly from Lemma 7.5.2. Hence suppose $D_+f(x) = 0$. Now, when $x \in Z^+$, we have as before $D_+h_Z(x) = 0$. Further, if $x \in N^+$, it is clear that $D_+f_N(x) = 0$, and if $x \in P^+$, it follows easily from the

Darboux property of $f$ that $D_+f_P(x) = 0$. Hence, when $x \in P^+$ or $N^+$, we have, by Lemma 7.5.2, $D_+h_P(x) = g'_+(y)D_+f_P(x) = 0$ or $D_+h_N(x) = g'_-(y)D_+f_N(x) = 0$, respectively. Consequently, $D_+h(x) = 0 = g'_+(y)D_+f(x)$. This establishes (9), and the assertions of the second part follow from it as before. □

In the case when the inner or outer function of a composite function is convex, we obtain the following result from the above theorem.

7.6.2. COROLLARY. *Suppose $X = I$.*

(a) *If $f$ is convex, $0 \notin Lf'(x)$, and $g$ has either a positive lower derivative or a negative upper derivative at $y = f(x)$, then $h = g \circ f$ has a semiderivative at $x$ similar to $Sg'(y)$ and (2) holds. The same holds when $0 \in Lf'(x)$ provided that $Sg'(y) \subset \mathbb{R}$.*

(b) *If $f$ is continuous, $g$ is convex, $0 \notin Lg'(y)$, and $f$ is $Lg'(y)$-semiderivable at $x$, then $h$ is lower derivable at $x$ and (2) holds.*

In the following theorem we deal with the exceptional case when $0 \in Sg'(y)$. It should be noted here that a function is always semiderivable at a point where it is unilaterally derivable from both sides. Also, it is interesting to note here that the inclusion (1) of Theorem 7.6.1 gets reversed in the following theorem.

7.6.3. THEOREM. *Suppose $X = I$, $f$ is continuous and unilaterally differentiable from both sides at $x$, $0 \notin Sf'(x)$, and $g$ has a zero semigradient at $y = f(x)$. Then $h = g \circ f$ has a semiderivative at $x$ similar to $Sg'(y)$ and*

$$0 \in Sh'(x) \subset Sg'(y)Sf'(x). \tag{10}$$

*Also, if $Sf'(x)$ and $Sg'(y)$ are unique, then so is $Sh'(x)$.*

*Further, if $Sg'(y) \geq 0$ or $\leq 0$ according to whether $Sf'(x)$ and $Sg'(y)$ are similar or dissimilar, then*

$$Sh'(x) = Sg'(y)Sf'(x). \tag{11}$$

*Moreover, the same holds when $0 \in Sf'(x)$ provided that $Sg'(y) \subset \mathbb{R}$.*

PROOF. It is enough to prove the result here when $Sf'(x)$ and $Sg'(y)$ are lower derivatives. On applying this result to $-f$ and $g_-$, the result would follow, with the help of Lemma 7.3.6, when $Sf'(x)$ is an upper derivative and $Sg'(y)$ is a lower derivative; and then on applying these two results to $f$ and $-g$ the result would follow when $Sg'(y)$ is an upper derivative.

Hence, suppose $f$ and $g$ are lower derivable at $x$ and $y$, respectively. Now, to prove the first two parts, it is enough as before to prove the results when $Lf'(x) > 0$; for on applying these results to $f_-$ and $g$ the results would follow when $Lf'(x) < 0$.

Hence, suppose $Lf'(x) > 0$. Then, according to the hypothesis, we have

$$0 < f'_-(x) \le f'_+(x) < +\infty \quad \text{and} \quad D^-g(y) \le 0 \le D_+g(y).$$

Hence it follows from Lemma 7.5.6 that

$$D^-h(x) = f'_-(x)D^-g(y) \le 0 \le f'_+(x)D_+g(y) = D_+h(x).$$

Consequently, $h$ is lower derivable at $x$ and

$$0 \in Lh'(x) = [f'_-(x)D^-g(y), f'_+(x)D_+g(y)]. \tag{12}$$

Further, it is clear that

$$Lg'(y)Lf'(x) = [f'_+(x)D^-g(y), f'_+(x)D_+g(y)]. \tag{13}$$

Hence it follows from (12) that (10) holds. Also, when $Lf'(x)$ and $Lg'(y)$ are unique, it follows from (12) that so is $Lh'(x)$.

Next, to prove the second part, suppose $Lg'(y) \ge 0$. Then since $0 \in Lg'(y)$, we have $D^-g(y) = 0$. Hence in this case it is clear from (12) and (13) that the identity (11) holds.

Finally, to prove the last part, suppose $0 \in Lf'(x)$ and $Lg'(y) \subset \mathbb{R}$. Then $-\infty < f'_-(x) \le 0 \le f'_+(x) < +\infty$ and $-\infty < D^-g(y) \le 0 \le D_+g(y) < +\infty$. We claim that

$$D^-h(x) = f'_-(x)D_+g(y) \le 0 \le f'_+(x)D_+g(y) = D_+h(x). \tag{14}$$

The middle two inequalities follow here from the relations just mentioned. The first equality follows on the other hand from Lemma 7.5.6 when $f'_-(x) > 0$, and from Lemma 7.5.7 when $f'_-(x) = 0$. The last equality also follows similarly from these two lemmas.

It is now clear from (14) that $h$ is lower derivable at $x$ and that

$$0 \in Lh'(x) = [f'_-(x)D_+g(y), f'_+(x)\ D_+g(y)]. \tag{15}$$

Further, it is easy to see that

$$\begin{aligned} &Lg'(y)Lf'(x) \\ &\quad = [f'_+(x)D^-g(y) \wedge f'_-(x)D_+g(y), f'_-(x)\ D^-g(y) \vee f'_+(x)D_+g(y)]. \end{aligned}$$

Hence it follows from (15) that (10) holds again.

Further, when $Lf'(x)$ is unique, it is clear from (15) that so is $Lh'(x)$; and if $Lg'(y) \ge 0$, then since $D^-g(y) = 0$, it follows from (15) and the last identity that (11) holds. □

When the inner function $f$ is convex, we obtain the following result from the above theorem in the case when $0 \in Sg'(y)$.

7.6.4. Corollary. *Suppose $X = I$, $f$ is convex, $0 \notin Lf'(x)$, and $g$ has a zero semigradient at $y$. Then $h$ has a semiderivative at $x$ similar to $Sg'(y)$ and* (10) *holds. Also, if $Lf'(x)$ and $Sg'(y)$ are unique, then so is $Sh'(x)$.*

*Further, if $Sg'(y) \geq 0$ or $\leq 0$ according to whether $Sg'(y)$ is a lower or upper derivative, then* (11) *holds.*

*Moreover, the same holds when $0 \in Lf'(x)$ provided that $Sg'(y) \subset \mathbb{R}$.*

Next, to obtain the chain rule for weak derivative, we need the following lemma.

7.6.5. Lemma. *Suppose $f$ is continuous at $x$. If $c \in M_+^* f(x)$ and $c > 0$, then $D_+h(x) \leq' cD^+g(y)$ and $cD_+g(y) \leq' D^+h(x)$.*

*Further, if $g$ is $D$-finite at $y$ and $0 \in M_+^* f(x)$, then $0 \in M_+^* h(x)$.*

*Moreover, the same holds on replacing $+$ everywhere by $-$.*

Proof. We will give the proof here for the first two parts; the proof of the last part is quite similar.

First, suppose $c \in M_+^* f(x)$ and $c > 0$. Then there is a sequence of points $\{x_n\}$ in $X$ which converges to $x$ from the right such that $f(x_n) > f(x) = y$ for each $n$ and

$$c = \lim_n Qf(x, x_n). \tag{16}$$

Hence, if the product $cD^+g(y)$ is not indeterminate, we obtain

$$\begin{aligned} D_+h(x) &\leq \liminf_{n\to\infty} Qh(x, x_n) \\ &= \liminf_{n\to\infty} \{Qg(y, f(x_n)) \cdot Qf(x, x_n)\} \leq cD^+g(y). \end{aligned}$$

The other inequality of the first part is obtained similarly.

Next, to prove the second part, suppose $g$ is $D$-finite at $y$ and $0 \in M_+^* f(x)$. Then there exist $M, \varepsilon \in \mathbb{R}_+$ such that $|Qg(y, z)| < M$ whenever $z \in f(X)$ and $0 < |z - y| < \varepsilon$. Also, since $f$ is continuous at $x$, there is a $\delta > 0$ such that $|f(t) - f(x)| < \varepsilon$ whenever $t \in X$ and $|t - x| < \delta$. Now, there is a decreasing sequence of points $\{x_n\}$ in $X \cap (x, x + \delta)$ such that (16) holds with $c = 0$. If $Z$ includes some subsequence $\{x_{n_i}\}$ of $\{x_n\}$, then since $h(x_{n_i}) = h(x)$ for each $i$, it is clear that $0 \in M_+^* h(x)$. Otherwise, we can assume that $x_n \notin Z$ for any $n$. Then for each $n$ we have

$$|Qh(x, x_n)| = |Qg(y, f(x_n)) \cdot Qf(x, x_n)| \leq M|Qf(x, x_n)|,$$

and so it follows from (16) with $c = 0$ that $0 \in M_+^* h(x)$. □

As we will see later in Remark 7.6.8(b), the chain rule does not hold in general for weak derivative. However, when the composite function is known to be weakly derivable at the point in question, the following conditional chain rule remains valid for weak derivative.

7.6.6. THEOREM. *Suppose $X = I$, $f$ is continuous, $f$ and $g$ have nonzero unique weak derivatives at $x$ and $y = f(x)$, respectively, and $h = g \circ f$ is weakly derivable at $x$. Then*

$$Wg'(y)Wf'(x) \in Wh'(x). \tag{17}$$

*Further, if $f$ and $g$ have either $-Wf'(x)$-compatible unilateral derivatives, or $-Wg'(y)$-compatible semiderivatives, at $x$ and $y$, respectively, then*

$$Wh'(x) = Wg'(y)Wf'(x). \tag{18}$$

*Also,* (18) *holds when either* (i) *$Wf'(x) = 0$ and $g$ is D-finite at $y$, or* (ii) *$f$ is D-finite at $x$, $g$ is continuous, and $Wg'(y) = 0$.*

PROOF. Let $c = Wg'(y)$ and $d = Wf'(x)$. First, to prove the first two parts, suppose $c \neq 0$ and $d \neq 0$. Then it is enough to prove the results when $c > 0$ and $d > 0$. On applying these results to $-f$ and $g_{-}$, the results are obtained when $c > 0$ and $d < 0$ (see Lemma 7.3.6); and then on applying these two results to $f$ and $-g$, the results are obtained when $c < 0$ and $d \neq 0$.

Hence, suppose $c > 0$ and $d > 0$. We need to consider here four distinct cases separately.

**Case (i):** $c = Sg'(y)$ and $d = Sf'(x)$. When these semiderivatives are similar, the result follows directly from Theorem 7.6.1.

To consider dissimilar semiderivatives, first suppose $c = Lg'(y)$ and $d = Uf'(x)$. Then $D^{-}g(y) = D_{+}g(y) = c$ and $D^{+}f(x) = D_{-}f(x) = d$. To obtain (18) in this case, it is enough to prove here that $cd \in M_b h(x)$, for then $Wh'(x)$ is unique and equal to $cd$ (see §4.2).

Since $D_{-}f(x)$ and $D^{-}g(y)$ are positive, by the last part of Lemma 7.5.1 we obtain

$$D_{-}h(x) = D_{-}h_P(x) \leq D^{-}g(y)D_{-}f_P(x) = D^{-}g(y)\ D_{-}f(x) = cd;$$

and since $D^{+}f(x)$ and $D_{+}g(y)$ are positive, by Lemma 7.5.1(b) we obtain

$$D^{+}h(x) \geq D^{+}h_P(x) \geq D_{+}g(y)D^{+}f_P(x) = D_{+}g(y)\ D^{+}f(x) = cd.$$

Further, since $x \in P^{-}$, by Lemma 7.5.5 we obtain $D^{-}h(x) \geq D^{-}g(y)D_{-}f(x) = cd$. It remains to show that $D_{+}h(x) \leq cd$, and this holds trivially when $c = +\infty$. When $c < +\infty$, since $D^{+}f(x) > 0$, it follows from Lemma 7.5.5(a) that $D_{+}h(x) \leq D_{+}g(y)D^{+}f(x) = cd$. Hence $cd \in M_b h(x)$.

In the other case of dissimilar semiderivatives, the result is obtained on applying the above result to $-f_-$ and $-g_-$.

**Case (ii):** $c = g'_u(y)$ and $d = f'_u(x)$. When these unilateral derivatives are symmetric, say right derivatives, we have $c = g'_+(y) > 0$ and $d = f'_+(x) > 0$, and so by Lemma 7.5.1(b) we obtain $cd = h'_+(x) \in Wh'(x)$.

To consider asymmetric unilateral derivatives, first suppose $c = g'_-(y)$ and $d = f'_+(x)$. Then since $f$ and $g$ are by hypothesis nonangular at $x$ and $y$, respectively, $d \in M_-f(x)$ and $c \in M_+g(y)$. Hence, when $d < +\infty$, from Lemma 7.5.6 we obtain

$$D_+h(x) = f'_+(x)D_+g(y) \leq dc \leq f'_+(x)D^+g(y) = D^+h(x);$$

and when $d = +\infty$, it is clear that $D_+h(x) \leq +\infty = cd$. Also, from Lemma 7.5.5 we obtain $D^+h(x) \geq D^+g(y)f'_+(x) = +\infty$, so that $D^+h(x) = +\infty = cd$. Hence in any case $cd \in M_+h(x)$. Further, it follows from the Darboux property of $f$ that $d \in M^*_-f(x)$, and hence from Lemma 7.6.5 we obtain $D_-h(x) \leq dg'_-(y) \leq D^-h(x)$. Thus $cd \in M_bh(x)$, so that (18) holds.

In the other case of asymmetric unilateral derivatives, the result follows on applying the above to $-f_-$ and $-g_-$.

**Case (iii):** $c = Sg'(y)$ and $d = f'_u(x)$. First, suppose $c = Lg'(y)$ and $d = f'_+(x)$. Then $d \in M_-f(x)$ as before, and $D^-g(y) = D_+g(y) = c$. Now since $d \in M^*_-f(x)$ as before, this time we obtain the following from Lemma 7.6.5: $D_-h(x) \leq dD^-g(y) = cd$. Further, when $d < +\infty$, from Lemma 7.5.6 we obtain $D_+h(x) = f'_+(x)D_+g(y) = cd$; and when $d = +\infty$, from Lemma 7.5.1(b) we obtain $D_+h(x) \geq f'_+(x)D_+g(y) = +\infty$. Hence, in any case, $D_+h(x) = cd$. Consequently,

$$cd = \max\{D_+h(x), D_-h(x)\} \in Wh'(x), \tag{19}$$

so that (17) holds.

A similar argument holds when $d = f'_-(x)$; and on applying these results to $-f_-$ and $-g_-$, the result is obtained when $c = Ug'(y)$.

**Case (iv):** $c = g'_u(y)$ and $d = Sf'(x)$. First, suppose $c = g'_+(y)$ and $d = Lf'(x)$. Then $c \in M_-g(y)$ and $D^-f(x) = D_+f(x) = d$. If $c < +\infty$, then since $D_+f(x)$ and $g'_+(y)$ are positive, by Lemma 7.5.2, $D_+h(x) = g'_+(y)D_+f(x) = cd$; and when $c = +\infty$, by Lemma 7.5.1(a), $D_+h(x) \geq g'_+(y)D_+f(x) = +\infty$. Hence, in any case, $D_+h(x) = cd$.

Further, to obtain (19) in this case, we need to show that $D_-h(x) \leq cd$. Since $D^-f(x) > 0$, there exists $x' \in X$, $x' < x$, such that $f(x') < f(x) = y$. Choose an increasing sequence of points $\{y_n\}$ in $J \equiv [f(x'), y]$ which converges to $y$ such that $Qg(y, y_n)$ converges to $D_-g(y)$ as $n \to \infty$. Let $x_n$ denote for each $n$ the supremum of the nonempty closed set $J \cap f^{-1}(y_n)$. Now it follows by an argument similar to the one used earlier for Lemma

7.5.5 that $\{x_n\}$ is an increasing sequence of points in $J$ which converges to $x$. Further,

$$D_-h(x) \le \liminf_{n\to\infty} Qh(x, x_n) = \liminf_{n\to\infty} \{Qg(y, y_n) \cdot Qf(x, x_n)\},$$

where $Qf(x, x_n) = (y_n - y)/(x_n - x) > 0$ for each $n$. Hence when $D_-g(y) < 0$, we have $D_-h(x) \le 0 < cd$; otherwise $D_-h(x) \le D_-g(y)D^-f(x) \le cd$. Thus (19) holds again, and hence (17) holds.

A similar argument holds when $c = g'_-(y)$; and on applying these results to $-f_-$ and $-g_-$, the result is obtained when $d = Uf'(x)$.

The first two parts are now established. To deal with the last part, first suppose $d = 0$ and $g$ is $D$-finite at $y$. Then since $f$ is continuous and $Wf'(x)$ is unique, it is clear that $0 \in M_b^* f(x)$. Hence, by Lemma 7.6.5, $0 \in M_b h(x)$. Consequently, $Wh'(x) = 0 = cd$, so that (18) holds.

Next, suppose $c = 0$, $f$ is $D$-finite at $x$, and $g$ is continuous. In this case, $0 \in M_b^* g(y)$. When $x \in Z^+$, it is clear that $D_+h(x) \le (h_Z)'_+(x) = 0 \le D^+h(x)$. Otherwise, $x \in P^+$ or $N^+$, and so it follows from Lemma 7.5.5 with $c = 0$ that $0 \in M_+h(x)$. It is proved similarly that $0 \in M_-h(x)$, and this implies (18) as before. □

It may be added here that (18) holds also when $h$ is nonangular at $x$, for then $Wh'(x)$ is unique, and so (17) becomes an equality. Further, on combining case (i) in the last part of the above theorem with Theorem 7.6.1, we obtain the following result which may be compared with Theorems 7.5.3 and 7.5.8.

7.6.7. COROLLARY. *Suppose $X = I$, $f$ is continuous, $x \in I^\circ$, $y = f(x)$, and $h = g \circ f$.*

(a) *If $f$ has a nonzero unique semiderivative at $x$ and $g$ has a nonzero derivative at $y$, then $h$ has a unique semiderivative at $x$, $g'(y)$-compatible with $Sf'(x)$, and the identity* (5) *of Theorem* 7.5.3 *holds.*

(b) *If $f$ has a nonzero derivative at $x$ and $g$ has a nonzero unique semiderivative at $y$, then $h$ has a unique semiderivative at $x$ similar to $Sg'(y)$ and the inclusion* (9) *of Theorem* 7.5.8 *holds.*

7.6.8. REMARK. We conclude this section with some examples to show the necessity of the hypotheses of the three theorems of this section, and that they are best possible in their direction.

(a) Firstly, the reverse inclusions of the inclusions in Theorems 7.6.1, 7.6.3, and 7.6.6 do not hold in general. First, let $f(x) = x$ or $|x|$ according to whether $x \in \mathbb{R}$ is rational or irrational; and $g(x) = |x|$ or $x$ according to whether $x \in f(\mathbb{R})$ is rational or irrational. Then $f$ is continuous at 0 and $y = f(0) = 0$. Clearly, each of $f$ and $g$ has a unique lower derivative equal to 1 at 0. However, if $h = g \circ f$, it is easy to see that $h(x) = |x|$, $x \in \mathbb{R}$, and so $Lh'(0) = [-1, 1]$. Hence the inclusions (1) and (17) of Theorems 7.6.1 and 7.6.6 cannot be reversed in general. This example shows further that $Sh'(x)$ need not be unique when $Sf'(x)$ and $Sg'(y)$ are unique, and even finite.

Next, the second inclusion of (10) in Theorem 7.6.3 cannot be reversed in general even when $f$ and $g$ are convex functions. For example, let $f(x) = x$ or $2x$ according to whether $x \in \mathbb{R}$ is $< 0$ or $\geq 0$; and $g(x) = |x|$, $x \in \mathbb{R}$. Then $f$ and $g$ are convex, $y = f(0) = 0$, $Lf'(0) = [1, 2]$, and $Lg'(0) = [-1, 1]$, so that $Lf'(0)Lg'(0) = [-2, 2]$. However, if $h = g \circ f$, then it is easy to see that $Lh'(0) = [-1, 2]$, so that the reverse inclusion of (10) does not hold in this case.

(b) Next, when the semiderivatives $Sf'(x)$ and $Sg'(y)$ lack the compatibility mentioned in Theorem 7.6.1, the composite function $h$ is not always semiderivable at $x$, or even weakly derivable at $x$. For example, let $f(x) = 2x$ or $x$ according to whether $x \in \mathbb{R}$ is $< 0$ or $\geq 0$; and $g(x) = x + \sqrt{|x|}\sin^2(1/x)$, $g(0) = 0$. Then $f$ and $g$ are continuous functions, $y = f(0) = 0$, and $f$ and $g$ are upper and lower derivable respectively at 0, where $Uf'(0) = [1, 2]$ and $Lg'(0) = 1$. However, if $h = g \circ f$, then

$$h(x) = 2x + \sqrt{|2x|}\sin^2(1/2x), \quad 0 \quad \text{or} \quad x + \sqrt{|x|}\sin^2(1/x)$$

according to whether $x < 0$, $= 0$ or $> 0$ respectively; hence it is easy to see that $h$ is upper knotted at 0, $D^-h(0) = 2$ and $D_+h(0) = 1$, so that $h$ is not even weakly derivable at 0.

(c) Finally, it is clear from the example (b) of Remark 7.5.9 that Theorems 7.6.1 and 7.6.3 do not hold in general when $0 \in Sf'(x)$, and Theorem 7.6.6 does not hold in turn when $Wf'(x) = 0$. Similarly, it is clear from example (a) of that remark that Theorem 7.6.1 does not hold in general when $0 \in Sg'(y)$, and Theorem 7.6.6 does not hold in turn when $Wg'(y) = 0$.

## §7.7 Calculus of semiderivatives on normed vector spaces.

In this section we obtain the results of the earlier sections of this chapter on semiderivatives for functions defined on a normed vector space. Due to the variation in the definitions of semiderivatives in the two cases, independent proofs are needed in this case. However, the results will be seen to assume a simpler and somewhat more elegant form in this setting.

To avoid hard analysis in terms of $\varepsilon$ and $\delta$, we will use here the Definition 4 of lower and upper gradients presented in §4.1. Also, we will use the notions of similarity and compatibility of semiderivatives defined in §7.1.

We will assume throughout this section, unless stated otherwise, that $X$ is a real normed vector space, $f, g : X \to \mathbb{R}$, and $x \in X$. As we saw in the earlier sections, the results of this section can also be obtained for functions defined on a subset of $X$. However, for the sake of simplicity, we will assume $f$ and $g$ to be defined on the whole of $X$.

The following version of Theorem 7.1.1 follows directly from the definitions of lower and upper derivatives in the present case (see §4.1).

7.7.1. THEOREM. *Suppose $f : X \to \mathbb{R}$, $E \subset X$ and $x \in E \cap E^d$. If $f$ is lower or upper derivable at $x$, then so is $f_E$ and*

$$Sf'(x) \subset S(f_E)'(x).$$

7.7.2. THEOREM. *If $f$ and $g$ are lower derivable at $x$, then so is $f + g$ and*

$$L(f+g)'(x) \supset Lf'(x) + Lg'(x). \tag{1}$$

*Further, if $f$ or $g$ is derivable at $x$, then* (1) *becomes an identity.*

PROOF. Suppose $f$ and $g$ are lower derivable at $x$, and let $x' \in Lf'(x)$ and $y' \in Lg'(x)$. Then

$$f(y) \geq f(x) + (x', y - x) + o(\|y - x\|),$$
$$g(y) \geq g(x) + (y', y - x) + o(\|y - x\|).$$

Consequently, we have

$$f(y) + g(y) \geq f(x) + g(x) + (x' + y', y - x) + o(\|y - x\|).$$

This proves that $x' + y'$ is a lower gradient of $f + g$ at $x$, and the first part is thus established.

Now, to obtain the second part, suppose for the sake of definiteness that $f$ is derivable at $x$. Then since $f + g$ and $-f$ are lower derivable at $x$ by the first part, it follows from (1) that

$$Lg'(x) \supset L(f + g)'(x) - f'(x),$$

which yields the reverse inclusion of (1). □

The above theorem clearly yields a similar result on upper derivative. With the help of that result we obtain the following as in Corollary 7.3.3.

7.7.3. COROLLARY. *If $f$ and $g$ are lower derivable at $x$ and $f + g$ is upper derivable at $x$, then $f$ and $g$ are derivable at $x$.*

In the following lemma we deal with scalar products of functions.

7.7.4. LEMMA. *Let $c \in \mathbb{R}$. If $f$ is semiderivable at $x$, then so is $cf$; and if so, then*

$$S(cf)'(x) = cSf'(x), \tag{2}$$

*where $S(cf)'(x)$ is $c$-compatible with $Sf'(x)$.*

*Consequently, if $f$ is lower or upper derivable at $x$, then $-f$ is upper or lower derivable, respectively, at $x$ and $S(-f)'(x) = -Sf'(x)$.*

PROOF. It is enough to consider here the case when $f$ is lower derivable at $x$; on applying this result to $-f$ the result is obtained when $f$ is upper derivable at $x$. Further, the result holds trivially when $c = 0$. We will prove the result when $c > 0$; when $c < 0$, the result is obtained again on applying this result to $-f$ and $-c$.

Hence, suppose $f$ is lower derivable at $x$ and $c > 0$. Let $x' \in Lf'(x)$. Then

$$f(y) \geq f(x) + (x', y - x) + o(\|y - x\|).$$

Hence we obtain

$$cf(y) \geq cf(x) + (cx', y - x) + o(\|y - x\|),$$

which proves that $cx'$ is a lower gradient of $cf$ at $x$. This proves the inclusion $\supset$ in (2). The reverse inclusion follows on the other hand on applying this result to $cf$ and $1/c$.

The last part follows directly from above on choosing $c = -1$. □

The following result on linear combinations of functions follows directly on combining the above lemma with Theorem 7.7.2 and its similar version on upper derivative.

7.7.5. THEOREM. *Let $h = af + bg$, where $a, b \in \mathbb{R}$ and $a \neq 0$. Suppose $f$ and $g$ have ab-compatible semiderivatives at $x$. Then $h$ has a semiderivative at $x$, a-compatible with $Sf'(x)$, and*

$$Sh'(x) \supset aSf'(x) + bSg'(x). \tag{3}$$

*Further, if $f$ or $g$ is derivable at $x$, then* (3) *becomes an identity.*

The following theorem deals with the product rule for semiderivative.

7.7.6. THEOREM. *Suppose $f$ is continuous at $x$, $f(x) \neq 0$, and $f$ and $g$ have $f(x)g(x)$-compatible semiderivatives at $x$. Then $fg$ has a semiderivative at $x$, $f(x)$-compatible with $Sg'(x)$, and*

$$S(fg)'(x) \supset f(x)Sg'(x) + g(x)Sf'(x). \tag{4}$$

*Further, in case $f$ or $g$ is derivable at $x$, then* (4) *becomes an identity.*

*Moreover, when $f(x) = 0$, the same holds with identity in* (4) *provided that $f$ is derivable at $x$ and $g$ is continuous at $x$.*

PROOF. To prove the first two parts, it is enough to consider the case when $f(x) > 0$. When $f(x) < 0$, the result follows on applying this result to $-f$ and $g$. Further, it is enough to consider the case when $g(x) \geq 0$ and $f$ and $g$ have similar semiderivatives at $x$; on applying this result to $f$ and $-g$ the result is obtained when $g(x) \leq 0$ and $f$ and $g$ have dissimilar semiderivatives at $x$.

Hence, suppose $f(x) > 0$ and $g(x) \geq 0$ and $f$ and $g$ have similar semiderivatives at $x$. Then since $f$ is continuous at $x$, given $\varepsilon > 0$, there is a $\delta > 0$ such that $f(y) > 0$ and $|f(y) - f(x)| < \varepsilon$ whenever $\|y - x\| < \delta$.

First, suppose $f$ and $g$ are lower derivable at $x$. Let $x' \in Lf'(x)$ and $y' \in Lg'(x)$, and set $z' = f(x)y' + g(x)x'$. We need to show here that $z' \in L(fg)'(x)$.

In this case we have

$$f(y) - f(x) \geq (x', y - x) + o(\|y - x\|),$$
$$g(y) - g(x) \geq (y', y - x) + o(\|y - x\|).$$

Hence when $\|y - x\| < \delta$, we obtain

$$\begin{aligned}
&f(y)g(y) - f(x)g(x) \\
&\quad = f(y)\{g(y) - g(x)\} + g(x)\{f(y) - f(x)\} \\
&\quad \geq f(y)\{(y', y - x) + o(\|y - x\|)\} + g(x)\{(x', y - x) + o(\|y - x\|)\} \\
&\quad = (z', y - x) + \{f(y) - f(x)\}\{(y', y - x) + o(\|y - x\|)\} \\
&\qquad + g(x)\, o(\|y - x\|) \\
&\quad \geq (z', y - x) - \varepsilon\|y'\|\,\|y - x\| - \varepsilon\, o(\|y - x\|) + g(x)\, o(\|y - x\|).
\end{aligned}$$

Now since $\varepsilon \to 0$ as $\|y - x\| \to 0$, this proves that $f(y)g(y) \geq (z', y - x) + o(\|y - x\|)$. Hence $fg$ is lower derivable at $x$ and $z' \in L(fg)'(x)$.

In the case when $f$ and $g$ are upper derivable at $x$, the result is obtained by a similar argument. The first part is thus established.

To prove the second part, first suppose $f$ is derivable at $x$. Then $1/f$ is also derivable at $x$, and so on applying the first part to $1/f$ and $fg$ we obtain

$$\begin{aligned}
Sg'(x) &\supset \{1/f(x)\}S(fg)'(x) + f(x)g(x)(1/f)'(x) \\
&= \{1/f(x)\}S(fg)'(x) - \{g(x)/f(x)\}f'(x).
\end{aligned}$$

The reverse inclusion of (4) follows clearly from this inclusion.

When $g$ is derivable at $x$ and $g(x) \neq 0$, the result is obtained by a similar argument. Hence suppose $g$ is derivable at $x$ and $g(x) = 0$. Let $x' \in S(fg)'(x)$, and set $y' = x'/f(x)$. Then it is enough to show that $y' = g'(x)$. Suppose, to be definite, that $x' \in L(fg)'(x)$. Then since $g(x) = 0$, we have

$$f(y)g(y) \geq (x', y - x) + o(\|y - x\|).$$

Now, since the function $f(y)/f(x)$ of $y$ is continuous at $x$, given $\varepsilon > 0$, there is a $\delta > 0$ such that $|f(y)/f(x) - 1| < \varepsilon \wedge 1$ for $\|y - x\| < \delta$. Hence for such $y$, $(1 - \varepsilon)f(x) < f(y) < (1 + \varepsilon)f(x)$.

Now, when $g(y) \geq 0$, from the last inequality we obtain

$$(1 + \varepsilon)f(x)g(y) \geq f(y)g(y) \geq (x', y - x) + o(\|y - x\|),$$

so that

$$\begin{aligned}
g(y) &\geq (1 + \varepsilon)^{-1}(x'/f(x), y - x) + \{(1 + \varepsilon)f(x)\}^{-1}o(\|y - x\|) \\
&= (y', y - x) - \{1 - (1 + \varepsilon)^{-1}\}(y', y - x) + o(\|y - x\|) \\
&\geq (y', y - x) - \varepsilon\|y'\|\,\|y - x\| + o(\|y - x\|).
\end{aligned}$$

When $g(y) < 0$, we have $(1 - \varepsilon)f(x)g(y) > f(y)g(y)$, and so we obtain the same by a similar argument with $\varepsilon$ in the place of $-\varepsilon$. Now since $\varepsilon \to 0$ as $\|y - x\| \to 0$, we thus obtain in any case $g(y) \geq (y', y - x) + o(\|y - x\|)$, which proves that $y' \in Lg'(x)$. Consequently, $y' = g'(x)$.

The second part is thus established.

Finally, to prove the last part, suppose $f$ is derivable at $x$, $g$ is continuous at $x$, and $f(x) = 0$. In case $g(x) \neq 0$, the result follows from the first two parts on inverting the roles of $f$ and $g$. Hence, suppose $g(x) = 0$, and let $x' = f'(x)$. Then since $g$ is continuous at $x$, and $f(x) = g(x) = 0$, we have

$$\lim_{\substack{y \to x \\ y \neq x}} \frac{|f(y)g(y)|}{\|y - x\|} = \lim_{\substack{y \to x \\ y \neq x}} \frac{|f(y)|}{\|y - x\|} \cdot \lim_{y \to x} |g(y)| \leq \|x'\| \cdot 0 = 0.$$

Now since $(0, y - x) = 0$ for every $y \in X$, it is thus clear that the zero element of $X'$ is the derivative of $fg$ at $x$. Both parts follow now from this fact in the present case. □

Next, to obtain the quotient rule for semiderivative, we first deal with the function $1/f$ in the following lemma.

7.7.7. LEMMA. *Suppose $f$ is continuous at $x$ and $f(x) \neq 0$. Then $1/f$ is semiderivable at $x$ iff $f$ is so; and if so, then the two semiderivatives at $x$ are dissimilar and*

$$S(1/f)'(x) = -\{1/f(x)\}^2 Sf'(x). \tag{5}$$

PROOF. It is enough to prove here the sufficiency part, for the necessity follows from it on applying it to $1/f$. Further, it is enough to consider the case when $f(x) > 0$; when $f(x) < 0$ the result is obtained on applying this result to $-f$.

Set $g = 1/f$, $c = \{f(x)\}^{-2}$, and $y' = -cx'$. Since $f$ is continuous at $x$, the function $h(y) = f(x)/f(y)$ is defined in a neighborhood of $x$ and is continuous at $x$. Hence, given $\varepsilon > 0$, we can choose a $\delta > 0$ such that $|h(y) - 1| < \varepsilon \wedge 1$ whenever $\|y - x\| < \delta$.

First, suppose $f$ is lower derivable at $x$, and let $x' \in Lf'(x)$. We need to show here that $-cx' \in Ug'(x)$. In this case we have

$$f(y) - f(x) \geq (x', y - x) + o(\|y - x\|).$$

Hence when $\|y - x\| < \delta$, then since $ch(y) > 0$, we obtain

$$\begin{aligned}
g(y) - g(x) &= 1/f(y) - 1/f(x) = -\{f(y) - f(x)\}/\{f(y)f(x)\} \\
&= -ch(y)\{f(y) - f(x)\} \\
&\le -ch(y)(x', y - x) - ch(y)\, o(\|y - x\|) \\
&= -c(x', y - x) + \{1 - h(y)\}c(x', y - x) - ch(y)\, o(\|y - x\|) \\
&\le (-cx', y - x) + |1 - h(y)|\, c\|x'\|\, \|y - x\| + c|h(y)|\, o(\|y - x\|) \\
&\le (-cx', y - x) + \varepsilon c\|x'\|\, \|y - x\| + 2c \cdot o(\|y - x\|).
\end{aligned}$$

Now, since $\varepsilon \to 0$ as $\|y - x\| \to 0$, this proves that $g(y) - g(x) \le (-cx', y - x) + o(\|y - x\|)$. Hence $-cx' \in Ug'(x)$, which proves the inclusion $\supset$ of (5). When $f$ is upper derivable at $x$, this inclusion of (5) is obtained by a similar argument.

The inclusion $\subset$ of (5) is obtained on the other hand by applying the above result to $1/f$. □

Now, on combining the above lemma with Theorem 7.7.6 we obtain the following theorem on quotients.

7.7.8. THEOREM. *Suppose g is continuous at x,* $g(x) \neq 0$, *and f and g have* $-f(x)g(x)$*-compatible semiderivatives at x. Then f/g has a semiderivative at x,* $g(x)$*-compatible with* $Sf'(x)$, *and*

$$(6) \qquad S(f/g)'(x) \supset \{1/g(x)\}^2\{g(x)Sf'(x) - f(x)Sg'(x)\}.$$

*Further, if f or g is derivable at x, then* (6) *becomes an identity.*

Finally, in the following theorem we obtain a chain rule similar to Theorem 7.5.8 for lower derivative in the present setting. A similar result holds of course for upper derivative.

7.7.9. THEOREM. *Suppose* $f: X \to Y$, *where X and Y are normed vector spaces, and* $g: f(X) \to \mathbb{R}$. *If f is derivable at* $x \in X$ *and g is lower derivable at* $f(x)$, *then* $g \circ f$ *is lower derivable at x and*

$$(7) \qquad L(g \circ f)'(x) \supset Lg'(f(x)) \circ f'(x).$$

*Further, if f is one-to-one and its inverse is derivable at* $f(x)$, *then* (7) *becomes an identity.*

PROOF. Suppose $f$ is derivable at $x_0 \in X$ and $g$ is lower derivable at $y_0 = f(x_0)$, and let $h = g \circ f$. Let $\varphi = f'(x_0)$, which is in this case a continuous linear mapping from $X$ to $Y$, and let $y' \in Lg'(y_0)$. Then

$$(8) \qquad f(x) - f(x_0) = (\varphi, x - x_0) + \psi(x - x_0),$$

where $\psi(x - x_0) = o(\|x - x_0\|)$. Further, given $\varepsilon > 0$, there is a $\delta_0 > 0$ such that whenever $0 < \|y - y_0\| < \delta_0$, we obtain

$$(9) \qquad g(y) - g(y_0) > (y', y - y_0) - \varepsilon\|y - y_0\|.$$

Also, since $f$ is obviously continuous at $x_0$, we can find a $\delta > 0$ such that

$$(10) \qquad \|f(x) - f(x_0)\| < \delta_0 \quad \text{when} \quad \|x - x_0\| < \delta.$$

Hence when $\|x - x_0\| < \delta$, with the help of the above three inequalities we obtain

$$\begin{aligned} h(x) &- h(x_0) \\ &= g(f(x)) - g(f(x_0)) \\ &> (y', f(x) - f(x_0)) - \varepsilon\|f(x) - f(x_0)\| \\ &\geq (y', \varphi(x - x_0)) + (y', \psi(x - x_0)) - \varepsilon\|\varphi(x - x_0)\| - \varepsilon\|\psi(x - x_0)\| \\ &\geq (y' \circ \varphi, x - x_0) - \|y'\|\,\|\psi(x - x_0)\| - \varepsilon\|\varphi\|\,\|x - x_0\| - \varepsilon\|\psi(x - x_0)\|. \end{aligned}$$

It is further clear from (9) and (10) that $\varepsilon \to 0$ as $\|x - x_0\| \to 0$. Hence we obtain $h(x) - h(x_0) \geq (y' \circ \varphi, x - x_0) - o(\|x - x_0\|)$, which proves that $h$ is lower derivable at $x_0$ and that $y' \circ \varphi \in Lh'(x_0)$. The inclusion (7) is thus established.

Next, to prove the second part, suppose that $f$ is one-to-one and that $f^{-1}$ is derivable at $y_0$. Then since $h = g \circ f$, it is clear that $g = h \circ f^{-1}$. Hence if $h$ is lower derivable at $x_0$, it follows from the first part that $g$ is lower derivable at $y_0$ and

$$Lg'(y_0) \supset Lh'(x_0) \circ (f^{-1})'(y_0) = Lh'(x_0) \circ \{f'(x)\}^{-1}.$$

The reverse inclusion of (7) follows clearly from this inclusion. □

7.7.10. Remark. To see that the results of this section are the best possible (e.g., the inclusions of these results cannot be replaced by identity in general), the examples presented in the earlier sections remain applicable in the present setting (see Remarks 7.2.2, 7.3.12, 7.4.11 and 7.6.8).

7.7.11. Problem. In the particular case when $X \subset \mathbb{R}$, we have been able to obtain four chain rules for semiderivatives as presented in Theorems 7.5.3, 7.5.8, 7.6.1, and 7.6.3. Out of these, the second theorem was extended in Theorem 7.7.9 to a general normed vector space $X$. The other three theorems do not seem to hold in general for such a space $X$. It will be interesting to obtain conditions under which these three theorems become valid for a general normed vector space $X$.

## §7.8 Developments in other fields related with new derivatives.

After the announcement of a good part of the results of Parts II and III in [G12] and [G13], it is heartening to note that the notions of lower and upper derivatives have been used by some authors on Banach spaces, with different terminologies, to tackle some of the problems of their fields.

**A. Use of lower and upper derivatives by others.** First, in connection with an evolution problem (e.g., the heat equation), these notions were considered by Giorgi, Marino, and Tosque [GMT] using the Definition 3 of §4.1. They used in fact the usual canonical extensions of the lower and upper derivatives to the domain of the function and called them *subdifferential* and *superdifferential*, respectively. Let $X$ be a Banach space. Given a function $f: X \to \mathbb{R}$ and $x \in X$, they used the notations $\partial^- f(x)$ and $\partial^+ f(x)$ to denote the subdifferential and superdifferential, respectively, of $f$ at $x$. They defined $\partial^- f(x)$ to be $Lf'(x)$ when $f$ is lower derivable at $x$, and otherwise they defined $\partial^- f(x) = \emptyset$. Thus $\partial^- f$ is the canonical extension of $Lf'$ to $X$, and $f$ is lower derivable at $x$ iff $\partial^- f(x) \neq \emptyset$. The superdifferential $\partial^+ f(x)$ is defined similarly in terms of upper derivative of $f$.

The above two differentials were used later successfully by Crandall and Lions (see [CL], [CEL], and [Li]) to obtain "viscosity solutions" of Hamilton-Jacobi equations. See further Deville, Godefroy and Zizler [DGZ, pp. 345–350] for some related results in infinite dimensions.

Later, Mirică, Staicu, and Angelescu [MSA] called these two differentials "Fréchet subdifferential" and "Fréchet superdifferential", respectively; and they used the term "Fréchet semidifferential" for both of these two differentials. The first two of these terms were also used subsequently by Deville, Godefroy and Zizler in [DGZ, p. 339]. Next, in a book of Aubin and Frankowska on set-valued analysis [AF, p. 220], they called these two differentials "local subdifferential" and "local superdifferential", respectively. Further, in a book of Rockafellar and Wets on variational analysis [RW], they used the Definition 4 of §4.1 and called these two differentials "regular subgradient" and "regular supergradient," respectively. Note that we have used the term "gradient" in the present work for various elements of set-valued derivatives, whereas they use this term for the derivative itself.

Apart from the use of different terminologies by these authors, they have also used totally different notations for the subdifferential of a function $f$ at a point $x$—namely, $\partial^- f(x)$ in [GMT], $\partial_F^- f(x)$ in [MSA], $\partial_- f(x)$ in [AF], and $\hat{\partial} f(x)$ in [RW]. Similar notations have been used by them for the superdifferential of $f$ at $x$ which are again totally different.

**B Relations with other subdifferentials used in nonsmooth analysis.*** We include here some other known notions of subdifferentials used in nonsmooth analysis which are related with the above subdifferential. The corresponding superdifferentials are related similarly with the above superdifferential.

*We take this opportunity to express our gratitude to Professor Jonathan Borwein and his student Wang Xianfu for bringing to our attention some of the above-mentioned developments in nonsmooth analysis which are related to new derivatives.

*(a) Ioffe's Dini subdifferential.* Ioffe defined in [I1] the *lower Dini derivative* of $f$ at $x$ *in the direction* $y$ ($\in X$) to be the following:

$$d^- f(x;y) = \liminf_{\substack{t\to 0^+ \\ u\to y}} t^{-1}\{f(x+tu) - f(x)\};$$

and in terms of this notion he defined the *Dini subdifferential* of $f$ at $x$ to be the following:

$$\partial^- f(x) = \{x' \in X' : (x', y) \le d^- f(x;y) \text{ for every } y\}.$$

In the case when $X = \mathbb{R}^n$, it was proved later by Aubin and Frankowska in [AF, p. 253] that this Dini subdifferential is equal to the above considered subdifferential of [GMT] for which the same notation was used by them.

*(b) Ioffe's approximate subdifferential.* Let $X = \mathbb{R}^n$. In this case Ioffe defined further in [I2] the notion of *approximate subdifferential* of $f$ at $x$, denoted by $\partial_a f(x)$, to be in a sense the upper limit of $\partial^- f(y)$ as $y \to x$ and $f(y) \to f(x)$. Thus it is clear that $\partial^- f(x) \subset \partial_a f(x)$.

*(c) Clarke's generalized gradient.* Let $X = \mathbb{R}^n$. Suppose $f: X \to \mathbb{R}$ is Lipschitz near $x$, and $y \in X$. Then Clarke [Cl] defined the *generalized directional derivative* of $f$ at $x$ *in the direction* $y$ to be the following:

$$f^0(x;y) = \limsup_{\substack{u\to x \\ t\to 0^+}} t^{-1}\{f(u+ty) - f(u)\};$$

and in terms of this notion the *Clarke's generalized gradient* of $f$ at $x$ is defined to be the following:

$$\partial_c f(x) = \{x' \in X' : (x', y) \le f^0(x;y) \text{ for every } y\}.$$

It is clear from the above definition that $\partial_c f$ is in fact a generalization of strong derivative defined in §2.5, and so its *USC* and the nicer properties of its calculus seem to come from this fact (see for example Chapter 14).

Ioffe obtained the following relations between $\partial_a f$ and $\partial_c f$ in [I2, Theorem 2]: If $f$ is *LSC*, then $\partial_a f(x) \subset \partial_c f(x)$; and if $\partial_c f(x) \ne \varnothing$, then so is $\partial_a f(x)$. Also, if $f$ is further Lipschitz near $x$, then $\partial_c f(x) = \overline{\text{co}}\,\partial_a f(x)$. Thus in this particular case, $\partial_c f$ is also indirectly related with lower derivative of $f$.

*(d) The #-subdifferential of Borwein and Preiss.* Given a Banach space $X$, let # be a given *bornology* on $X$, namely, a nonempty family of bounded sets in $X$. Then Borwein and Preiss [BP] defined a function $f: X \to \mathbb{R}$ to be

*#-subdifferentiable* at $x \in X$ if there is an $x' \in X'$ such that for each set $S$ in the #-bornology the Definition 2 of §4.1 holds for $y \in S$. Clearly, this notion provides a generalization of lower derivability.

**C. Extension of first part of the First Derivability Theorem.** In the case when $X = \mathbb{R}^n$, since the subdifferential $\partial^- f$ of [GMT] coincides with the Dini subdifferential of Ioffe as mentioned above, it follows from Proposition 2 of Ioffe [I1, p. 519] that the first part of the First Derivability Theorem 5.1.2 remains valid for *LSC* functions on $X$. This extension has also been obtained by Mirică, Staicu, and Angelescu [MSA] in their Proposition 4.6 in the case when $f$ is continuous.

Further, given a bornology # on a Banach space $X$, it has been proved in this direction by Borwein and Preiss [BP, p. 521, Theorem 3.1] that if $X$ has a #-smooth renorm and $f: X \to \mathbb{R}$ is *LSC*, then $f$ is #-subdifferentiable at a dense set of points in its graph. (See Preiss [Pr2] for a related result on the differentiability of locally Lipschitz functions on Banach spaces.)

**D. Calculus of lower and upper derivatives.** In the case when $X = \mathbb{R}^n$, using some of the characterizations of lower and upper derivatives obtained by Mirică, Staicu, and Angelescu in [MSA], these authors obtained parts of Theorem 7.7.2, Lemma 7.7.4, and Theorems 7.7.6 and 7.7.9 in terms of subdifferentials. However, their proofs do not extend to a general normed vector space $X$. As for Theorem 7.7.6, they obtained this result for continuous functions $f$ and $g$ on $\mathbb{R}^n$ when they are upper derivable at $x$ and $f(x) > 0$ and $g(x) > 0$; they somehow used the inequality $\geq$ in the place of $>$ due to an oversight. See Chapter 10 of [RW] for some variants of Theorems 7.7.2 and 7.7.9 and for a detailed study of the calculus of various other subdifferentials.

However, in the case of optimization theory it is the opposite inclusions of the inclusions (1) and (7) in Theorems 7.7.2 and 7.7.9 which are found more useful. In [I1] Ioffe developed a calculus for Dini subdifferentials in that direction which he described as a "fuzzy calculus." The rules of this calculus are more in the direction of existence of elements in every neighborhood of the point in question for which the desired relation holds. The notion of approximate subdifferentials was developed by him in [I2] for that very reason where he obtained the actual opposite inclusions for these subdifferentials.

7.8.1. Remark. It may be mentioned here that since nonsmooth analysts use frequently what is called directional Dini derivates, they may find the theory of unilateral derivates presented here in Part I useful. Further, the results of the theory of new derivatives presented in Parts II and III in the case of functions defined on the real line are also in a sense in the same direction, and so may be found useful.

7.8.2. PROBLEM. (a) First, in the theory of distribution developed by L. Schwartz, every continuous function has a derivative in the distributional sense. But such a derivative is not defined at any particular point. It will thus be interesting to investigate if there is any relation between the new derivatives presented here and the derivative in the distributional sense.

(b) As mentioned above, the first part of the First Derivability Theorem 5.1.2 holds also for any finite-dimensional space $X$. It will be worthwhile to investigate conditions under which this part holds for an infinite dimensional Banach space $X$.

(c) The dense set of points in (b) where $f$ is lower differentiable may be only countable as seen in Remark 5.1.4 in the case of the Weierstrass' nonderivable function. Hence it should be investigated whether the notion of nonangularity can be extended to functions on $\mathbb{R}^n$, and whether the second part of Theorem 5.1.2 can also be extended in this case to obtain a $c$-dense set of points where $f$ is lower differentiable.

(d) It is indeed the Third Derivability Theorem 6.2.2 which plays a more basic role in the development of the present theory of new derivatives. Hence its extension to $\mathbb{R}^n$ should be explored.

# *PART III*

# THEORY OF NEW DERIVATIVES (CONTINUED)

**Introduction to Part III.** In this part we deal with the remaining aspects of the theory of new derivatives. Namely, we obtain mean value theorems and related results on new derivatives, along with several monotonicity theorems and other properties of functions in terms of new derivatives. Further, we obtain various properties of new derivatives and define and investigate various Denjoy and Perron integrals corresponding to new derivatives.

First, in Chapter 8, we obtain a General Mean Value Theorem which holds in terms of semiderivative without any derivability hypothesis, and a similar Second Mean Value Theorem is obtained in terms of lower derivative for upper nonangular functions. Also, in the case of a function that is lower derivable everywhere, the Third Mean Value Theorem is obtained under a weaker continuity hypothesis.

Further, we deduce from the first two mean value theorems a General Cauchy's Mean Value Theorem and a General l'Hôpital's Rule in terms of lower and semi- derivatives without any derivability hypothesis on the function in the numerator. Also, Taylor's formulae are obtained for lower and weak derivatives; and results on term-by-term differentiation of sequences and series of functions are obtained for unilateral, strong, semi- and weak derivatives.

Next, in Chapter 9, we first obtain three monotonicity theorems in terms of new derivatives which hold without any derivability hypothesis. We deduce from one of these theorems and the Second Mean Value Theorem generalized versions of the well-known first and second derivative tests for points of extrema, as well as the concavity test, in terms of new derivatives without any derivability hypothesis at points other than the point in question.

Also, some inequalities similar to the ones obtained by Saks for functions with property $T_2$ in terms of ordinary derivative, are obtained here for arbitrary functions in terms of lower and upper derivatives. Further, we obtain in this chapter five median theorems: Three of them deal with the properties of various medians at nonmonotonicity points (i.e., points in no neighborhood of which the function is nondecreasing), and two of them deal with monotonicity conditions in terms of medians. From the last two median theorems we deduce some monotonicity theorems for functions that are lower or weakly derivable n.e., which in turn include the Goldowski-Tonelli theorems for various new derivatives.

Next, in Chapter 10, we deduce from the Fifth Derivate Theorem results on the Borel class and measurability of various new derivability sets. Also, we obtain the Zahorski property of these sets which was obtained originally by Zahorski for ordinary derivability sets. Further, we investigate here the Baire class and measurability of various new derivatives and medians as multifunctions.

Further, we obtain the Darboux property of new derivatives without any derivability hypothesis. Also, the Denjoy property of ordinary and lower derivatives is obtained for nonangular functions that are lower derivable n.e.; and this property of ordinary, semi- and weak derivatives is obtained for nonangular functions that are weakly derivable n.e. Also, characterizations of stationary sets of various new derivatives are obtained.

Finally, in Chapter 11, we deal with the problem of reconstruction of a primitive function from its various new derivatives. For this purpose we define lower, upper, and weak Perron integrals which provide a solution of this problem for lower, upper, and weak derivatives, respectively. We obtain here various results on these new Perron integrals similar to the ones known for ordinary Perron integral. Further, we present in this chapter descriptive definitions of these new Perron integrals, and we prove theorems similar to the Alexandroff-Looman and Hake's theorems to obtain the equivalence of the two integrals in each case.

## *CHAPTER 8*

# Mean Value Theorems and Related Results in Terms of New Derivatives

In this chapter we obtain some mean value theorems along with l'Hôpital's rule in terms of new derivatives. Also, Taylor's formulae are obtained for new derivatives. Further, theorems are obtained for term-by-term differentiation in terms of unilateral, strong, and new derivatives.

First, in §8.1, we obtain the three mean value theorems mentioned in the introduction of Part III in terms of new derivatives. Then, in §8.2, we deduce from the first two mean value theorems a General Cauchy's Mean Value Theorem and a General l'Hôpital's Rule in terms of lower and semi-derivatives without any derivability hypothesis on the function in the numerator.

Next, in §8.3, we deduce from the above mean value theorems some characterizations of strong derivative in terms of limits of lower and semi-derivatives. Then, in §8.4, from the results of §§8.1 and 8.3 we deduce Taylor's formulae for lower and weak derivatives.

Finally, in §8.5, we obtain results on term-by-term differentiation of sequences and series of functions for unilateral and strong derivatives, and in §8.6 similar results are obtained for weak and semi- derivatives.

It may be pointed out here that according to our convention it will be understood in §§8.1 to 8.3 that $f: I \to \mathbb{R}$, where $I$ is a subinterval of $\mathbb{R}$.

**§8.1 Mean value theorem in terms of new derivatives.** In this section we obtain three mean value theorems in terms of new derivatives.

First, in Theorem 8.1.2, we obtain a general mean value theorem in terms of semiderivative which holds without any derivability hypothesis. Clearly, this theorem holds also in terms of weak derivative. Next, in the case of upper nonangular functions we obtain, in Theorem 8.1.4, a mean value theorem in terms of lower derivative which also holds without any derivability hypothesis. Finally, in the case of a function that is lower derivable everywhere, we obtain, in Theorem 8.1.5, a mean value theorem in terms of lower derivative under a weaker continuity hypothesis.

We call these three theorems the General, Second, and Third Mean Value Theorems, respectively.

We begin with the general mean value theorem in terms of semiderivative; the term "general" is used here since it holds without any derivability hypothesis. This somewhat unexpected fact does have some deeper implications. As we will see later in Chapters 16 and 17, for most of the known generalizations of derivative the mean value theorem follows directly from this theorem, although generalized derivatives are not comparable with semiderivative in general.

Let us recall here that we call a function $f: I \to \mathbb{R}$, $I^*C$ if it is simultaneously $WC$ and inherently continuous (see §§1.2 and 1.3 for definitions). Also, $Qf(a, b)$ is used to denote the difference quotient $\{f(b) - f(a)\}/(b - a)$. We begin with a lemma on $I^*C$ functions.

8.1.1. LEMMA. *Suppose $f$ is $I^*C$. Then if $f$ is not monotone, then there is a nondegenerate subinterval $J$ of $I$ on which $f$ is continuous but not monotone.*

PROOF. Suppose $f$ is not monotone, and let $F$ denote the set of points $x \in I$ such that $f$ is not continuous in any neighborhood of $x$. Then $F$ is closed relative to $I$; and since $f$ is $IC$, $F$ is nowhere dense. Suppose $F \neq \emptyset$, for otherwise $I$ can be chosen to be $J$. Now, let $\{I_n\}$ be the sequence of contiguous intervals of $F$ in $I$. Suppose, if possible, that $f$ is monotone on each $I_n$. Since $f$ is $IC$, there exists an open subinterval $I_0$ of $I$ such that $F \cap I_0 \neq \emptyset$ and $f$ is continuous relative to $F \cap I_0$. Now, since $f$ is $WC$, it is clearly continuous on $\bar{I}_n$ for each $n$. Hence it is clear that $f$ is continuous on $I_0$, which contradicts the fact that $F \cap I_0 \neq \emptyset$. Consequently, $f$ is not monotone on some $I_n$, which can be chosen to be $J$. □

8.1.2. THEOREM (GENERAL MEAN VALUE). *Suppose $I = [a, b]$ and $f$ is $I^*C$. Then there is a point $x \in I^\circ$ such that $f$ is semidifferentiable at $x$ and $Qf(a,b) \in Sf'(x)$.*

PROOF. Let $c = Qf(a, b)$. It is enough to prove the result here, as usual, for $c = 0$; on applying this result to $f_{-c}$ the result is obtained in the general case with the help of Theorem 7.3.1.

Hence, suppose $c = 0$, that is, $f(b) = f(a)$. Suppose further that $f$ is not monotone; otherwise $f$ is constant in the present case, and so the result holds trivially. Then by the above lemma there is a nondegenerate subinterval $J$ of $I$ on which $f$ is continuous but not monotone. Hence we can find $x_1, x_2, x_3 \in J$, $x_1 < x_2 < x_3$, such that either $f(x_2) < f(x_1) \wedge f(x_3)$ or $f(x_2) > f(x_1) \vee f(x_3)$. Consequently, $f$ has a relative minimum or maximum at some point $x$ in $(x_1, x_3)$. Now it is clear that $x \in I^\circ$ and that $f$ has a zero semigradient at $x$. □

In the case of a function with symmetrical derivates, since such a function is derivable at every point where it is semiderivable, the above theorem yields the following extended version of the mean value theorem in terms of ordinary derivative which has been obtained by W. H. Young and G. C. Young [YY] for continuous functions.

8.1.3. COROLLARY (YOUNG-YOUNG). *Suppose $I = [a, b]$ and $f$ is $I^*C$. If $f$ has symmetrical derivates, then there is a point $x \in I^\circ$ such that $f$ is differentiable at $x$ and $f(b) - f(a) = f'(x)(b - a)$.*

It should be noted here that the above mean value theorem does not hold in such generality in terms of lower or upper derivative. If $f(x) = -|x|$, $x \in \mathbb{R}$, it is easy to see that the mean value theorem does not hold for $f$ in terms of lower derivative; similarly, this theorem does not hold for $-f$ in terms of upper derivative.

Further, it should be noted here that the inclusion in Theorem 8.1.2 becomes an identity in the case when $f$ is nonangular. However, in that case, as we see in the following theorem, $Qf(a, b)$ can indeed be chosen to be lower or upper derivative of $f$ at some point of $I^\circ$. The result for upper derivative follows on applying the following theorem to $-f$.

8.1.4. THEOREM (SECOND MEAN VALUE). *Suppose $I = [a, b]$ and $f$ is $I^*C$. If $f$ is upper nonangular, then there is a point $x \in I^\circ$ such that $f$ is lower differentiable at $x$ and $Qf(a, b) \subset Lf'(x)$.*

*Consequently, if $f$ is nonangular, then there is a point $x \in I^\circ$ such that $f$ is uniquely lower differentiable at $x$ and $f(b) - f(a) = Lf'(x)(b - a)$.*

PROOF. Suppose $f$ is upper nonangular. As in the case of Theorem 8.1.2, it is enough to consider the case when $Qf(a, b) = 0$ and $f$ is not monotone. Then, by Lemma 8.1.1, there is a subinterval $J$ of $I$ on which $f$ is continuous but not monotone. Hence as seen in the proof of Theorem 8.1.2, there is a point $x \in J^\circ$ where $f$ has a relative extremum.

When $f$ is lower derivable at $x$, it is clear that $0 \in Lf'(x)$; for if $f$ has a relative maximum at $x$, then $f$ is indeed derivable at $x$ and $f'(x) = 0$. Hence suppose $f$ is not lower derivable at $x$. Then $f$ has a relative maximum at $x$. Now, since $f$ is upper nonangular, it has a unique upper derivative equal to zero at $x$; that is, $D_-f(x) = D^+f(x) = 0$.

Further, since $f$ is not lower derivable at $x$, we have either (i) $D^-f(x) > 0$ or (ii) $D_+f(x) < 0$. In the case (i), since $D_-f(x) = 0$, it is easy to see that there exist $x_1, x_2 \in J$, $x_1 < x_2 < x$, such that $f(x_2) < f(x) \wedge f(x_1)$; and in case (ii), since $D^+f(x) = 0$, there similarly exist $x_1, x_2 \in J$, $x < x_2 < x_1$, such that $f(x_2) < f(x) \wedge f(x_1)$. Hence, in any case, since $f$ is continuous on $J$, it thus has a relative minimum at some point $y \in J^\circ$. Consequently, $f$ is lower differentiable at $y$ and $0 \in Lf'(y)$.

The last part is of course a direct consequence of the first. □

The hypothesis of upper nonangularity is not dispensable in the above theorem as it is clear from the function $f(x) = -|x|$, $x \in \mathbb{R}$. However, in the case when $f$ is lower derivable at every point of $I^\circ$, it automatically becomes upper nonangular. Hence, the classical mean value theorem for ordinary derivative remains valid for lower derivative in terms of a lower gradient.

In the following theorem (see its last part) we obtain this version of the mean value theorem for lower derivative under a weaker continuity hypothesis. To obtain the weakening of continuity, a totally different argument is used here.

8.1.5. Theorem (Third Mean Value). *Suppose $I = [a, b]$ and $f$ is WC and inherently QLSC. If $f$ is lower derivable n.e., then there is a point $x \in I^\circ$ such that $f$ is semidifferentiable at $x$ and $Qf(a, b) \in Sf'(x)$.*

*Consequently, if $f$ is lower derivable on $I^\circ$, then there is a point $x \in I^\circ$ such that $f$ is lower differentiable at $x$ and $Qf(a, b) \in L'f(x)$; and if $f$ is uniquely lower derivable on $I^\circ$, then $f(b) - f(a) = Lf'(x)(b - a)$.*

Proof. First, to prove the first part, suppose $f$ is lower derivable n.e. As in the case of Theorem 8.1.2, it is enough to consider the case when $Qf(a, b) = 0$, that is, $f(b) = f(a)$. We assume further, as before, that $f$ is not monotone; otherwise $f$ is constant, and so the result holds trivially.

Next, suppose $f$ does not have a relative extremum at any point of $I^\circ$; the result holds trivially otherwise. Let $N$ denote the set of points $x \in I$ such that $f$ is not monotone in any neighborhood of $x$. Then since $f$ is $WC$ and not monotone, it follows from Lemma 5.3 of [G11, p. 323] that $N$ is nonempty and perfect, and that $f$ is nowhere monotone on $N$.

Now, $f$ is by hypothesis $QLSC$ on some portion $N_0$ of $N$. Hence, it follows from the Third Derivate Theorem 2.4.1 that $D_+f \leq 0$ r.e. in $N_0$, and on applying this theorem to $f_-$ it follows that $D^-f \geq 0$ r.e. in $N_0$. Now, since $f$ is lower derivable n.e., there thus exists a point $x \in N_0 \cap I^\circ$ where $f$ is lower derivable and $D_+f(x) \leq 0 \leq D^-f(x)$. Then it is clear that $f$ has a zero (unique) lower derivative at $x$, which completes the proof of the first part.

The second part is of course a direct consequence of the first. □

In the remaining sections of this chapter we present some applications of the mean value theorems of this section. More applications will appear in the following chapters as well.

**§8.2 Cauchy's mean value theorem and l'Hôpital's rule in terms of new derivatives.** In this section we deduce from the mean value theorems of the last section extended versions of various l'Hôpital's rules in terms of lower and semi- derivatives without any derivability hypothesis on the function in the numerator. Also, an extended version of Cauchy's mean value theorem is obtained in the same direction for this purpose.

We will call the results obtained here, as in the case of the General Mean Value Theorem, the "General Cauchy's Mean Value Theorem" and "General l'Hôpital's Rule" since they hold for a general function $f$ without any derivability hypothesis on it. Other versions of these two results will be obtained later in Chapters 16–18 in terms of various notions of generalized derivatives.

We begin with the extension of Cauchy's mean value theorem.

8.2.1. THEOREM (GENERAL CAUCHY'S MEAN VALUE). *Let $f, g : I \to \mathbb{R}$, where $I = [a, b]$. Suppose* (i) *$f$ is $I^*C$ and $g$ is continuous on $I$, and* (ii) *$g$ is differentiable on $I^\circ$ and $g'(x) \neq 0$ for $x \in I^\circ$. Then there is a point $x \in I^\circ$ such that $f$ is semidifferentiable at $x$ and*

$$\frac{f(b) - f(a)}{g(b) - g(a)} \in \frac{Sf'(x)}{g'(x)}. \tag{1}$$

*Further, when $f$ is upper nonangular, the same holds in terms of lower derivative of $f$; and if $f$ is nonangular, then identity holds in place of* (1) *in terms of $Lf'(x)$.*

PROOF. It is clear from (ii) and the General Mean Value Theorem 8.1.2 that $g(a) \neq g(b)$. Hence the ratio on the left side of (1) is defined. Let $c$ denote this ratio, and define

$$h(x) = f(x) - f(a) - c\{g(x) - g(a)\}, \qquad x \in I. \tag{2}$$

Then $h(a) = 0 = h(b)$, and it is clear from (i) that $h$ is $I^*C$ on $I$. Hence by Theorem 8.1.2 there exists a point $x \in I^\circ$ such that $h$ is semidifferentiable at $x$ and $0 \in Sh'(x)$. Now, since $g$ is differentiable by hypothesis at $x$, it follows from Theorem 7.3.10(b) and (2) that $f$ is semidifferentiable at $x$ and $Sh'(x) = Sf'(x) - cg'(x)$. Consequently, $cg'(x) \in Sf'(x)$, which proves (1). The first part is thus established.

Next, to prove the second part, suppose $f$ is upper nonangular. Then since $g$ is differentiable on $I^\circ$, it is clear from (2) that $h$ also is upper nonangular. Hence the result follows as above from the Second Mean Value Theorem 8.1.4. The last part is of course an immediate consequence of this result. □

When $f$ has symmetrical derivates (e.g., when $f$ is smooth), the above theorem yields the following direct extension of Cauchy's mean value theorem in terms of ordinary derivative. This extension was obtained by W. H. Young and G. C. Young [YY] in the case when $f$ also is continuous.

8.2.2. COROLLARY (YOUNG-YOUNG). *Let $f, g : I \to \mathbb{R}$, where $I = [a, b]$. Suppose* (i) *$f$ is $I^*C$ and $g$ is continuous on $I$,* (ii) *$f$ has symmetrical derivates, and* (iii) *$g$ is differentiable on $I^\circ$ and $g'(x) \neq 0$ for $x \in I^\circ$. Then there is a point*

*$x \in I^\circ$ such that f is differentiable at x and*

$$\frac{f(b)-f(a)}{g(b)-g(a)} = \frac{f'(x)}{g'(x)}.$$

Next, we deduce from the above version of Cauchy's mean value theorem a general version of l'Hôpital's rule in terms of lower and semi- derivatives. Since these derivatives are set-valued in general, we need here the following notion of limit which is related to *USC* of multifunctions (see §1.5).

Suppose $\varphi : X \to 2^{\overline{\mathbb{R}}}$, where $X \subset \mathbb{R}$. Then we define $c \in \overline{\mathbb{R}}$ to be the *U-limit* (or upper limit) of $\varphi$ at $x \in X$ if for each neighborhood $V$ of $c$ in $\overline{\mathbb{R}}$ there is a neighborhood $U$ of $x$ in $X$ such that $\varphi(y) \subset V$ whenever $y \in U$, $y \neq x$. The notions of unilateral $U$-limits of $\varphi$ at $x$, as well as $U$-limit relative to a set, are defined by usual modifications in this definition.

It should be noted here that if $\varphi(x) = \{c\}$, then $\varphi$ has $U$-limit $c$ at $x$ iff it is *USC* at $x$. Also, when $\varphi$ is a single-valued function, the above notion of $U$-limit coincides with the ordinary limit of a function.

We state the following theorem in terms of right limit; the result on left limit, including the one when $x \to \infty$, is obtained on applying this theorem to $f_-$ and $g_-$; and then on combining these two results the result on bilateral limit is obtained.

It may be recalled here that $\lim_{x \to a+}\{f(x)/g(x)\}$ is said to be of *indeterminate form* $0/0$ if $f(x)$ and $g(x)$ have limit zero as $x$ approaches $a$ from the right, and to be of *indeterminate form* $\infty/\infty$ if $f(x)$ and $g(x)$ have infinite limits $+\infty$ or $-\infty$ as $x$ approaches $a$ from the right. The following proof is an adaptation of the proof in the case of ordinary derivative (see, e.g., Flatto [Fla, pp. 44, 45]).

8.2.3. THEOREM (GENERAL L'HÔPITAL'S RULE). *Let $f, g : I \to \mathbb{R}$, where $I = (a, b)$, $-\infty \le a < b < \infty$. Suppose* (i) *$f$ is $I^*C$ on $I$, and* (ii) *$g$ is differentiable and $g'(x) \neq 0$ for $x \in I$. Let $E$ denote the set of points in $I$ where $f$ is semidifferentiable. Then if the following limit on the left side is of indeterminate form $0/0$ or $\infty/\infty$, then*

$$\lim_{x\to a+} \frac{f(x)}{g(x)} = \underset{x\in E;\, x\to a+}{U\text{-lim}} \frac{Sf'(x)}{g'(x)}, \tag{3}$$

*provided that the limit on the right exists (finite or infinite).*

*Further, when $f$ is upper nonangular, the same holds in terms of lower derivative of $f$ with $E = \underline{\Delta}^*(f)$; and if $f$ is nonangular, then* (3) *holds for $Lf'$ in terms of ordinary limit.*

PROOF. We first prove the first part. Suppose the limit on the right side of (3) exists, and let $L$ denote this limit. We need to consider here some cases separately.

(a) First, suppose the limit on the left side of (3) is of indeterminate form $0/0$. We need to consider in this case the following two cases separately: (a1) $a$ is finite; (a2) $a = -\infty$.

**Case (a1):** Suppose $a$ is finite. Define $F(x) = f(x)$ and $G(x) = g(x)$ for $x \in I$, and also define $F(a) = G(a) = 0$. Then $F$ and $G$ are continuous at $a$. Let $x \in I$. Then it is clear that $F$ is $I^*C$ and $G$ is continuous on $J \equiv [a, x]$, and that $G$ is differentiable on $J^\circ$ and $G'(x) \neq 0$ for $x \in J^\circ$. Hence it follows from Theorem 8.2.1 that there is a point $y \in J^\circ$ such that $F$ is semidifferentiable at $y$ and

$$\frac{f(x)}{g(x)} = \frac{F(x)}{G(x)} = \frac{F(x) - F(a)}{G(x) - G(a)} \in \frac{SF'(y)}{G'(y)} = \frac{Sf'(y)}{g'(y)}.$$

Now since $y \to a$ as $x \to a$, this clearly proves (3) in the present case.

**Case (a2):** Suppose $a = -\infty$. Given any neighborhood $V$ of $L$ in $\overline{\mathbb{R}}$, there is a point $c \in I$ such that for each $z \in E \cap (a, c)$ we have $Sf'(z)/g'(z) \subset V$. Let $a < y < x < c$. Then by Theorem 8.2.1 there is a point $z \in (y, x)$ such that

$$\frac{f(x) - f(y)}{g(x) - g(y)} \in \frac{Sf'(z)}{g'(z)} \subset V.$$

Now since $f(x)$ and $g(x)$ converge to zero as $x \to a+$, we obtain

$$\frac{f(x)}{g(x)} = \lim_{y \to a+} \frac{f(x) - f(y)}{g(x) - g(y)} \subset \overline{V}.$$

This again proves (3) in this case.

(b) Next suppose the limit on the left side of (3) is of indeterminate form $\infty/\infty$. We may assume here without loss of generality that $\lim_{x \to a+} f(x) = \lim_{x \to a+} g(x) = +\infty$; for if $\lim_{x \to a+} f(x) = -\infty$ or $\lim_{x \to a+} g(x) = -\infty$, then the result will follow on replacing $f$ or $g$ by $-f$ or $-g$ respectively. We need to consider in this case the following two cases separately: (b1) $L$ is finite; (b2) $L$ is infinite.

**Case (b1):** Suppose $L$ is finite. Then, given $\varepsilon > 0$, there is a point $c \in I$ such that $Sf'(z)/g'(z) \subset (L - \varepsilon, L + \varepsilon)$ for $z \in E \cap (a, c)$. Let $a < x < y < c$. Then by Theorem 8.2.1 there is a point $z \in E \cap (x, y)$ such that

$$k \equiv \frac{f(x) - f(y)}{g(x) - g(y)} \in \frac{Sf'(z)}{g'(z)}.$$

Consequently, $|k - L| < \varepsilon$.

Now, since $f(x) - f(y) = k\{g(x) - g(y)\}$ and $|k| \le |k - L| + |L| < |L| + \varepsilon$, we obtain

$$\left|\frac{f(x)}{g(x)} - L\right| \le \left|\frac{f(y)}{g(x)}\right| + \left|\frac{g(y)}{g(x)}\right|(|L| + \varepsilon) + |k - L|.$$

Further, since $\lim_{x\to a+} g(x) = +\infty$, there is a point $d \in (a, c)$ such that for each $x \in (a, d)$, $|f(y)/g(x)| < \varepsilon$ and $|g(y)/g(x)|(|L| + \varepsilon) < \varepsilon$. Then for each $x \in (a, d)$, $|f(x)/g(x) - L| < 3\varepsilon$, so that $\lim_{x\to a+} f(x)/g(x) = L$.

**Case (b2):** We give the proof here for $L = +\infty$; a similar argument holds when $L = -\infty$. Then given $M \in \mathbb{R}_+$, there is a point $c \in I$ such that for each $z \in E \cap (a, c)$ we have $Sf'(z)/g'(z) > M$. Hence if $a < x < y < c$, it follows from Theorem 8.2.1 as before that $k \equiv \{f(x) - f(y)\}/\{g(x) - g(y)\} > M$. Thus

$$\frac{f(x)}{g(x)} = \frac{f(y)}{g(x)} + k\left\{1 - \frac{g(y)}{g(x)}\right\},$$

and since $\lim_{x\to a+} g(x) = +\infty$, there is a point $d \in (a, c)$ such that for each $x \in (a, d)$, $|f(y)/g(x)| < M/4$ and $|g(y)/g(x)| < 1/2$. Now since $k > M$, we obtain

$$\frac{f(x)}{g(x)} > k\left\{1 - \left|\frac{g(y)}{g(x)}\right|\right\} - \left|\frac{f(y)}{g(x)}\right| > \frac{M}{2} - \frac{M}{4} = \frac{M}{4},$$

which proves that $\lim_{x\to a+} f(x)/g(x) = +\infty$.

This completes the proof of the first part.

The second part is obtained by an analogous argument from the second part of Theorem 8.2.1. □

When $f$ has symmetrical derivates, the above theorem yields the following direct extension of l'Hôpital's rule in terms of ordinary derivative.

8.2.4. COROLLARY. *Let $f, g : I \to \mathbb{R}$, where $I = (a, b)$, $-\infty \le a < b < \infty$. Suppose* (i) *$f$ has symmetrical derivates and is $I^*C$ on $I$, and* (ii) *$g$ is differentiable and $g'(x) \ne 0$ for $x \in I$. Let $E$ denote the set of points in $I$ where $f$ is differentiable. Then if the following limit on the left side is of indeterminate form $0/0$ or $\infty/\infty$, then*

$$\lim_{x\to a+} \frac{f(x)}{g(x)} = \lim_{x\in E;\, x\to a+} \frac{f'(x)}{g'(x)},$$

*provided that the limit on the right exists (finite or infinite).*

**§8.3 Strong derivative in terms of limits of new derivatives.** In this section we obtain with the help of the mean value theorems of §8.1 some characterizations of strong derivative in terms of $U$-limits of lower and semi- derivatives (see Theorems 8.3.1, 8.3.5, and 8.3.7).

We need here some terminology. First, suppose $f: I \to \mathbb{R}$ is $I^*C$. Then it is clear from the General Mean Value Theorem 8.1.2 that the set $\Delta_s(f)$ is dense in $I$. Hence using the notion of $U$-limit defined in the last section, we will call $c \in \overline{\mathbb{R}}$ to be the *U- limit* of $Sf'$ at $x \in I$ if $c$ is the $U$-limit of $Sf'(y)$ as $y$ approaches $x$ through $\Delta_s(f)$. It should be noted here that we have defined $U$-limit only to be a singleton set, or a unique element of $\overline{\mathbb{R}}$.

Further, in this case when $x \in \Delta_{su}(f)$, we will call the semiderivative of $f$ *USC* at $x$ if it is so relative to $\Delta_s(f)$, or, equivalently, if $Sf'(x)$ is the $U$-limit of $Sf'$ at $x$. Similar terminology will be used for lower derivative of $f$ when $\underline{\Delta}(f)$ is dense in $I$.

8.3.1. THEOREM. *Suppose $f$ is $I^*C$ and $x \in I$. Then $f$ is strongly derivable at $x$ iff the semiderivative of $f$ has a (unique) U-limit at $x$; and if so, then $f'_*(x)$ is the U-limit of $Sf'$ at $x$.*

PROOF. Given $c \in \overline{\mathbb{R}}$, it is clearly enough to prove here that $f$ has strong derivative $c$ at $x$ iff $c$ is $U$-limit of $Sf'$ at $x$.

First, to prove the necessity part, suppose $f$ has strong derivative $c$ at $x$. Let us recall here that we use $M_*f(x)$ to denote the strong median of $f$ at $x$, namely, the interval $[\underline{D}_*f(x), \overline{D}_*f(x)]$. It is clear that $Sf'(y) \subset M_*f(y)$ for each $y \in \Delta_s(f)$. Hence the necessity of the condition follows directly from the *USC* of $M_*f$ which is easy to prove (or see Theorem 14.1.1).

Next, to prove the sufficiency, suppose $c \in \overline{\mathbb{R}}$ is the $U$-limit of $Sf'$ at $x$. Let $V$ be any subinterval of $\overline{\mathbb{R}}$ open relative to $\overline{\mathbb{R}}$ such that $c \in V$. Then there is a $\delta > 0$ such that $Sf'(t) \subset V$ whenever $t \in \Delta_s(f)$ and $0 < |t - x| < \delta$. Now let $y, z \in I$, $y < z$, and suppose $|y - x| \vee |z - x| < \delta$. Set $d = Qf(y, z)$. Then it is enough to show that $d \in V$, for this will prove that $c$ is the strong derivative of $f$ at $x$.

First, suppose $z \leq x$. Then there is by Theorem 8.1.2 a point $t \in \Delta_s(f)$, $y < t < z$, such that $d \in Sf'(t)$. Now since $0 < x - t < \delta$, it follows that $d \in V$. A similar argument holds when $y \geq x$.

Next, suppose $y < x < z$. Set $a = Qf(x, z)$ and $b = Qf(x, y)$. Then it is easy to see that

$$d = a(z - x)/(z - y) + b(x - y)/(z - y),$$

which is a convex sum of $a$ and $b$. Since $a, b \in V$ by above, it thus follows again that $d \in V$. □

We include here a few interesting consequences of the above theorem, some of which will be used later.

8.3.2. Corollary. *Suppose $f$ is $I^*C$ and $x \in I^\circ$. Then $f$ is strongly derivable at $x$ iff it is uniquely semiderivable at $x$ and $Sf'$ is USC at $x$, or, equivalently, iff $f$ is uniquely weakly derivable at $x$ and $Wf'$ is USC at $x$.*

The first part follows directly from the above theorem, and the second part follows from the first on taking into account the fact that $\Delta_s(f) \subset \Delta_w(f)$ and $Wf'$ coincides with $Sf'$ on $\Delta_s(f)$.

In the case of a function with symmetrical derivates, the following characterization of strong derivative is obtained from the above theorem in terms of ordinary derivative.

8.3.3. Corollary. *Suppose $f$ is $I^*C$. If $f$ has symmetrical derivates, then it is strongly derivable at $x \in I$ iff it is derivable at $x$ and $f'$ is continuous at $x$ relative to $\Delta(f)$.*

Further, when a function $f$ is strongly differentiable at a point $x$, it is easy to see that there is a neighborhood of $x$ in which $f$ is Lipschitz, and thus continuous. Hence from the above theorem we obtain the following characterization of strong differentiability.

8.3.4. Corollary. *The function $f$ is strongly differentiable at $x \in I$, and $f'_*(x) = c$, iff $f$ is continuous in some neighborhood of $x$ and $c$ is the U-limit of $Sf'$ at $x$.*

Next, we deduce from the Second Mean Value Theorem the following characterization of strong derivative in terms of lower derivative.

8.3.5. Theorem. *Suppose $f$ is $I^*C$ and upper nonangular. Then $f$ is strongly derivable at $x \in I$ iff the lower derivative of $f$ has a (unique) U-limit at $x$, and if so, then $f'_*(x)$ is the U-limit of $Lf'$ at $x$.*

Proof. We omit the proof here since this result is obtained by an argument similar to the one used for Theorem 8.3.1, using this time Theorem 8.1.4 in the place of Theorem 8.1.2. □

In the case when $f$ is nonangular, we thus obtain the following simpler characterization.

8.3.6. Corollary. *Suppose $f$ is $I^*C$ and nonangular. Then $f$ is strongly derivable at $x \in I$ iff it is lower derivable at $x$ and $Lf'$ is continuous at $x$.*

The following characterization in terms of lower derivative is obtained similarly from the Third Mean Value Theorem 8.1.5.

8.3.7. Theorem. *If $f$ is WC, inherently QLSC, and lower derivable on $I^\circ$, then $f$ is strongly derivable at $x \in I^\circ$ iff $Lf'$ is unique and USC at $x$.*

Finally, with help of Theorem 4.5.2 we obtain the following from this theorem.

8.3.8. COROLLARY. *If $f$ is uniquely lower differentiable on $J = [a, b] \subset I^\circ$ and $Lf'$ is continuous on $J$, then $f$ is continuous and strongly differentiable on $J$.*

**§8.4 Taylor's formulae for new derivatives.** In this section we deduce from the mean value theorems of §8.1, with the help of the results of the last section, Taylor's formulae for weak and lower derivatives.

When a function $f: I \to \mathbb{R}$ is $n$ times ($n > 1$) uniquely weakly differentiable on $I$, we will use $Wf^{(n)}$ to denote the $n$th weak derivative of $f$; and when $Wf^{(n)}$ is weakly derivable at some point $x \in I$, we will use, similarly, $Wf^{(n+1)}(x)$ to denote the weak derivative of $Wf^{(n)}$ at $x$. The lower derivatives $Lf^{(n)}$ and $Lf^{(n+1)}(x)$ are defined similarly.

We begin with the Taylor's formula for weak derivative.

8.4.1. THEOREM (TAYLOR'S FORMULA). *Let $f: I \to \mathbb{R}$ and $J = [a, b] \subset I^\circ$. Suppose $f$ is $I^*C$ on $J$, it is $n$ times uniquely weakly differentiable on $J$, and its first $n$ weak derivatives are continuous on $J$. Then $f$ has an $(n+1)$th semigradient $c$ at some point of $J^\circ$ such that*

$$(1) \qquad f(b) = f(a) + (b-a)Wf'(a) + \cdots + \frac{(b-a)^n}{n!}Wf^{(n)}(a) + \frac{(b-a)^{n+1}c}{(n+1)!}.$$

*Consequently, if $Wf^{(n)}$ is nonangular on $J^\circ$, then there is a point $x \in J^\circ$ such that $c = Wf^{(n+1)}(x)$.*

PROOF. Let $c$ be the real number for which (1) holds. Then we need to show that $c$ is a semigradient of $Wf^{(n)}$ at some point of $J^\circ$.

Define, for $x \in J$,

$$(2) \qquad g(x) = f(b) - f(x) - (b-x)Wf'(x) - \cdots - \frac{(b-x)^n}{n!}Wf^{(n)}(x) - \frac{(b-x)^{n+1}c}{(n+1)!}.$$

Then since the first $n$ weak derivatives of $f$ are continuous on $J$, it is clear from (2) that $g$ is $WC$ and $IC$. Further, it follows from (1) that $g(a) = 0 = g(b)$. Hence, by the General Mean Value Theorem 8.1.2, $g$ has a zero semigradient at some point $t \in J^\circ$.

Now, since the weak derivatives of $f, Wf', \ldots$ and $Wf^{(n-1)}$ are continuous on $J$, it follows from Corollary 8.3.2 that these functions are strongly differentiable on $J^\circ$. Hence all the functions on the right side of (2), except for the last but one, are differentiable on $J^\circ$. Now, since $-(b-t)^n/n! < 0$, it follows form (2) and Theorems 7.3.1 and 7.4.1 that $Wf^{(n)}$ has a semiderivative at $t$ dissimilar to $Sg'(t)$ and

$$\begin{aligned} Sg'(t) &= -f'(t) + Wf'(t) - (b-t)Wf^{(2)}(t) + (b-t)Wf^{(2)}(t) \\ &\quad - \frac{(b-t)^2}{2!}Wf^{(3)}(t) + \cdots + \frac{(b-t)^{n-1}}{(n-1)!}Wf^{(n)}(t) \\ &\quad - \frac{(b-t)^n}{n!}Sf^{(n+1)}(t) + \frac{(b-t)^n c}{n!} \\ &= \frac{(b-t)^n}{n!}\{c - Sf^{(n+1)}(t)\}. \end{aligned}$$

Now, since $0 \in Sg'(t)$, this implies that $c \in Sf^{(n+1)}(t)$, which completes the proof. □

Next, we deduce from the above theorem the following Taylor's formula for lower derivative.

8.4.2. Theorem (Taylor's Formula). *Let $f: I \to \mathbb{R}$ and $J = [a,b] \subset I^\circ$. Suppose $f$ is $n$ times uniquely lower differentiable on $J$ and $Lf^{(n)}$ is continuous on $J$. Then $f$ has an $(n+1)$th semigradient $c$ at some point of $J^\circ$ such that*

$$f(b) = f(a) + (b-a)Lf'(a) + \cdots + \frac{(b-a)^n}{n!}Lf^{(n)}(a) + \frac{(b-a)^{n+1}c}{(n+1)!}. \tag{3}$$

*Further, if $Lf^{(n)}$ is upper nonangular on $J^\circ$, then $c$ can be chosen to be an $(n+1)$th lower gradient of $f$ at some point of $J^\circ$. Consequently, if $Lf^{(n)}$ is nonangular on $J^\circ$, then there is a point $x \in J^\circ$ such that $c = Lf^{(n+1)}(x)$.*

Proof. Since $Lf^{(n)}$ is continuous on $J$, it follows from Corollary 8.3.8 that $Lf^{(n-1)}$ is continuous on $J$. Similarly, it follows from that corollary by induction that $Lf^{(n-2)}, \ldots, Lf'$ and $f$ are continuous on $J$. Hence the first part follows directly from Theorem 8.4.1.

Next, to prove the second part, suppose $Lf^{(n)}$ is upper nonangular on $J^\circ$. Let $c$ be the real number for which (3) holds, and let $g$ be defined as before by (2) using this time lower derivatives in the place of weak derivatives. Further, define the following for $x \in J$:

$$h(x) = f(b) - f(x) - (b-x)Lf'(x) - \cdots - \frac{(b-x)^{n-1}}{(n-1)!}Lf^{(n-1)}(x) - \frac{(b-x)^{n+1}}{(n+1)!}c. \tag{4}$$

Then, by (2), we obtain

$$g(x) = h(x) - \frac{(b-x)^n}{n!} Lf^{(n)}(x), \qquad x \in J. \tag{5}$$

Now, let $x \in J^\circ$. Then since $Lf', \ldots, Lf^{(n)}$ are continuous on $J$, it follows from Corollary 8.3.2 that each of the functions on the right side of (4) is differentiable at $x$. Hence $h$ is differentiable at $x$. Now, since $-(b-x)^n/n! < 0$, it is clear from (5) that

$$D^- g(x) = h'(x) + \frac{(b-x)^{n-1}}{(n-1)!} Lf^{(n)}(x) - \frac{(b-x)^n}{n!} D_-(Lf^{(n)}(x)),$$

$$D_+ g(x) = h'(x) + \frac{(b-x)^{n-1}}{(n-1)!} Lf^{(n)}(x) - \frac{(b-x)^n}{n!} D^+(Lf^{(n)}(x)).$$

It is now clear from these two relations that $g$ is lower nonangular at $x$.

Further, it follows from (2) and (3) that $g(a) = 0 = g(b)$, and it is clear that $g$ is continuous on $J$. Hence it follows from the Second Mean Value Theorem 8.1.4 that $g$ has a zero upper gradient at some point $t \in J^\circ$. It follows now by an argument similar to the one used in the last part of the proof of Theorem 8.4.1 that $c$ is a lower gradient of $Lf^{(n)}$ at $t$.

The last part is of course a direct consequence of the result just established. □

From each of the above two theorems we obtain directly the following extended version of the classical Taylor's formula for ordinary derivative.

8.4.3. COROLLARY. *Let $f : I \to \mathbb{R}$ and $J = [a, b] \subset I^\circ$. Suppose $f$ is $n$ times differentiable on $J$ and $f^{(n)}$ is continuous on $J$. If $f^{(n)}$ has symmetrical derivates on $J^\circ$, then there is a point $x \in J^\circ$ such that*

$$f(b) = f(a) + (b-a)f'(a) + \cdots + \frac{(b-a)^n}{n!} f^{(n)}(a) + \frac{(b-a)^{n+1}}{(n+1)!} f^{(n+1)}(x).$$

**§8.5 Term-by-term differentiation for unilateral and strong derivatives.** In this section we obtain some results on term-by-term differentiation of sequences and series of functions in terms of unilateral, ordinary, and strong derivatives. The results for new derivatives will be considered in the next section, where some applications of the results of this section are also included.

We assume throughout this section that $\{f_n\}$ is a sequence of real-valued functions defined on $X \subset \mathbb{R}$ which converges pointwise to $f$.

Some terminology is needed here to describe the hypotheses needed in these two sections. It may be recalled here that we use $\mathbb{N}$ to denote the set of all positive integers.

Following Hahn [Ha2, p. 247] (or see Hobson [H2, p. 112]), we call the sequence $\{f_n\}$ *uniformly convergent at* $x \in X$ if for every $\varepsilon > 0$ there is a $\delta > 0$ and a $k \in \mathbb{N}$ such that, for each $n$, $i > k$, we have

$$(1) \qquad |f_n(y) - f_i(y)| < \varepsilon \quad \text{for} \quad y \in X \cap (x - \delta, x + \delta).$$

Further, $\{f_n\}$ is said to be *continuously convergent at* $x$ (see, e.g., Hobson [H2, p. 113]) if for every $\varepsilon > 0$ there is a $\delta > 0$ and a $k \in \mathbb{N}$ such that, for each $n$, $i > k$, we have

$$(2) \qquad |f_n(y) - f_i(z)| < \varepsilon \quad \text{for} \quad y, z \in X \cap (x - \delta, x + \delta).$$

Applying these notions to the difference quotients of $f_n$'s, we will call the sequence $\{Qf_n\}$ *uniformly convergent at* $x$ if for every $\varepsilon > 0$ there is a $\delta > 0$ and a $k \in \mathbb{N}$ such that, for each $n,i > k$, we have

(3)
$$|Qf_n(x, y) - Qf_i(x, y)| < \varepsilon \quad \text{for} \quad y \in X \cap (x - \delta, x + \delta),\, y \neq x.$$

On replacing the interval $(x - \delta, x + \delta)$ in this definition by $(x, x + \delta)$ or $(x - \delta, x)$ we obtain, on the other hand, the notions of *uniform convergence on the right* or *left*, respectively, at $x$.

Further, $\{Qf_n\}$ will be said to be *continuously convergent at* $x$ if for every $\varepsilon > 0$ there is a $\delta > 0$ and a $k \in \mathbb{N}$ such that, for each $n$, $i > k$, we have

$$(4) \quad |Qf_n(y, z) - Qf_i(y, z)| < \varepsilon \quad \text{for} \quad y, z \in X \cap (x - \delta, x + \delta),\, y \neq z.$$

It should be noted here that the notions of continuous convergence of $\{f_n\}$ and $\{Qf_n\}$ at $x$ are stronger than those of their uniform convergence at $x$.

We will in fact use here the following two weaker forms of uniform and continuous convergence of $\{f_n\}$ and $\{Qf_n\}$ at $x$.

In the case when for every $\varepsilon > 0$ there is a $k \in \mathbb{N}$ such that to each $n > k$ there corresponds a $\delta > 0$ so that for each $i > n$, (1) or (2) holds, then we will call $\{f_n\}$ *quasiuniformly convergent* or *quasicontinuously convergent*, respectively, at $x$. On replacing the interval $(x - \delta, x + \delta)$ in (1) by $(x, x + \delta)$ or $(x - \delta, x)$ we obtain, on the other hand, the notions of *quasiuniform convergence on the right* or *left*, respectively, of $\{f_n\}$ at $x$.

Further, in the case when for every $\varepsilon > 0$ and $k \in \mathbb{N}$ there is an $n > k$ and a $\delta > 0$ so that for each $i > n$, (1) or (2) holds, then $\{f_n\}$ will be said to be *weakly uniformly convergent* or *weakly continuously convergent*, respectively, at $x$.

The quasi- and weakly uniform, and continuous, convergences of $\{Qf_n\}$ are defined similarly using (3) and (4) in the place of (1) and (2), respectively.

In each of these two cases, it should be noted here that weakly uniform convergence is weaker than quasiuniform convergence; and, similarly, weakly continuous convergence is weaker than quasicontinuous convergence. Further, quasicontinuous convergence is stronger than quasiuniform convergence; and weakly continuous convergence is stronger than weakly uniform convergence.

We begin with the following basic lemma which deals with the relations between the unilateral derivates of $f$ and $f_n$'s.

8.5.1. LEMMA. *Let $x \in X^+$. (a) If $\{Qf_n\}$ is quasiuniformly convergent on the right at x, then*

(5) $$D_+f(x) = \lim_n D_+f_n(x), \qquad D^+f(x) = \lim_n D^+f_n(x).$$

(b) *If $\{Qf_n\}$ is weakly uniformly convergent on the right at x, then*

(6) $$\liminf_{n\to\infty} D_+f_n(x) \le D_+f(x) \le \limsup_{n\to\infty} D_+f_n(x);$$

*and the same holds on replacing $D_+$ everywhere by $D^+$.*

PROOF. To prove (a), suppose its hypothesis holds. It is enough to prove here the first equality of (5), for its second equality follows on applying the first to $\{-f_n\}$ and $-f$. Hence it is enough to show that

(7) $$D_+f(x) \le \liminf_{n\to\infty} D_+f_n(x) \le \limsup_{n\to\infty} D_+f_n(x) \le D_+f(x).$$

To prove the first inequality of (7), suppose it does not hold. Then we can choose $a$ and $b$ such that $\liminf_{n\to\infty} D_+f_n(x) < a < b < D_+f(x)$. Hence there exists an $\eta > 0$ such that $Qf(x, y) > b$ for $y \in X \cap (x, x + \eta)$. Now let $k$ be the positive integer determined by the hypothesis for $\varepsilon = b - a$. Then, given $n > k$, there is by hypothesis a $\delta > 0$ such that (3) holds for each $i > n$ and $y \in X \cap (x, x + \delta)$. Clearly, we can assume $\delta < \eta$. Then for such $i$ and $y$ we have $Qf_n(x, y) > Qf_i(x, y) - \varepsilon$. However, since $Qf(x, y) > b$ and $\{f_n\}$ converges to $f$ at $x$ and $y$, on choosing $i$ large enough we obtain $Qf_i(x, y) > b$. Hence $Qf_n(x, y) > b - \varepsilon = a$. This proves that $D_+f_n(x) \ge a$ for each $n > k$, which contradicts the choice of $a$. The middle inequality of (7) is of course obvious.

Next, suppose the last inequality of (7) does not hold. Then we can choose $c$ and $d$ such that $D_+f(x) < c < d < \limsup_{n\to\infty} D_+f_n(x)$. This time let $k$ be the positive integer determined by the hypothesis for $\varepsilon = d - c$. Then, given $n > k$, there is by the hypothesis a $\delta > 0$ such that (3) holds for $i > n$ and $y \in X \cap (x, x + \delta)$. Further, since $D_+f(x) < c$, there is a sequence of points

$\{y_j\}$ in $X \cap (x, x+\delta)$ which converges to $x$ from the right such that $Qf(x, y_j) < c$ for each $j$. Now for each $j$, and $i > n$, $Qf_n(x, y_j) < Qf_i(x, y_j) + \varepsilon$. But since $Qf(x, y_j) < c$, on choosing $i$ large enough we obtain $Qf_i(x, y_j) < c$. Hence $Qf_n(x, y_j) < c + \varepsilon = d$. This proves that $D_+ f_n(x) \le d$ for each $n > k$, which contradicts the choice of $d$. Part (a) is thus established.

Next, to prove (b), suppose $\{Qf_n\}$ is weakly uniformly convergent on the right at $x$. It is enough to prove here (6) for $D_+$, for on applying this result to $\{-f_n\}$ and $-f$ the result is obtained as before for $D^+$.

First, suppose the left inequality of (6) does not hold. Choose $a$ and $b$ such that $D_+ f(x) < a < b < \liminf_{n\to\infty} D_+ f_n(x)$. Then there is a $k \in \mathbb{N}$ such that $D_+ f_n(x) > b$ for each $n > k$. Now, choosing $\varepsilon = b - a$, there is by the hypothesis an $n > k$ and a $\delta > 0$ such that (3) holds for $i > n$ and $y \in X \cap (x, x+\delta)$. Further, since $D_+ f_n(x) > b$, there is an $\eta > 0$, $\eta < \delta$, such that $Qf_n(x, y) > b$ for $y \in X \cap (x, x+\eta)$. Hence for such $y$ we have by (3), for each $i > n$,

$$Qf_i(x, y) > Qf_n(x, y) - \varepsilon > b - \varepsilon = a.$$

Hence $Qf(x, y) = \lim_i Qf_i(x, y) \ge a$. This proves that $D_+ f(x) \ge a$, which is a contradiction.

Next, to prove the second inequality of (6), suppose it does not hold. Choose $c$ and $d$ such that $\limsup_{n\to\infty} D_+ f_n(x) < c < d < D_+ f(x)$. Then there is a $k \in \mathbb{N}$ such that $D_+ f_n(x) < c$ for each $n > k$. Now, choosing $\varepsilon = d - c$, there is by the hypothesis an $n > k$ and a $\delta > 0$ such that (3) holds for $i > n$ and $y \in X \cap (x, x+\delta)$. Further, since $D_+ f(x) > d$, there is an $\eta > 0$, $\eta < \delta$ such that $Qf(x, y) > d$ for $y \in X \cap (x, x+\eta)$. Hence for such $y$ and $i > n$, we have by (3), $Qf_n(x, y) > Qf_i(x, y) - \varepsilon$. However, on choosing $i$ large enough we have $Qf_i(x, y) > d$. Hence we obtain

$$Qf_n(x, y) > Qf(x, y) - \varepsilon > d - \varepsilon = c.$$

This proves that $D_+ f_n(x) \ge c$, which is a contradiction. □

The following theorem deals with term-by-term differentiation for the right derivative.

8.5.2. THEOREM. *Let $x \in X^+$ and suppose each $f_n$ is right derivable at $x$. If $\{Qf_n\}$ is quasiuniformly convergent on the right at $x$, then $f$ is right derivable at $x$ and $f'_+(x) = \lim_n (f_n)'_+(x)$.*

*Further, if $\{(f_n)'_+(x)\}$ is convergent in $\overline{\mathbb{R}}$, then the same holds provided that $\{Qf_n\}$ is weakly uniformly convergent on the right at $x$.*

PROOF. The first part follows clearly from the part (a) of the above lemma. To prove the second part, suppose $\{(f_n)'_+(x)\}$ is convergent in $\overline{\mathbb{R}}$ and that $\{Qf_n\}$ is weakly uniformly convergent on the right at $x$. Then it follows from the two parts of Lemma 8.5.1(b) that $D_+f(x) = \lim_n(f_n)'_+(x) = D^+f(x)$, which proves the result. □

We include here some significant consequences of the above theorem.

Firstly, on applying the above theorem to the sequence of partial sums of a series of functions, we obtain the following result on series of functions.

8.5.3. COROLLARY. *Suppose $f$ is the pointwise limit of a series of functions $\Sigma_n g_n$ on $X$, $x \in X^+$, each $g_n$ is right derivable at $x$, and $\Sigma_{i \le n}(g_i)'_+(x)$ is defined for each $n$. Let $f_n = \Sigma_{i \le n} g_i$ for each $n$.*

*Then if $\{Qf_n\}$ is quasiuniformly convergent on the right at $x$, then $f$ is right derivable at $x$ and $f'_+(x) = \Sigma_n(g_n)'_+(x)$.*

*Further, if $\Sigma_n(g_n)'_+(x)$ is convergent in $\overline{\mathbb{R}}$, then the same holds provided that $\{Qf_n\}$ is weakly uniformly convergent on the right at $x$.*

Next, on applying Theorem 8.5.2 to the sequence $\{(f_n)_-\}$, and its pointwise limit $f_-$, a similar result is obtained on the left derivative. On combining Theorem 8.5.2 with this result we obtain, on the other hand, the following result on the ordinary derivative.

8.5.4. COROLLARY. *Let $x \in X'$ and suppose each $f_n$ is derivable at $x$. If $\{Qf_n\}$ is quasiuniformly convergent at $x$, then $f$ is derivable at $x$ and $f'(x) = \lim_n f'_n(x)$.*

*Further, if $\{f'_n(x)\}$ is convergent in $\overline{\mathbb{R}}$, then the same holds provided that $\{Qf_n\}$ is weakly uniformly convergent at $x$.*

Clearly, a result similar to Corollary 8.5.3 can be deduced from this corollary in the case of a series of functions on ordinary derivative in terms of quasiuniform convergence. This result has, however, been known for some time in a somewhat different form in the case when $X = I$ and the functions are differentiable at the point in question (see, e.g., [H2, pp. 336, 337]).

Finally, in the following theorem we deal with the strong derivative.

8.5.5. THEOREM. *Let $x \in X'$ and suppose each $f_n$ is strongly derivable at $x$. If $\{Qf_n\}$ is quasicontinuously convergent at $x$, then $f$ is strongly derivable at $x$ and $f'_*(x) = \lim_n(f_n)'_*(x)$. Further, if $\{(f_n)'_*(x)\}$ is convergent in $\overline{\mathbb{R}}$, then the same holds provided that $\{Qf_n\}$ is weakly continuously convergent at $x$.*

PROOF. First, to prove the first part, suppose $\{Qf_n\}$ is quasicontinuously convergent at $x$. Then it follows from the first part of Corollary 8.5.4 that $f$ is derivable at $x$ and $f'(x) = \lim_n(f_n)'_*(x)$. Hence it is enough to show here that $f$ is strongly derivable at $x$.

First, suppose $f'(x)$ is finite. Then, given $\varepsilon > 0$, it is clear from the hypothesis that there is an $n \in \mathbb{N}$ and a $\delta > 0$ such that $|(f_n)'_*(x) - f'(x)| < \varepsilon$

and (4) holds for each $i > n$. Thus $(f_n)'_*(x)$ is finite, and so there is further an $\eta > 0$, $\eta < \delta$ such that

$$|Qf_n(y,z) - (f_n)'_*(x)| < \varepsilon \quad \text{for} \quad y, z \in X \cap (x - \eta,\ x + \eta), y \neq z.$$

Hence, for such $y$ and $z$, on choosing $i$ large enough we obtain

$$\begin{aligned} |Qf(y,z) - f'(x)| &\leq |Qf(y,z) - Qf_i(y,z)| + |Qf_i(y,z) - Qf_n(y,z)| \\ &\quad + |Qf_n(y,z) - (f_n)'_*(x)| + |(f_n)'_*(x) - f'(x)| < 4\varepsilon, \end{aligned}$$

which proves that $f$ is strongly derivable at $x$.

Next, suppose $f'(x) = +\infty$; the proof is similar when $f'(x) = -\infty$. Then, given $M \in \mathbb{R}_+$, there is an $n \in \mathbb{N}$ and a $\delta > 0$ such that $(f_n)'_*(x) > M + 2$ and (4) holds for $\varepsilon = 1$ and $i > n$. Further, there is an $\eta > 0$, $\eta < \delta$ such that $Qf_n(y,z) > M + 2$ for $y, z \in X \cap (x - \eta, x + \eta)$, $y \neq z$. Hence, for such $y$ and $z$, on choosing $i$ large enough we obtain

$$\begin{aligned} Qf(y,z) &= \{Qf(y,z) - Qf_i(y,z)\} + \{Qf_i(y,z) - Qf_n(y,z)\} + Qf_n(y,z) \\ &> -1 - 1 + (M + 2) = M. \end{aligned}$$

This proves that $f$ has a strong derivative $+\infty$ at $x$. The first part is thus established.

The second part is obtained, on the other hand, from the second part of Theorem 8.5.2 by similar arguments. □

Once again, it is clear that a result similar to Corollary 8.5.3 can be deduced from the above theorem in the case of a series of functions on strong derivative in terms of quasicontinuous convergence.

**§8.6 Term-by-term differentiation for new derivatives.** In this section we investigate term-by-term differentiation for weak and semi- derivatives. First, in Theorem 8.6.2, we deal with a sequence of functions, and then in Theorem 8.6.3 we obtain the result for a series of functions in terms of lower derivative.

Finally, in Theorems 8.6.4, 8.6.6, and 8.6.7, we present some applications of the first two theorems and of the results obtained in the last section.

In this section we continue to assume, unless stated otherwise, that $\{f_n\}$ is a sequence of functions on $X \subset \mathbb{R}$ which converges pointwise to $f$.

We first state a simple lemma on the convergence of closed intervals in $2^{\overline{\mathbb{R}}}$ which will be used here frequently. This lemma can be verified without difficulty from the definition of Vietoris topology on $2^{\overline{\mathbb{R}}}$.

8.6.1. Lemma. *Let $I_n = [a_n, b_n]$ for each $n$, where $a_n, b_n \in \overline{\mathbb{R}}$. Then $\{I_n\}$ is convergent in the Vietoris topology of $2^{\overline{\mathbb{R}}}$ iff $\{a_n\}$ and $\{b_n\}$ are convergent in $\overline{\mathbb{R}}$;*

*and if so, then*

$$\lim_n I_n = \left[\lim_n a_n, \lim_n b_n\right].$$

In the following theorem we deal with term-by-term differentiation for all the forms of new derivatives introduced in §§4.2 and 4.3 in the case of a sequence of functions.

8.6.2. THEOREM. *Let $x \in X^b$ and suppose each $f_n$ is weakly derivable at $x$. If $\{Qf_n\}$ is quasiuniformly convergent at $x$, then $f$ is weakly derivable at $x$ and $Wf'(x) = \lim_n Wf_n'(x)$ in the Vietoris topology of $2^{\overline{\mathbb{R}}}$; and if each $f_n$ is lower or upper derivable, nonangular, or lower or upper knotted at $x$, then so is $f$.*

*Further, if $\{Wf_n'(x)\}$ is convergent in $2^{\overline{\mathbb{R}}}$, then the same holds provided that $\{Qf_n\}$ is weakly uniformly convergent at $x$.*

PROOF. First, suppose $\{Qf_n\}$ is quasiuniformly convergent at $x$. Then, by Lemma 8.5.1(a), we have

$$(1) \qquad D_+f(x) = \lim_n D_+f_n(x), \qquad D^+f(x) = \lim_n D^+f_n(x).$$

Further, on applying this lemma to $\{(f_n)_-\}$ and $f_-$, we obtain

$$(2) \qquad D_-f(x) = \lim_n D_-f_n(x), \qquad D^-f(x) = \lim_n D^-f_n(x).$$

Now, since $\overline{D}_w f_n(x) \le \underline{D}_w f_n(x)$ for each $n$, it follows from the continuity of the operations $\wedge$ and $\vee$ that

$$\overline{D}_w f(x) = \lim_n \overline{D}_w f_n(x) \le \lim_n \underline{D}_w f_n(x) = \underline{D}_w f(x).$$

Hence $f$ is weakly derivable at $x$, and it follows from Lemma 8.6.1 that

$$Wf'(x) = \left[\overline{D}_w f(x), \underline{D}_w f(x)\right] = \lim_n \left[\overline{D}_w f_n(x), \underline{D}_w f_n(x)\right] = \lim_n Wf_n'(x).$$

Further, when each $f_n$ is lower or upper derivable, nonangular, or lower or upper knotted at $x$, it follows from (1) and (2) by similar arguments that the same holds for $f$.

Next, to prove the second part, suppose $\{Wf_n'(x)\}$ is convergent in $2^{\overline{\mathbb{R}}}$ and $\{Qf_n\}$ is weakly uniformly convergent at $x$. Then by Lemma 8.5.1(b), and its analogous version for the left side, we have

$$(3) \qquad \liminf_{n\to\infty} D_\pm f_n(x) \le D_\pm f(x) \le \limsup_{n\to\infty} D_\pm f_n(x),$$

$$(4) \qquad \liminf_{n\to\infty} D^\pm f_n(x) \le D^\pm f(x) \le \limsup_{n\to\infty} D^\pm f_n(x).$$

Now, let us recall that

$$\underline{D}_w f(x) = \vee D_{\pm} f(x), \qquad \overline{D}_w f(x) = \wedge D^{\pm} f(x),$$

and the same holds on replacing $f$ by $f_n$ for each $n$. Hence, using the continuity of the operations $\vee$ and $\wedge$, we obtain the following from (3) and (4):

$$\liminf_{n\to\infty} \underline{D}_w f_n(x) \le \underline{D}_w f(x) \le \limsup_{n\to\infty} \underline{D}_w f_n(x),$$

$$\liminf_{n\to\infty} \overline{D}_w f_n(x) \le \overline{D}_w f(x) \le \limsup_{n\to\infty} \overline{D}_w f_n(x).$$

Further, since $\{Wf_n'(x)\}$ is convergent by the hypothesis, it follows from Lemma 8.6.1 that the sequences $\{\underline{D}_w f_n(x)\}$ and $\{\overline{D}_w f_n(x)\}$ are convergent in $\overline{R}$. Hence from the last two relations we obtain

$$(5) \qquad \underline{D}_w f(x) = \lim_n \underline{D}_w f_n(x), \qquad \overline{D}_w f(x) = \lim_n \overline{D}_w f_n(x).$$

Now, since each $f_n$ is weakly derivable at $x$, we have $\overline{D}_w f_n(x) \le \underline{D}_w f_n(x)$ for each $n$. Hence it follows clearly from (5) that $f$ is weakly derivable at $x$, and further from Lemma 8.6.1 that $Wf'(x) = \lim_n Wf_n'(x)$.

Further, if each $f_n$ is nonangular at $x$, then $\underline{D}_w f_n(x) \le \overline{D}_w f_n(x)$ for each $n$, and so it follows from (5) that $\underline{D}_w f(x) \le \overline{D}_w f(x)$, that is, $f$ is nonangular at $x$.

Next, suppose each $f_n$ is lower derivable at $x$. Then $Wf_n'(x) = [D^- f_n(x), D_+ f_n(x)]$ for each $n$. Further, by Lemma 8.6.1, $\lim_n Wf_n'(x) = [a, b]$, where $a = \lim_n D^- f_n(x)$ and $b = \lim_n D_+ f_n(x)$. Hence it follows from (3) and (4) that $D^- f(x) = a$ and $D_+ f(x) = b$. Consequently, $f$ is lower derivable at $x$ and $Wf'(x) = [a, b] = \lim_n Wf_n'(x)$. A similar argument holds when $f_n$'s are upper derivable at $x$.

Finally, suppose each $f_n$ is lower knotted at $x$. Then $D^- f_n(x) = +\infty$ and $D_+ f_n(x) = -\infty$ for each $n$, and so it follows clearly from (3) and (4) that $D^- f(x) = +\infty$ and $D_+ f(x) = -\infty$, that is, $f$ is lower knotted at $x$. A similar argument holds when $f_n$'s are upper knotted at $x$. □

Since weak derivative is not additive in general, the above theorem does not yield a corresponding result for weak derivative on series of functions. However, in the case of lower derivative we still have the following result on series of functions. But since lower derivative is also not totally additive in general, this theorem does not follow directly from the above theorem.

8.6.3. THEOREM. *Suppose $f$ is the pointwise limit of a series of functions $\Sigma_n g_n$ on $X$, $x \in X^b$, each $g_n$ is lower derivable at $x$, and $\Sigma_{i=1}^n Lg_i'(x)$ is defined for each $n$. Let $f_n = \Sigma_{i=1}^n g_i$ for each $n$.*

*Then if $\{Qf_n\}$ is quasiuniformly convergent at $x$, then $f$ is lower derivable at $x$, $\Sigma_n Lg_n'(x)$ is convergent in $2^{\overline{\mathbb{R}}}$, and*

$$\sum_n Lg_n'(x) \subset Lf'(x). \tag{6}$$

*Further, if $\Sigma_n Lg_n'(x)$ is convergent in $2^{\overline{\mathbb{R}}}$, then the same holds provided that $\{Qf_n\}$ is weakly uniformly convergent at $x$.*

PROOF. Given $n > 1$, it follows from Theorem 7.3.2, by induction, that $f_n$ is lower derivable at $x$ and $\Sigma_{i=1}^n Lg_i'(x) \subset Lf_n'(x)$. Further, since $D^-g_i(x) \le D_+g_i(x)$ for each $i$, it is clear that

$$D^-f_n(x) \le \sum_{i=1}^n D^- g_i(x) \le \sum_{i=1}^n D_+g_i(x) \le D_+f_n(x); \tag{7}$$

and it is also clear that

$$\sum_{i=1}^n Lg_i'(x) = \left[\sum_{i=1}^n D^-g_i(x), \sum_{i=1}^n D_+g_i(x)\right]. \tag{8}$$

First, suppose $\{Qf_n\}$ is quasiuniformly convergent at $x$. Then since each $f_n$ is lower derivable at $x$, it follows from Theorem 8.6.2 that $f$ is lower derivable at $x$ and $Lf'(x) = \lim_n Lf_n'(x)$.

Now, given $\varepsilon > 0$, there is by the hypothesis a $k \in \mathbb{N}$ and a $\delta > 0$ corresponding to $k$ such that, for each $n > k$ and $y \in X \cap (x - \delta, x + \delta)$, $y \neq x$,

$$|Q(f_n - f_k)(x, y)| = |Qf_n(x, y) - Qf_k(x, y)| < \varepsilon.$$

Hence, for each $n > k$ we obtain, as in (7),

$$-\varepsilon \le D^-(f_n - f_k)(x) \le \sum_{i=k+1}^n D^-g_i(x) \le \sum_{i=k+1}^n D_+g_i(x)$$
$$\le D_+(f_n - f_k)(x) \le \varepsilon.$$

Now since this holds for each $n > k$, it is thus clear that the series $\Sigma_n D^-g_n(x)$ and $\Sigma_n D_+g_n(x)$ are convergent in $\overline{\mathbb{R}}$. Hence it follows from Lemma 8.6.1 and (8) that the series $\Sigma_n Lg_n'(x)$ converges in $2^{\overline{\mathbb{R}}}$ to the interval $[\Sigma_n D^-g_n(x), \Sigma_n D_+g_n(x)]$. Further, it is clear from Lemma 8.5.1(a) and (7) that

$$D^-f(x) = \lim_n D^-f_n(x) \le \sum_n D^-g_n(x) \le \sum_n D_+g_n(x)$$
$$\le \lim_n D_+f_n(x) = D_+f(x).$$

Consequently, we obtain (6). The first part is thus established.

Next, to prove the second part, suppose $\Sigma_n Lg_n'(x)$ is convergent in $2^{\overline{\mathbb{R}}}$ and $\{Qf_n\}$ is weakly uniformly convergent at $x$. Then it follows from Lemma 8.6.1 and (8) that the series $\Sigma_n D^- g_n(x)$ and $\Sigma_n D_+ g_n(x)$ are convergent in $\overline{\mathbb{R}}$ and

$$(9) \qquad \sum_n Lg_n'(x) = \left[\sum_n D^- g_n(x), \sum_n D_+ g_n(x)\right].$$

Further, as seen earlier in the proof of Theorem 8.6.2, we have the inequalities (3) and (4) in the present case. Hence, with the help of (3), (4) and (7), we obtain

$$\begin{aligned} D^- f(x) &\le \limsup_{n\to\infty} D^- f_n(x) \le \sum_n D^- g_n(x) \le \sum_n D_+ g_n(x) \\ &\le \liminf_{n\to\infty} D_+ f_n(x) \le D_+ f(x). \end{aligned}$$

Consequently, $f$ is lower derivable at $x$, and (6) is obtained now with the help of (9). □

Next, to obtain some applications of the results of the present and the last section, we present two concrete situations in which the convergence hypotheses of these two sections on $\{Qf_n\}$ become valid.

First, let $\Lambda$ denote the linear space of real-valued Lipschitz functions on the compact interval $I = [a, b]$ under the norm

$$\|f\| = |f(a)| + \sup\{|Qf(x, y)| : x, y \in I,\ x \ne y\},\ f \in \Lambda.$$

It is well known that $\Lambda$ is a Banach space under this norm (see, e.g., Rudin [Ru, p. 114]). We will call this norm *Lipschitz norm* and denote it by $\|\ \|_\lambda$.

Then for every sequence $\{f_n\}$ in $\Lambda$ which is Cauchy in the Lipschitz norm, it is clear that the sequence $\{Qf_n\}$ is continuously convergent at every point of $I$.

Hence, for such a sequence it follows from Lemma 8.5.1(a) that each of the unilateral derivates is stable under convergence in $\|\ \|_\lambda$.

Further, under this convergence, we obtain the following result directly from Theorems 8.5.2, 8.5.6, and 8.6.2.

8.6.4. THEOREM. *Suppose $\{f_n\}$ is a sequence of Lipschitz functions on $I = [a, b]$ which converges in the Lipschitz norm to $f$.*

(a) *If each $f_n$ is right derivable at $x \in I^+$, then so is $f$ and $f_+'(x) = \lim_n (f_n)_+'(x)$.*

(b) *If each $f_n$ is strongly derivable at $x \in I$, then so is $f$ and $f_*'(x) = \lim_n (f_n)_*'(x)$.*

(c) *If each $f_n$ is weakly, lower, or upper derivable at $x \in I^\circ$, then so is $f$ and $Wf'(x) = \lim_n Wf_n'(x)$.*

Next, in the following lemma we investigate the relation between continuous convergence of $\{Qf_n\}$ and uniform convergence of derivatives, or lower derivatives, of $f_n$'s.

8.6.5. LEMMA. *Suppose $f_n : I \to \mathbb{R}$ is continuous and uniquely lower differentiable on $I$ for each $n$. If $\{Lf_n'\}$ is uniformly, quasiuniformly, or weakly uniformly convergent at $x \in I$, then $\{Qf_n\}$ is continuously, quasicontinuously, or weakly continuously convergent, respectively, at $x$.*

PROOF. First, let $n, i \in \mathbb{N}$ and $y, z \in I$, $y \neq z$. Then since $f_n - f_i$ is continuous on $I$, it follows from the General Mean Value Theorem 8.1.2 that there is a point $t$ between $y$ and $z$ where $f_n - f_i$ is semidifferentiable and

$$Qf_n(y, z) - Qf_i(y, z) = Q(f_n - f_i)(y, z) \in S(f_n - f_i)'(t).$$

Now, since $f_n$ and $f_i$ are, by the hypothesis, uniquely lower differentiable at $t$, it follows from Theorem 7.3.11 that $S(f_n - f_i)'(t) = Lf_n'(t) - Lf_i'(t)$. Hence, $S(f_n - f_i)'(t)$ is unique and

$$Qf_n(y, z) - Qf_i(y, z) = Lf_n'(t) - Lf_i'(t). \tag{10}$$

First, suppose $\{Lf_n'\}$ is uniformly convergent at $x$. Then, given $\varepsilon > 0$, there is a $\delta > 0$ and a $k \in \mathbb{N}$ such that, for each $n, i > k$, we have

$$|Lf_n'(t) - Lf_i'(t)| < \varepsilon \quad \text{for} \quad t \in I \cap (x - \delta, x + \delta). \tag{11}$$

Hence, if $n$, $i > k$ and $y$, $z \in I \cap (x - \delta, x + \delta)$, $y \neq z$, it follows clearly from (10) and (11) that $|Qf_n(y, z) - Qf_i(y, z)| < \varepsilon$. Consequently, $\{Qf_n\}$ is continuously convergent at $x$.

A similar argument holds in each of the other two cases. □

Next, we deduce from the above lemma the following result on term-by-term differentiation in the case of a sequence of uniquely lower differentiable functions. In part (b) of this theorem we do not assume the pointwise convergence of $\{f_n\}$, but obtain it as a consequence.

8.6.6. THEOREM. *Suppose $I = (a, b)$, and $f_n : I \to \mathbb{R}$ is continuous and uniquely lower differentiable on $I$ for each $n$.*

(a) *If $\{f_n\}$ converges pointwise to $f$ and $\{Lf_n'\}$ is quasiuniformly convergent at $x \in I$, then $f$ is uniquely lower differentiable at $x$ and*

$$Lf'(x) = \lim_n Lf_n'(x); \tag{12}$$

*and if each $f_n$ is derivable or strongly derivable at $x$, then so is $f$.*

(b) *If $\{f_n\}$ converges at some point in $I$ and $\{Lf_n'\}$ is uniformly convergent on $I$, then $\{f_n\}$ converges uniformly to a continuous, uniquely lower differentiable function $f$ on $I$ and $Lf'(x) = \lim_n Lf_n'(x)$ for each $x$. Moreover, if $Lf_n'$ is continuous for each $n$, then $f$ is strongly differentiable on $I$.*

PROOF. First, to prove (a), suppose its hypothesis holds. Then, by the above lemma, $\{Qf_n\}$ is quasicontinuously convergent at $x$. Hence it follows from Theorem 8.6.2 that $f$ has a unique lower derivative at $x$ and that (12) holds. Also, it is clear from the quasiuniform convergence of $\{Lf_n'\}$ at $x$ that $Lf'(x)$ is finite. The remaining part of (a) follows on the other hand from Corollary 8.5.4 and Theorem 8.5.5.

Next, to prove (b), suppose $\{f_n\}$ converges at some point $c$ in $I$ and $\{Lf_n'\}$ is uniformly convergent on $I$. Then, given $\varepsilon > 0$, $\eta = \varepsilon/(b-a)$, there is a $k \in \mathbb{N}$ such that for each $n$, $i > k$, we have

$$|f_n(c) - f_i(c)| < \varepsilon, \quad |Lf_n'(t) - Lf_i'(t)| < \eta \quad \text{for} \quad t \in I.$$

Hence, for such $n$ and $i$ it follows from (10) as in the proof of Lemma 8.6.5 that

$$|Qf_n(c,x) - Qf_i(c,x)| < \eta \quad \text{for} \quad x \in I,\ x \neq c.$$

Hence if $n$, $i > k$ and $x \neq c$, we obtain

$$\begin{aligned}|f_n(x) - f_i(x)| &\leq |\{f_n(x) - f_n(c)\} - \{f_i(x) - f_i(c)\}| + |f_n(c) - f_i(c)|\\ &\leq |Qf_n(c,x) - Qf_i(c,x)|\,|x-c| + \varepsilon\\ &< \eta(b-a) + \varepsilon = 2\varepsilon.\end{aligned}$$

This proves that $\{f_n\}$ is uniformly convergent on $I$, and so it converges uniformly to a continuous function $f$ on $I$. Now, it follows from (a) that $f$ is uniquely lower differentiable on $I$ and that (12) holds for each $x \in I$.

Further, if each $Lf_n'$ is continuous on $I$, then by Corollary 8.3.6 each $f_n$ is strongly derivable on $I$, and hence it follows from (a) that $f$ is strongly differentiable on $I$. □

Finally, in the following theorem we present two interesting consequences of the last theorem.

8.6.7. THEOREM. *Let $\{f_n\}$ be a sequence of functions on $I = (a,b)$ which converges pointwise to $f$. Suppose each $f_n$ is either* (i) *uniquely lower differentiable on $I$, or* (ii) *$I^*C$ and uniquely weakly differentiable on $I$.*

(a) *If $Wf_n'$ is continuous for each $n$ and $\{Wf_n'\}$ has a finite limit at each point of $I$, then there is a residual set of points $x$ in $I$ where $f$ is strongly differentiable and $f_*'(x) = \lim_n Wf_n'(x)$.*

(b) *If $\{Wf_n'\}$ is pointwise bounded and equicontinuous on $I$, then $f$ is strongly differentiable on $I$, and for every compact subinterval $J$ of $I$ there is a subsequence $\{f_{n_i}\}$ of $\{f_n\}$ which converges uniformly to $f$ on $J$ and such that $f_*'(x) = \lim_i Wf_{n_i}'(x)$ for $x \in J$.*

PROOF. Let us first observe that in each of the parts (a) and (b), $Wf_n'$ is continuous by the hypothesis for each $n$. Hence, for each $n$, it follows from Theorems 4.5.2 and 8.3.5(b) in case (i), and from Corollary 8.3.2 in case (ii), that $f_n$ is strongly differentiable on $I$. Consequently, each $f_n$ is also continuous on $I$.

Now, to prove (a), suppose its hypothesis holds. Then, using the $\sigma$-compactness of $I$, it follows from a theorem of Osgood (see [H2, p. 137]) that $\{(f_n)'_*\}$ is uniformly convergent at a residual set of points $E$ in $I$. Hence, for each $x \in E$, it follows from Theorem 8.6.6(a) that $f$ is strongly differentiable at $x$ and $f'_*(x) = \lim_n (f_n)'_*(x)$.

Next, to prove (b), suppose its hypothesis holds, and let $J$ be any compact subinterval of $I$. Then there exists another compact subinterval $K$ of $I$ such that $J \subset K^\circ$. Now, on account of Ascoli's theorem (see, e.g., [Roy, p. 179]), there is a subsequence $\{f_{n_i}\}$ of $\{f_n\}$ such that $\{(f_{n_i})'_*\}$ converges uniformly on $K$. Hence it follows from Theorem 8.6.6(b) that $\{f_{n_i}\}$ converges to $f$ uniformly on $K^\circ$, $f$ is strongly differentiable on $K^\circ$, and $f'_*(x) = \lim_i (f_{n_i})'_*(x)$ for $x \in K^\circ$. Also, since $I$ is a regular space, it is further clear that $f$ is strongly differentiable at each point of $I$. □

# *CHAPTER 9*

# Monotonicity and Other Properties in Terms of New Derivatives

In this chapter we obtain several monotonicity theorems in terms of new derivatives, along with some median theorems which lead to Goldowski-Tonelli theorems for new derivatives.

First, in §9.1, we obtain three monotonicity theorems in terms of new derivatives without any derivability hypothesis, which we call the Second, Third, and Fifth Monotonicity Theorems. In the Fourth Monotonicity Theorem we deduce from the Third a known monotonicity result in terms of a unilateral derivate. Also, some inequalities similar to the ones obtained by Saks for functions with property $(T_2)$ in terms of ordinary derivative, are obtained here for arbitrary functions in terms of lower derivative.

Next, in §9.2, we deduce from the Second Monotonicity and Second Mean Value Theorems generalized versions of the well-known first and second derivative tests for points of extrema, along with the concavity test, in terms of lower derivative without any derivability hypothesis at points other than the point in question. All the results of these two sections hold also for the other new derivatives.

Next, in §9.3, we obtain a generalized version of the above-mentioned theorem of Saks and deduce from it monotonicity conditions for functions with properties $(T_2)$ and $(N)$ in terms of ordinary derivative. Then, in §9.4, we obtain three median theorems dealing with properties of various medians at nonmonotonicity points, which we call the First, Second, and Third Median Theorems. Further, in §9.5, we deduce from the First and Second Derivate Theorems properties of functions that are lower or weakly derivable n.e.

Finally, in §9.6, we deduce from the Third Median Theorem the Fourth and Fifth Median Theorems in which we obtain monotonicity conditions in terms of medians. From these median theorems we deduce in turn some monotonicity theorems for functions that are lower or weakly derivable n.e., including the First and Second Goldowski-Tonelli Theorems in terms of various new derivatives.

Throughout this chapter it will be understood according to our convention, unless stated otherwise, that $f: I \to \mathbb{R}$ where $I$ is a subinterval of $\mathbb{R}$.

**§9.1 Monotonicity in terms of new derivatives.** In this section we obtain three monotonicity theorems in terms of new derivatives without any derivability hypothesis, and we deduce from one of them another known monotonicity theorem in terms of a unilateral derivate. We call these theorems the Second, Third, Fourth, and Fifth Monotonicity Theorems. We present the results here in terms of lower derivative; they hold, however, also for upper, semi- and weak derivatives as seen later in Remark 9.1.12.

First, in the Second Monotonicity Theorem 9.1.3 we obtain some monotonicity conditions in terms of the sets of points where the function has a finite negative or positive lower gradient. Next, in Theorem 9.1.5, we deduce from the Normal Level Structure Theorem some general inequalities in terms of normal lower derivative; these inequalities have been obtained by Saks [S4] in the case of functions with property $(T_2)$ in terms of ordinary derivative.

Then from the last theorem we deduce in the Third Monotonicity Theorem 9.1.7 a general monotonicity condition in terms of lower regularity and lower singularity relative to lower derivative; this result does not hold in terms of ordinary derivative in general. As an immediate consequence of this theorem we obtain in the Fourth Monotonicity Theorem 9.1.8 a well known monotonicity result in terms of a unilateral derivate.

Finally, in the Fifth Monotonicity Theorem 9.1.10, we obtain a monotonicity result similar to the First Monotonicity Theorem 3.5.3 in terms of a single value of lower gradients.

It may be pointed out here that each of the three mean value theorems of §8.1 leads as usual to a corresponding monotonicity theorem in terms of lower or semi- derivative. The monotonicity theorems obtained in this section are however more general and require independent arguments.

We begin with a reduction lemma similar to Lemma 3.5.1. It should be recalled here that we call a function $I^*LSC$ (or strongly inherently $LSC$) if it is $WLSC$ and $ILSC$.

9.1.1. Lemma (Second Reduction). *Let $P$ be any function-theoretic property. If every $LSC$ function on an open interval with property $P$ is nondecreasing, then so is every $I^*LSC$ function on an open interval which has property $P$.*

Proof. Suppose $P$ satisfies the given hypothesis, and let $f$ be any function on an open interval $I$ which is $I^*LSC$ and has property $P$. Then it is enough to show that the set $E$ of points where $f$ is not $LSC$ is empty.

Suppose $E \neq \emptyset$. Then since $f$ is $ILSC$, $E$ is nowhere dense in $I$, and so is in turn its closure $\overline{E}$ relative to $I$. Also, due to the same property, there exists an open subinterval $J$ of $I$ such that $\overline{E} \cap J \neq \varnothing$ and $f$ is $LSC$ relative to $\overline{E} \cap J$. Choose $x \in E \cap J$. Now, it is clearly enough to show that $f$ is $LSC$ at $x$. However, since $f$ is $I^*LC_+$, it follows by the arguments used earlier in the proof of Lemma 3.5.1 that $f$ is $LC_+$ at $x$. Hence we need to show here only that $f$ is $UC_-$ at $x$.

First, suppose $x \notin E^-$. Then $x$ is the right endpoint of some contiguous interval $(y, x)$ of $E$, and so $f$ is by hypothesis nondecreasing on $(y, x)$. Now since $f$ is $WUC_-$, it follows that $f$ is $UC_-$ at $x$.

Next, suppose $x \in E^-$ and let $\varepsilon > 0$. Then since $f$ is $LSC$ relative to $\overline{E} \cap J$, there exists a point $y \in \overline{E} \cap J$ such that $y < x$ and

$$(1) \qquad f(t) > f(x) - \varepsilon \quad \text{for} \quad t \in \overline{E} \cap (y, x).$$

When $t \in (y, x) \sim \overline{E}$, $t$ is in some contiguous interval $(a, b)$ of $\overline{E}$ such that $a \in \overline{E} \cap (y, x)$. Then $f$ is clearly nondecreasing on $(a, b)$, and since $f$ is $WLSC$, it is indeed nondecreasing on $[a, b)$. Hence it follows from (1) that $f(t) \geq f(a) > f(x) - \varepsilon$. Thus (1) holds for every $t$ in $(y, x)$, which proves again that $f$ is $UC_-$ at $x$. □

We include here another lemma obtained earlier in [G11] which will be used here repeatedly. Given a function $f: I \to \mathbb{R}$, we used there $N(f)$ to denote the set of points $x$ in $I$ such that $f$ is not nondecreasing in any neighborhood of $x$. This set is clearly closed in $I$, and it will be called the *nonmonotonicity set* of $f$. Further, we will call the points of $N(f)$ the *nonmonotonicity points* of $f$. Note that $N(-f)$ is also in a sense a nonmonotonicity set of $f$, but for the sake of simplicity we use this term only for $N(f)$. Following is the lemma proved in [G11, p. 316].

9.1.2. LEMMA. *If $f$ is $WLC$, then the set $N(f)$ is perfect in $I$.*

9.1.3. THEOREM (SECOND MONOTONICITY). *Suppose $I$ is open and $f$ is $I^*LSC$. Let $E$ be the set of points where $f$ has a finite negative (or positive) lower gradient. Then $f$ is nondecreasing (or nonincreasing) whenever one of the following holds*: (a) *$E$ is empty*; (b) *$f$ is $WC_-$ (or $WC_+$) and $E$ is finite*; *or* (c) *$f$ is lower nonangular and* card $(E) < c$.

PROOF. It is enough to prove here the result for $f$ to be nondecreasing, for the other result follows on applying it to $f_-$ (see Lemma 1.1.2). Further, on account of Lemma 9.1.1, it is enough to prove this result when $f$ is $LSC$. Also, we can assume here that $\underline{D}f \geq 0$ a.e., for otherwise it follows from Theorem 5.5.2 that $f$ has a finite negative derivative at a set of points of positive measure, which contradicts each of the hypotheses (a), (b), and (c).

Hence, suppose that $f$ is $LSC$ and $\underline{D}f \geq 0$ a.e. and that $f$ is not nondecreasing. Then there exist $a,b \in I$, $a < b$, such that $f(a) > f(b)$. Hence $r \equiv Qf(a, b) < 0$. We claim that for each $c \in (r, 0)$ there is a point $x_c$ in $E$ where $f$ has a lower gradient $c$.

Given $r < c < 0$, it is clear that $f_{-c}(a) > f_{-c}(b)$. Now since $\overline{D}f_{-c} \geq -c > 0$ a.e., there exist two points $s, t \in I$ such that $b < s < t$ and $f_{-c}(s) < f_{-c}(t)$. Thus $a < b < s < t$ and

$$d \equiv \inf\{f_{-c}(x) : a \leq x \leq t\} \leq f_{-c}(b) \wedge f_{-c}(s) < f_{-c}(a) \wedge f_{-c}(t).$$

Now since $f_{-c}$ is $LSC$, it follows that there is a point $x_c$ in $(a, t)$ such that $f_{-c}(x_c) = d$. Then $f_{-c}$ has clearly a relative minimum at $x_c$, and so $f$ has a lower gradient $c$ at $x_c$. Thus since $x_c \in E$, this proves the sufficiency of (a).

Next, let $N = N(f)$. Clearly, $E \subset N$, and on applying the above result to restrictions of $f$ to open subintervals of $I$ which meet $N$, it follows that $E$ is dense in $N$. Now, if $f$ is $WLC_-$, it is indeed $WLC$. Hence, by Lemma 9.1.2, $N$ is dense in itself, which implies that $E$ is infinite. The sufficiency of (b) is thus established.

Finally, to prove the sufficiency of (c), suppose $f$ is lower nonangular. Then the lower derivative of $f$ is unique at each point of $E$. Hence, if $r < c < c' < 0$, it is clear that $x_c \neq x_{c'}$. Consequently, the subset $\{x_c : r < c < 0\}$ of $E$ has cardinality $\mathfrak{c}$, which proves the sufficiency of (c). □

It is now easy to see that the following results on strong monotonicity and Lipschitz property follow from the two cases of Theorem 9.1.3(a). Similar results can be deduced from parts (b) and (c) of this theorem as well.

9.1.4. COROLLARY. *Suppose $I$ is open and $f$ is $I^*LSC$. Let $G$ be the set of all finite lower gradients of $f$ wherever they exist. Then*

(a) *if* $\inf G > 0$ *or* $-\infty$, *then $f$ is strongly increasing or lower Lipschitz, respectively; and*

(b) *if $G$ is bounded by $c \in \mathbb{R}_+$ or it consists of only* 0, *then $f$ is $c$-Lipschitz or constant, respectively.*

Next, we deduce from the Normal Level Structure Theorem 6.2.2 the following theorem in terms of lower and upper derivatives which may be compared with the theorem of Saks [S4] (or see [S8, p. 280]) where this result was obtained for continuous functions with property $(T_2)$ in terms of ordinary derivative; (see Theorem 9.3.3 for an extension of Saks' theorem).

9.1.5. THEOREM. *Suppose $I = [a, b]$ and let $E_-$ and $E_+$ denote the sets of points where $f$ has a negative or positive normal lower (or upper) derivative, respectively. Then if $f$ is $R^*LI$ (or $L^*LI$), then*

$$f(a) - f(b) \leq |f(E_-)|; \tag{2}$$

*and if $f$ is $L^*UI$ (or $R^*UI$), then*

$$f(b) - f(a) \leq |f(E_+)|. \tag{3}$$

PROOF. It is enough to prove (2) here, for (3) follows from (2) on applying it to $f_-$. Further, it is enough to prove (2) in the case when $f$ is $R^*LI$, for on applying this result to $-f_-$ the result is obtained when $f$ is $L^*LI$.

Hence, suppose $f$ is $R^*LI$ and that $f(a) > f(b)$, for (2) holds trivially otherwise. Then, since $f$ is lower Darboux by Lemma 6.2.1, the interval $J \equiv (f(b), f(a))$ is included in $f((a, b))$. Let $H$ denote the set of elements $c$ of $J$ for which the level set $f^{-1}(c)$ is differentially normal. Then, by Theorem 6.1.2, $|J \sim H| = 0$, that is, $|H| = f(a) - f(b)$. Hence it is enough to show that $H \subset f(E_-)$.

Given $c \in H$, let $x$ be the infimum of the nonempty level set $f^{-1}(c)$. Then since $f(a) > c$ and $\underline{f}(x-0) \leq f(x) \leq \underline{f}(x+0)$, it follows easily from the lower Darboux property of $f$ that $a < x < b$ and $f(x) = c$. Now, since $f(t) \geq c$ for $a < t < x$, we have $D^- f(x) \leq 0$; and since $f^{-1}(c)$ is differentially normal, it follows that $f$ has a negative normal lower derivative at $x$. Thus $x \in E_-$, and hence $c \in f(E_-)$, which proves that $H \subset f(E_-)$. □

Next, on combining the above theorem with Theorem 6.7.3 we obtain the following inequalities in terms of integrals.

9.1.6. COROLLARY. *Suppose $I = [a, b]$ and $f$ is LSC, USI, and L-regular. Then*

$$\int_{E_-} Lf'\, dm \leq f(b) - f(a) \leq \int_{E_+} Lf'\, dm,$$

*where $E_-$ and $E_+$ are the sets of points where $f$ has a finite negative or positive normal lower derivative, respectively.*

In the following theorem we deduce from Theorem 9.1.5 the Third Monotonicity Theorem in terms of lower regularity and lower singularity with respect to lower and upper derivatives. It should be noted here that each of the hypotheses of this theorem is also necessary for the function to be nondecreasing. The second part of this theorem was obtained earlier in [G7] in a somewhat cruder form in the case of functions that are $L^*LI$, which is included in case (iii).

9.1.7. THEOREM (THIRD MONOTONICITY). *Suppose $f$ is either* (i) *$I^*LC$, or* (ii) *$LI_-$ and $I^*LC_+$. If $f$ is lower L-regular and lower L-singular, then it is nondecreasing.*

*Similarly, if $f$ is either* (i) *$I^*LC$, or* (iii) *$LI_+$ and $I^*LC_-$, and it is lower U-regular and lower U-singular, then it is nondecreasing.*

PROOF. It is enough to prove here the first part, for the second part follows on applying the first to $-f_-$.

Further, since $f$ is clearly $WLC$ in each of cases (i) and (ii), there is no loss of generality in assuming $I$ to be open. Also, in each of these cases it follows from Lemma 3.5.1 that it is enough to prove the result when $f$ is $LI_-$ and $LC_+$, that is, when $f$ is $R^*LI$.

Hence, suppose that $I$ is open and $f$ is $R^*LI$, and that the hypothesis of the first part holds. Let $E_-$ denote the set of points where $f$ has a negative normal lower derivative, and let $A$ and $B$ be the sets of points in $E_-$ where $Lf'$ is finite or infinite, respectively.

Now, since $f$ is lower $L$-singular, $|A| = 0$, and since $f$ has, by Corollary 6.1.4, property $(N)$ on $A$, it follows that $|f(A)| = 0$. Further, since $f$ is lower $L$-regular, $|f(B)| = 0$. Consequently, $|f(E_-)| = 0$. Hence it follows clearly from Theorem 9.1.5 that $f$ is nondecreasing. □

As an immediate consequence of the second part of the above theorem we obtain the following monotonicity theorem in terms of the derivate $D^+f$. This theorem is known in the case when $f$ is $L^*LI$ (i.e., $LC_-$ and $LI_+$), and seems to have an interesting past (see Remark 9.1.12).

9.1.8. Theorem (Fourth Monotonicity). *Suppose $f$ is either* (i) *$I^*LC$, or* (ii) *$I^*LC_-$ and $LI_+$. Then if $D^+f \geq 0$ a.e. and $D^+f > -\infty$ n.e., then $f$ is nondecreasing.*

Next, to obtain the fifth monotonicity theorem we need here the following lemma on the existence of points of relative minima which seems to be known only in the case of continuous nowhere monotone functions.

9.1.9. Lemma. *Suppose $I$ is open and $f$ is $I^*LSC$. If $f$ is nowhere nonincreasing and not increasing on $I$, then it has a relative minimum at some point of $I$.*

Proof. When $f$ is nowhere nonincreasing, it is also nowhere constant, and so $f$ is increasing on $I$ iff it is nondecreasing on $I$. Hence, on choosing $P$ to be the property of being nowhere nonincreasing and having nowhere a relative minimum, it follows from Lemma 9.1.1 that it is enough to prove the result when $f$ is $LSC$.

Hence suppose that $f$ is $LSC$ and nowhere nonincreasing and that it is not nondecreasing on $I$. Then there exist $a, b \in I$, $a < b$, such that $f(a) > f(b)$, and there exist further $c, d \in I$ such that $b < c < d$ and $f(c) < f(d)$. Thus $a < b < c < d$ and

$$t \equiv \inf\{f(x) : a \leq x \leq d\} < f(b) \wedge f(c) < f(a) \wedge f(d).$$

Now, since $f$ is $LSC$, it follows that there is a point $x$ in $(a, d)$ where $f(x) = t$. Clearly, $x$ is a point of relative minimum of $f$. □

The following monotonicity theorem in terms of a single value of lower gradients is somewhat similar to the First Monotonicity Theorem 3.5.3 of Morse, but its proof is much simpler.

9.1.10. THEOREM (FIFTH MONOTONICITY). *Suppose $I$ is open and $f$ is $I^*LSC$ and nowhere decreasing. Let $E$ be the set of points where $f$ has a zero lower gradient. Then $f$ is increasing whenever one of the following holds:* (a) *$E$ is empty; or* (b) *$f$ is $WLC_-$ and $E$ is finite.*

PROOF. Let $A$ denote the set of points where $f$ has a relative minimum. Then $A \subset E$, and since $E$ is finite in each of the two cases, $f$ is clearly nowhere constant. Hence $f$ is nowhere nonincreasing. Now the sufficiency of (a) follows directly from the above lemma.

Next, suppose (b) holds but $f$ is not increasing on $I$. Then $f$ is clearly not nondecreasing, and hence the set $N = N(f)$ is nonempty. Further, it is clear that $A \subset N$, and it follows from the above lemma that $A$ is dense in $N$. Also, since $f$ is clearly $WLC$, $N$ is dense in itself by Lemma 9.1.2. Consequently, $A$ is infinite, and so in turn is $E$, which contradicts (b). □

Finally, using arguments similar to the ones used earlier for Corollary 3.5.4, we obtain from part (a) of the above theorem the following necessary and sufficient conditions for an $I^*LSC$ function to be nondecreasing. Similar necessary and sufficient conditions can be deduced from its part (b).

9.1.11. COROLLARY. *Suppose $I$ is open and $f$ is $I^*LSC$. Then $f$ is nondecreasing whenever one of the following holds:*

(a) *$f$ is nowhere strongly decreasing and there exists an arbitrarily small $c > 0$ such that $f$ has nowhere a lower gradient $-c$; or*

(b) *$f$ is lower singular and there exists $c \in \mathbb{R}_+$ such that $f$ has nowhere a lower gradient $-c$.*

9.1.12. REMARK. As the semi- and weak derivatives are only generalizations of lower (or upper) derivative, it is clear that the results of this section hold also for these derivatives without any modifications. In the case of upper derivative, the results follow on the other hand on applying the results of this section to $-f_-$. This, however, involves modifications in continuity hypotheses, some of which have been indicated in the text of this section.

As for the origins of the Fourth Monotonicity Theorem 9.1.8, this theorem seems to have been obtained first by Carathéodory [Ca, p. 580] in the case of continuous functions (or see Saks [S6, p. 140]). In the more general case when $f$ is $L^*LI$ (i.e., $LC_-$ and $LI_+$), this theorem follows directly from a theorem of Zygmund [S8, p. 203] with the help of Saks' theorem obtained in Corollary 6.1.4. The somewhat surprising continuity hypothesis $L^*LI$ seems to have been discovered by Saks; the Zygmund's unpublished theorem first appeared in Saks's paper [S3, p. 248] in the case of continuous functions, and the hypothesis $L^*LI$ was introduced by Saks without this nomenclature in [S8, p. 203]. Subsequently, this theorem was rediscovered in this case by Gál [Ga] with an independent proof.

## §9.2 Points of extrema and concavity in terms of new derivatives.

In this section we deduce from the Second Monotonicity Theorem of the last section and the Second Mean Value Theorem of §8.1 generalized versions of the two commonly known derivative tests for points of relative extrema in terms of lower derivative. These generalized versions hold without any derivability hypothesis at points other than the point in question, and thus they are found to be much more general than the known tests.

Also, from the Second Mean Value Theorem we obtain a concavity test in terms of ordinary derivative of lower derivative relative to the set of points where it exists. This test is also found to be much more general than the known test since it holds without any hypothesis of first derivability.

Although the results of this section are presented in terms of lower derivative, as we see later in Remark 9.2.6 they hold also for upper, semi-, and weak derivatives.

Let $f : I \to \mathbb{R}$ and $x \in I^\circ$. If $f$ has a relative extremum at $x$, it clearly has a zero semigradient at $x$. Hence we will call $x$ a *critical point* of $f$ if $f$ has a zero semigradient at $x$.

9.2.1. THEOREM (FIRST DERIVATIVE TEST). *Suppose $f$ is $I^*LSC$ and $x \in I^\circ$ is a critical point of $f$. Let $E_-$ and $E_+$ denote the sets of points where $f$ has a finite negative or positive lower gradient, respectively. Then*

(a) *if there is a $\delta > 0$ such that $E_+ \cap (x - \delta, x)$ and $E_- \cap (x, x + \delta)$ are empty, then $f$ has a relative minimum at $x$; and*

(b) *if $f$ is WC at $x$ and there is a $\delta > 0$ such that $E_- \cap (x - \delta, x)$ and $E_+ \cap (x, x + \delta)$ are empty, then $f$ has a relative maximum at $x$.*

*Further, if $f$ is WC or lower nonangular, then the same holds provided that the sets considered in* (a) *and* (b) *are finite or countable, respectively.*

PROOF. First, suppose the hypothesis of (a) holds; we can clearly assume here that $(x - \delta, x + \delta) \subset I$. Then it follows from the Second Monotonicity Theorem 9.1.3(a) that $f$ is nonincreasing on $(x - \delta, x)$ and nondecreasing on $(x, x + \delta)$. Now since $f$ is *WLSC*, it is clear that $f$ has a relative minimum at $x$. A similar argument holds for (b).

The last two parts are obtained similarly from parts (b) and (c) of the Second Monotonicity Theorem 9.1.3. □

In terms of the ordinary derivative, the last part of the above theorem yields the following generalization of the commonly known first derivative test (see, e.g., Swokowski [Swo, p. 149]).

9.2.2. COROLLARY. *Suppose $f$ is $I^*LSC$ and derivable everywhere, and $x \in I^\circ$ is a critical point of $f$. Then if there is a $\delta > 0$ such that $f' \le 0$ n.e. in $(x - \delta, x)$ and $\ge 0$ n.e. in $(x, x + \delta)$, then $f$ has a relative minimum at $x$; and if $f$ is WC at $x$ and $f' \ge 0$ n.e. in $(x - \delta, x)$ and $\le 0$ n.e. in $(x, x + \delta)$, then $f$ has a relative maximum at $x$.*

Next, suppose $f$ is uniquely weakly differentiable at $x \in I^\circ$. Extending the usual notions of local concavity (see, e.g., [Swo, p. 154]) we define $Gf$, the graph of $f$, to be *concave upward* at $Gf(x) \equiv (x, f(x))$ if there is a $\delta > 0$ such that

$$(1) \qquad f(y) > f(x) + Wf'(x)(y-x) \quad \text{for} \quad 0 < |y-x| < \delta;$$

and if the reverse inequality of the first inequality of (1) holds, then $Gf$ will be said to be *concave downward* at $Gf(x)$.

When $Gf$ is concave upward or downward at each point of $Gf$, then $Gf$ will be said as usual to be so on $I$.

Further, when $f$ is $I^*C$ and nonangular, it follows from the Second Mean Value Theorem 8.1.4 that $f$ is uniquely lower differentiable at a dense set of points in $I$. Hence, if $x \in \underline{\Delta}^*(f)$, the lower derivative $Lf'$ will be said to be derivable at $x$ if it is so relative to $\underline{\Delta}^*(f)$, and if so, then we will use $(Lf')'(x)$ to denote this derivative.

In the following theorem we deduce from the Second Mean Value Theorem a generalized version of the well-known test for concavity (see, e.g., [Swo, p. 155]). It should be noted here that the following result does not involve any derivability hypothesis, except for nonangularity, at points other than the point in question.

9.2.3. THEOREM (TEST FOR CONCAVITY). *Let $f$ be $I^*C$ and nonangular. Suppose $f$ is lower differentiable at $x \in I^\circ$ and $Lf'$ is derivable at $x$ relative to $\underline{\Delta}^*(f)$. Then if $(Lf')'(x) > 0$, then $Gf$ is concave upward at $Gf(x)$; and if $(Lf')'(x) < 0$, then $Gf$ is concave downward at $Gf(x)$.*

PROOF. First, suppose $(Lf')'(x) > 0$. Then there is a $\delta > 0$ such that $J \equiv (x-\delta, x+\delta) \subset I$ and

$$(2) \qquad Q(Lf')(x,y) > 0 \quad \text{whenever} \quad y \in \underline{\Delta}^*(f) \cap J,\ y \neq x.$$

Now define

$$(3) \qquad g(y) = f(y) - f(x) - Lf'(x)(y-x), \qquad y \in J.$$

Then, according to (1), it is enough to show that $g(y) > 0$ for $y \in J$, $y \neq x$.

Now, given $y \in J$, $y \neq x$, there exists by the Second Mean Value Theorem 8.1.4 a point $t$ between $x$ and $y$ such that $t \in \underline{\Delta}^*(f)$ and $f(y) - f(x) = Lf'(t)(y-x)$. Hence it follows from (3) that

$$g(y) = \{Lf'(t) - Lf'(x)\}(y-x) = Q(Lf')(x,t)(t-x)(y-x).$$

Now since $(t-x)$ and $(y-x)$ are of the same sign, their product is positive, and hence it follows from (2) that $g(y) > 0$.

A similar argument holds in the other case. □

As for the global concavity, the above theorem yields the following test in terms of lower derivative in place of the ordinary derivative used in the commonly known test.

9.2.4. COROLLARY. *Suppose $I$ is open, $f$ is $I^*C$ and uniquely lower differentiable on $I$ and that $Lf'$ is derivable at each point of $I$. Then if $(Lf')'(x) > 0$ for each $x$, then $Gf$ is concave upward on $I$; and if $(Lf')'(x) < 0$ for each $x$, then $Gf$ is concave downward on $I$.*

Finally, from Theorem 9.2.3 we deduce the following generalization of the second derivative test for points of extrema. Once again, the following test does not involve any derivability hypothesis, except for nonangularity, at points other than the point in question.

9.2.5. THEOREM (SECOND DERIVATIVE TEST). *Suppose $f$ is $I^*C$ and nonangular, $x \in I^\circ$, and $Lf'(x) = 0$. Then if $Lf'$ has a positive or negative derivative at $x$ relative to $\underline{\Delta}^*(f)$, then $f$ has a relative minimum or maximum, respectively, at $x$.*

PROOF. First, suppose $(Lf')'(x) > 0$. Then, by Theorem 9.2.3, $Gf$ is concave upward at $Gf(x)$, and since $Lf'(x) = 0$, it follows from (1) that $f$ has a (proper) relative minimum at $x$. A similar argument holds in the other case. □

9.2.6. REMARK. As the semi- and weak derivatives are only generalizations of lower derivative, taking into account the proof of Theorem 9.2.3 it is easy to see that the results of this section hold also in terms of these derivatives without any modifications.

This section's results hold also for upper derivative as it follows easily on applying them to $-f$; however, in this case the continuity hypotheses need some obvious modifications.

**§9.3 Monotonicity under properties $(T_2)$ and $(N)$ in terms of ordinary derivative.** In this section we deduce from the Normal Level Structure Theorem an extended version of a theorem of Saks [S4] on functions with property $(T_2)$ in terms of ordinary derivative which is similar to Theorem 9.1.5 (see Theorem 9.3.3). This theorem was obtained originally by Saks for continuous functions, and it was later extended to Darboux functions of $B_1$ by Bruckner [Br1]. We extend it here further to lower internal functions of $B_1$, and we deduce from it two monotonicity results for functions with properties $(T_2)$ and $(N)$.

We need here two lemmas. It is easy to see that a lower Darboux function on an interval is always lower internal. In the following lemma we obtain its converse for a function of $B_1$. The last part of this lemma has been obtained by Sen [Se] in the case when $f$ is real-valued.

9.3.1. LEMMA. *Suppose $f : I \to \overline{\mathbb{R}}$ is of $B_1$. Then $f$ is lower or upper Darboux iff it is lower or upper internal, respectively. Consequently, $f$ is a Darboux function iff it is internal.*

PROOF. It is clearly enough to prove here the result for lower Darboux property, and as mentioned above, we need to prove only its sufficiency part.

Hence, suppose $f$ is lower internal but it is not lower Darboux. Then there exist $a$, $b \in I$, $a < b$, and a real number $c$ such that $f(a) > c > f(b)$ and $(a, b) \cap f^{-1}(c) = \varnothing$. Let $J = (a, b)$, and set

$$E_- = \{x \in J : f(x) < c\}, \qquad E_+ = \{x \in J : f(x) > c\}.$$

Further, let $\mathscr{I}_-$ and $\mathscr{I}_+$ denote the families of maximal open intervals that are contained in $E_-$ and $E_+$ respectively. Since $f$ is lower internal, if an open interval $(x, y) \in \mathscr{I}_-$, then $f(x) < c$, and if $(x, y) \in \mathscr{I}_+$, then $f(y) > c$. Hence neither of the families $\mathscr{I}_-$ and $\mathscr{I}_+$ contains any abutting intervals. Also, if $(x_1, y_1) \in \mathscr{I}_+$ and $(x_2, y_2) \in \mathscr{I}_-$, then $y_1 \neq x_2$.

Now, let $\mathscr{I}$ be the family of open intervals obtained from $\mathscr{I}_- \cup \mathscr{I}_+$ on replacing each pair of abutting intervals in the family by the minimal open interval that contains both of them. Then the intervals in $\mathscr{I}$ are nonabutting, and $P \equiv J \sim \bigcup \mathscr{I}$ is thus a dense in itself $G_\delta$-set. Further, since $f(a) > c$ and $f(b) < c$, it is clear from above that $J = (a, b) \notin \mathscr{I}$. Hence, $P$ is nonempty.

Next, we claim that the sets $P_- = P \cap E_-$ and $P_+ = P \cap E_+$ are both dense in $P$. Let $x$, $y$ be any pair of points in $P$ such that $x < y$ and $P \cap (x, y) \neq \varnothing$. If $P \cap [x, y] \subset E_+$, then it is clear from above that $\mathscr{I}_-$ does not contain any subinterval of $[x, y]$, i.e., $[x, y] \sim P \subset \bigcup \mathscr{I}_+ \subset E_+$, and so $[x, y] \subset E_+$. But this contradicts the fact that $P \cap (x, y) \neq \varnothing$. Consequently, $P_-$ is dense in $P$. A similar argument holds for $P_+$.

Now, since $P \cap f^{-1}(c) = \varnothing$ and $f$ is of $B_1$, it is clear that $P_-$ and $P_+$ are disjoint $G_\delta$-sets each of which is dense in $P$. However, this contradicts the fact that $P$ is nonmeager in itself. □

Next, in the following lemma we prove a result which is somewhat stronger than the theorem that follows. This lemma will be needed again in the next section, and its proof is an adaptation of the arguments used by Saks and Bruckner in their proofs of the following theorem.

9.3.2. LEMMA. *Suppose $f : [a, b] \to \mathbb{R}$ is lower internal and of $B_1$, and it has property ($T_2$). If $f(a) > f(b)$, then almost every level set of $f$ contains a point where $f$ is continuous and has a negative derivative.*

PROOF. Let $I = [a, b]$. Since $f$ is lower Darboux by the above lemma, and it has property ($T_2$), it follows from the Second Level Structure Theorem 6.4.1 that there is a set $H \subset f(I)$ such that $|f(I) \sim H| = 0$ and, for each $c \in H$, $f^{-1}(c)$ is countable and $f$ has a nonzero derivative at each isolated point of $f^{-1}(c)$.

Let $C$ be the set of points where $f$ is continuous, and set

$$E_- = \{x \in \Delta(f) : f'(x) < 0\}, \qquad E_+ = \{x \in \Delta(f) : f'(x) > 0\}.$$

Now, since $E_- \sim C$ is countable by Theorem 4.5.5, so is $f(E_- \sim C)$. Thus it is enough to show that $H \subset f(E_-)$. Hence, let $c \in H$. Then $S \equiv I^\circ \cap f^{-1}(c)$ is a nonempty, countable $G_\delta$-set. It is now enough to show that $S \cap E_- \neq \varnothing$.

Suppose $S \cap E_- = \varnothing$. Let $S_0$ be the set of isolated points of $S$. Then since $c \in H$, $S_0 \subset E_+$, and since $f$ is lower Darboux and $f(a) > f(b)$, it is easy to see that the following facts hold: (i) $S$ consists of more than one point, and (ii) if $x, y \in S_0$, $x < y$, then $S \cap (x, y) \neq \varnothing$.

We claim further that (iii) $S_0$ does not contain any portion of $S$ which consists of more than one point. Suppose $S_0$ contains such a portion $P$ of $S$. Then $P$ is infinite by (ii). Let $\{x_n\}$ be an enumeration of the countable set $P$. Now, for each $n$, since $x_n \in S_0 \subset E_+$, there is a maximal open interval $(a_n, b_n)$ such that $P \cap (a_n, b_n) = \{x_n\}$. Then, by (ii), $a_n, b_n \notin P$, and since $x_n \in E_+$, it is clear that $f(a_n) < c < f(b_n)$. Now set

$$F = \bar{P} \sim P = \bar{P} \sim \bigcup_n (a_n, b_n).$$

Then $F$ is closed, and each $x \in F$ is the limit of some subsequence $\{x_{n_i}\}$ of $\{x_n\}$, and thus also of the corresponding sequences $\{a_{n_i}\}$ and $\{b_{n_i}\}$. The sets $\{x \in F : f(x) < c\}$ and $\{x \in F : f(x) > c\}$ are thus both dense in $F$. Now since $F \cap S = \varnothing$, the restriction of $f$ to $S$ is thus not continuous at any point of $S$, which contradicts the fact that $f$ is of $B_1$. Hence (iii) holds.

Now, it follows from (i), (ii), and (iii) that the set $S_1 = S \sim S_0 = S \cap S'$ is nonempty. This set is further dense in itself; for if $S_1$ has some isolated point $x$, then since $x \in S'$, there is an infinite portion of $S$ on one side of $x$ which is contained in $S_0$, and this contradicts (iii). Now, since $S_0$ is countable, $S_1$ is thus a dense in itself $G_\delta$-set, and thus is uncountable. But this contradicts the fact that $S$ is countable. □

Now, from the above two lemmas we obtain directly the following extended version of the theorem of Saks and Bruckner. This theorem was obtained originally by Saks [S4] (or see [S8, p. 280]) for continuous $f$, and it was extended later by Bruckner [Br1] to a Darboux function of $B_1$.

9.3.3. THEOREM (SAKS-BRUCKNER). *Suppose $I = [a, b]$, $f$ is lower internal and of $B_1$, and $f$ has property $(T_2)$. Then*

$$f(a) - f(b) \le |f(E)|,$$

*where $E$ is the set of points where $f$ has a negative derivative.*

Next, on combining the above theorem, and the result obtained on applying it to $-f$, with Theorem 6.7.3 we obtain the following inequalities in terms of integrals.

9.3.4. Corollary. *Suppose $I = [a, b]$, $f$ is internal and of $B_1$, and it has property $(T_2)$. Then if $f$ is regular, then*

$$(1) \qquad \int_{E_-} f'\,dm \le f(b) - f(a) \le \int_{E_+} f'\,dm,$$

*where $E_-$ and $E_+$ are the sets of points where $f$ has a finite negative or positive derivative, respectively.*

It may be observed here that for the validity of the left and right inequalities of (1), $f$ needs to be only lower or upper internal, respectively.

Finally, we obtain the following monotonicity theorem on functions with properties $(T_2)$ and $(N)$, which may be compared with the Third Monotonicity Theorem 9.1.7. Part (b) of this theorem is known in the case of continuous functions (see [S8, p. 286]).

9.3.5. Theorem (Sixth Monotonicity). *Suppose $f$ is lower internal and of $B_1$. Then $f$ is nondecreasing whenever one of the following holds:*

(a) *$f$ has property $(T_2)$ and it is lower regular and lower singular;*

(b) *$f$ has property $(N)$ and it is lower singular.*

Proof. Part (a) follows clearly from Theorem 9.3.3 with the help of Corollary 6.1.4. Part (b) is obtained on the other hand from (a) with the help of the Banach-Ellis Theorem 6.6.4. □

9.3.6. Remark. In connection with Theorem 6.6.1, it is natural to ask here whether Lemma 9.3.1 or Theorem 9.3.3 holds also for a function which is nearly of $B_1$. This is not true in general. If $f$ is the characteristic function of the set of rational numbers on $[0, \sqrt{2}\,]$, it is clear that $f$ is internal, nearly of $B_1$, and has property $(T_2)$, but none of the two results holds for $f$ since it is not lower or upper Darboux.

## §9.4 Medians at nonmonotonicity points.

In this section we investigate various medians of a function $f$ at the points of its nonmonotonicity set $N(f)$. We obtain here three theorems in this direction which we call the First, Second, and Third Median Theorems.

First, in the First Median Theorem 9.4.1, we deduce from the Third Derivate Theorem strengthened versions of two known results on medians of nowhere monotone functions. Then in the Second Median Theorem 9.4.2 we deal with lower singular functions that are nowhere nondecreasing. In the Third Median Theorem 9.4.6 we investigate lower median at the points of $N(f)$ for a general $f$.

The Third Median Theorem will be used subsequently in §9.6 to obtain extended versions of the Goldowski-Tonelli theorem for various new derivatives. Some results similar to the ones obtained here were obtained earlier in [G11, §4] in the case of ordinary median.

We begin with an introduction of two known results on bilateral medians.

Let $f: X \to \mathbb{R}$. Then $f$ is called *nowhere monotone* [G2] if it is not monotone on any portion of $X$. Clearly, if $f$ is so, then $X$ is dense in itself. We will, similarly, call $f$ *nowhere strongly monotone* if it is not strongly monotone (i.e., strongly increasing or strongly decreasing) on any portion of $X$ (see §2.1 for definitions). It should be noted here that nowhere strongly monotone functions are more general than nowhere monotone functions.

Further, we will call $f$ *strongly nowhere monotone* if $f_{+c}$ is nowhere monotone for every $c \in \mathbb{R}$, or, equivalently, if $f$ is nowhere of monotonic type (see [G15]). Such a function was called earlier in [G4] a nowhere monotone function of the "second species."

Now suppose $f$ is a continuous function on an interval $I$. Then if $f$ is nowhere monotone, then it was proved in [G2, p. 176] that $0 \in M_b f$ r.e. in $I$; and in the case when $f$ is strongly nowhere monotone, it was proved in [G4, p. 87] that $f$ is biknotted (or $\overline{\mathbb{R}} \subset M_b f$) r.e. in $I$.

In the following theorem we obtain strengthened versions of these two results from the Third Derivate Theorem 2.4.1.

9.4.1. THEOREM (FIRST MEDIAN). *Let $f: X \to \mathbb{R}$ and $E \subset X$. Suppose $f$ is QLSC (or QBSC) relative to $E$. Then if $f$ is nowhere strongly monotone on $E$, then $0 \in \underline{M}f$ (or $M_b f$) r.e. in $E$; and if $f$ is strongly nowhere monotone on $E$, then $f$ is lower knotted (or biknotted) r.e. in $E$.*

PROOF. It is enough to prove each of the two results when $f$ is *QLSC* relative to $E$. For, on combining this result with the one obtained on applying it to $-f$, the result follows when $f$ is *QBSC* relative to $E$.

Hence, suppose $f$ is *QLSC* relative to $E$. In case $f$ is nowhere strongly monotone on $E$, then it follows from Theorem 2.4.1 that $D_+ f \leq 0$ r.e. in $E$, and from the result obtained on applying this result to $f_-$ (see Lemma 1.1.2) that $D^- f \geq 0$ r.e. in $E$. Consequently, $0 \in \underline{M}f$ r.e. in $E$.

Next, suppose $f$ is strongly nowhere monotone on $E$. Then $f$ is clearly nowhere lower or upper Lipschitz on $E$, and so the result follows this time from Corollary 2.4.2 and from the result obtained on applying it to $f_-$. □

In the following theorem we deal with lower singular functions.

9.4.2. THEOREM (SECOND MEDIAN). *Suppose $f$ is lower singular and nowhere nondecreasing. If $f$ is QLSC (or QBSC), then $[-\infty, 0] \subset \underline{M}f$ (or $M_b f$) r.e. in $I$.*

PROOF. It is enough to consider here as before the case when $f$ is *QLSC*. Then since $f$ is nowhere nondecreasing, it follows from Theorem 2.4.1, as before, that $D^- f \geq 0$ r.e. in $I$. Hence we need to show only that $D_+ f = -\infty$ r.e. in $I$. Suppose it is not so. Then it follows from Corollary 2.4.2 that $f$ is lower Lipschitz on some subinterval $J$ of $I$. Thus $f$ is *LAC* on $J$ (see [G11, p. 310] or §11.1 for definition of *LAC*), and hence it follows from Proposition 2.24 of [G11] that $f$ is nondecreasing on $J$, which contradicts the hypothesis.

□

Next, to obtain the Third Median Theorem we need the following generalization of lower continuity. The following condition $(C)$ is somewhat complicated in its nature, but is used here since it synthesizes several interesting cases as we will see later in the proofs of Theorems 9.6.1 and 9.6.2.

Suppose $f: I \to \mathbb{R}$, $N = N(f)$ and let $E$ be any countable subset of $I$. Then $f$ and $E$ will be said to satisfy the *condition* $(C)$ if the following four conditions hold:

$(C_1)$ $f$ is $WLC_-$,

$(C_2)$ $f$ is $LI_+$ at each point of $E$ and it is of $UB_1$ on $\bar{E}$,

$(C_3)$ $f$ is $LC_+$ at each point of $N \sim (E \cup N^-)$, and

$(C_4)$ $f(x) \leq \vee \underline{f}(x \pm 0)$ for each $x \in I^\circ \cap N^- \sim E$.

Also, $f$ will be said to satisfy condition $(C)$ when there exists some countable subset $E$ of $I$ such that $f$ and $E$ satisfy this condition.

It should be noted here that $f$ satisfies condition $(C)$ automatically when it is $WLC_-$ and $LC_+$ (choose $E = \varnothing$), and thus, in particular, when $f$ is $R^*LI$, or $LC$. The proof of the next median theorem becomes considerably simpler in these particular cases.

We will need here a few lemmas.

9.4.3. LEMMA. *Suppose $E \subset I$ is countable and $f$ is $QLSC$ at the points of $I \sim E$. If $f$ is of $UB_1$ on $\bar{E}$ and $f(x) \leq \bar{f}(x)$ for $x \in E$, then $f$ is $QLSC$.*

PROOF. Suppose the hypotheses hold for $f$ and $E$. Then, given $x \in E$, it is enough to show that $f$ is $QLSC$ at $x$.

Now, given $\varepsilon, \delta > 0$, let $J = I \cap (x - \delta, x + \delta)$. Then it is enough to show that there is a point $y \in J \sim E$ such that $f(y) > f(x) - \varepsilon$. For then, since $f$ is $QLSC$ at $y$, there exists an open interval $U \subset J$ such that $f(t) > f(y) - \varepsilon > f(x) - 2\varepsilon$ for $t \in U$, which would imply that $f$ is $QLSC$ at $x$.

Let $0 < \eta < \varepsilon$, and set $A = \{t \in J : f(t) > f(x) - \eta\}$. Then $A \neq \varnothing$ since $x \in A$. When $A$ contains any point outside $E$, we can choose that point as $y$. Hence, suppose $A \subset E$. Then for each $t \in A$, $\bar{f}(t) \geq f(t) > f(x) - \eta$; that is, in every neighborhood of $t$ there is a point $s \neq t$ such that $f(s) > f(x) - \eta$. Hence $A$ is dense in itself, and $\bar{A}$ is consequently perfect in $I$.

Now, since $\bar{A} \subset \bar{E}$ and $f$ is of $UB_1$ on $\bar{E}$, it follows from Theorem 1.4.1 that there is a residual set of points $B$ in $\bar{A}$ where $f$ is $USC$ relative to $\bar{A}$. Then since $B$ is uncountable, it contains a point $y \in J \sim E$; and since $y \in A'$, it follows that $f(y) \geq f(x) - \eta > f(x) - \varepsilon$. This proves the existence of the required point $y$. □

9.4.4. LEMMA. *Suppose $f$ and $E \subset I$ satisfy condition $(C)$. Then $N = N(f)$ is perfect in $I$; and for every $x \in N^+$ and $\varepsilon, \delta > 0$ there is a point $y \in N^+ \sim E$ such that $x \leq y < x + \delta$ and $f(y) > f(x) - \varepsilon$.*

Proof. The set $N$ is obviously closed in $I$. Suppose, if possible, that $N$ contains an isolated point $x$. First, let $x \in I^\circ$. Then $x$ is a common endpoint of two contiguous intervals $(a, x)$ and $(x, b)$ of $N$, and $f$ is nondecreasing on each of these intervals. Hence $f$ is regulated at $x$; and since $f$ is $WLC_-$, $f(x-0) \leq f(x)$. Further, since $x \notin N^-$, it follows from the hypotheses $(C_2)$ and $(C_3)$ that $f(x) \leq f(x+0)$. Thus $f$ is nondecreasing on $(a, b)$, which contradicts the fact that $x \in N$. A similar argument holds when $x$ is an endpoint of $I$. This proves that $N$ is perfect in $I$.

Next, to prove the other part, suppose $x \in N^+$ and let $\varepsilon, \delta > 0$. We can clearly assume that $x + \delta \in I$. Let $0 < \eta < \varepsilon$, and set

$$(1) \qquad A = \{t \in N^+ : x \leq t < x + \delta, f(t) > f(x) - \eta\}.$$

Then $A \neq \emptyset$ since $x \in A$. When $A$ contains some point outside $E$, we can choose $y$ to be that point. Hence, suppose $A \subset E$.

We need to prove here first that $A$ is dense in itself. Suppose $A$ contains some isolated point $u$. Then since $u \in N^+$, where $N$ is perfect, there is a point $v \in N^b$ such that

$$(2) \qquad u < v < x + \delta \quad \text{and} \quad A \cap (u, v) = \emptyset.$$

Further, since $u \in E$, it follows from $(C_2)$ and (1) that $\bar{f}(u+0) \geq f(u) > f(x) - \eta$. Hence there is a point $w$ in $(u, v)$ such that $f(w) > f(x) - \eta$. Now since $w \notin A$ by (2), it follows from (1) that $w \notin N^+$. Hence there is a contiguous interval $(a, b)$ of $N$ such that $u < a \leq w < b < v$. Then we have, as before, $f(b-0) \leq f(b)$. Thus in case either (i) $w > a$, or (ii) $w = a$ and $f(a) \leq f(a+0)$, we obtain

$$f(b) \geq f(b-0) \geq f(w) > f(x) - \eta,$$

so that $b \in A \cap (u, v)$, which contradicts (2). Otherwise, $w = a$ and $f(a) > f(a+0)$, so that $a \notin E$ by $(C_2)$, and hence it follows from $(C_4)$ that

$$\underline{f}(a-0) \geq f(a) = f(w) > f(x) - \eta.$$

Now since $a \in N^-$, there is thus a point $t \in N^+ \cap (u, a)$ such that $f(t) > f(x) - \eta$, and this contradicts (2) again. Hence $A$ is dense in itself.

The set $\bar{A}$ is thus perfect in $I$; and since $A \subset E$, $\bar{A} \subset \bar{E}$. Now since $f$ is of $UB_1$ on $\bar{E}$, there is by Theorem 1.4.1 a residual set of points $B$ in $\bar{A}$ where $f$ is $USC$ relative to $\bar{A}$. Now since $A \subset N$, $B \subset \bar{A} \subset N$, and since $B$ is uncountable, there is a point $y \in B \cap N^+ \cap (x, x+\delta) \sim E$. Further, since $y \in A'$, it follows from (1) that $f(y) \geq f(x) - \eta > f(x) - \varepsilon$. Hence $y$ is the required point. □

We include here another lemma proved earlier in [G11, p. 316] which will also be used subsequently.

9.4.5. LEMMA. *If $f$ is lower singular, then for every $c \in \mathbb{R}_+$, we have $N(f_{+c}) = N(f)$.*

Now we come to the main median theorem. Similar results on the ordinary median $Mf$ were obtained earlier in Theorems 4.7 and 4.11 of [G11].

9.4.6. THEOREM (THIRD MEDIAN). *Suppose $f$ satisfies condition $(C)$. Then $D_+f \leq 0$ r.e. in $N(f)$. Further,*

(a) *if $f$ is nowhere strongly decreasing, then $0 \in \underline{M}f$ r.e. in $N(f)$; and*

(b) *if $f$ is lower singular, then $[-\infty, 0] \subset \underline{M}f$ r.e. in $N(f)$.*

PROOF. There is by hypothesis a countable set $E \subset I$ such that $f$ and $E$ satisfy condition $(C)$. Suppose $N \equiv N(f)$ is nonempty, for there is nothing to prove otherwise. Then, by Lemma 9.4.4, $N$ is perfect in $I$.

First, let $A = \{x \in N \sim N^+ : f(x) > f(x-0)\}$, and set $P = N \sim (E \cup A)$. Then the set $N \sim P$ is clearly countable, and so it is meager in $N$. Hence to prove the first part it is enough to show that $D_+f \leq 0$ r.e. in $P$. Suppose it is not so. Then, since it is clear that $f$ is *QLSC* relative to $P$, it follows from Theorem 2.4.1 that $f$ is *SRI* on some portion of $P$. Hence there is an open interval $J$ with endpoints in $P$ such that $P \cap J \neq \varnothing$ and $f$ is *SRI* on $P \cap J$. Now, it follows easily from Lemma 9.4.4 that $f$ is *SRI* on $(P \cup N^+) \cap J$.

We now claim that $f$ is nondecreasing on the closure of each contiguous interval of $N$ in $J$. Let $(a, b)$ be any such interval. Then $f$ is nondecreasing on $(a, b)$; and since $f$ is $WLC_-$, $f(b-0) \leq f(b)$. Further, if $a \in P$, then it is clear that $f(a) \leq f(a+0) = f(a+0)$. Otherwise, $a \in A$, and so $f(a) \leq f(a+0)$ by hypothesis. Hence $f$ is nondecreasing on $[a, b]$. It is now easy to see that $f$ is nondecreasing on $N \cap J$, and hence it follows that $f$ is nondecreasing on $J$. Thus $N \cap J = \varnothing$, which contradicts the fact that $P \cap J \neq \varnothing$.

Next, to prove (a), suppose $f$ is nowhere strongly decreasing. Then due to the result just proved, it is enough to show that the set $B = \{x \in N : D^-f(x) < 0\}$ is meager in $N$. Suppose it is not so. Then there is an open interval $J \subset I$ such that $B$ is nonmeager in each portion of $N \cap J$. Now, if $N$ is dense in any open subinterval $J_0$ of $J$, then $J_0 \subset N$, and so $B \cap J_0$ is nonmeager, and since $f$ is then, by Lemma 9.4.3, *QLSC* on $J_0$, it is by Theorem 2.4.1 strongly decreasing on some subinterval $J_1$ of $J_0$, which contradicts the hypothesis. Hence $N$ is nowhere dense in $J$.

Now, let $Q = J \cap N^+ \sim E$. Then since $f$ is *QLSC* relative to $Q$ and $B \cap Q$ is nonmeager in $Q$, $f$ is, by Theorem 2.4.1, *SLD* on some portion of $Q$. Hence there is an open interval $J$ with endpoints in $Q$ and a $c \in \mathbb{R}_+$ such that $Q \cap J \neq \varnothing$ and $f_{+c}$ is *WLD* on $Q \cap J$. Then it follows from Lemma 9.4.4 that $f_{+c}$ is *WLD* on $J \cap N^+$. Now, let $(a, b)$ be any contiguous interval of $N$ in $J$, and let $a < x < b$. Then since $b \in N^+$, $f_{+c}(x) \geq f_{+c}(b)$, or $f(x) \geq f(b) + c(b-x) > f(b)$. Since this contradicts the fact that $f$ is nondecreasing on $(a, b]$, part (a) is thus established.

Finally, to prove (b), suppose $f$ is lower singular. Then $f$ is clearly nowhere strongly decreasing, and hence due to (a) it is enough to prove in this case that $D_+f = -\infty$ r.e. in $N$. Let $k$ be any positive integer. Then, by Lemma 9.4.5, $N(f_{+k}) = N$. Hence condition $(C)$ holds also for $f_{+k}$ and $E$. Consequently, by the first part there is a residual set of points $H_k$ in $N$ where $D_+f_{+k} \leq 0$, or where $D_+f \leq -k$. Now the set $H = \bigcap_k H_k$ is again residual in $N$, and $D_+f(x) = -\infty$ at each $x \in H$. □

On applying the above theorem to $-f_-$, a similar result is obtained on the upper median. The following result on bilateral median is obtained on combining this result with Theorem 9.4.6.

9.4.7. Corollary. *Let $f$ be WLC and $N = N(f)$. Suppose there is a countable set $E \subset I$ such that* (i) *$f$ is lower internal at the points of $E$ and it is of $B_1$ on $\overline{E}$,* (ii) *$f$ is $LC_-$ at the points of $N \sim (E \cup N^+)$ and $LC_+$ at the points of $N \sim (E \cup N^-)$, and* (iii) *$\wedge\bar{f}(x \pm 0) \leq f(x) \leq \vee\underline{f}(x \pm 0)$ for $x \in I^\circ \cap N \sim E$. Then*

(a) *if $f$ is nowhere strongly decreasing, then $0 \in M_b f$ r.e. in $N$; and*

(b) *if $f$ is lower singular, then $[-\infty, 0] \subset M_b f$ r.e. in $N$.*

**§9.5 Properties of lower and weakly derivable functions.** In this section we deduce from the first two derivate theorems of §2.2, and their consequences obtained in §5.4, some properties of functions that are lower or weakly derivable n.e.

First, in Theorem 9.5.1, we obtain some generalized properties of functions that are lower or weakly derivable n.e. Then in the next two theorems we deal with properties of functions that are lower or weakly derivable at a nonmeager set of points.

We begin with generalized properties of functions that are lower or weakly derivable n.e. on a set.

9.5.1. Theorem. *Suppose $f: X \to \mathbb{R}$ and $E \subset X$.*

(a) *If $f$ is lower (or weakly) derivable n.e. in $E$, then it is $G_\delta$-GL-MT$_*$ (or GMT) on $E$.*

(b) *If $f$ is lower (or weakly) differentiable n.e. in $E$, then it is F-GL-L$_*$ (or GL) on $E$.*

(c) *Given $c \in \mathbb{R}_+$, if $|Lf'| < c$ (or $|Wf'| < c$) n.e. in $E$, then $f$ is F-GL-L$_*(c)$ (or GL$(c)$) on $E$.*

Proof. We will use here the notations for derivability and knot sets of $f$ without reference to $f$.

The result on lower derivability in part (a) follows directly from Theorem 5.4.1. Also, if $f$ is weakly derivable n.e. in $E$, then it is clear that there is a countable set $C \subset E$ such that $E = C \cup (E \sim \underline{K}) \cup (E \sim \overline{K})$, and so the result follows again from Theorem 5.4.1.

The results on lower differentiability and lower derivative in (b) and (c) follow on the other hand directly from Theorem 5.4.2 since $Lf'$ is unique n.e. Further, if $f$ is weakly differentiable n.e. in $E$, then there is a countable subset $C$ of $E$ such that $E = C \cup \underline{\Delta}^* \cup \overline{\Delta}^* \cup \Delta^*_+ \cup \Delta^*_-$. Hence, in this case the results in (b) and (c) follow easily from the results just obtained and Theorem 2.2.7. □

In the case when $f$ is lower or weakly derivable at a nonmeager set of points in $E$, the following result is obtained similarly from Theorems 5.4.3, 5.4.4 and 2.4.4 and Corollary 2.4.3.

9.5.2. THEOREM. *Suppose $f : X \to \mathbb{R}$, $E \subset X$ and that $f$ is QLSC (or QBSC) relative to $E$.*

(a) *If $f$ is lower (or weakly) derivable at a nonmeager set of points in $E$, then it is $L$-$MT_*$ (or $MT$) on some portion of $E$.*

(b) *If $f$ is lower (or weakly) differentiable at a nonmeager set of points in $E$, then it is $L$-$L_*$ (or Lipschitz) on some portion of $E$.*

(c) *Given $c \in \mathbb{R}_+$, if $|Lf'| < c$ (or $|Wf'| < c$) at a nonmeager set of points in $E$, then $f$ is $L$-$L_*(c)$ (or $L(c)$) on some portion of $E$.*

In each of the above two theorems it should be noted here that the property $L$-$MT_*$ implies $L$-$BV_*$ (see Lemma 5.5.1), and $MT$ implies $BV$. Similarly, as will be seen later in §11.5, the properties $L$-$L_*$ and "Lipschitz" imply $L$-$AC_*$ and $AC$, respectively.

Further, on replacing "lower" derivability by "ordinary" derivability in these theorems, we obtain the properties $MT_*$ (or $BV_*$) and $L_*$ (or $AC_*$) in the place of their $L$ versions. Also, on applying the above theorems to $-f$, similar results are obtained on upper derivable functions.

As a prelude to the Goldowski-Tonelli theorems obtained in the next section for new derivatives, we state the following result in the direction of Theorem 9.5.2 which is obtained directly on applying the Second Median Theorem 9.4.2 to $f$ and $f_-$.

9.5.3. THEOREM. *Suppose $f$ is QLSC (or QBSC) and it is lower (or weakly) derivable at a nonmeager set of points in $I$. If $f$ is lower singular, then it is nondecreasing on some subinterval of $I$.*

## §9.6 Monotonicity in terms of medians and Goldowski-Tonelli theorems in terms of new derivatives.

In this section we deduce from the Third Median Theorem the Fourth and Fifth Median Theorems in which we obtain monotonicity conditions in terms of medians. From these three median theorems we deduce in turn some monotonicity theorems for functions that are lower or weakly derivable n.e., including the First and Second Goldowski-Tonelli Theorems in terms of various new derivatives.

First, in the Fourth Median Theorem 9.6.1, we deduce from the Third Median Theorem some monotonicity conditions in terms of lower median. Similar monotonicity conditions in terms of bilateral median follow from Corollary 9.4.7 with the help of Lemma 9.4.4; but a more general result is obtained in this direction in the Fifth Median Theorem 9.6.6.

Part (a) of the following theorem may be compared with Morse's Theorem 3.5.3 on unilateral derivates, and with Corollary 4.9 of [G11] on ordinary median. Also, for a result similar to part (b) on ordinary median see Corollary 4.12 of [G11].

9.6.1. THEOREM (FOURTH MEDIAN). *Suppose $I$ is open and $f$ is $WLC_-$. Let $A$ and $B$ denote the sets of points where $0 \in \underline{M}f$ or $[-\infty, 0] \subset \underline{M}f$, respectively.*

(a) *If $f$ is nowhere strongly decreasing and either* (i) *$A$ is empty,* (ii) *$A$ is countable and $f$ is $LC_+$ at the points of $A$, or* (iii) *$A$ is countable and $f$ is $LI_+$ at the points of $A$ and it is of $UB_1$ on $\bar{A}$, then $f$ is increasing.*

(b) *If $f$ is lower singular and one of the conditions* (i), (ii), *and* (iii) *holds for $B$ in the place of $A$, then $f$ is nondecreasing.*

PROOF. We give the proof here for part (a); part (b) is obtained by a similar argument from part (b) of Theorem 9.4.6.

First, we claim that $f$ satisfies condition $(C)$ under each of the hypotheses (i), (ii), and (iii). Let $N = N(f)$.

Let $x \in I \sim A$. Then since $0 \notin \underline{M}f(x)$, we have either $D^-f(x) < 0$ or $D_+f(x) > 0$. Hence, in any case, $f(x) \le \vee \underline{f}(x \pm 0)$. Further, if case $x \in N \sim N^-$, then $x$ is the right endpoint of some contiguous interval $(a, x)$ of $N$, and as $f$ is $WLC_-$, it is indeed nondecreasing on $(a, x]$. Hence $D^-f(x) \ge 0$, and so it follows from above that $D_+f(x) > 0$. Consequently, $f$ is $LC_+$ at $x$. It is now easy to see that condition $(C)$ holds for $f$ with $E = \varnothing$ in cases (i) and (ii) and with $E = A$ in case (iii).

Now, it follows from Lemma 9.4.4 that $N$ is perfect in $I$, and from part (a) of Theorem 9.4.6 that $A \cap N$ is residual in $N$. But since $A$ is countable in any case, this is not possible unless $N = \varnothing$. Hence $f$ is nondecreasing. Further, since $A$ is countable, it is clear that $f$ is nowhere constant. Consequently, $f$ is increasing. □

Next, we deduce from part (a) of the Third Median Theorem 9.4.6 the following monotonicity theorem for lower derivable functions.

9.6.2. THEOREM (SEVENTH MONOTONICITY). *Suppose $I$ is open and $f$ is $WLC_-$. If $f$ is nowhere strongly decreasing and the set of points where $f$ has a zero* *lower derivative is countable, then $f$ is increasing whenever one of the following holds:*

(a) *$f$ is lower derivable everywhere;*

(b) *$f$ is lower derivable n.e. and it is $LC_+$; or*

(c) *$f$ is lower derivable n.e. and it is $LI_+$ and of $UB_1$.*

PROOF. It is enough to prove here that $f$ satisfies condition $(C)$ of §9.4 in each of the three cases. For then $N \equiv N(f)$ is perfect in $I$ by Lemma 9.4.4, and since the set $\{x : 0 \in \underline{M}f(x)\}$ is clearly countable in each case, it follows from Theorem 9.4.6(a) that $N = \varnothing$. Hence $f$ is nondecreasing, and since it is clearly nowhere constant, it will follow that $f$ is increasing.

Condition $(C)$ holds clearly for $f$ in case (b) with $E = \varnothing$. Suppose $x \in \underline{\Delta}(f)$. Then it is clear that $f(x) \leq \vee f(x \pm 0)$. Further, if $x \in N \sim N^-$, then $x$ is the right endpoint of some contiguous interval $(a, x)$ of $N$, and since $f$ is $WLC_-$, $f$ is nondecreasing on $(a, x]$. Hence $D_+f(x) \geq D^-f(x) \geq 0$, and so $f$ is $LC_+$ at $x$. Thus conditions $(C_3)$ and $(C_4)$ hold at each point of $\underline{\Delta}(f)$, and so it is clear that condition $(C)$ holds for $f$ with $E = \varnothing$ in case (a) and with $E = I \sim \underline{\Delta}(f)$ in case (c). □

In the case when $f$ is derivable in the ordinary sense n.e., the above theorem has been obtained by Świątkowski [Sw, p. 196] under stronger hypotheses. However, in that case, as seen in [G11], Corollary 4.9, this result holds whenever $f$ is $WLC$.

Next, we deal with some versions of Goldowski-Tonelli theorem in terms of various new derivatives. The original theorem was obtained by Goldowski [Go] and Tonelli [Ton] in terms of ordinary derivative (see Saks [S8, p. 206]).

The following version of Goldowski-Tonelli theorem in terms of lower derivative is obtained from part (b) of the Third Median Theorem 9.4.6 by an argument similar to the one used for Theorem 9.6.2.

9.6.3. THEOREM (FIRST GOLDOWSKI-TONELLI). *Suppose $I$ is open and $f$ is $WLC_-$. If $f$ is lower singular, then it is nondecreasing whenever one of the following holds:*

(a) *$f$ is lower derivable everywhere;*

(b) *$f$ is lower derivable n.e. and it is $LC_+$; or*

(c) *$f$ is lower derivable n.e. and it is $LI_+$ and of $UB_1$.*

We include here the following consequence of the above theorem for a function to be Lipschitz or constant to indicate the hypotheses under which this result holds.

9.6.4. COROLLARY. *Suppose $I$ is open and that one of the following holds:*

(a) *$f$ is $WUSC$ and lower derivable everywhere;*

(b) *$f$ is $WUSC$, $LSI$ and of $UB_1$, and it is lower derivable n.e.; or*

(c) *$f$ is $WC_-$, or $WC_+$, or $WLSC$, or $WUSC$, and it is derivable everywhere.*

*Then if $f'$ is essentially bounded on $\Delta(f)$ by $c \in \mathbb{R}_+$, then $f$ is c-Lipschitz; and if $f$ is singular, then it is constant.*

Suppose $f$ is singular; a similar argument holds in the other case. Then in cases (a) and (b) the result follows on applying parts (a) and (c) of the above theorem, respectively, to $f$ and $f_-$. As for case (c), when $f$ is $WC_-$, the result is obtained on applying Theorem 9.6.3(a) to $f$ and $-f$; and on applying this result to $f_-$ the result is obtained when $f$ is $WC_+$; further, when $f$ is $WUSC$, the result follows from case (a), and on applying this result to $-f$ the result is obtained when $f$ is $WLSC$.

Next, we come to the extension of Goldowski-Tonelli theorem to functions that are weakly derivable n.e. This result follows clearly from Corollary 9.4.7 in the case when the function is lower internal and $LC$ n.e. We will, however, obtain a more general version of this result for lower internal functions of $B_1$. For this purpose we first obtain a modified version of a reduction theorem of Bruckner [Br1, p. 22].

Let $\mathcal{F}$ be any family of functions defined on intervals. We will call $\mathcal{F}$ *stable under restriction* if for each $f \in \mathcal{F}$, each restriction of $f$ to an interval is in $\mathcal{F}$.

9.6.5. LEMMA (THIRD REDUCTION). *Let $\mathcal{F}$ be any family of lower internal functions defined on intervals such that*

(a) *$\mathcal{F}$ is stable under restriction,*

(b) *every function in $\mathcal{F}$ which is $BV$ is nondecreasing, and*

(c) *every function in $\mathcal{F}$ which is of $B_1$ is $GBV$.*

*Then every function in $\mathcal{F}$ which is of $B_1$ is nondecreasing.*

PROOF. Suppose the hypothesis holds for $\mathcal{F}$, and let $f: I \to \mathbb{R}$ be a function in $\mathcal{F}$ of $B_1$. Suppose, if possible, that $f$ is not nondecreasing. Then $N \equiv N(f)$ is nonempty, and $N$ is perfect in $I$ by Lemma 9.1.2. Further, since $f$ is $GBV$ by (c), $N$ is the union of a sequence of sets $\{N_k\}$ on each of which $f$ is $BV$. Hence there is a $k$ such that $N_k$ is dense in some portion of $N$. Consequently, there is an interval $J = [a, b] \subset I$ such that $N \cap J^\circ \neq \varnothing$ and $N_k$ is dense in $N \cap J$.

We claim that $f$ is $BV$ on $J$. Let $a = x_0 < x_1 < \cdots < x_n = b$ be any partition of $J$. Set $S = \{1, 2, \ldots, n\}$, $S_- = \{i \in S : f(x_{i-1}) > f(x_i)\}$ and $S_+ = S \sim S_-$. Now, given $i \in S_-$, since $f$ has property $(T_2)$ by Theorem 6.5.7, it follows from Lemma 9.3.2 that there are two points $a_i$, $b_i$ in $(x_{i-1}, x_i)$ where $f$ is continuous and has negative derivatives such that $f(a_i) > f(x_{i-1}) - 1/n$ and $f(b_i) < f(x_i) + 1/n$. Then $a_i, b_i \in N \cap J$; and since $N_k$ is dense in $N \cap J$ and $f$ is continuous at $a_i$, $b_i$, there further exist two points $c_i, d_i \in N_k \cap (x_{i-1}, x_i)$ such that $|f(c_i) - f(a_i)| < 1/n$ and $|f(d_i) - f(b_i)| < 1/n$. Let, further, $V$ be the finite total variation of $f$ on $N_k$. Then we obtain

$$\begin{aligned}\sum_{i=1}^{n} |f(x_i) - f(x_{i-1})| &= \sum_{i \in S_+} \{f(x_i) - f(x_{i-1})\} + \sum_{i \in S_-} \{f(x_{i-1}) - f(x_i)\} \\ &= \sum_{i=1}^{n} \{f(x_i) - f(x_{i-1})\} + 2 \sum_{i \in S_-} \{f(x_{i-1}) - f(x_i)\} \\ &\leq f(b) - f(a) + 2 \sum_{i \in S_-} \{f(a_i) - f(b_i) + 2/n\} \\ &< f(b) - f(a) + 2 \sum_{i \in S_-} \{f(c_i) - f(d_i) + 4/n\} \\ &\leq f(b) - f(a) + 2V + 8.\end{aligned}$$

This proves that $f$ is $BV$ on $J$. Hence it follows from the hypotheses (a) and (b) that $f$ is nondecreasing on $J$. Consequently, $N \cap J^\circ = \varnothing$, which is a contradiction. □

Next, with the help of the above reduction lemma we obtain the following median theorem from the Third Median Theorem which deals with monotonicity conditions in terms of bilateral median.

9.6.6. THEOREM (FIFTH MEDIAN). *Suppose $f$ is lower internal and of $B_1$. Then $f$ is nondecreasing whenever one of the following holds:*

(a) *$f$ is nowhere strongly decreasing and the set of points where $0 \in M_b f$ is countable; or*

(b) *$f$ is lower singular and the set of point where $[-\infty, 0] \subset M_b f$ is countable.*

PROOF. We will obtain each part here with help of the above reduction lemma. First, let $\mathscr{F}$ be the family of all lower internal functions defined on intervals for which the above condition (a) holds. Then it is clear that $\mathscr{F}$ is stable under restriction. Further, if $f \in \mathscr{F}$ is $BV$, it is clearly lower continuous, and so, by Lemma 9.4.4, $N(f)$ is perfect in the domain of $f$. Hence if $f \in \mathscr{F}$, it follows from Corollary 9.4.7 that $f$ is nondecreasing. Next, if $f \in \mathscr{F}$, it follows clearly from condition (a) and Corollary 2.2.4 (on applying it to $f$ and $f_-$) that $f$ is $GMT$, and so it is $GBV$. Part (a) follows now from the Reduction Lemma 9.6.5.

Part (b) is obtained by a similar argument by replacing condition (a) in the definition of $\mathscr{F}$ by (b). □

From part (b) of the above median theorem we obtain directly the following version of the Goldowski-Tonelli theorem in terms of various new derivatives. In each of these cases the following theorem is somewhat stronger than the Goldowski-Tonelli theorem, for instead of assuming the derivative in question to be nonnegative a.e., the theorem holds as long as the "ordinary" derivative of the function is nonnegative at almost all the points where it exists. In the particular case of lower derivative, a slightly more general version of the following theorem has already been obtained in Theorem 9.6.3(c).

9.6.7. THEOREM (SECOND GOLDOWSKI-TONELLI). *Suppose that $f$ is lower internal and of $B_1$ and that it is derivable in one of the following senses n.e.*: (i) *weak derivative,* (ii) *lower derivative,* (iii) *upper derivative,* (iv) *semiderivative,* (v) *right derivative,* (vi) *left derivative,* (vii) *unilateral derivative,* (viii) *ordinary derivative.*

*Then if $f$ is lower singular, then it is nondecreasing.*

The above theorem yields, in particular, a generalization of the classical Goldowski-Tonelli theorem on ordinary derivative (see Saks [S8, p. 206]) from continuous functions to lower internal functions of $B_1$. This hypothesis is on the other hand necessary for a function to be nondecreasing.

As a consequence of the above theorem we obtain as usual the following characterizations of Lipschitz property and constant functions in terms of new derivatives. We state the result here in terms of weak derivative, but it holds in terms of each new derivative.

9.6.8. COROLLARY. *Suppose that $f$ is internal and of $B_1$ and that it is weakly derivable n.e. Then if $f'$ is essentially bounded on $\Delta(f)$ by $c \in \mathbb{R}_+$, then $f$ is c-Lipschitz; and if $f$ is singular, then it is constant.*

Finally, we deduce from part (a) of the Fifth Median Theorem 9.6.6 the following monotonicity theorem in terms of a single value of various new derivatives. We use here the fact that when a function $f$ is weakly derivable at a point $x$ where it has a bilateral median $c$, then $Wf'(x) = c$.

9.6.9. THEOREM (EIGHTH MONOTONICITY). *Let G-derivative denote any of the derivatives* (i) *to* (viii) *considered in Theorem* 9.6.7. *Suppose that $f$ is lower internal and of $B_1$ and that it is nowhere strongly decreasing and G-derivable n.e. Then*

(a) *if the set of points where $Gf' = 0$ is countable, then $f$ is increasing; and*

(b) *if there is an increasing sequence of negative numbers $\{c_n\}$ with limit 0 such that the set of points where $Gf' = c_n$ is countable for each $n$, then $f$ is nondecreasing.*

PROOF. To prove (a), suppose its hypothesis holds. Then it follows from Theorem 9.6.6(a) that $f$ is nondecreasing; and hence under the present hypothesis it is clear that $f$ is increasing.

Part (b) follows on the other hand directly from (a). □

It should be noted here that all the hypotheses involved in part (b) of the above theorem are also necessary for a function to be nondecreasing. This, however, is not true of part (a). Further, in the particular case of lower derivative a slightly more general version of part (a) was obtained in Theorem 9.6.2(b). Part (b) has been obtained on the other hand by Świątkowski [Sw] in the case of Darboux functions in terms of ordinary derivative. However, part (a) is clearly stronger than part (b).

9.6.10. Remark. We include here a few comments on the Third Reduction Lemma 9.6.5 of this section. Although the basic idea in the proof of this lemma is the same as that in the proof of the original reduction theorem of Bruckner [Br1, p. 22], the present proof is considerably simpler. The reduction theorem of Bruckner, although stated differently, follows from Lemma 9.6.5 on restricting the functions in $\mathscr{F}$ to be Darboux functions.

Bruckner used his reduction theorem to extend the Goldowski-Tonelli theorem on approximate derivative, obtained originally by Tolstoff [To] for approximately continuous functions, to Darboux functions of $B_1$ (see also Świątkowski [Sw]). With the help of the Third Reduction Lemma 9.6.5, this result can be extended further to lower internal functions of $B_1$ as will be seen in a subsequent paper. A nondecreasing function is on the other hand a Darboux function only if it is continuous.

# *CHAPTER 10*

# Properties of New Derivatives and New Derivability Sets

In this chapter we obtain the Borel class, measurability, and the Zahorski property of various new derivability sets, along with the Baire class and Darboux and Denjoy properties of new derivatives. Also, we investigate here the stationary sets of new derivatives.

First, in §10.1, we deduce from the Fifth Derivate Theorem results on the Borel class and measurability of various new derivability and knot sets. Then, in §10.2, we obtain another property of nonknotted new derivability sets which was obtained earlier by Zahorski for ordinary derivability sets.

Next, in §10.3, we investigate the Baire class and measurability of various new derivatives and medians as multifunctions. These properties of unilateral derivatives are also investigated here.

Next, in §10.4, we obtain the Darboux property of new derivatives without any derivability hypothesis. Also, the Denjoy property of ordinary and lower derivatives is obtained for nonangular functions that are lower derivable n.e.; and this property of ordinary, semi-, and weak derivatives is obtained for nonangular functions that are weakly derivable n.e. Finally, characterizations of stationary sets of various new derivatives are obtained in §10.5 from the Second and Fourth Monotonicity Theorems.

**§10.1 Borel class and measurability of new derivability and knot sets.** In this section we deduce from the Fifth Derivate Theorem results on the Borel class and measurability of various new derivability and knot sets of a function defined on a set $X \subset \mathbb{R}$.

First, in Theorem 10.1.2, we investigate these properties of various knot sets. Then, in Theorem 10.1.4, we deal with the properties of unilateral derivability sets. Finally, in Theorem 10.1.6, we obtain these properties for various lower, weak and semi- derivability sets of a function. Also, we include here some examples to show that the results obtained are the best possible in their direction (see Example 10.1.7).

Given a function $f: X \to \mathbb{R}$, it is well known that the ordinary derivability sets $\Delta(f)$ and $\Delta^*(f)$ are always $F_{\sigma\delta}$-sets in $X$ (see, e.g., [G11, p. 315]). Among the other derivability sets considered in §4.3, this result is found to hold in general only for the unilateral differentiability sets $\Delta^*_+(f)$ and $\Delta^*_-(f)$ (see Theorem 10.1.4). As we will see later in Example 10.1.7, the remaining derivability sets can be of any given additive or multiplicative class $\alpha > 1$ in $X$ without being of a lower class. Further, with the exception of $\Delta_+(f)$ and $\Delta_-(f)$ (see Theorem 10.1.4), the other derivability sets can even be nonmeasurable in $X$ (see Example 10.1.7).

We first consider the knot sets of $f$ (see §§1.1 and 4.3 for definitions). As for the ordinary knot set $K(f)$, it is known to be of the form $G_\delta \cup C$ in $X$, where $C$ is a countable set of points where $f$ is not $WC$ (see [G11, Corollary 3.7]). The other knot sets, e.g. $\underline{K}(f)$ and $K_+(f)$, are on the other hand not even measurable in $X$ in general (see Example 10.1.7(a)).

We will use in the proofs here the notations for various knot and derivability sets of $f$ introduced earlier in §§1.1 and 4.3 without reference to $f$. Further, if $A$ and $E$ are two subsets of $\mathbb{R}$, $A$ will be said as usual to have some property in $E$ if $A \cap E$ has that property relative to $E$. For example, $A$ is of $\Sigma_\alpha(E)$ iff $A \cap E$ is of that class. Similarly, as mentioned earlier, $f$ is said to be of $B_\alpha(E)$ iff its restriction to $E$ is of that class.

The results on $\overline{K}$ and $\overline{\Delta}$ will not be stated here explicitly since they follow on applying the results on $\underline{K}$ and $\underline{\Delta}$ to $-f$; similarly the results on $K_-$ and $\Delta_-$ follow on applying the results on $K_+$ and $\Delta_+$ to $f_-$.

10.1.1. Lemma. *Let $f: X \to \mathbb{R}$, $E \subset X$, $\underline{S} = \underline{S}_E(f)$ and $\overline{S} = \overline{S}_E(f)$. Then $\underline{K}(f)$ is of the form $G_\delta \cup (G_\delta \sim \underline{S})$ in $E$; and further the knot sets $K_+(f)$ and $K_b(f)$ are of the form $G_\delta \cup (G_\delta \sim \underline{S}) \cup (G_\delta \sim \overline{S}) \cup (G_\delta \sim \underline{S} \cup \overline{S})$ in $E$.*

Proof. On applying the Fifth Derivate Theorem 3.1.1 to $f$ and $f_-$, it is easy to see that the following holds relative to $E$:

$$E \cap \underline{K} = \{x \in E : D_+f(x) = -\infty\} \cap \{x \in E : D^-f(x) = +\infty\}$$
$$= \{G_\delta \cup (G_\delta \sim \underline{S})\} \cap \{G_\delta \cup (G_\delta \sim \underline{S})\} = G_\delta \cup (G_\delta \sim \underline{S}).$$

Similarly, on applying Theorem 3.1.1 to $f$ and $-f$, we obtain the following on $K_+$ relative to $E$:

$$E \cap K_+ = \{x \in E : D_+f(x) = -\infty\} \cap \{x \in E : D^+f(x) = +\infty\}$$
$$= \{G_\delta \cup (G_\delta \sim \underline{S})\} \cap \{G_\delta \cup (G_\delta \sim \overline{S})\},$$

which is clearly a set of the required form in $E$. Next, on applying this result to $f_-$, the same is obtained for $E \cap K_-$; and on combining these two results it follows that $E \cap K_b = (E \cap K_+) \cap (E \cap K_-)$ is of the same form in $E$. □

It is now easy to see that the following theorem on Borel class and measurability of various knot sets follows directly from the above lemma with the help of Lemma 3.2.3.

10.1.2. THEOREM. *Let $f : X \to \mathbb{R}$ and $E \subset X$.*
(a) *If $f$ is LSC on $E$, then $\underline{K}(f)$ is a $G_\delta$-set in $E$; and if $f$ is continuous on $E$, then the same holds for $K_+(f)$ and $K_b(f)$.*
(b) *Given $1 < \alpha < \Omega$, if $f$ is of $LB_\alpha(E)$, then $\underline{K}(f)$ is of $\Sigma_\alpha(E)$; and if $f$ is of $B_\alpha(E)$, then the same holds for $K_+(f)$ and $K_b(f)$.*
(c) *If $f$ is Lebesgue or Borel measurable on $E$, then each knot set of $f$ is a measurable or Borel set, respectively, in $E$.*

Now, with the help of Lemma 4.5.4, we obtain the following interesting characterization of measurability of a function from the above theorem.

10.1.3. COROLLARY. *Suppose $f : X \to \mathbb{R}$, $E \subset X$ and $|E \cap K_b(f)| = 0$. Then $f$ is measurable on $E$ iff $E \cap \underline{K}(f)$ (or $E \cap \overline{K}(f)$) is measurable in $E$.*

The necessity part follows from the above theorem. To see the sufficiency, suppose $A = E \cap \underline{K}$ is measurable in $E$, where $\underline{K} = \underline{K}(f)$, and so on. Let $A_1 = A \cap K_b$, $A_2 = A \sim \overline{K}$ and $B = E \sim \underline{K}$. Then $A = A_1 \cup A_2$, where $|A_1| = 0$ by the hypothesis. Hence $A_2$ is measurable in $E$. Now, it follows from Lemma 4.5.4 that $f$ is measurable on $A_2$ and $B$. Now since $B$ is also measurable in $E$, it is clear that $f$ is measurable on $E$.

Among the various general weak derivability sets, we first consider the unilateral derivability sets in the following theorem. Parts of this result hold more generally as in the case of ordinary derivability sets. The result on $\Delta_+(f)$ has been obtained by Filipczak [F, p. 11] in the case when $X$ is an interval.

10.1.4. THEOREM. *Let $f : X \to \mathbb{R}$. Then the unilateral derivability sets $\Delta_+(f)$ and $\Delta_u(f)$ are always measurable in $X$, and the unilateral differentiability sets $\Delta_+^*(f)$ and $\Delta_u^*(f)$ are $F_{\sigma\delta}$-sets in $X$.*

*Further, if $E \subset X$ and $f$ is of $B_\alpha(E)$, $\alpha > 1$, then $\Delta_+(f)$ and $\Delta_u(f)$ are of $\pi_\alpha(E)$.*

PROOF. It is enough to prove the results here for $\Delta_+$ and $\Delta_+^*$; on applying these results to $f_-$, similar results are obtained on $\Delta_-$ and $\Delta_-^*$, and then the results on $\Delta_u = \Delta_+ \cup \Delta_-$ and $\Delta_u^* = \Delta_+^* \cup \Delta_-^*$ follow clearly.

We will first prove the result for $\Delta_+^*$. Let $C$ denote the set of points in $X$ where $f$ is continuous, and let $Q$ be the set of all rational numbers. Then it is clear that

$$C \cap \Delta_+ = C \cap X^+ \sim \bigcup_{r,s \in Q; r<s} \left[\{x \in C : D_+f(x) \leq r\} \cap \{x \in C : D^+f(x) \geq s\}\right].$$

Now, according to Theorem 3.2.2, $D_+f$ is of $LB_1$ on $C$ and $D^+f$ is of $UB_1$ on $C$. Hence it follows clearly from above that $C \cap \Delta_+$ is an $F_{\sigma\delta}$-set in $C$. But since $C$ is a $G_\delta$-set in $X$, this proves that $C \cap \Delta_+$ is an $F_{\sigma\delta}$-set in $X$.

Next, it is clear that

$$C \cap \Delta_+^* = C \cap \Delta_+ \sim \left[\{x \in C : D_+f(x) = -\infty\} \cup \{x \in C : D^+f(x) = +\infty\}\right].$$

According to Theorem 3.1.1, each set within the square brackets is a $G_\delta$-set in $C$. Hence it follows that $C \cap \Delta_+^*$ is an $F_{\sigma\delta}$-set in $X$. Further, since $f$ is right continuous at each point of $\Delta_+^*$, it follows from Theorem 1.1.1 that the set $\Delta_+^* \sim C$ is countable. Consequently, $\Delta_+^*$ is an $F_{\sigma\delta}$-set in $X$.

Now, since $\Delta_+^*$ is measurable in $X$ by above, and $\Delta_+ \sim \Delta_+^*$ is by Corollary 5.5.3 of measure zero, it is clear that $\Delta_+$ is measurable in $X$.

Finally, to prove the last part, let $E \subset X$ and suppose $f$ is of $B_\alpha(E)$, where $\alpha > 1$. Then, by Theorem 3.2.4(b), $D_+f$ is of $UB_\alpha$ on $E$ and $D^+f$ is of $LB_\alpha$ on $E$. Now, since

$$E \cap \Delta_+ = E \cap X^+ \sim \bigcup_{r,s \in Q; r<s} \left[\{x \in E : D_+f(x) < r\} \cap \{x \in E : D^+f(x) > s\}\right],$$

it is clear that $E \cap \Delta_+$ is of $\pi_\alpha(E)$. □

Next, to deal with the other derivability sets we first prove the following lemma. It may be recalled here that we call all the lower, upper, unilateral, semi-, and weak derivability sets of a function $f$ the "general" weak derivability sets of $f$ (see §4.3 for definitions).

10.1.5. LEMMA. *Let $f : X \to \mathbb{R}$, $E \subset X$, $\underline{S} = \underline{S}_E(f)$ and $\overline{S} = \overline{S}_E(f)$. Then each lower derivability set of $f$ is of the form $(F_{\sigma\delta} \cap \underline{S}) \cup C$ in $E$, and each weak derivability set of $f$ is of the form $(F_{\sigma\delta} \cap \underline{S}) \cup (F_{\sigma\delta} \cap \overline{S}) \cup C$ in $E$, where $C$ is a countable set.*

*Consequently, if $\underline{S}$ is* (a) *of $\pi_\alpha(E)$, $\alpha > 1$,* (b) *of $\Sigma_\alpha(E)$, $\alpha > 2$, or* (c) *measurable in $E$, then the same holds for each lower derivability set of $f$ in $E$; and if* (a), (b), *or* (c) *holds for $\overline{S}$ also, then the same holds for each general weak derivability set of $f$ in $E$.*

PROOF. It is enough to prove here the first part, for the results of the second part follow directly from the corresponding results of the first part.

We will first consider the lower derivability set $\underline{\Delta}$ of $f$. Let $Q$ denote the set of all rational numbers. Then it is clear that

$$E \cap \underline{\Delta} = E \sim \bigcup_{r, s \in Q; r<s} \left[ \{x \in E : D_+ f(x) \leq r\} \cap \{x \in E : D^- f(x) \geq s\} \right].$$

Further, it follows clearly from Theorem 3.1.1 on applying it to $f$ and $f_-$ that each of the two sets within the square brackets is a set of the form $G_\delta \cup (G_\delta \sim \underline{S})$ in $E$, and so is in turn their intersection. Hence, relative to $E$, we obtain

$$E \cap \underline{\Delta} = E \sim \{G_{\delta\sigma} \cup (G_{\delta\sigma} \sim \underline{S})\} = F_{\sigma\delta} \cap (F_{\sigma\delta} \cup \underline{S}) = F_{\sigma\delta} \cup (F_{\sigma\delta} \cap \underline{S})$$
$$= (F_{\sigma\delta} \cap \underline{S}) \cup (F_{\sigma\delta} \sim \underline{S}).$$

Now, by Theorem 4.5.5, the set $E \cap \underline{\Delta} \sim \underline{S}$ is countable, and so is in turn the set $F_{\sigma\delta} \sim \underline{S}$ in the above representation of $E \cap \underline{\Delta}$. Hence $E \cap \underline{\Delta}$ is of the required form $(F_{\sigma\delta} \cap \underline{S}) \cup C$ in $E$.

Next, we consider the lower derivability sets $\underline{\Delta}_u$, $\underline{\Delta}_k$ and $\underline{\Delta}_n$. First, since $\underline{\Delta} \sim \underline{\Delta}_u$ is countable (see Theorem 4.4.1), it follows clearly from above that the set $E \cap \underline{\Delta}_u = E \cap \underline{\Delta} \sim (\underline{\Delta} \sim \underline{\Delta}_u)$ is of the required form in $E$. Hence, with the help of Lemma 10.1.1 we further obtain, relative to $E$,

$$E \cap \underline{\Delta}_k = (E \cap \underline{\Delta}_u) \cap (E \cap \overline{K})$$
$$= \{(F_{\sigma\delta} \cap \underline{S}) \cup C\} \cap \{G_\delta \cup (G_\delta \sim \overline{S})\}$$
$$= (F_{\sigma\delta} \cap \underline{S}) \cup \{F_{\sigma\delta} \cap \underline{S} \sim (\underline{S} \cap \overline{S})\} \cup C_0,$$

where $C_0 \subset C$. Now, since $\underline{S} \cap \overline{S}$ is the set of points in $E$ where $f$ is continuous relative to $E$, it is a $G_\delta$-set in $E$, and hence it is clear that the result holds also for $E \cap \underline{\Delta}_k$. Further, since the ordinary derivability set $\Delta$ of $f$ is always an $F_{\sigma\delta}$-set in $X$ (see, e.g., [G11], Theorem 3.12), we further obtain, relative to $E$,

$$E \cap \underline{\Delta}_n = (E \cap \Delta) \cup (E \cap \underline{\Delta}_k)$$
$$= F_{\sigma\delta} \cup \{(F_{\sigma\delta} \cap \underline{S}) \cup C\} = (F_{\sigma\delta} \cap \underline{S}) \cup C,$$

for $E \cap \underline{\Delta}_n \sim \underline{S}$ is countable by Theorem 4.5.5. Hence the result holds also for $E \cap \underline{\Delta}_n$.

Next, to deal with lower differentiability sets, we first observe that

$$E \cap \underline{\Delta}^* = E \cap \underline{\Delta} \sim \left[ \{x \in E : D_+ f(x) = -\infty\} \cup \{x \in E : D^- f(x) = +\infty\} \right].$$

Hence it follows once again from Theorem 3.1.1 that, relative to $E$,

$$\begin{aligned} E \cap \underline{\Delta}^* &= E \cap \underline{\Delta} \sim [\{G_\delta \cup (G_\delta \sim \underline{S})\} \cup \{G_\delta \cup (G_\delta \sim \underline{S})\}] \\ &= E \cap \underline{\Delta} \sim \{G_\delta \cup (G_\delta \sim \underline{S})\} \\ &= \{(F_{\sigma\delta} \cap \underline{S}) \cup C\} \cap F_\sigma \cap (F_\sigma \cup \underline{S}) = (F_{\sigma\delta} \cap \underline{S}) \cup C. \end{aligned}$$

This proves the result for $\underline{\Delta}^*$. The result on $\underline{\Delta}^*_u$ follows on the other hand from the equation

$$E \cap \underline{\Delta}^*_u = (E \cap \underline{\Delta}_u) \cap (E \cap \underline{\Delta}^*).$$

Similar argument holds for $E \cap \underline{\Delta}^*_k$ and $E \cap \underline{\Delta}^*_n$. The result has thus been established for all the lower derivability sets of $f$ in $E$.

Now, the results on various upper derivability sets follow clearly on applying the above results to $-f$. Further, as for the semiderivability sets of $f$ in $E$, it follows clearly from above that, relative to $E$,

$$E \cap \Delta_s = (E \cap \underline{\Delta}) \cup (E \cap \overline{\Delta}) = \{(F_{\sigma\delta} \cap \underline{S}) \cup C\} \cup \{(F_{\sigma\delta} \cap \overline{S}) \cup C\},$$

which is clearly of the required form in $E$. A similar argument holds for the other seven semiderivability sets of $f$ in $E$. The results on unilateral derivability sets have already been established in Theorem 10.1.4.

Finally, since

$$E \cap \Delta_w = (E \cap \Delta_s) \cup (E \cap \Delta_+) \cup (E \cap \Delta_-),$$

it follows from the above results that this set is of the required form in $E$. The results for $\Delta_{wu}$, $\Delta^*_w$, and $\Delta^*_{wu}$ are obtained by similar arguments. □

The following theorem on Borel class and measurability of various lower and general weak derivability sets follows now from the last part of the above lemma with the help of Lemma 3.2.3.

10.1.6. Theorem. *Let $f: X \to \mathbb{R}$, $E \subset X$ and $1 < \alpha < \Omega$.*

(a) *If $f$ is of $LB_\alpha(E)$, then each lower derivability set of $f$ is of $\pi_\alpha(E)$; and if $f$ is of $B_\alpha(E)$, then the same holds for each general weak derivability set of $f$.*

(b) *If $f$ is Lebesgue or Borel measurable on $E$, then each general weak derivability set of $f$ is a measurable or Borel set, respectively, in $E$.*

10.1.7. Example. We include here some examples to show that the results of this section are the best possible in their direction; that is, the order of the Borel class of various derivability and knot sets in these results cannot be reduced in general. We will assume here that $E = X = I$.

(a) First, we show that the results of Theorems 10.1.2 and 10.1.6 are best possible. Let $A$ and $B$ be any two mutually disjoint countable dense sets in $\mathbb{R}$, and let $P$ be an arbitrary subset of $\mathbb{R} \sim (A \cup B)$. Set $C = (P \cup A)$ and $D = \mathbb{R} \sim (P \cup A \cup B)$, and define $f(x) = 0$, 1 or 2 according to whether $x \in C$, $D$, or $B$, respectively. Then it is easy to see that each of the lower derivability sets of $f$ coincides with $C$, whereas each of the weak and semiderivability sets of $f$ coincides with $C \cup B$.

Now, on choosing $P$ to be a nonmeasurable set we obtain a function $f$ for which the lower, weak and semi- derivability sets are nonmeasurable. Also, on choosing $P$ to be a measurable set which is not a Borel set, we obtain a measurable function $f$ for which none of these derivability sets is a Borel set. Next, given $1 < \alpha < \Omega$, choose $P$ to be a set of $\Delta_\alpha$ which is not of any additive or multiplicative class lower than $\alpha$. Then since $C$, $D$, $B$ are of $\Delta_\alpha$, it is clear that $f$ is of $B_\alpha$, whereas none of the above-considered derivability sets is of an additive or multiplicative class lower than $\alpha$.

It is further clear in the present example that $\underline{K}(f) = \mathbb{R} \sim C$ and $K_+(f) = D$. Hence, when $P$ is nonmeasurable, we obtain a function $f$ for which these knot sets are not measurable. Similar statements hold in the other two cases considered above, which in turn show that the results of Theorem 10.1.2 are best possible.

(b) Next, in connection with Theorem 10.1.4, we claim that the result on $\Delta_+^*$ does not extend to $\Delta_+$ in general. The set $\Delta_+$ is of course measurable as seen in that theorem. Let $g$ be any continuous, increasing, singular function that maps $I = [0, 1]$ onto itself (see, e.g., [S8, p. 101]). Then $E = \{x : g'(x) = +\infty\}$ is known to be an $F_{\sigma\delta}$-set (see, e.g., [G11, Theorem 3.4]). Further, $E$ is of measure zero, and as it follows from Corollaries 6.2.4 and 5.3.1, $E$ is $c$-dense in $I$. Now, let $A$ be an arbitrary dense subset of $E$, let $B$ be a countable dense set in $I$ which is disjoint with $E$, and let $C = I \sim (A \cup B)$. Define $f(x) = g(x)$, 2, or 1 according to whether $x \in A$, $B$, or $C$, respectively. Then it is clear that $\Delta_+(f) = A$.

Now, since $E$ has cardinality $c$, its power set is of cardinality $2^c$, whereas the cardinality of the family of all Borel sets in $I$ is only $c$. Hence $A$ can be chosen to be a set which is not a Borel set. This shows that $\Delta_+$ is not always a Borel set. Further, given $1 < \alpha < \Omega$, we can choose $A$ to be a set of $\Delta_\alpha$ which is not of any additive or multiplicative class lower than $\alpha$. Then we obtain a function $f$ of $B_\alpha$ for which $\Delta_+(f)$ is not of any additive or multiplicative class lower than $\alpha$.

## §10.2 Zahorski property of new derivability sets.

In this section we obtain another property of various (nonknotted) lower and semi- derivability sets, and weak derivability sets. This property was obtained originally by Zahorski for the two ordinary derivability sets, and it was later extended by Filipczak to unilateral differentiability sets.

First, in Theorem 10.2.3, we obtain this property of the new (nonknotted) derivability sets. Then, in Theorem 10.2.4, we deduce extended versions of the results of Zahorski and Filipczak from the results of this section. At the end of this section some open problems are indicated that arise naturally in this connection.

Suppose $A \subset E \subset \mathbb{R}$. Then we will say that $A$ has *property* $(Z_1)$ *in* $E$ if there is an $F_\sigma$-set $H$ in $E$ such that $A \subset H$ and $|H \sim A| = 0$. Further, when $A$ has property $(Z_1)$ and it is further an $F_{\sigma\delta}$-set in $E$, then $A$ will be said to have *property* $(Z_2)$ *in* $E$. Also, when $H$ and $A$ are $F_\sigma$- and $F_{\sigma\delta}$-sets respectively in $\mathbb{R}$, the reference to $E$ in this terminology will be dropped as usual.

In terms of the notion of measurable cover used in measure theory (see, e.g., Halmos [Hal, p. 50]), $A$ thus has property $(Z_1)$ in $E$ iff it has an $F_\sigma$-cover in $E$. Further, it is easy to see that $A$ has property $(Z_2)$ in $E$ iff it is of the form $F_\sigma \sim G_{\delta\sigma}$ in $E$ where the $G_{\delta\sigma}$-set is of measure zero; or, equivalently, iff its complement $E \sim A$ is of the form $G_\delta \cup G_{\delta\sigma}$ in $E$ where the $G_{\delta\sigma}$-set is of measure zero. It is indeed this property of $E \sim A$, with $E = I$, which was used by Zahorski [Z1, Z2] to characterize the set of points where a continuous function on $I$ is not derivable in the ordinary sense.

We begin in the following lemma with some basic facts about properties $(Z_1)$ and $(Z_2)$ which follow easily from their definitions. These facts will be used repeatedly in this section.

10.2.1. LEMMA. *Let $A \subset E \subset \mathbb{R}$. (a) If $A$ has property $(Z_1)$ in $E$, then the same holds for every subset $B$ of $A$ for which $|A \sim B| = 0$.*

*(b) If two sets $A$ and $B$ have property $(Z_1)$ or $(Z_2)$ in $E$, then the same holds for $A \cup B$ and $A \cap B$.*

It should be recalled here that we call the sets $\underline{\Delta}$, $\underline{\Delta}^*$, $\underline{\Delta}_u$, $\underline{\Delta}^*_u$, $\underline{\Delta}_n$, and $\underline{\Delta}^*_n$ "nonknotted" lower derivability sets, and, similarly, we call $\Delta_s$, $\Delta^*_s$, $\Delta_{su}$, $\Delta^*_{su}$, $\Delta_{sn}$, and $\Delta^*_{sn}$ "nonknotted" semiderivability sets (see §4.3 for their definitions).

10.2.2. LEMMA. *Let $f : X \to \mathbb{R}$, $E \subset X$, $\underline{S} = \underline{S}_E(f)$, and $\overline{S} = \overline{S}_E(f)$.*

*(a) If $\underline{S}$ (or $\overline{S}$) has property $(Z_1)$ or $(Z_2)$ in $E$, then the same holds for $E \sim \underline{K}$ (or $E \sim \overline{K}$) in $E$.*

*(b) If $\underline{S}$ has property $(Z_1)$ or $(Z_2)$ in $E$, then each of the nonknotted lower derivability sets of $f$ has that property in $E$.*

*(c) If both $\underline{S}$ and $\overline{S}$ have property $(Z_1)$ or $(Z_2)$ in $E$, then each of the weak derivability sets and nonknotted semiderivability sets of $f$ has that property in $E$.*

PROOF. To prove (a), first suppose $\underline{S}$ has property $(Z_1)$ in $E$. Then $\underline{S} = F_\sigma \sim H$ in $E$, where $|H| = 0$. Hence, relative to $E$, it follows from Lemma 10.1.1 that

$$
\begin{aligned}
E \sim \underline{K} &= E \sim E \cap \underline{K} = E \sim \{G_\delta \cup (G_\delta \sim \underline{S})\} \\
&= F_\sigma \cap (F_\sigma \cup \underline{S}) = F_\sigma \cup (F_\sigma \cap \underline{S}) \\
&= F_\sigma \cup F_\sigma \cap (F_\sigma \sim H) = F_\sigma \cup (F_\sigma \sim H) = F_\sigma \sim H_0,
\end{aligned}
$$

where $H_0 \subset H$. Hence it is clear that $E \sim \underline{K}$ has property $(Z_1)$ in $E$. Further, if $\underline{S}$ has property $(Z_2)$ in $E$, then it is an $F_{\sigma\delta}$-set in $E$, and so it follows clearly from Lemma 10.1.1 that $E \sim \underline{K}$ also is such a set. Hence then $E \sim \underline{K}$ has property $(Z_2)$. The other part of (a) follows on the other hand on applying the results just proved to $-f$.

Next, we give the proof of (b) for $\underline{\Delta}$; the same argument holds on replacing this set by $\underline{\Delta}^*$, $\underline{\Delta}_u$, $\underline{\Delta}^*_u$, $\underline{\Delta}_n$, and $\underline{\Delta}^*_n$.

First, suppose $\underline{S}$ has property $(Z_1)$ in $E$. Set $A = E \cap \underline{\Delta}$ and $B = E \sim \underline{K}$. Then it is clear that $A \subset B$, and, by (a), $B$ has property $(Z_1)$ in $E$. Further, since $B \sim A = E \sim (\underline{K} \cup \underline{\Delta})$, it follows from the Denjoy-Young-Saks Theorem 5.5.2 that $|B \sim A| = 0$. Hence it follows from Lemma 10.2.1(a) that $A$ has property $(Z_1)$ in $E$. Further, if $\underline{S}$ has property $(Z_2)$ in $E$, then since $\underline{S}$ is an $F_{\sigma\delta}$-set in $E$, it follows from Lemma 10.1.5 that $A$ also is such a set. Hence then $A$ has property $(Z_2)$ in $E$.

Next, we give the proof of (c) for $\Delta_w$; the same argument holds on replacing this set with the other weak derivability sets and nonknotted semiderivability sets of $f$.

First, suppose $\underline{S}$ and $\overline{S}$ have property $(Z_1)$ in $E$. Set this time $A = E \cap \Delta_w$ and $B = E \sim K_b$. Then it is clear that $A \subset B$; and since $K_b = \underline{K} \cap \overline{K}$ we have $B = (E \sim \underline{K}) \cup (E \sim \overline{K})$. Hence it follows from (a) and Lemma 10.2.1(b) that $B$ has property $(Z_1)$ in $E$. Further, since $B \sim A = E \sim (K_b \cup \Delta_w)$, it follows from Theorem 5.5.2 that $|B \sim A| = 0$. Hence, by Lemma 10.2.1(a), $A$ has property $(Z_1)$ in $E$. Further, if $\underline{S}$ and $\overline{S}$ have property $(Z_2)$ in $E$, then since they are $F_{\sigma\delta}$-sets in $E$, it follows from Lemma 10.1.5 that $A$ also is such a set. Hence then $A$ has property $(Z_2)$ in $E$. □

10.2.3. THEOREM. *Let $f: X \to \mathbb{R}$ and $E \subset X$.*

(a) *If $f$ is LSC relative to $E$ a.e. in $E$, then each of the nonknotted lower derivability sets of $f$ has property $(Z_1)$ in $E$; and if $f$ is further of $LB_2$ on $E$, then these derivability sets have property $(Z_2)$ in $E$.*

(b) *If $f$ is continuous relative to $E$ a.e. in $E$, then each of the weak derivability sets and nonknotted semiderivability sets of $f$ has property $(Z_1)$ in $E$; and if $f$ is further of $B_2$ on $E$, then these derivability sets have property $(Z_2)$ in $E$.*

PROOF. To prove (a), suppose $f$ is $LSC$ relative to $E$ a.e. in $E$. Then it is clear that $\underline{S}_E(f)$ has property $(Z_1)$ in $E$, and hence it follows from Lemma 10.2.2(b) that each nonknotted lower derivability set of $f$ has this property in $E$. Further, if $f$ is also of $LB_2$ on $E$, then each of these derivability sets is by Theorem 10.1.6(a) an $F_{\sigma\delta}$-set in $E$, and hence in this case it has property $(Z_2)$ in $E$.

Part (b) is obtained by a similar argument. □

In the following theorem we obtain extended versions of the above-mentioned results of Zahorski and Filipczak (see Remark 10.2.6 for details).

10.2.4. THEOREM. *Let $f: X \to \mathbb{R}$. Then each of the ordinary derivability sets $\Delta$ and $\Delta^*$ of $f$, as well as the unilateral differentiability sets $\Delta^*_+$, $\Delta^*_-$, and $\Delta^*_u$, has property $(Z_2)$ in $X$.*

PROOF. First, let $A = \Delta^*_+$ and $B$ be the set of points where the right derivates of $f$ are finite. Then $A \subset B$, and according to Theorem 2.3.2, $B$ is of the form $F_\sigma \sim C$ in $X$, where $C$ is countable. Hence $B$ has property $(Z_1)$

in $X$, and $|B \sim A| = 0$ by Theorem 5.5.2. Consequently, by Lemma 10.2.1, $A$ has property $(Z_1)$ in $X$, and so it follows from Theorem 10.1.4 that it has property $(Z_2)$ in $X$.

Now, the result on $\Delta^*_-$ is obtained on applying the above result to $f_-$, and then the results on $\Delta^*$ and $\Delta^*_u$ follow clearly from these two results with the help of Lemma 10.2.1(b).

Next, to obtain this property for $\Delta$, let $A = X \sim K(f)$. Then it follows from Corollary 3.7 of [G11] that there is an $F_\sigma$-set $B$ in $X$ such that $A \subset B$ and $B \sim A$ is countable. Also, by Lemma 4.5.4, $f$ is continuous n.e. in $A$, and thus also n.e. in $B$. Hence it follows from Theorem 10.2.3(a) that $\underline{\Delta}$ and $\overline{\Delta}$ have property $(Z_2)$ in $B$, and so $\Delta = \underline{\Delta} \cap \overline{\Delta}$ has that property in $B$ by Lemma 10.2.1. Now since $B$ is an $F_\sigma$-set in $X$, it is clear that $\Delta$ has property $(Z_2)$ in $X$. □

In the following corollary we include an interesting consequence of the above two theorems which is obtained with the help of the Denjoy-Young-Saks Theorem 5.5.2, and it may be compared with Theorem 9.5.2.

10.2.5. COROLLARY. *Let $f : X \to \mathbb{R}$ and $E \subset X$.*

(a) *If $f$ is LSC relative to $E$ a.e. in $E$ and it is lower derivable at a nonmeager set of points in $E$, then $f$ is normally lower differentiable a.e. in some portion of $E$.*

(b) *If $f$ is continuous relative to $E$ a.e. in $E$ and it is weakly derivable at a nonmeager set of points in $E$, then $f$ is normally semidifferentiable a.e. in some portion of $E$.*

(c) *If $f$ is unilaterally differentiable at a nonmeager set of points in $E$, then it is differentiable a.e. in some portion of $E$.*

10.2.6. REMARK. It should be noted here that the two parts of Theorem 10.2.3 and parts (a) and (b) of Corollary 10.2.5 do not hold without their continuity hypotheses. This can be seen easily from the function defined in Example 10.1.7(a) on choosing $P$ to be any residual set in $\mathbb{R}$ which is of measure zero.

The result in Theorem 10.2.4 on the ordinary derivability sets $\Delta$ and $\Delta^*$ was obtained first by Zahorski [Z1, Z2] for a continuous function $f$ on an interval $I$; it was extended later by Brudno [Bru] to an arbitrary function on $I$, and then it was extended in [G11, p. 315] to a function on an arbitrary domain $X \subset \mathbb{R}$. The result on $\Delta^*_+$ and $\Delta^*_-$ was obtained on the other hand by Filipczak [F, p. 11] for an arbitrary function on an interval. His proof does not extend to the general case and the present proof is of course much simpler.

What is further interesting is that each of these derivability sets is also characterized by property $(Z_2)$. Given any set $E$ with this property in $I$, Zahorski [Z1, Z2] proved that there exist continuous functions on $I$ for which $E = \Delta$ or $\Delta^*$. The same result was obtained later by Filipczak [F, p. 47] for $\Delta^*_+$, $\Delta^*_-$, and $\Delta^*_u$.

10.2.7. PROBLEM. As seen in Theorem 10.2.3, property $(Z_2)$ is shared by all the nonknotted lower and semiderivability sets and weak derivability sets, at least when the function is continuous. This property does not characterize, however, any of these derivability sets. For, according to Theorems 5.1.2 and 6.2.2, each of these sets is always dense in the domain of the function. It will be interesting to obtain characterizations of these derivability sets. This, however, will involve investigation of further properties of these sets.

Further, in connection with Theorem 10.2.4, it is not known if its result extends to the general unilateral derivability sets $\Delta_+$ and $\Delta_-$. These sets are of course not $F_{\sigma\delta}$-sets in general as seen in Example 10.1.7(b). However, when the function is of $B_2$, these two sets become $F_{\sigma\delta}$-sets by Theorem 10.1.6. So the question is indeed whether, or when, $\Delta_+$ and $\Delta_-$ have property $(Z_1)$. Incidentally, the converse of this has already been obtained by Filipczak [F, p. 47].

**§10.3 Baire class and measurability of new derivatives and medians.** In this section we investigate the Baire class and measurability of new derivatives and medians along with that of unilateral derivatives.

First, in Theorems 10.3.1 and 10.3.3, these properties are investigated for lower and weak derivatives as multifunctions. Then, in Theorem 10.3.4, we obtain these properties of unilateral derivatives, and in Theorem 10.3.5 we investigate the Baire class and measurability of lower median. Also, we include here some examples to show that the results obtained here are the best possible (see Example 10.3.8).

The ordinary derivative of a function $f: X \to \mathbb{R}$ is known to be always of $B_1$ relative to the set of points where it is defined (see, e.g., [G11, p. 322]). The lower derivative is found on the other hand to be of $B_2$ in general. The weak and semi- derivatives are, however, not even measurable in general (see Example 10.3.8(b)). The unilateral derivatives are found to be Lebesgue measurable in general (see Theorem 10.3.4), but not always Borel measurable (see Example 10.3.8(c)). We obtain here hypotheses under which these derivatives are measurable or of a given Baire class.

Following our earlier convention, any derivative of a function $f: X \to \mathbb{R}$ will be said to have some property on a set $E \subset X$ if it has that property relative to the set of points in $E$ where it is defined. Also, in the use of notations of derivability and knot sets in the proofs, the reference to $f$ will be dropped as in §10.1.

We begin with lower derivative.

10.3.1. THEOREM. *Let $f: X \to \mathbb{R}$. Then $Lf'$ is always of $B_2$, and it is of $LB_1$ on $\underline{S}(f) \cap \underline{\Delta}(f)$.*

*Consequently, $Lf'$ is of $LB_1$ on $\underline{\Delta}^*(f)$, and of $B_1$ on $\underline{\Delta}^*_u(f)$.*

PROOF. Let $\varphi = Lf'$ on $\underline{\Delta} \equiv \underline{\Delta}(f)$. Then $\varphi_* = D^-f$ and $\varphi^* = D_+f$. According to Theorem 4.5.5, $f$ is $LSC$ n.e. in $\underline{\Delta}$. Hence it follows from Theorem 3.2.2(b) on applying it to $f$ and $f_-$ that $D_+f$ and $D^-f$ are of $B_2$ on $\underline{\Delta}$. Consequently, by Theorem 1.5.1, $Lf'$ is of $B_2$ on $\underline{\Delta}$.

Next, let $A = \underline{S}(f) \cap \underline{\Delta}$. Then by part (a) of Theorem 3.2.2, $D_+f$ is of $LB_1$ on $A$, and $D^-f$ is of $UB_1$ on $A$. Hence, by Theorem 1.5.1, $Lf'$ is of $LB_1$ on $A$. Further, since $\underline{\Delta}^* \subset A$ by Theorem 4.5.2(b), it is clear that $Lf'$ is of $LB_1$ on $\underline{\Delta}^*$. The last part is now obvious since $Lf'$ is single-valued on $\underline{\Delta}^*_u$. □

We include here the following interesting consequence of the above theorem which follows with the help of Corollaries 1.5.2 and 1.5.3.

10.3.2. COROLLARY. *Let $f: X \to \mathbb{R}$. If $f$ is LSC, then $Lf'$ has a selector of $B_1$, and every subset $E$ of $\underline{\Delta}(f)$ contains a residual set of points where $Lf'$ is LSC relative to $E$.*

The following theorem deals with the weak derivative. This theorem holds also for semiderivative, for $Sf'$ is only a restriction of $Wf'$.

Similar to the notation $\underline{S}_E(f)$, we will use here $C_E(f)$ to denote the set of points in $E$ where $f$ is continuous relative to $E$. Also, $C(f)$ will be used to denote $C_X(f)$.

10.3.3. THEOREM. *Let $f: X \to \mathbb{R}$, $E \subset X$ and $C = C_E(f)$. Then*
*(a) $Wf'$ is of $LB_1$ on $C \cap \Delta_w(f)$, and of $B_1$ on $C \cap \Delta_{wu}(f)$;*
*(b) if $f$ is of $B_\alpha(E)$, $\alpha > 1$, then so is $Wf'$; and*
*(c) if $f$ is Lebesgue or Borel measurable on $E$, then so is $Wf'$.*

PROOF. Let $\varphi = Wf'$ on $\Delta_w$. Then $\varphi_* = \overline{D}_w f$ and $\varphi^* = \underline{D}_w f$ on $\Delta_w$. First, to prove (a), let $A = C \cap \Delta_w$. Then, according to Theorem 3.2.2(a), $D_+ f$ and $D_- f$ are of $LB_1$ on $A$, and $D^+ f$ and $D^- f$ are of $UB_1$ on $A$. Hence $\varphi_*$ is of $UB_1$ on $A$ and $\varphi^*$ is of $LB_1$ on $A$. Consequently, it follows from Theorem 1.5.1 that $\varphi$ is of $LB_1$ on $A$. The second part of (a) is now obvious as before.

Next, to prove (b), suppose $f$ is of $B_\alpha(E)$, where $\alpha > 1$. Let $B = E \cap \Delta_w$, $\underline{S} = \underline{S}_E(f)$ and $\overline{S} = \overline{S}_E(f)$. Then according to Theorem 3.2.3(a), $\underline{S}$ and $\overline{S}$ are of $\Pi_\alpha(E)$. We claim that these sets are of $\Delta_\alpha(B)$.

We clearly have

$$B \cap \underline{S} = \left(B \cap \underline{S} \cap \overline{S}\right) \cup \left\{B \cap \left(\underline{S} \cup \overline{S}\right) \sim B \cap \overline{S}\right\}.$$

Now, $\underline{S} \cap \overline{S}$ is clearly a $G_\delta$-set in $E$. Further, since $f$ is not biknotted at any point of $B$, $B \sim (\underline{S} \cup \overline{S})$ is countable by Lemma 4.5.4. Hence, relative to $B$, $B \cap \underline{S}$ is of the form $G_\delta \cup (G_\delta \sim \Pi_\alpha)$. This proves that $\underline{S}$ is of $\Sigma_\alpha(B)$, and hence it is of $\Delta_\alpha(B)$. On reversing the roles of $\underline{S}$ and $\overline{S}$ in this argument, it follows that the same holds for $\overline{S}$.

Now, it follows from Lemma 3.2.1(c) that each unilateral derivate of $f$ is of $B_\alpha$ on $B$. Hence it follows from Theorem 1.5.1 as before that $Wf'$ is of $B_\alpha$ on $B$.

Part (c) is obtained directly from Theorems 3.2.4(c) and 1.5.1 as in (a). □

Next, in the following theorem we consider unilateral derivatives.

10.3.4. THEOREM. *Let $f: X \to \mathbb{R}$. Then $f'_+$ and $f'_u$ are always measurable. Further,*
*(a) $f'_+$ is of $B_1$ on $C(f) \cap \Delta_+(f)$, and if $f$ is nonangular at the points of $\Delta_u$, then $f'_u$ is of $B_1$ on $C(f) \cap \Delta_u(f)$; and*
*(b) if $E \subset X$ and $f$ is of $B_\alpha(E)$, $\alpha > 1$, then $f'_+$ and $f'_u$ are also of $B_\alpha(E)$.*

PROOF. We will prove here parts (a) and (b) first.

First, let $A = C(f) \cap \Delta_+$. Then, by Theorem 3.2.2(a), $D_+f$ is of $LB_1$ on $A$, and $D^+f$ is of $UB_1$ on $A$. But since these two derivates coincide with $f'_+$ on $A$, this proves that $f'_+$ is of $B_1$ on $A$. The result on $f'_u$ in (a) follows, on the other hand, from Theorem 10.3.3(a).

Next, to prove (b), suppose $E \subset X$ and $f$ is of $B_\alpha(E)$, where $\alpha > 1$. Set $B = E \cap \Delta_u$ and $C = B \cap \Delta_{wu}$. Then $f'_u = Wf'$ on $C$. Hence it follows from Theorem 10.3.3(b) that $f'_u$ is of $B_\alpha$ on $C$. Now, since $B \sim C$ is countable (see Theorem 4.4.1), it is easy to see that $f'_u$ is of $B_\alpha$ on $B$ as well. A similar argument holds for $f'_+$.

Finally, since $f$ is right continuous at each point of $\Delta^*_+$, it is, by Theorem 1.1.1, continuous n.e. in $\Delta^*_+$. Hence $f$ is of $B_2$ on $\Delta^*_+$, and so, by (b), $f'_+$ is of $B_2$ on $\Delta^*_+$. Further, since $|\Delta_u \sim \Delta^*_u| = 0$ by Corollary 5.5.3, it is thus clear that $f'_+$ is measurable on $\Delta_+$. A similar argument holds for $f'_u$. □

In the following theorem we consider finally the Baire class and measurability of the lower median. This theorem should be compared with Theorem 3.3.2 obtained earlier on unilateral and bilateral medians.

10.3.5. THEOREM. *Let $f: X \to \mathbb{R}$ and $E \subset X$. If $f$ is of $LB_0$, $LB_1$, or $LB_\alpha$ ($\alpha > 1$) on $E$, then $\underline{M}f$ is of $UB_1$, $B_2$, or $LB_\alpha$, respectively, on $E$. Further, if $f$ is Lebesgue or Borel measurable on $E$, then so is $\underline{M}f$.*

PROOF. Let $A$ be the set of points in $E$ where $\underline{M}f$ is defined. First, suppose $f$ is $LSC$ on $E$. Then by Theorem 3.2.2(a), $D_+f$ is of $LB_1(E)$ and $D^-f$ is of $UB_1(E)$. Hence it follows from the definition of $\underline{M}f$ and Theorem 1.5.1 that $\underline{M}f$ is of $UB_1$ on $A$. In each of the remaining four cases the result is obtained similarly from Theorems 3.2.4 and 1.5.1. □

We include here two consequences of the above theorem which may be compared with Corollaries 3.3.3 and 3.3.4. The following result follows directly from the above theorem with the help of Corollary 1.5.3.

10.3.6. COROLLARY. *Let $f: X \to \mathbb{R}$. If $f$ is of $LB_\alpha$, $\alpha > 1$, then $\underline{M}f$ has a selector of $B_\alpha$; and if $f$ is measurable, then $\underline{M}f$ has a measurable selector.*

The following result is obtained, on the other hand, with the help of Corollary 1.5.2 and the definition of $USC$ multifunctions.

10.3.7. COROLLARY. *Suppose $f: X \to \mathbb{R}$ is $LSC$ relative to $E \subset X$. Then $\underline{M}f$ is $USC$ relative to $E$ r.e. in $E$; and given $c \in \overline{\mathbb{R}}$, each of the sets $\{x : c \in \underline{M}f(x)\}$ and $\underline{K}(f)$ is residual in every subset of $E$ in which it is dense.*

A result similar to Theorem 3.3.6 on lower median can also be deduced from Corollary 2.5.2 in this direction under $QLSC$.

10.3.8. EXAMPLE. (a) First, we show that unlike the ordinary derivative $f'$ which is always of $B_1$, the derivatives $Lf'$, $Sf'$, and $Wf'$ are not in general of $B_1$ even when $f$ is continuous; the unilateral derivative $f'_+$ was seen, however, in Theorem 10.3.4 to be of $B_1$ when $f$ is continuous.

Let $P$ be any nowhere dense perfect set in $I = [0, 1]$ with extreme points 0 and 1, and let $\{I_n\}$ be the sequence of contiguous intervals of $P$ in $I$, where $I_n = (a_n, b_n)$ for each $n$. Define $f(x) = 0$ for $x \in P$, and on each $I_n$ define

$$f(x) = (b_n - a_n)^{-2}\sqrt{(b_n - x)(x - a_n)}, \qquad x \in I_n.$$

Then it is easy to see that $f$ is a continuous function on $I$ which is lower derivable everywhere in $I^\circ$. Further, if $x \in P^b$, then $Lf'(x) = \{0\}$, and for each $n$, $Lf'(a_n) = [0, +\infty]$ and $Lf'(b_n) = [-\infty, 0]$. Hence, the restriction of $Lf'$ to $P$ is not $USC$ at any point of $P^b$ which is of course nonmeager in $P$. Hence, according to Corollary 1.5.2, $Lf'$ is not of $UB_1$, and thus not of $B_1$. The same holds, of course, for $Wf'$ and $Sf'$.

(b) Next, we will show that the results in parts (b) and (c) of Theorem 10.3.3 are best possible for $Wf'$ and $Sf'$. Let $A$ and $B$ be any two mutually disjoint, countable, dense sets in $I = (0, 1)$. Given an arbitrary subset $P$ of $I$, set $C = (P \cup A) \sim B$ and $D = I \sim C$. Now define $f(x) = x$ or 0 according to whether $x \in C$ or $D$, respectively. Then it is easy to see that $f$ is knottedly semidifferentiable everywhere, where $Sf'(x) = 1$ or 0 according to whether $x \in C$ or $D$, respectively.

Now, on choosing $P$ to be a nonmeasurable set we obtain a function $f$ for which $Wf'$ and $Sf'$ are not measurable. Also, on choosing $P$ to be a measurable set which is not a Borel set, we obtain a measurable function $f$ for which $Wf'$ and $Sf'$ are not Borel measurable. Next, given $1 < \alpha < \Omega$, choose $P$ to be a set of $\Delta_\alpha$ which is not of any additive or multiplicative class lower than $\alpha$. Then $f$ is of $B_\alpha$, but $Wf'$ and $Sf'$ are not of any (lower or upper) Baire class lower than $\alpha$.

(c) Finally, we consider Theorem 10.3.4 on unilateral derivatives. As we saw in that theorem, $f'_+$ and $f'_u$ are always measurable. Let $g$ be any continuous, increasing, singular function that maps $I = [0, 1]$ onto itself. Then, as seen earlier in Example 10.1.7(b), the set $E = \{x \in I^\circ : g'(x) = +\infty\}$ is an $F_{\sigma\delta}$-set which is $c$-dense in $I$, and $|E| = 0$. Let $A$ and $B$ be two arbitrary, mutually disjoint, dense subsets of $E$, and set $C = I \sim (A \cup B)$. Define $f(x) = g(x) - 1$, $1 - g(x)$, or 0 according to whether $x \in A$, $B$, or $C$, respectively. Then it is easy to see that $A \cup B = \Delta_+(f) = \Delta_u(f)$, where $f'_+(x) = +\infty$ or $-\infty$ according to whether $x \in A$ or $B$, respectively.

Now, on choosing $A$ to be a non-Borel set, as seen earlier in Example 10.1.7(b), and $B = E \sim A$, we obtain a measurable function $f$ for which $f'_+$ and $f'_u$ are not Borel measurable. Further, given $1 < \alpha < \Omega$, choose $A$ and $B$ to be two mutually disjoint, dense subsets of $E$ which are of $\Delta_\alpha$ but not of an additive or multiplicative class lower than $\alpha$ (see [K1, pp. 348, 350] for the existence of such sets). Then we obtain a function $f$ of $B_\alpha$ for which $f'_+$ and $f'_u$ are not of any (lower or upper) Baire class lower than $\alpha$.

## §10.4 Darboux and Denjoy properties of new derivatives.

In this section we obtain the Darboux and Denjoy properties of various new derivatives.

First, in Theorem 10.4.1, we obtain the Darboux property of weak, semi-, and lower derivatives without any derivability hypothesis. Further, similar to the well-known result on ordinary derivative, the Darboux property of lower derivative is obtained in Theorem 10.4.3 under a weaker continuity hypothesis in the case when the function is lower derivable everywhere. Both of these theorems are deduced from the mean value theorems established earlier for these derivatives in §8.1.

Next, in Theorem 10.4.6, the Denjoy property of ordinary and lower derivatives are obtained for a nonangular function that is lower derivable n.e. Finally, in Theorem 10.4.8, the Denjoy property of ordinary, semi-, and weak derivatives are obtained for a nonangular function which is weakly derivable n.e. These two theorems are in turn deduced from the Third Derivate Theorem of §2.4 and two monotonicity theorems of §9.5.

Since the new derivatives are set-valued in general, to obtain their Darboux property we will employ the following extension of that property defined earlier in §6.2 to multifunctions.

Given a set $E \subset \mathbb{R}$, a multifunction $\varphi : E \to 2^{\overline{\mathbb{R}}}$ will be said to have *Darboux property* if for every connected set $C$ in $\mathbb{R}$, the set $\bigcup\{\varphi(x) : x \in E \cap C\}$ is connected in $\overline{\mathbb{R}}$.

When $E$ is connected and $\varphi$ is single-valued, it is clear that the above property coincides with the usual Darboux property which is weaker than continuity. Also, when $E$ is connected and $\varphi$ is a connected-valued multifunction, it is easy to see that $\varphi$ has the Darboux property whenever it is $LSC$ or $USC$. When $E$ is on the other hand disconnected, the Darboux property of $\varphi$ is not comparable with continuity in general.

We begin with the Darboux property of semi- and weak derivatives.

10.4.1. THEOREM. *Suppose $f : I \to \mathbb{R}$ is $I^*C$. Then $Sf'$ and $Wf'$ possess the Darboux property. Moreover, if $f$ is upper nonangular, then $Lf'$ also possesses the Darboux property.*

PROOF. We will consider here $Wf'$ first. Given an arbitrary subinterval $J$ of $I$, let $A = J \cap \Delta_w(f)$ and set $H = \bigcup\{Wf'(x) : x \in A\}$. To show that $H$ is connected, let $c \in \text{co}(H)$. Then to obtain the Darboux property of $Wf'$, it is enough to show that $c \in H$.

Suppose $c \notin H$. Then there exist $a, b \in H$ such that $a < c < b$. Hence there exist $x,y \in A$ such that $a \in Wf'(x)$ and $b \in Wf'(y)$. If $x = y$, then since $Wf'(x)$ is connected, it is clear that $c \in Wf'(x) \subset H$. Hence $x \neq y$. Let $g$ denote the restriction of $f_{-c}$ to $J$. We claim that $g$ is not monotone (on $J$).

According to Theorem 7.3.1, $g$ is weakly derivable at each point of $A$, and $Wg'(x) = Wf'(x) - c$ and $Wg'(y) = Wf'(y) - c$. Now, since $c \notin Wf'(x)$, $0 \notin Wg'(x)$, and since $a \in Wf'(x)$, $g$ has a negative weak gradient $a - c$ at $x$. But since $Wg'(x)$ is connected, this implies that $Wg'(x) < 0$; that is, $\wedge D^{\pm} g(x) \le \vee D_{\pm} g(x) < 0$. Hence $g$ is not nondecreasing on $J$ even in case $x$ is an extreme point of $J$. Also, since $b \in Wf'(y)$, it is seen similarly that $g$ is not nonincreasing on $J$. Then $g$ is not monotone on $J$.

Hence, by Lemma 8.1.1, there is a nondegenerate subinterval $J_0$ of $J$ on which $g$ is continuous but not monotone. Consequently, there exist $x_1, x_2, x_3 \in J_0$ such that $x_1 < x_2 < x_3$ and either $g(x_2) < g(x_1) \wedge g(x_3)$ or $g(x_2) > g(x_1) \vee g(x_3)$. Now, since $g$ is continuous on $J_0$, it is easy to see in each of these two cases that there exist $s \in (x_1, x_2)$ and $t \in (x_2, x_3)$ such that

$g(s) = g(t)$. Hence it follows from the General Mean Value Theorem 8.1.2 that there exists a point $z$ in $(s, t)$ such that $0 \in Sg'(z)$. Then it is clear that $z \in A$ and $c \in Sf'(z) = Wf'(z) \subset H$, which is a contradiction. This proves the Darboux property of $Wf'$.

The Darboux property of $Sf'$ is proved by a similar argument. Also, in the case when $f$ is upper nonangular, the Darboux property of $Lf'$ is obtained by a similar argument from the Second Mean Value Theorem 8.1.4. □

Since the weak derivative of a function with symmetrical derivates coincides with ordinary derivative, from the first part of the above theorem we obtain the following extended version of the classical result on the Darboux property of ordinary derivative.

10.4.2. COROLLARY. *Suppose $f: I \to \mathbb{R}$ is $I^*C$. If $f$ has symmetrical derivates, then its ordinary derivative possesses the Darboux property.*

In the following theorem we obtain the Darboux property of lower derivative under a weaker continuity hypothesis.

10.4.3. THEOREM. *Suppose $f: I \to \mathbb{R}$ is $WC$ and inherently $QLSC$. If $f$ is lower derivable n.e., then $Sf'$ and $Wf'$ possess the Darboux property. Consequently, if $f$ is lower derivable everywhere, then $Lf'$ also possesses the Darboux property.*

PROOF. We consider, once again, $Wf'$ first. Given an arbitrary subinterval $J$ of $I$, let $A$ and $H$ be defined as in the proof of the last theorem. Then, given $c \in \mathrm{co}(H)$, it is enough to show that $c \in H$.

Suppose $c \notin H$, and let $g$ be the restriction of $f_{-c}$ to $J$. Then as was seen in the proof of the last theorem, $g$ is not monotone on $J$. Hence it follows by an argument used earlier in the proof of Theorem 8.1.5 that there exists a point $x \in J$ where $g$ has a unique semiderivative equal to zero. Thus $c = Sf'(x) \subset H$, which is a contradiction. This proves the Darboux property of $Wf'$.

The Darboux property of $Sf'$ is proved by a similar argument. The last part on the Darboux property of $Lf'$ is of course a direct consequence of the above result. □

In the case of a function $f$ which is uniquely lower differentiable everywhere, it follows from parts (b) and (c) of Theorem 4.5.2 that $f$ is $LSC$ and internal. Hence in that case the following result is obtained from the last part of the above theorem, which is in the direction of the classical result on finite ordinary derivative.

10.4.4. COROLLARY. *If $f: I \to \mathbb{R}$ is uniquely lower differentiable everywhere, then $Lf'$ possesses the Darboux property.*

Next, we investigate the Denjoy property of new derivatives. We will consider here only nonangular functions, and so each of these derivatives becomes single-valued in this case.

Given a measurable set $E$ in $\mathbb{R}$, a function $g : E \to \overline{\mathbb{R}}$ is said to have *Denjoy property* if for every $a, b \in \mathbb{R}$, $a < b$, the set $\{x \in E : a < g(x) < b\}$ has a positive measure if it is not empty.

We need here the following lemma.

10.4.5. LEMMA. *Suppose $f : I \to \mathbb{R}$ is WC and nonangular. Given $a, b \in \mathbb{R}$, $a < b$, let $N = N(-f_{-a}) \cap N(f_{-b})$.*

(a) *Then $N$ is perfect in $I$; and if $N = \varnothing$, then either $\overline{D}f \leq a$ everywhere or $\underline{D}f \geq b$ everywhere.*

(b) *If $f_{-a}$ is strongly increasing on some portion $P$ of $N$, then it is so on $co(P)$.*

(c) *If $f_{-b}$ is strongly decreasing on some portion $P$ of $N$, then it is so on $co(P)$.*

PROOF. The set $N$ is clearly closed in $I$. To see that it is dense in itself, let $\mathscr{I}_-$ and $\mathscr{I}_+$ denote the families of contiguous intervals of $N(-f_{-a})$ and $N(f_{-b})$, respectively, in $I$. Then $\overline{D}f \leq a$ at the points of each interval of $\mathscr{I}_-$, and $\underline{D}f \geq b$ at the points of each interval of $\mathscr{I}_+$. The intervals of the family $\mathscr{I} = \mathscr{I}_- \cup \mathscr{I}_+$ are thus mutually disjoint, and they constitute the contiguous intervals of $N$ in $I$. Now, since $f$ is $WC$, the intervals of $I_-$ are nonabutting, and so are the intervals of $\mathscr{I}_+$. If, on the other hand, $(x, y) \in \mathscr{I}_-$ and $(y, z) \in \mathscr{I}_+$, then $f_{-a}$ is nonincreasing on $[x, y]$ and $f_{-b}$ is nondecreasing on $[y, z]$. Hence then $D^-f(y) \leq a < b \leq D_+f(y)$, which contradicts the nonangularity of $f$. Similarly, it is not possible that $(x, y) \in \mathscr{I}_+$ and $(y, z) \in \mathscr{I}_-$. Hence the intervals of $\mathscr{I}$ are nonabutting, and $N$ is consequently dense in itself.

Next, suppose $N = \varnothing$. Then $I \in \mathscr{I}$, that is, $I \in \mathscr{I}_-$ or $\mathscr{I}_+$. Hence either $\overline{D}f \leq a$ everywhere or $\underline{D}f \geq b$ everywhere. Part (a) is thus established.

Next, to prove (b), suppose $f_{-a}$ is strongly increasing on some portion $P$ of $N$. Then there exists $c > a$ such that $f_{-c}$ is increasing on $P$. Given $x, y \in P$, $x < y$, it is then enough to show that $f_{-c}$ is increasing on $J \equiv [x, y]$. Let $(s, t)$ be any contiguous interval of $N$ in $J$. Then $f_{-c}(s) < f_{-c}(t)$, so that $f_{-a}(s) < f_{-a}(t)$, and so $-f_{-a}$ is not nondecreasing on $(s, t)$. Hence $(s, t) \in \mathscr{I}_+$, and consequently $f_{-b}$ is nondecreasing on $(s, t)$. The function $f_{-a}$ is thus increasing on the closure of each contiguous interval of $N$ in $J$, and hence it is increasing on $J$.

Part (c) is proved by a similar argument. □

Earlier, in Theorem 5.10 of [G11], the Denjoy property (see [D3]) of the ordinary derivative was obtained for any $WC$, nonangular function $f : I \to \mathbb{R}$ which is derivable n.e. In the following theorem we obtain similarly, under suitable continuity hypotheses, the Denjoy property of ordinary and lower derivatives of a nonangular function which is lower derivable n.e.

10.4.6. THEOREM. *Let $f: I \to \mathbb{R}$ be WC, LSI, inherently QLSC, and of $UB_1$. Suppose f is nonangular and lower derivable n.e. Then if $a, b \in \mathbb{R}$, $a < b$, and $E \equiv \{x : a < f'(x) < b\}$ is of measure zero, then either $\overline{D}f \leq a$ everywhere or $\underline{D}f \geq b$ everywhere.*

*Consequently, both $f'$ and $Lf'$ possess the Denjoy property.*

PROOF. It is enough to prove here the first part, for the second part is a direct consequence of the first.

Hence, given $a, b \in \mathbb{R}$, $a < b$, suppose $|E| = 0$, and let $N$ be defined as in the last lemma. Then according to part (a) of that lemma, $N$ is perfect in $I$, and it is enough to show that $N = \varnothing$.

Suppose $N \neq \varnothing$. Then we claim that $f_{-a}$ is nowhere strongly increasing on $N$. Suppose, if possible, that it is strongly increasing on some portion $P$ of $N$, and let $x, y, z \in P$, $x < y < z$. Then by part (b) of the above lemma, $f_{-a}$ is strongly increasing on $J \equiv [x, z]$. But since $|E| = 0$, this clearly implies that $f_{-b}$ is lower singular on $J$. Hence it follows from Theorem 9.5.5(c) that $f_{-b}$ is nondecreasing on $J^\circ$, which contradicts the fact that $y \in N$. Hence the function $g \equiv f_{-a}$ is nowhere strongly increasing on $N$.

It is deduced similarly from Lemma 10.4.5(c) and Theorem 9.5.5(c) by applying the latter to $g_-$ that the function $h \equiv f_{-b}$ is nowhere strongly decreasing on $N$. Now, since $g$ and $h$ are inherently $QLSC$, there exists a portion $Q$ of $N$ on which they are $QLSC$. Hence it follows from the Third Derivate Theorem 2.4.1 on applying it to $g$ and $h_-$ that $D_+g \leq 0$ and $D^-h \geq 0$ r.e. in $Q$. Consequently, $D_+f \leq a < b \leq D^-f$ r.e. in $Q$, which contradicts the fact that $f$ is lower derivable n.e. □

The following result is obtained from the second part of the above theorem with the help of Theorem 4.5.2 as before.

10.4.7. COROLLARY. *If $f: I \to \mathbb{R}$ is uniquely lower differentiable everywhere, then $f'$ and $Lf'$ possess the Denjoy property.*

Analogous to the above theorem, in the following theorem we obtain, under suitable continuity hypotheses, the Denjoy property of ordinary, semi-, and weak derivatives of a nonangular function which is weakly derivable n.e.

10.4.8. THEOREM. *Let $f: I \to \mathbb{R}$ be internal, inherently QBSC and of $B_1$. Suppose f is nonangular and weakly derivable n.e. Then if $a, b \in \mathbb{R}$, $a < b$ and $E \equiv \{x : a < f'(x) < b\}$ is of measure zero, then either $\overline{D}f \leq a$ everywhere or $\underline{D}f \geq b$ everywhere.*

*Consequently, $f'$, $Sf'$, and $Wf'$ possess the Denjoy property.*

PROOF. It is enough to prove here again the first part, for the second part is a direct consequence of the first.

Hence, given $a, b \in \mathbb{R}$, $a < b$, suppose $|E| = 0$, and let $N$ be defined as in Lemma 10.4.5. Then according to part (a) of that lemma, $N$ is perfect in $I$, and it is enough to show that $N = \varnothing$.

Suppose $N \neq \varnothing$. Then it follows this time from parts (b) and (c) of Lemma 10.4.5 and Theorem 9.5.9(b) by an argument similar to the one used in the proof of Theorem 10.4.6 that the function $g \equiv f_{-a}$ is nowhere strongly increasing on $N$, and the function $h \equiv f_{-b}$ is nowhere strongly decreasing on $N$. Now, there is a portion $Q$ of $N$ on which $g$ and $h$ are $QBSC$. Hence it follows from Theorem 2.4.1 on applying it to $g$ and $-g_{-}$ that $D_{\pm}g \leq 0$ r.e. in $Q$, and on applying it to $-h$ and $h_{-}$ that $D^{\pm}h \geq 0$ r.e. in $Q$. Consequently, $D_{\pm}f \leq a < b \leq D^{\pm}f$ r.e. in $Q$, which contradicts the fact that $f$ is weakly derivable n.e. □

Finally, with the help of Theorem 4.5.2(c) we obtain the following result from the second part of the above theorem.

10.4.9. COROLLARY. *Suppose $f: I \to \mathbb{R}$ is inherently QBSC and of $B_1$. If $f$ is uniquely weakly differentiable everywhere, then $f'$, $Sf'$, and $Wf'$ possess the Denjoy property.*

10.4.10. REMARK. Given $f: I \to \mathbb{R}$, let $Gf$ be any generalized derivate or derivative of $f$ which coincides with $f'$ where it exists and such that $\underline{D}f \leq Gf \leq \overline{D}f$ at every point where $Gf$ is defined. Then it follows clearly from the first parts of Theorems 10.4.6 and 10.4.8 that if $f$ satisfies the hypothesis of any of these two theorems, then $Gf$ possesses the Denjoy property. Hence under each of these two hypotheses, each unilateral, bilateral, approximate, and symmetric derivate of $f$, any selection of each of its medians, and the approximate, symmetric, unilateral, lower, upper, semi-, and weak derivatives of $f$ possess the Denjoy property.

## §10.5 Stationary sets of new derivatives.

In this section we deduce from the Eighth and Ninth Monotonicity Theorems of §9.6 characterizations of stationary sets of various new derivatives. This characterization has been known for some time for the stationary sets of ordinary derivatives.

Let $\mathscr{F}$ be any class of functions or multifunctions defined on $\mathbb{R}$ with values in $\overline{\mathbb{R}}$ or $2^{\overline{\mathbb{R}}}$, respectively. Generalizing the notion of stationary sets introduced by Boboc and Marcus in [BMa] for functions, we will call a set $E \subset \mathbb{R}$ a *stationary set* of $\mathscr{F}$ if every element of $\mathscr{F}$ which is constant on $E$ is constant on $\mathbb{R}$.

Part (a) of the following theorem was obtained originally by Boboc and Marcus [BMa] for bounded derivatives; Marcus [M4] later extended it to ordinary derivatives of continuous functions, and then in [G11, p. 327] it was extended to derivatives of weakly continuous functions.

10.5.1. THEOREM. *For each of the following classes of derivatives a set* $E \subset \mathbb{R}$ *is a stationary set iff its complement is of inner measure zero*:

(a) *ordinary derivatives of functions that are* $WC_-$, $WC_+$, $WLSC$, *or* $WUSC$;

(b) *lower derivatives of functions that are* $WUSC$; *and*

(c) *weak derivatives of functions that are internal and of* $B_1$.

PROOF. For each of the classes (a), (b), and (c), the necessity part is obvious. For, given $E \subset \mathbb{R}$, if the inner measure of $\mathbb{R} \sim E$ is positive, then as shown by Zahorski [Z4], there exists a differentiable function $f$ on $\mathbb{R}$ for which $f'$ is constant on $E$ but not on $\mathbb{R}$.

To prove the sufficiency part, suppose $E$ is a subset of $\mathbb{R}$ for which the inner measure of $\mathbb{R} \sim E$ is zero, and let $\varphi$ be a derivative in any of the classes (a), (b), and (c), say of $f$, which is constant on $E$. Then since $E$ is uncountable, it is clear that $\varphi$ is single-valued on $E$ (see Theorem 4.4.1). Let $c$ be the constant value of $\varphi$ on $E$. Now, since $\Delta \equiv \Delta(f)$ is an $F_{\sigma\delta}$-set in $\mathbb{R}$ (see [G11, Theorem 3.12]) and $f'$ is of $B_1$ on $\Delta$ (see [G11, Theorem 5.1]), it is clear that the set $A = \{x \in \Delta : f'(x) \neq c\}$ is measurable. Further, since $A \subset \mathbb{R} \sim E$ in each of the three cases, it follows that $|A| = 0$. Hence the function $f_{-c}$ is singular.

Now, in cases (a) and (b) it follows from parts (c) and (a), respectively, of Corollary 9.6.4 that $f_{-c}$ is constant; and in case (c) the same follows from Corollary 9.6.8. Hence, in each case, $f$ has an ordinary derivative equal to $c$ everywhere, and $\varphi$ is consequently constant on $\mathbb{R}$. □

10.5.2. REMARK. In connection with part (c) of Theorem 10.5.1, it may be pointed out here that this result has been obtained by Leonard [Le, Theorem 9] for right derivatives of Darboux functions of $B_1$.

# *CHAPTER 11*

# Denjoy and Perron Integrals Corresponding to New Derivatives

In this final chapter of Part III we deal with the problem of reconstruction of a primitive function from its various new derivatives. For this purpose we define here lower, upper, and weak Perron integrals which provide a solution of this problem for lower, upper, and weak derivatives, respectively.

Further, we present descriptive definitions of these three new Perron integrals in terms of corresponding Denjoy-Perron integrals, and we prove theorems similar to the Alexandroff-Looman and Hake's theorems to obtain the equivalence of the two integrals in each case.

First, in §11.1, we define lower, upper, and weak Perron integrals and investigate their interrelations. Then, in §§11.2 and 11.3, we obtain various properties of these integrals similar to the ones known for ordinary Perron integral.

Next, in §11.4, we define properties $L\text{-}AC_*$ and $U\text{-}AC_*$ similar to the property $AC_*$ and obtain several results dealing with these two properties. Also, in §11.5, we obtain criteria in terms of derivates under which a function has these properties. Then, in §11.6, we present descriptive definitions of lower and upper Perron integrals in terms of $GL\text{-}AC_*$ and $GU\text{-}AC_*$ functions and obtain the equivalence of the two definitions in each case.

Finally, in §§11.7 and 11.8, we present the descriptive definition of weak Perron integral in terms of a similar property $GS\text{-}AC_*$ and obtain the equivalence of the two definitions.

**§11.1 Perron integrals corresponding to new derivatives.** To deal with the problem of reconstruction of a primitive function from its new derivatives, in this section we define lower, upper, and weak Perron integrals and obtain some results on their compatibility and comparability.

We will use throughout this chapter $I = [a, b]$ to denote an arbitrary compact subinterval of $\mathbb{R}$. Given $f : I \to \overline{\mathbb{R}}$, let $U$ and $V$ be any "continuous" real-valued functions defined on $I$. Then we will call $U$ a *lower major function* of $f$ and $V$ a *lower minor function* of $f$ if the following conditions hold:

(a) $U(a) = 0$, (a′) $V(a) = 0$,

(b) $D_+U \geq f$ a.e., (b′) $D^-V \leq f$ a.e.,

(c) $D_+U > -\infty$ n.e., (c′) $D^-V < +\infty$ n.e.

We call $f$ *lower Perron integrable* if it has at least one lower major function and at least one lower minor function and if $\inf U(b) = \sup V(b)$, where inf and sup are taken over all lower major and lower minor functions, respectively, of $f$. Further, if it is so, then this common bound will be called the *lower Perron integral* of $f$ and denoted by $(LP)\int_a^b f(x)\,dx$.

The *upper major* and *upper minor functions* of $f$ are defined similarly by replacing the derivates $D_+U$ and $D^-V$ in the above definitions by $D_-U$ and $D^+V$, respectively. Further, the *upper Perron integrability* of $f$ and the *upper Perron integral* of $f$ are defined as above in terms of upper major and upper minor functions of $f$.

Finally, the *weak major* and *weak minor functions* of $f$ are defined by replacing the derivates $D_+U$ and $D^-V$ in the above definitions by the weak derivates $\underline{D}_wU$ and $\overline{D}_wV$, respectively. Further, the *weak Perron integrability* of $f$ and the *weak Perron integral* of $f$ are defined as above in terms of weak major and weak minor functions of $f$.

Further, for the sake of brevity, we will also call a lower, upper, or weak major function of $f$ an *L-major*, *U-major*, or *W-major function* (respectively) of $f$. Similarly, a lower, upper or weak minor function of $f$ will be called an *L-minor*, *U-minor*, or *W-minor function* of $f$. Further, when $f$ is lower, upper or weak Perron integrable, we will call it *LP-*, *UP-*, or *WP-integrable*, and the lower, upper, or weak Perron integral of $f$ will be called simply the *LP-*, *UP-*, or *WP-integral* of $f$, and denoted by $(LP)\int_a^b f(x)\,dx$, $(UP)\int_a^b f(x)\,dx$, or $(WP)\int_a^b f(x)\,dx$, respectively. The first two of these integrals were considered earlier by Ridder [Ri1] and Ionescu-Tulcea [I-T] (see Remark 11.6.12).

On replacing the derivates $D_+U$ and $D^-V$ in the above definitions by the bilateral derivates $\underline{D}U$ and $\overline{D}V$, respectively, we obtain on the other hand the notions of ordinary *major* and *minor functions* of $f$, and then the ordinary *Perron integrability* and the ordinary *Perron integral* of $f$ are defined in terms of ordinary major and minor functions of $f$ as above.

It should be pointed out here that in the classical definition of major or minor functions of $f$ the inequalities in the conditions (b), (c), ($b'$), and ($c'$) were assumed to hold everywhere. These two definitions of ordinary Perron integral are, however, known to be equivalent.

We begin in the following theorem with the relation between lower and upper Perron integrals.

11.1.1. THEOREM. *A function $f : I \to \overline{\mathbb{R}}$ is UP-integrable iff $-f$ is LP-integrable; and if so, then $(UP)\int_a^b f(x)\,dx = -(LP)\int_a^b (-f)(x)\,dx$.*

PROOF. Let $\overline{\mathscr{U}}$ and $\overline{\mathscr{V}}$ denote the classes of upper major and minor functions of $f$, and, similarly, let $\underline{\mathscr{U}}$ and $\underline{\mathscr{V}}$ denote the classes of lower major and minor functions of $-f$. Now, for each $U, V: I \to \mathbb{R}$, we have clearly $D_-U = -D^-(-U)$ and $D^+V = -D_+(-V)$. Hence it is easy to see that $U$ is an upper major function of $f$ iff $-U$ is a lower minor function of $-f$, and, similarly, $V$ is an upper minor function of $f$ iff $-V$ is a lower major function of $-f$. Thus $\overline{\mathscr{U}} = -\underline{\mathscr{V}}$ and $\overline{\mathscr{V}} = -\underline{\mathscr{U}}$. Hence we obtain

$$\inf_{U \in \overline{\mathscr{U}}} U(b) = -\sup_{V \in \underline{\mathscr{V}}} V(b), \qquad \sup_{V \in \overline{\mathscr{V}}} V(b) = -\inf_{U \in \underline{\mathscr{U}}} U(b).$$

Now the result follows clearly from these two relations. □

It is clear from the above theorem that it is enough to consider here the lower and weak Perron integrals. The results on the upper Perron integral will follow directly from the corresponding results on the lower Perron integral with the help of the above theorem.

Next, we obtain compatibility of lower and upper Perron integrals.

Given any two generalized integrals $G_1$- and $G_2$-, we will call them *compatible* if for every function $f: I \to \overline{\mathbb{R}}$ defined on any compact interval $I$ in $\mathbb{R}$, if $f$ is simultaneously $G_1$- and $G_2$-integrable, then the values of these two integrals of $f$ on $I$ are equal. Further, we will call the $G_1$-integral *weaker than* $G_2$-integral, or the $G_2$-integral *stronger than* $G_1$-integral, if these two integrals are compatible and if every function $f: I \to \overline{\mathbb{R}}$ which is $G_2$-integrable is also $G_1$-integrable. Also, we will call these two integrals *equivalent* if each of them is weaker than the other.

11.1.2. THEOREM. *The lower and upper Perron integrals are compatible. Consequently, if a function $f: I \to \overline{\mathbb{R}}$ is simultaneously lower and upper Perron integrable, then*

$$(1) \qquad (LP)\int_a^b f(x)\,dx = (UP)\int_a^b f(x)\,dx.$$

PROOF. Suppose $f: I \to \overline{\mathbb{R}}$ is simultaneously lower and upper Perron-integrable. Let $\underline{\mathscr{U}}$ and $\underline{\mathscr{V}}$ denote the classes of lower major and minor functions of $f$, and let $\overline{\mathscr{U}}$ and $\overline{\mathscr{V}}$ denote the classes of upper major and minor functions of $f$. First, suppose $U \in \underline{\mathscr{U}}$ and $V \in \overline{\mathscr{V}}$. Then $D_+U \geq f$ a.e. and $> -\infty$ n.e., and $D^+V \leq f$ a.e. and $< +\infty$ n.e. Hence it is easy to see that

$$D^+(U - V) \geq D^+U - D^+V \geq 0 \text{ a.e.} \quad \text{and} \quad > -\infty \text{ n.e.}$$

Thus $U - V$ is nondecreasing on $I$ by Theorem 9.1.8. Hence $U(b) - V(b) \geq U(a) - V(a) = 0$, and so $U(b) \geq V(b)$. Also, when $U \in \overline{\mathscr{U}}$ and $V \in \underline{\mathscr{V}}$, it is proved similarly that $U(b) \geq V(b)$. Hence we obtain

$$\inf_{U \in \underline{\mathscr{U}}} U(b) = \sup_{V \in \underline{\mathscr{V}}} V(b) \leq \inf_{U \in \overline{\mathscr{U}}} U(b) = \sup_{V \in \overline{\mathscr{V}}} V(b) \leq \inf_{U \in \underline{\mathscr{U}}} U(b).$$

The relation (1) follows now from the equality of these four bounds, and this proves the result. □

In connection with the above theorem we will see later in §11.6 that a function is Perron integrable in the ordinary sense iff it is simultaneously lower and upper Perron integrable.

Next, to deal with the comparison of the new Perron integrals, we need the following lemma. This lemma plays a basic role in the theory of these integrals.

11.1.3. Lemma. *If $U$ and $V$ are weak major and minor functions of some function $f: I \to \overline{\mathbb{R}}$, then $U - V$ is nondecreasing on $I$. The same holds when $U$ and $V$ are lower major and minor functions of $f$.*

Proof. First, suppose $U$ and $V$ are weak major and minor functions of $f$. Let $W = U - V$, and define $u = \underline{D}_w U$ and $v = \overline{D}_w V$. Now let $E$ denote the set of points in $I$ where $U, V$, and $W$ are nonangular. Then $I \sim E$ is countable by Theorem 2.3.1. Hence there is a set $A \subset E$ such that $I \sim A$ is countable and $u > -\infty$ and $v < +\infty$ at each point of $A$. Further, there is a subset $B$ of $A$ such that $|I \sim B| = 0$ and $u \geq f \geq v$ at each point of $B$.

We claim that $D^+W(x) \geq u(x) - v(x)$ for each $x \in A$. Let $x \in A$. We need to consider here four cases separately. If $u(x) = D_+U(x)$ and $v(x) = D^+V(x)$, then it is clear that

$$D^+W(x) \geq D^+U(x) - D^+V(x) \geq D_+U(x) - D^+V(x) = u(x) - v(x).$$

Also, when $u(x) = D_-U(x)$ and $v(x) = D^-V(x)$, then since $W$ is nonangular at $x$, we obtain

$$\begin{aligned} D^+W(x) &\geq D_-W(x) \geq D_-U(x) - D_-V(x) \geq D_-U(x) - D^-V(x) \\ &= u(x) - v(x). \end{aligned}$$

Further, if $u(x) = D_+U(x)$ and $v(x) = D^-V(x)$, then from the nonangularity of $V$ at $x$ we obtain

$$\begin{aligned} D^+W(x) &\geq D_+W(x) \geq D_+U(x) - D_+V(x) \geq D_+U(x) - D^-V(x) \\ &= u(x) - v(x). \end{aligned}$$

Finally, when $u(x) = D_-U(x)$ and $v(x) = D^+V(x)$, from the nonangularity of $U$ at $x$ we obtain

$$D^+W(x) \geq D^+U(x) - D^+V(x) \geq D_-U(x) - D^+V(x) = u(x) - v(x).$$

This proves the claim. Hence it is clear that $D^+W(x) > -\infty$ for each $x \in A$. Further, let $x \in B$. Then when $f(x)$ is finite, it is clear that $D^+W(x) \geq u(x) - v(x) \geq f(x) - f(x) = 0$. If $f(x)$ is on the other hand infinite, then since $u(x) > -\infty$ and $v(x) < +\infty$, it is easy to see that $D^+W(x) = +\infty$. Hence $D^+W(x) \geq 0$ a.e. Now since $W$ is continuous, it follows from Theorem 9.1.8 that $W$ is nondecreasing.

The second part follows from the first since every lower major and minor functions of $f$ are also weak major and minor functions of $f$. □

Finally, in the following theorem we compare the lower and upper Perron integrals with the ordinary and weak Perron integrals.

11.1.4. THEOREM. *The lower and upper Perron integrals are weaker than the Perron integral, but each of them is stronger than the weak Perron integral.*

PROOF. We give the proof here for the lower Perron integral; a similar argument holds for the upper Perron integral.

First, suppose $f: I \to \overline{\mathbb{R}}$ is Perron integrable in the ordinary sense. Let $\mathscr{U}$ and $\mathscr{V}$ denote the classes of ordinary major and minor functions of $f$, and let $\underline{\mathscr{U}}$ and $\underline{\mathscr{V}}$ denote the classes of lower major and minor functions of $f$. Then it is clear that $\mathscr{U} \subset \underline{\mathscr{U}}$ and $\mathscr{V} \subset \underline{\mathscr{V}}$. Further, if $U \in \underline{\mathscr{U}}$ and $V \in \underline{\mathscr{V}}$, then according to the above lemma $U - V$ is nondecreasing, and so $U(b) \geq V(b)$. Hence we obtain

$$(2) \qquad \inf_{U \in \underline{\mathscr{U}}} U(b) \leq \inf_{U \in \mathscr{U}} U(b) = \sup_{V \in \mathscr{V}} V(b) \leq \sup_{V \in \underline{\mathscr{V}}} V(b) \leq \inf_{U \in \underline{\mathscr{U}}} U(b).$$

This proves that all the bounds in (2) are equal. Hence $f$ is also $LP$-integrable and the $LP$-integral of $f$ coincides with its $P$-integral.

The fact that the $LP$-integral is stronger than the $WP$-integral is deduced from the above lemma by a similar argument using the fact that the lower major and minor functions of $f$ are also weak major and minor functions, respectively, of $f$. □

**§11.2 Elementary properties of new Perron integrals.** In this section we present some elementary properties of lower, upper, and weak Perron integrals.

Later, in Chapter 21, we develop an axiomatic theory of generalized Perron integrals which also applies to the present new Perron integrals.

Hence, to avoid repetition, we present the results in the present section and the next section without proof (with few exceptions). The results of the present section follow directly from the ones obtained in §21.3 as proved in Theorem 21.5.2. As observed earlier, it is enough to consider here lower and weak Perron integrals.

We begin with the following theorem establishing that the lower and weak Perron integrals provide the solution of the problem of reconstruction of a primitive function from its lower and weak derivatives, respectively.

11.2.1. THEOREM (RECONSTRUCTION). *Suppose a continuous function* $F: I \to \mathbb{R}$ *is weakly differentiable* n.e. *and* $f: I \to \overline{\mathbb{R}}$. *Then if* $f(x) = WF'(x)$ *for* n.e. $x$, *then* $f$ *is WP-integrable and*

$$F(b) - F(a) = (WP)\int_a^b f(x)\,dx.$$

*Similarly, if* $F$ *is lower differentiable* n.e. *and* $f(x) = LF'(x)$ *for* n.e. $x$, *then* $f$ *is LP-integrable and*

$$F(b) - F(a) = (LP)\int_a^b f(x)\,dx.$$

PROOF. First, suppose $F$ is weakly differentiable n.e. and $f(x) = WF'(x)$ for n.e. $x$. Define $G(x) = F(x) - F(a)$, $x \in I$. Then since $F$ and $G$ have the same derivates at each point of $I$, it is clear that $G$ is simultaneously a $W$-major and $W$-minor function of $f$. Hence it follows easily from Lemma 11.1.3 that $f$ is $WP$-integrable and that $(WP)\int_a^b f(x)\,dx = G(b) = F(b) - F(a)$. A similar argument holds in the other case. □

We present the remaining results of this section in terms of weak Perron integral, except for Theorem 11.2.7 which holds only for lower and upper Perron integrals. All the results of this section hold for lower and upper Perron integrals as well, except that the Theorem 11.2.6 holds for these integrals only for $c \geq 0$.

The following lemma is of course obvious.

11.2.2. LEMMA. *A function* $f: I \to \overline{\mathbb{R}}$ *is WP-integrable iff for every* $\varepsilon > 0$ *there exist W-major and W-minor functions* $U$ *and* $V$ *of* $f$ *such that* $U(b) - V(b) < \varepsilon$.

11.2.3. THEOREM. *Let* $a < c < b$. *If* $f$ *is WP-integrable on* $[a, b]$, *then it is also so on* $[a, c]$ *and* $[c, b]$ *and*

$$(WP)\int_a^b f(x)\,dx = (WP)\int_a^c f(x)\,dx + (WP)\int_c^b f(x)\,dx.$$

*Conversely, if $f$ is WP-integrable on $[a, c]$ and $[c, b]$, then it is also so on $[a, b]$.*

11.2.4. THEOREM. *If $f$ and $g$ are WP-integrable and $f \leq g$ a.e., then*

$$(WP)\int_a^b f(x)\,dx \leq (WP)\int_a^b g(x)\,dx.$$

11.2.5. THEOREM. *If $f$ is WP-integrable and $g = f$ a.e., then $g$ also is WP-integrable and*

$$(WP)\int_a^b g(x)\,dx = (WP)\int_a^b f(x)\,dx.$$

11.2.6. THEOREM. *Suppose $f$ is WP-integrable. Then for each $c \in \mathbb{R}$, $cf$ also is WP-integrable and*

$$(WP)\int_a^b cf(x)\,dx = c(WP)\int_a^b f(x)\,dx.$$

*In the case of LP-integral this holds only for $c \geq 0$.*

The following additivity theorem docs not hold on the other hand in the case of $WP$-integral. This is not entirely unexpected since the weak derivative is not additive in general.

11.2.7. THEOREM. *Suppose $f$ and $g$ are LP-integrable and $f + g$ is defined* a.e. *Then $f + g$ also is LP-integrable and*

$$(LP)\int_a^b (f+g)(x)\,dx - (LP)\int_a^b f(x)\,dx + (LP)\int_a^b g(x)\,dx.$$

Finally, in the following theorem we obtain relations between the new Perron integrals of the functions $f$ and $f_-$. It is somewhat surprising to note in its part (b) that although the roles of lower and upper derivatives are preserved by the function $f_-$ (see Lemma 7.3.6), the roles of lower and upper Perron integrals are somehow reversed by $f_-$.

11.2.8. THEOREM. *Let $f: I \to \overline{\mathbb{R}}$. Then* (a) *$f$ is WP-integrable iff $f_-$ is so; and if so, then*

$$(WP)\int_a^b f(x)\,dx = (WP)\int_{-b}^{-a} f_-(x)\,dx. \tag{1}$$

(b) *$f$ is LP-integrable iff $f_-$ is UP-integrable; and if so, then*

$$(LP)\int_a^b f(x)\,dx = (UP)\int_{-b}^{-a} f_-(x)\,dx. \tag{2}$$

*Also, the same holds on inverting the roles of LP and UP.*

PROOF. First, suppose $f$ is *WP-integrable*. Given $\varepsilon > 0$, choose weak major and minor functions $U$ and $V$ of $f$ such that $U(b) - V(b) < \varepsilon$. Then

(i) $\underline{D}_w U \geq f \geq \overline{D}_w V$ a.e.,

(ii) $\underline{D}_w U > -\infty$ n.e., and

(iii) $\overline{D}_w V < +\infty$ n.e.

Let $J = [-b, -a]$, and define for each $t \in J, U^*(t) = U(b) - U_-(t)$ and $V^*(t) = V(b) - V_-(t)$. We claim that $U^*$ and $V^*$ are weak major and minor functions of $f_-$. Clearly, $U^*$ and $V^*$ are continuous on $J$, and $U^*(-b) = V^*(-b) = 0$. Further, given $x \in I^0$, we obtain from Lemma 1.1.2(c),

$$\underline{D}_w U^*(-x) = \vee D_{\pm} U^*(-x) = \vee D_{\mp} U(x) = \underline{D}_w U(x);$$

and similarly $\overline{D}_w V^*(-x) = \overline{D}_w V(x)$. Hence it follows from (i) that $\underline{D}_w U^*(-x) \geq f(x) = f_-(-x)$ for a.e. $x$, and, similarly, $\overline{D}_w V^*(-x) \leq f_-(-x)$ for a.e. $x$. Also, it follows from (ii) and (iii) that $\underline{D}_w U^* > -\infty$ and $\overline{D}_w V^* < +\infty$ n.e. This proves the claim.

Further, it is clear that

$$\begin{aligned} U^*(-a) - V^*(-a) &= U(b) - U_-(-a) - \{V(b) - V_-(-a)\} \\ &= U(b) - V(b) < \varepsilon, \end{aligned}$$

and so $f_-$ is *WP*-integrable by Lemma 11.2.2. Also,

$$\begin{aligned} (WP)\int_{-b}^{-a} f_-(x)\,dx &= \inf U^*(-a) = \inf U(b) - U_-(-a) \\ &= \inf U(b) = (WP)\int_a^b f(x)\,dx, \end{aligned}$$

which proves the relation (1). The necessity part of (a) is thus established. Its converse follows on the other hand on applying it to the function $f_-$, for $(f_-)_- = f$.

Next, to obtain part (b), suppose $f$ is *LP*-integrable. Given $\varepsilon > 0$, choose this time lower major and minor functions $U$ and $V$ of $f$ such that $U(b) - V(b) < \varepsilon$. Then

(i) $D_+ U \geq f \geq D^- V$ a.e.,

(ii) $D_+ U > -\infty$ n.e., and

(iii) $D^- V < +\infty$ n.e.

Let $U^*$ and $V^*$ be defined as above. Then given $x \in I^0$, we obtain from Lemma 1.1.2(c), $D_-U^*(-x) = D_+U(x)$ and $D^+V^*(-x) = D^-V(x)$. Hence from (i) we obtain, as before, $D_-U^* \geq f_- \geq D^+V^*$ a.e., and from (ii) and (iii) we obtain $D_-U^* > -\infty$ and $D^+V^* < +\infty$ n.e. Hence $U^*$ and $V^*$ are this time upper major and minor functions of $f_-$. Further, it follows by an argument similar to above that $f_-$ is *UP*-integrable and that relation (2) holds.

This establishes the necessity part of (b). Further, on applying this result to $-f$, a similar result is obtained with the help of Theorem 11.1.1 when $f$ is *UP*-integrable. The sufficiency parts of these two results follow now on applying these results to $f_-$. □

11.2.9. Remark. In the case of the ordinary Perron integral all the results of this section, except for Theorem 11.2.8, follow directly from the axiomatic theory developed in Chapter 21 (see in particular §21.3). However, it is interesting to note here that the results on the *P*-integral can also be deduced directly from the corresponding results on the *LP*-integral.

For, as we will see later in Theorem 11.6.9 from the descriptive definitions of lower and upper Perron integrals, any function $f: I \to \overline{\mathbb{R}}$ is *P*-integrable iff it is simultaneously lower and upper Perron integrable; and if so, then these three integrals of $f$ are equal on $I$. Hence if any result holds for the *LP*-integral, then it follows clearly from Theorems 11.1.1 and 11.6.9 that it holds also for the *P*-integral. Consequently, Theorem 11.2.8 also holds for the ordinary *P*-integral. The same applies to the results of the next section.

## §11.3 Properties of new indefinite Perron integrals.

In this section we present various properties of indefinite lower, upper, and weak Perron integrals.

First, suppose $f: I \to \overline{\mathbb{R}}$ is *WP*-integrable. Then according to Theorem 11.2.3, $f$ is *WP*-integrable on $[a, x]$ for $a < x \leq b$. We further define

$$(WP)\int_a^a f(x)\,dx = 0$$

and call the function

$$F(x) = (WP)\int_a^x f(t)\,dt, \qquad a \leq x \leq b,$$

the *indefinite weak Perron integral* of $f$. The indefinite lower and upper Perron integrals of *LP*- and *UP*-integrable functions are defined similarly.

Once again, we state the results in this section without proof since they follow directly from the results of §21.4 as proved in Theorem 21.5.2. The results stated in this section for the *WP*-integral (in 11.3.1 to 11.3.6) hold also for lower and upper Perron integrals, but the results stated for the *LP*-integral (in 11.3.7 to 11.3.10) do not hold for the *WP*-integral although they do hold for the *UP*-integral.

11.3.1. LEMMA. *Suppose $F$ is the indefinite WP-integral of some function $f$. Then if $U$ and $V$ are W-major and W-minor functions of $f$, then $U - F$ and $F - V$ are nondecreasing on $I$, and $U(x) \geq F(x) \geq V(X)$ for each $x \in I$.*

11.3.2. THEOREM. *The indefinite WP-integral of every WP-integrable function is continuous.*

The following fundamental theorem is obtained on the other hand from Theorem 21.4.3 with the help of the Denjoy-Young-Saks Theorem 5.5.2.

11.3.3. THEOREM (FUNDAMENTAL). *If $F$ is the indefinite WP-integral of some function $f$, then $F$ is normally semidifferentiable* a.e. *and $f(x) = SF'(x)$ for* a.e. $x$.

*Similarly, if $F$ is the indefinite LP-integral of $f$, then $F$ is normally lower differentiable* a.e. *and $f(x) = LF'(x)$ for* a.e. $x$.

11.3.4. COROLLARY. *If $f$ is WP-integrable, then every W-major and W-minor functions of $f$ are normally semidifferentiable* a.e. *Similarly, if $f$ is LP-integrable, then every L-major and L-minor functions of $f$ are normally lower differentiable* a.e.

Following is another consequence of the above theorem which is obtained with the help of Theorems 11.3.2 and 10.3.3.

11.3.5. COROLLARY. *Every WP-integrable function is measurable and finite* a.e.

The following theorem deals with a relation between the *WP*-integral and the Lebesgue integral. We use here $(L)$ for Lebesgue integral.

11.3.6. THEOREM. *If $f$ is WP-integrable and non-negative* a.e., *then it is Lebesgue integrable and*

$$(WP)\int_a^b f(x)\,dx = (L)\int_a^b f(x)\,dx.$$

Following is a corollary of the above theorem in the case of the *LP*-integral. This corollary does not hold for the *WP*-integral since Theorem 11.2.7 is not valid for this integral.

11.3.7. COROLLARY. *If $f$ is LP-integrable and there is a Lebesgue integrable function $g$ such that $g \leq f$* a.e., *then $f$ is Lebesgue integrable.*

The following two theorems deal with the extension of two well-known convergence theorems on Lebesgue integral to lower Perron integral.

11.3.8. THEOREM. *Suppose $\{f_n\}$ is a nondecreasing sequence of LP-integrable functions whose LP-integrals are bounded from above. Then if $-f_1$ is also LP-integrable, then so is the pointwise limit $f$ of $\{f_n\}$ and*

$$(1) \qquad (LP)\int_a^b f(x)\,dx = \lim_n (LP)\int_a^b f_n(x)\,dx.$$

11.3.9. THEOREM. *Let $\{f_n\}$ be a sequence of LP-integrable functions which converges in measure to $f$. Then if $-f_1$ is also LP-integrable and there is a Lebesgue integrable function $g$ such that for each $n$, $|f_n - f_1| \le g$ a.e., then $f$ is LP-integrable and (1) holds.*

Next, we deal with the integration by parts formula in the case of *LP*- and *UP*-integrals. The following theorem has been obtained by Bullen [Bu2, p. 199] in this direction.

11.3.10. THEOREM (INTEGRATION BY PARTS). *Suppose $f$ is LP-integrable and that $F$ is the indefinite LP-integral of $f$. Then for each increasing function $G$ on $I$ with $G(a) = 0$, the function $fG$ is also LP-integrable and*

$$(2) \qquad (LP)\int_a^b f(x)G(x)\,dx = F(b)G(b) - F(a)G(a) - \int_a^b F\,dG,$$

*where the right side of (2) is to be interpreted as follows:*

$$G(a) = G(a+), \qquad G(b) = G(b-),$$

$$\int_a^b F\,dG = \lim_{\alpha \to a+,\ \beta \to b-} (RS)\int_\alpha^\beta F\,dG.$$

*Moreover, the same holds on replacing LP everywhere by UP.*

Finally, as we will see in the next two theorems, the rule regarding integration by substitution also extends to new Perron integrals when the function in terms of which the substitution is made is monotone. These two theorems have been obtained by McShane [Mc2, pp. 329, 331] for his notion of Perron integral, and the main idea behind the following proof is based on his proof. In the case of the ordinary Perron integral, the following two theorems follow from their following versions as seen in Remark 11.2.9.

11.3.11. THEOREM (SUBSTITUTION RULE). *Suppose $g$ is a continuous nondecreasing function on $J = [c, d]$ such that $g(c) = a$ and $g(d) = b$. Then if $f : I = [a, b] \to \overline{R}$ is WP-integrable and $g$ is D-finite* n.e., *then the function*

*$f(g(y))g'(y)$ is WP-integrable on $J$ and*

$$(2)\qquad (WP)\int_a^b f(x)\,dx = (WP)\int_c^d f(g(y))g'(y)\,dy.$$

*Moreover, the same holds on replacing WP by LP or UP.*

PROOF. We give the proof here for the weak Perron integral; in the case of lower or upper Perron integral the proof is similar and somewhat simpler.

Let $E$ denote the set of points in $J$ where $g$ is differentiable, and define $F(y) = f(g(y))g'(y)$ for $y \in E$, and define $F(y) = 0$ for $y \in J \sim E$. Given $\varepsilon > 0$, choose weak major and minor functions $U$ and $V$ of $f$ such that $U(b) - V(b) < \varepsilon$. Now define

$$(3)\qquad U^*(y) = U(g(y)),\qquad V^*(y) = V(g(y)),\qquad y \in J.$$

We claim that $U^*$ and $V^*$ are weak major and minor functions of $F$ on $J$. We give the proof here for $U^*$.

It is clear that $U^*$ is continuous and that $U^*(c) = U(a) = 0$. Let $\{I_n\}$ denote the sequence of maximal closed intervals in $J$ on which $g$ is constant. Set $A = \bigcup_n I_n^0$ and let $C_0$ denote the set of endpoints of the intervals $I_n$. Then $C_0$ is countable. Let $y \in A$. Then $g$ is constant on some neighborhood of $y$, and so $U^*$ is also constant on that neighborhood. Hence $g$ and $U^*$ have a zero derivative at $y$. Thus $F(y) = 0 = \underline{D}_w U^*(y)$, and so the conditions (b) and (c) hold for $U^*$ at the points of $A$.

Now let $B = I \sim (A \cup C_0)$. Then $g$ is clearly increasing on $B$. Set

$$C_1 = \{y \in B : \underline{D}_w U(g(y)) = -\infty\},\qquad C_2 = \{y \in B : \overline{D}g(y) = +\infty\}.$$

Then since $U$ is a $W$-major function of $f$ and $g$ is $D$-finite n.e., the sets $C_1$ and $C_2$ are countable. Hence the set $C = C_1 \cup C_2$ is also countable. Now set $B_0 = B \sim C$. Let $y \in B_0$. Then it is clear that $Qg(y, z) > 0$ for every $z \neq y$. Hence from Lemma 7.5.1(b) we obtain

$$(4)\qquad D_\pm U^*(y) \geq D_\pm U(g(y))D_\pm g(y) \wedge D_\pm U(g(y))D^\pm g(y).$$

Since $y \notin C$, $\underline{D}_w U(g(y)) > -\infty$ and $\overline{D}g(y) < +\infty$, and thus it is clear that $\underline{D}_w U^*(y) > -\infty$. Hence the condition (c) holds for $U^*$ on $B$ and thus, on $J$.

Next, to obtain the condition (b) for $U^*$, it is clearly enough to show that $\underline{D}_w U^*(y) \geq F(y)$ for a.e. $y$ in $B_0$. Let $S = \{x \in I : \underline{D}_w U(x) \geq f(x)\}$, and set $T = I \sim S$. Then it is clear that $|T| = 0$. Further, let $E_0 = \{y \in E : g'(y) = 0\}$, and set $E_1 = E \sim E_0$ and $G = g(E_1)$. Then since $g'(y) > 0$ for each $y \in E_1$, $g$ is increasing on $E_1$, and so it has an inverse function $h = g^{-1}$ on $G$. Further, for each $x \in G$, $h$ is differentiable at $x$ (relative to $G$) and $h'(x) = 1/g'(h(x))$, which is finite. Hence it follows from Corollary 6.1.4 that $h$ has the property $(N)$. Hence $|h(T)| = 0$. Set $E_2 = E_1 \cap h(T)$, so that $|E_2| = 0$.

Now, set $B_1 = B_0 \cap E \sim E_2$. Then it is clear that $|B_1| = |B_0|$. Now let $y \in B_1$. Then since $y \notin h(T)$, $g(y) \in S$, and since $g$ is differentiable at $y$, it follows from (3) that

$$\underline{D}_w U^*(y) = \vee D_\pm U^*(y) \geq \vee D_\pm U(g(y))g'(y) \geq f(g(y))g'(y) = F(y).$$

This proves the condition (b) for $U^*$, and so $U^*$ is a $W$-major function of $F$.

It is proved by a similar argument that $V^*$ is a $W$-minor function of $F$. Further, since

$$U^*(d) - V^*(d) = U(g(d)) - V(g(d)) = U(b) - V(b) < \varepsilon,$$

this proves by Lemma 11.2.2 that $F$ is $WP$-integrable on $J$. Also, we have

$$U(b) = U^*(d) \geq (WP)\int_c^d F(y)\,dy \geq V^*(d) = V(b),$$

where $U(b) \geq (WP)\int_a^b f(x)\,dx \geq V(b)$ and $U(b) - V(b) < \varepsilon$. Hence it is clear that

$$\left|(WP)\int_a^b f(x)\,dx - (WP)\int_c^d F(y)\,dy\right| < \varepsilon,$$

which proves the relation (2). □

As we see next, the above theorem also holds in the case when $g$ is nonincreasing except for a change of sign on the right side of (2).

11.3.12. THEOREM. *Suppose $g$ is a continuous nonincreasing function on $J = [c, d]$ such that $g(c) = b$ and $g(d) = a$. Then if $f : [a, b] \to \overline{\mathbb{R}}$ is WP-integrable and $g$ is D-finite n.e., then the function $f(g(y))g'(y)$ is WP-integrable on $J$ and*

$$(5) \qquad (WP)\int_a^b f(x)\,dx = -(WP)\int_c^d f(g(y))g'(y)\,dy.$$

*Moreover, the same holds on replacing WP by LP or UP.*

PROOF. First, note that $g_-$ is a continuous nondecreasing function on $-J = [-d, -c]$, where $g_-(-d) = a$ and $g_-(-c) = b$. Let $E$ denote the set of points in $J$ where $g$ is differentiable. Then for each $y \in E$, $g_-$ is differentiable at $-y$ and $(g_-)'(-y) = -g'(y)$. For each $y \in -E$, define $F(y) = f(g_-(y))(g_-)'(y) = -f(g(-y))g'(-y)$, and $F(y) = 0$ elsewhere in $-J$.

First, suppose $f$ is *WP*-integrable. Then it follows on applying the above theorem to $f$ and $g_-$ with the help of Theorem 11.2.6 that $F$ is *WP*-integrable on $-J$ and

$$(WP)\int_a^b f(x)\,dx = (WP)\int_{-d}^{-c} F(y)\,dy = -(WP)\int_{-d}^{-c} f(g(-y))g'(-y)\,dy.$$

Hence we obtain the relation (5) with the help of Theorem 11.2.8.

Next, suppose $f$ is *LP*-integrable. Then it follows as above with the help of Theorem 11.1.1 that $F$ is *LP*-integrable on $-J$ and

$$(LP)\int_a^b f(x)\,dx = (LP)\int_{-d}^{-c} F(y)\,dy = -(UP)\int_{-d}^{-c} f(g(-y))g'(-y)\,dy.$$

Hence with the help of Theorem 11.2.8 we obtain

$$(LP)\int_a^b f(x)\,dx = -(LP)\int_c^d f(g(y))g'(y)\,dy.$$

The result on the *UP*-integral is obtained by a similar argument. □

11.3.13. REMARK. In connection with Theorem 11.3.10, it may be noted here that it was stated in [Bu2, p. 199] due to an oversight that this theorem holds for any function $G$ of bounded variation. However, this does not hold in general. Suppose $f$ is *LP*-integrable but not *P*-integrable (see Example 11.6.11 and the comments before it). Then as we see later in Theorem 11.6.9, $f$ is not *UP*-integrable, and so $-f$ is not *LP*-integrable. Hence Theorem 11.3.10 does not hold in this case for the function $G(x) = -1$, $x \in I$.

## §11.4 Lower and upper $AC$, $L\text{-}AC_*$, $U\text{-}AC_*$, and $AC_*$ functions.

In this section we define properties $L\text{-}AC_*$ and $U\text{-}AC_*$ on a set in terms of which the descriptive definitions of lower and upper Perron integrals are presented later in §11.6. These properties are related with the properties $L\text{-}BV_*$ and $U\text{-}BV_*$ defined earlier in §5.2. Also, we define here lower and upper versions of these two properties along with the ones of ordinary $AC$ and $AC_*$.

First, in Theorems 11.4.2 and 11.4.3, we obtain relations between the properties $LAC$ and $BV$, and between $LL\text{-}AC_*$ and $L\text{-}BV_*$. Then in Theorem 11.4.5 we obtain a characterization of the property $L\text{-}BV_*$. Finally, in Theorem 11.4.7 we obtain characterizations of the properties $LL\text{-}AC_*$, $UL\text{-}AC_*$, and $L\text{-}AC_*$.

Given a function $f : X \to \mathbb{R}$, where $X \subset \mathbb{R}$, let $E \subset X$. We begin with the definitions of various variations of $f$ on $E$.

The (weak) *variation* of $f$ on $E$ is defined [S8, p. 221] to be

$$V(f;E) = \sup \sum_n |f(b_n) - f(a_n)|,$$

where the sup is taken over all finite sequences of nonoverlapping intervals $\{[a_n, b_n]\}$ whose endpoints are in $E$. When $V(f;E) < +\infty$, $f$ is said to be of *bounded variation* (or *BV*) on $E$. Further, the *lower* and *upper variations* of $f$ on $E$ are defined [S8, p. 98] similarly to be

$$\underline{V}(f;E) = \sup \sum_n \{f(a_n) - f(b_n)\}, \qquad \overline{V}(f;E) = \sup \sum_n \{f(b_n) - f(a_n)\},$$

where the suprema are taken as before over all finite sequences of nonoverlapping intervals $\{[a_n, b_n]\}$ whose endpoints are in $E$. The variations $\underline{V}(f;E)$, $\overline{V}(f;E)$, and $V(f;E)$ are also known as the negative, positive, and absolute variations of $f$ on $E$. The above definition of $\underline{V}(f;E)$ differs, however, with the one used by Saks in its sign.

Next, we consider ordinary absolute continuity and its lower and upper versions on a set. The function $f$ is called *absolutely continuous* (or *AC*) on $E$ if for every $\varepsilon > 0$ there is a $\delta > 0$ such that for each finite sequence of nonoverlapping intervals $\{[a_n, b_n]\}$ with endpoints in $E$, if $\Sigma_n(b_n - a_n) < \delta$, then $\Sigma_n|f(b_n) - f(a_n)| < \varepsilon$. The *lower* or *upper absolute continuity* (or *LAC* or *UAC*, respectively) of $f$ on $E$ is defined similarly by replacing the last sum by $\Sigma_n\{f(a_n) - f(b_n)\}$ or $\Sigma_n\{f(b_n) - f(a_n)\}$, respectively.

It is now easy to see that $f$ is *AC* on $E$ iff it is *LAC* and *UAC* on $E$. Also, it is easy to see that $f$ is *LAC* or *UAC* on $E$ iff for every $\varepsilon > 0$ there is a $\delta > 0$ such that for each finite sequence of nonoverlapping intervals $\{I_n\}$ with endpoints in $E$ and $\Sigma_n|I_n| < \delta$, $\Sigma_n\underline{V}(f; E \cap I_n)$ or $\Sigma_n\overline{V}(f; E \cap I_n)$, respectively, is $< \varepsilon$.

Now the generalized properties *GAC*, *GLAC*, and *GUAC* of $f$ on $E$ are defined as in §1.3. It should be noted here that in the definition of *GAC* in Saks [S8] $f$ is further assumed to be continuous.

Next, we define the properties *L-AC*$_*$ and *U-AC*$_*$ and their lower and upper versions on a set. First, let $I = [a, b]$, where $a, b \in X$. Then we define the following variants of the lower and upper oscillations considered in §5.2:

$$0_-(f;I) = f(a) - \inf_{x \in I \cap X} f(x), \qquad 0_+(f;I) = f(b) - \inf_{x \in I \cap X} f(x),$$

$$0^-(f;I) = \sup_{x \in I \cap X} f(x) - f(a), \qquad 0^+(f;I) = \sup_{x \in I \cap X} f(x) - f(b).$$

Here $0_-(f;I)$ and $0_+(f;I)$ may be called the *left* and *right lower oscillations*, respectively, of $f$ on $I$; and, similarly, $0^-(f;I)$ and $0^+(f;I)$ may be called the *left* and *right upper oscillations*, respectively, of $f$ on $I$.

Now it is clear from the above definitions that we have the following relations between these oscillations and the lower, upper, and ordinary oscillations of $f$ on $I$ defined in §5.2:

$$\underline{0}(f;I) = \max\{0_-(f;I), 0_+(f;I)\}, \tag{1}$$

$$\overline{0}(f;I) = \max\{0^-(f;I), 0^+(f;I)\}, \tag{2}$$

$$0(f;I) = 0_-(f;I) + 0^-(f;I) = 0_+(f;I) + 0^+(f;I). \tag{3}$$

Now, given $E \subset X$, we will call $f$ $L\text{-}AC_*$, $LL\text{-}AC_*$, or $UL\text{-}AC_*$ on $E$ if for every $\varepsilon > 0$ there is a $\delta > 0$ such that for each finite sequence of nonoverlapping intervals $\{I_n\}$ with endpoints in $E$ and $\Sigma_n |I_n| < \delta$,

$$\sum_n \underline{0}(f;I_n), \quad \sum_n 0_-(f;I_n), \quad \text{or} \quad \sum_n 0_+(f;I_n),$$

respectively, is $< \varepsilon$. The properties $U\text{-}AC_*$, $LU\text{-}AC_*$, and $UU\text{-}AC_*$ of $f$ on $E$ are defined similarly in terms of the upper oscillations $\overline{0}(f;I_n)$, $0^+(f;I_n)$, and $0^-(f;I_n)$, respectively. Note the reversal in the role of signs here.

It is now easy to see from the definitions of various oscillations considered above that if $f$ is $LL\text{-}AC_*$ or $LU\text{-}AC_*$ on $E$, then it is $LAC$ on $E$. Similarly, if $f$ is $UL\text{-}AC_*$ or $UU\text{-}AC_*$ on $E$, then it is $UAC$ on $E$. Hence if $f$ is $L\text{-}AC_*$ or $U\text{-}AC_*$ on $E$, then it is $AC$ on $E$. Further, it is clear from relation (1) that $f$ is $L\text{-}AC_*$ on $E$ iff it is $LL\text{-}AC_*$ and $UL\text{-}AC_*$ on $E$; similarly, it is clear from (2) that $f$ is $U\text{-}AC_*$ on $E$ iff it is $LU\text{-}AC_*$ and $UU\text{-}AC_*$ on $E$.

Finally, we will call $f$ *lower* $AC_*$ (or $LAC_*$) on $E$ if it is simultaneously $LL\text{-}AC_*$ and $LU\text{-}AC_*$ on $E$; similarly, we will call $f$ *upper* $AC_*$ (or $UAC_*$) on $E$ if it is simultaneously $UL\text{-}AC_*$ and $UU\text{-}AC_*$ on $E$. However, the properties $LAC_*$ and $UAC_*$ should not be confused with the properties $L\text{-}AC_*$ and $U\text{-}AC_*$.

The generalized versions of all the above properties are defined now as usual as in §1.3.

We begin with the following lemma which will be needed later.

11.4.1. Lemma. *Let $f: X \to \mathbb{R}$ and $E \subset X$. Then $f$ is $AC_*$ on $E$ iff any of the following equivalent conditions holds:*

(a) *$f$ is $L\text{-}AC_*$ and $U\text{-}AC_*$ on $E$,*

(b) *$f$ is $LL\text{-}AC_*$ and $UU\text{-}AC_*$ on $E$,*

(c) *$f$ is $UL\text{-}AC_*$ and $LU\text{-}AC_*$ on $E$,*

(d) *$f$ is $LAC_*$ and $UAC_*$ on $E$.*

Proof. We will prove here that each of the conditions (a) to (d) is equivalent to the proposition (e) that $f$ is $AC_*$ on $E$. The equivalence of (a) with (e) follows clearly from the relation (1) of §5.2. Further, the equivalence of (b) and (c) with (e) follows from the above relations (3). Now the equivalence of (d) with (e) follows from the definitions of $LAC_*$ and $UAC_*$ with the help of the equivalence of (e) with (b) and (c). □

In the following theorem we deal with the relation between $LAC$ and $BV$. In the case when $E$ is an interval, this relation was deduced earlier in [G16, p. 38] from a more general result.

11.4.2. THEOREM. *Suppose $f: X \to \mathbb{R}$ and $E \subset X$ is compact. Then if $f$ is $LAC$ or $UAC$ on $E$, then it is $BV$ on $E$.*

PROOF. Suppose $f$ is $LAC$ on $E$. Then there is a $\delta > 0$ such that if $\{[a_n, b_n]\}$ is any finite sequence of nonoverlapping intervals with endpoints in $E$ and $\Sigma_n(b_n - a_n) < \delta$, then $\Sigma_n\{f(a_n) - f(b_n)\} < 1$.

We will first prove that $f$ is bounded on $E$. First, suppose $f$ is not bounded from above on $E$. Then there is a sequence of points $\{x_n\}$ in $E$ such that the sequence $\{f(x_n)\}$ increases to $+\infty$. Now since $E$ is compact, there is a subsequence $\{x_i\}$ of $\{x_n\}$ which converges to some point $x$ of $E$ from some fixed side. Now, if $\{x_i\}$ converges to $x$ from the left, then we can choose an $i$ such that $x - \delta < x_i < x$ and $f(x_i) > f(x) + 1$. But this contradicts the choice of $\delta$. Further, in case $\{x_i\}$ converges to $x$ from the right, then we can clearly choose $i_1$ and $i_2$ such that $x < x_{i_2} < x_{i_1} < x + \delta$ and $f(x_{i_2}) > f(x_{i_1}) + 1$, which again contradicts the choice of $\delta$. This proves that $f$ is bounded from above on $E$. It is proved by a similar argument that $f$ is also bounded from below on $E$.

Now, let $M = \sup\{|f(x)| : x \in E\}$, and set $J = \operatorname{co}E$. Choose a finite decomposition $\{J_i : i = 1, \dots, k\}$ of $J$ such that $|J_i| < \delta$ for each $i$. Now, given any finite sequence of points $x_0 < x_1 < \cdots < x_p$ in $E$, let $N_+$ and $N_-$ denote the sets of positive integers $n \le p$ for which $f(x_n) - f(x_{n-1}) \ge 0$ or $< 0$, respectively. Further, for each $i \le k$, let $N_i$ denote the set of integers $n \in N_-$ for which $[x_{n-1}, x_n] \subset J_i$, and finally let $N_0 = N_- \sim \bigcup_{i \le k} N_i$. Clearly, $N_0$ cannot have more than $k$ elements. Hence we have

$$\begin{aligned}
&\sum_{n \in N_-} \{f(x_{n-1}) - f(x_n)\} \\
&\qquad = \sum_{i \le k} \sum_{n \in N_i} \{f(x_{n-1}) - f(x_n)\} + \sum_{n \in N_0} \{f(x_{n-1}) - f(x_n)\} \\
&\qquad < \sum_{i \le k} 1 + \sum_{n \in N_0} 2M \le (2M + 1)k,
\end{aligned}$$

and so we obtain

$$\begin{aligned}
&\sum_{n \le p} |f(x_n) - f(x_{n-1})| \\
&\qquad = \sum_{n \in N_+} \{f(x_n) - f(x_{n-1})\} + \sum_{n \in N_-} \{f(x_{n-1}) - f(x_n)\} \\
&\qquad = \sum_{n \le p} \{f(x_n) - f(x_{n-1})\} + 2 \sum_{n \in N_-} \{f(x_{n-1}) - f(x_n)\} \\
&\qquad < f(x_n) - f(x_0) + 2(2M + 1)k \\
&\qquad \le 2M + 2(2M + 1)k < (2M + 1)(2k + 1).
\end{aligned}$$

This proves that $V(f; E) \le (2M+1)(2k+1)$, and so $f$ is $BV$ on $E$.

When $f$ is $UAC$ on $E$, the result follows on applying the above result to $-f$. □

Next, in the following theorem we deal with the relation between the properties $LL\text{-}AC_*$ and $L\text{-}BV_*$ on a set.

11.4.3. THEOREM. *Let $f: X \to \mathbb{R}$ and $E \subset X$. If $E$ is bounded and $f$ is bounded and $LL\text{-}AC_*$ on $E$, and $f$ is also bounded from below on $X \cap (\operatorname{co} E)$, then $f$ is $L\text{-}BV_*$ on $E$. Consequently, if $f$ is either $GLL\text{-}AC_*$ or $GUL\text{-}AC_*$ on $E$, then it is $GL\text{-}BV_*$ on $E$.*

PROOF. First, to prove the first part, suppose $E$ is bounded, $f$ is $LL\text{-}AC_*$ on $E$, and

$$c = \sup\{|f(x)| : x \in E\} < +\infty, \qquad d = \inf\{f(x) : x \in X \cap (\operatorname{co} E)\} > -\infty.$$

Then there is a $\delta > 0$ such that if $\{I_n\}$ is any finite sequence of nonoverlapping intervals with endpoints in $E$ and $\Sigma_n |I_n| < \delta$, then $\Sigma_n 0_-(f; I_n) < 1$. Set $J = \overline{\operatorname{co}} E$, and let $\{J_i : i = 1, \dots, k\}$ be a finite decomposition of $J$ such that $|J_i| < \delta$ for each $i$.

Now let $x_0 < x_1 < \cdots < x_p$ be any finite sequence of points in $E$. Set $I_n = [x_{n-1}, x_n]$ for $n = 1, \dots, p$. Further, let $N_i$ denote for each $i \le k$ the set of integers $n \le p$ for which $I_n \subset J_i$, and let $N_0$ denote the set of integers $n \le p$ which are not in $N_i$ for any $i \le k$. Then it is clear that $N_0$ has at most $k-1$ elements. Hence we have

$$\sum_{n \le p} 0_-(f; I_n) = \sum_{i \le k} \sum_{n \in N_i} 0_-(f; I_n) + \sum_{n \in N_0} 0_-(f; I_n)$$
$$< \sum_{i \le k} 1 + \sum_{n \in N_0} (c-d) \le (c-d+1)k.$$

Further,

$$\sum_{n \le p} 0_+(f; I_n) = \sum_{n \le p} \Big\{ f(x_n) - \inf_{t \in X \cap I_n} f(t) \Big\}$$
$$= \sum_{n \le p} \{ f(x_n) - f(x_{n-1}) \} + \sum_{n \le p} \Big\{ f(x_{n-1}) - \inf_{t \in X \cap I_n} f(t) \Big\}$$
$$= f(x_p) - f(x_0) + \sum_{n \le p} 0_-(f; I_n) \le 2c + (c-d+1)k.$$

Consequently, we obtain by (1),

$$\sum_{n \le p} \underline{0}(f; I_n) \le \sum_{n \le p} 0_-(f; I_n) + \sum_{n \le p} 0_+(f; I_n) < 2c + 2(c-d+1)k.$$

This proves that $L\text{-}V_* (f; E) \le 2c + 2(c - d + 1)k < +\infty$, and so $f$ is $L\text{-}BV_*$ on $E$.

Next, suppose $f$ is $LL\text{-}AC_*$ on an arbitrary set $E \subset X$, and let $\delta$ be chosen as above. It is now easy to see that we can choose a sequence of sets $\{E_n\}$ in $E$ such that $E = \bigcup_n E_n$ and such that for each $n$, $E_n$ includes its two bounds and $|\text{co}\, E_n| < \delta$. Now, given $n$, let $I_n \equiv [a_n, b_n] = \text{co}\, E_n$. Then $0_-(f; I_n) < 1$, and so $\inf\{f(x): x \in X \cap I_n\} > f(a_n) - 1 > -\infty$. Further, for each $x \in E_n$, $f(x) - f(b_n) \le 0_-(f; [x, b_n]) < 1$, so that $f(x) \le f(b_n) + 1$. This proves that $f$ is bounded on $E_n$, and so it follows from the above result that $f$ is $L\text{-}BV_*$ on $E_n$. Consequently, $f$ is $GL\text{-}BV_*$ on $E$.

It is now clear that $f$ is $GL\text{-}BV_*$ on $E$ when it is $GLL\text{-}AC_*$ on this set. Also, on applying this result to the function $f_-$, the result is obtained in the case when $f$ is $GUL\text{-}AC_*$ on $E$. □

Next, we obtain a characterization of the property $L\text{-}BV_*$ on a compact set similar to a known characterization of the property $BV_*$ (see [S8, p. 232]). We need the following lemma for this purpose.

11.4.4. Lemma. *Suppose $f: X \to \mathbb{R}$, $E \subset X$ is compact, $I = \text{co}(E)$ and $\{I_n\}$ is the sequence of contiguous intervals of $E$ in $I$. Then*

$$0_-(f; I) \le \underline{V}(f; E) + \min\left\{\sum_n 0_-(f; I_n), \sum_n 0_+(f; I_n)\right\} \tag{4}$$

*and*

$$\underline{0}(f; I) \le V(f; E) + \min\left\{\sum_n 0_-(f; I_n), \sum_n 0_+(f; I_n)\right\}. \tag{5}$$

Proof. Let $c$ and $d$ denote the infima of $f$ on $I \cap X$ and $E$, respectively. Then $c \le d$. The above two inequalities hold trivially in the case when $c = d$, for then $\underline{0}(f; I) = 0$. Hence suppose $c < d$. Let $I = [a, b]$. Then it is clear that

$$0_-(f; I) = f(a) - d + d - c \le \underline{V}(f; E) + d - c. \tag{6}$$

Further, for each real number $r$ in the interval $(c, d)$, there is clearly a point $x \in I \cap X \sim E$ such that $f(x) < r$, and hence there is an integer $n_r$ such that $x \in I_{n_r}$. Hence we obtain

$$d - c \le \min\{0_-(f; I_{n_r}), 0_+(f; I_{n_r})\}.$$

It is thus clear that

$$d - c \le \min\left\{\sum_n 0_-(f; I_n), \sum_n 0_+(f; I_n)\right\}. \tag{7}$$

The inequality (4) follows now on combining (6) and (7). Inequality (5) is obtained on the other hand on combining (7) with the following inequality, which is obvious:

$$\underline{0}(f;I) = \max\{f(a), f(b)\} - d + d - c \leq V(f;E) + d - c. \quad \square$$

11.4.5. THEOREM. *Suppose $f: X \to \mathbb{R}$, $E \subset X$ is compact, $I = \operatorname{co}(E)$ and $\{I_n\}$ is the sequence of contiguous intervals of $E$ in $I$. Then $f$ is $L$-$BV_*$ on $E$ iff it is $BV$ on $E$ and one (or both) of the series $\Sigma_n 0_-(f; I_n)$ and $\Sigma_n 0_+(f; I_n)$ is convergent.*

PROOF. The necessity part follows clearly from Theorem 11.4.2 and (1). To prove the sufficiency, suppose $f$ is $BV$ on $E$ and the first of the two series is convergent. Let $\{J_i\}$ be any finite sequence of nonoverlapping intervals with endpoints in $E$. Then it is clear from the second inequality of the above lemma that

$$\sum_i \underline{0}(f; J_i) \leq \sum_i V(f; E \cap J_i) + \sum_n 0_-(f; I_n) \leq V(f;E) + \sum_n 0_-(f; I_n).$$

Thus $L$-$V_*(f;E) < +\infty$, which proves that $f$ is $L$-$BV_*$ on $E$. The same argument holds in the other case as well. $\square$

Next, in the following theorem we deal with the converse of Theorem 11.4.3.

11.4.6. THEOREM. *Let $f: X \to \mathbb{R}$ and suppose $E \subset X$ is compact and $f$ is $L$-$BV_*$ on $E$. Then if $f$ is LAC, UAC, or AC on $E$, then it is $LL$-$AC_*$, $UL$-$AC_*$, or $L$-$AC_*$, respectively, on $E$.*

PROOF. It is enough to prove the result here in the case when $f$ is $LAC$ on $E$: On applying this result to the function $f_-$, the result is obtained in the case of $UAC$, and the result on $AC$ is obtained on combining these two results.

Hence, suppose $f$ is $LAC$ on $E$, and let $\{I_n\}$ be the sequence of contiguous intervals of $E$ in $\operatorname{co} E$. Then since $f$ is $L$-$BV_*$ on $E$, the series $\Sigma_n 0_-(f; I_n)$ is convergent by Theorem 11.4.5. Hence, given $\varepsilon > 0$, there is an integer $p$ such that $\Sigma_{n>p} 0_-(f; I_n) < \varepsilon/2$. Further, since $f$ is $LAC$ on $E$, there is a $\delta > 0$ such that if $\{J_i\}$ is any finite sequence of nonoverlapping intervals with endpoints in $E$ such that $\Sigma_i |J_i| < \delta$, then $\Sigma_i \underline{V}(f; E \cap J_i) < \varepsilon/2$.

Now let $\eta = \min\{\delta, |I_n|: n = 1, \ldots, p\}$, and let $\{J_i\}$ be any finite sequence of nonoverlapping intervals with endpoints in $E$ such that $\Sigma_i |J_i| < \eta$. Then for each $i$, $J_i$ does not include any of the intervals $\{I_n : n = 1, \ldots, p\}$. Hence it follows clearly from Lemma 11.4.4 that

$$\sum_i 0_-(f; J_i) \leq \sum_i \underline{V}(f; E \cap J_i) + \sum_{n>p} 0_-(f; I_n) < \varepsilon.$$

This proves that $f$ is $LL$-$AC_*$ on $E$. $\square$

Now, on combining the above theorem with Theorem 11.4.3, we obtain the following characterizations of the properties $LL\text{-}AC_*$, $UL\text{-}AC_*$, and $L\text{-}AC_*$, and their generalized versions.

11.4.7. THEOREM. *Let $I = [a, b]$, and suppose $f: I \to \mathbb{R}$ is continuous and $E \subset I$ is closed. Then f is $LL\text{-}AC_*$, $UL\text{-}AC_*$, or $L\text{-}AC_*$ on E iff it is simultaneously $L\text{-}BV_*$ and LAC, UAC, or AC, respectively, on E.*

*Consequently, f is $F\text{-}GLL\text{-}AC_*$, $F\text{-}GUL\text{-}AC_*$, or $F\text{-}GL\text{-}AC_*$ on E iff it is simultaneously $F\text{-}GL\text{-}BV_*$ and F-GLAC, F-GUAC, or F-GAC, respectively, on E.*

PROOF. To prove the first part, it is enough to consider here, as in the last theorem, the result on $LL\text{-}AC_*$. First, suppose $f$ is $LL\text{-}AC_*$ on $E$. Then it follows clearly from Theorem 11.4.3 that $f$ is $L\text{-}BV_*$ on $E$. The $LAC$ of $f$ on $E$ follows on the other hand from the fact that for each interval $J = [c, d]$ with endpoints in $E$ we have $f(c) - f(d) \leq O_-(f; J)$. The converse has already been established in the last theorem.

The second part is of course a direct consequence of the first. □

Now, in the case of a continuous function, the following result in terms of inherent properties follows clearly from the above theorem with the help of Theorem 1.3.1.

11.4.8. COROLLARY. *Let $I = [a, b]$, and suppose $f: I \to \mathbb{R}$ is continuous and $E \subset I$ is closed. Then f is $GLL\text{-}AC_*$, $GUL\text{-}AC_*$, or $GL\text{-}AC_*$ on E iff it is simultaneously $IL\text{-}BV_*$ and ILAC, IUAC, or IAC, respectively, on E.*

**§11.5 Criteria for $GL\text{-}AC_*$ and $GL\text{-}BV_*$ in terms of derivates.** In this section we deduce from the First Derivate Theorem of §2.2 criteria in terms of derivates under which a function is generalized $LL\text{-}AC_*$, $L\text{-}AC_*$, $LAC_*$, or $L\text{-}BV_*$ on a set. These results lead in turn to strengthened versions of some known theorems of Denjoy on the properties $GBV_*$, $GAC_*$, $GBV$, and $GAC$ on a set (see Theorems 11.5.4 and 11.5.5).

First, in the following lemma we establish the relations between the properties $LL\text{-}AC_*$, $UL\text{-}AC_*$, and $AC_*$ on a set with the Lipschitz properties $RLL$, $LUL$, $RL$, and $LL$ on that set defined in §2.1. It should be recalled here that a function $f: X \to \mathbb{R}$ is defined in §5.4 to be $L\text{-}L_*$ (or *lower strongly Lipschitz*}) on a set $E \subset X$ if it is simultaneously $RLL$ and $LUL$ on $E$.

11.5.1. LEMMA. *Let $f: X \to \mathbb{R}$ and $E \subset X$.*

(a) *If f is RLL or LUL on E, then it is $LL\text{-}AC_*$ or $UL\text{-}AC_*$, respectively, on E.*

(b) *If f is $L\text{-}L_*$ on E, then it is $L\text{-}AC_*$ on E.*

(c) *If f is right or left Lipschitz on E, then it is $AC_*$ on E.*

PROOF. Re (a). First, suppose $f$ is *RLL* on $E$. Then there is a $c \in \mathbb{R}_+$ such that the function $f_{+c}$ is right increasing on $E$. Given $\varepsilon > 0$, choose $\delta = \varepsilon/c$, and let $\{I_n\}$ be any finite sequence of nonoverlapping intervals with endpoints in $E$ such that $\Sigma_n|I_n| < \delta$. Now, given $n$, let $I_n = [a_n, b_n]$. Then for each point $x \in X \cap I_n$ we have $f_{+c}(a_n) \leq f_{+c}(x)$, and so

$$f(a_n) - f(x) \leq c(x - a_n) \leq c|I_n|.$$

Thus it is clear that $0_-(f; I_n) \leq c|I_n|$ for each $n$. Hence we obtain

$$\sum_n 0_-(f; I_n) \leq c\sum_n |I_n| < c\delta = \varepsilon,$$

which proves that $f$ is *LL-AC*$_*$ on $E$.

In the case when $f$ is *LUL* on $E$, the result is obtained on applying the above result to the function $f_-$.

Re (b). This part follows clearly on combining the two results in (a).

Re (c). First, suppose $f$ is right Lipschitz on $E$. Then it is clearly *RLL* and *RUL* on $E$, and so it follows from part (a) that $f$ is *LL-AC*$_*$ and *UU-AC*$_*$ on $E$. Hence it follows from Lemma 11.4.1 that $f$ is *AC*$_*$ on $E$. A similar argument holds in the other case. □

Next, in the following theorem we deduce from the First Derivate Theorem generalized properties *LL-AC*$_*$, *UL-AC*$_*$, *L-AC*$_*$, and *LAC*$_*$ in terms of derivates.

11.5.2. THEOREM. *Let* $f : X \to \mathbb{R}$ *and* $E \subset X$.

(a) *If* $D_+f > -\infty$ *or* $D^-f < +\infty$ n.e. *in* $E$, *then* $f$ *is* $G_\delta$-*GLL-AC*$_*$ *or* $G_\delta$-*GUL-AC*$_*$, *respectively, on* $E$, *and in case both of these inequalities hold* n.e. *in* $E$, *then* $f$ *is* $G_\delta$-*GL-AC*$_*$ *on* $E$. *Further, if* $f$ *is LSC relative to* $E$, *then each of these results holds with "F-" in the place of "$G_\delta$-".*

(b) *If* $\underline{D}f > -\infty$ n.e. *in* $E$, *then* $f$ *is* $G_\delta$-*GLAC*$_*$ *on* $E$, *and if* $f$ *is continuous relative to* $E$, *then it is F-GLAC*$_*$ *on* $E$.

PROOF. To obtain part (a), first suppose $D_+f > -\infty$ n.e. in $E$. Then the result follows directly from Corollary 2.2.3 with the help of part (a) of the above lemma. Also, when $f$ is *LSC* relative to $E$, it follows from the same results that $f$ is *F-GLL-AC*$_*$ on $E$. Further, on applying these results to the function $f_-$ the corresponding results are obtained in the case when $D^-f < +\infty$ n.e. in $E$. Finally, on combining these two results the results are obtained clearly in the case when the two inequalities hold simultaneously n.e. in $E$.

Next, to obtain (b), suppose $\underline{D}f > -\infty$ n.e. in $E$. Then on applying part (a) to the functions $f$ and $-f_-$ it follows that $f$ is $G_\delta$-*GLL-AC*$_*$ and $G_\delta$-*GLU-AC*$_*$ on $E$. Hence it is clear that $f$ is $G_\delta$-*GLAC*$_*$ on $E$. Also, in the case when $f$ is continuous relative to $E$, the result is obtained by the same argument. □

Now since various major and minor functions are continuous by definition, we obtain their following inherent properties from the above theorem with the help of Theorem 1.3.1.

11.5.3. COROLLARY. *Let* $f: I \to \overline{\mathbb{R}}$ *where* $I = [a, b]$. *Then every L-major or L-minor function of f is* $ILL\text{-}AC_*$ *or* $IUL\text{-}AC_*$, *respectively. Similarly, every U-major or U-minor function of f is* $ILU\text{-}AC_*$ *or* $IUU\text{-}AC_*$, *respectively.*

*Furthermore, every ordinary major or minor function of f is* $ILAC_*$ *or* $IUAC_*$, *respectively.*

Next, we deduce from the above theorem, and another consequence of the First Derivate Theorem obtained in §5.4, the following strengthened versions of some well-known theorems of Denjoy [D4] (or see Saks [S8, pp. 234–239]) on the properties $GBV_*$, $GBV$, and $GAC$ of a function in terms of its derivates. The result on $GAC$ has been obtained by Denjoy, however, in terms of approximate derivates, and the result on $GL\text{-}BV_*$ is of course new.

11.5.4. THEOREM. *Let* $f: X \to \mathbb{R}$ *and* $E \subset X$.

(a) *If the knot set* $\underline{K}(f)$, $K(f)$, *or* $K_b(f)$ *is countable in E, then f is* $G_\delta\text{-}GL\text{-}BV_*$, $G_\delta\text{-}GBV_*$, *or* $GBV$, *respectively, on E.*

(b) *If at nearly every point of E at least one lower derivate of f is* $> -\infty$ *and at least one upper derivate of f is* $< +\infty$, *then f is GAC on E; and if this condition holds for fixed derivates of f, then f is* $G_\delta\text{-}GAC$ *on E.*

*Moreover, if f is continuous relative to E, then each of these results holds also on replacing "G" or "*$G_\delta$*-G" by "F-G".*

PROOF. Re (a). First, suppose $\underline{K}(f)$ is countable in $E$. Then according to Theorem 5.4.1, $f$ is $G_\delta\text{-}GL\text{-}MT_*$ on $E$, and so it follows easily from Lemma 5.5.1 that $f$ is $G_\delta\text{-}GL\text{-}BV_*$ on $E$.

Next, suppose $K(f)$ is countable in $E$. Then $\underline{K}(f)$ and $\overline{K}(f)$ are both countable in $E$, and thus on applying the first result to $f$ and $-f$ it follows easily that $f$ is $G_\delta\text{-}GBV_*$ on $E$.

Next, suppose $K_b(f)$ is countable in $E$. Set $A = E \sim \underline{K}(f)$ and $B = E \sim A$. Then since $B \subset \underline{K}(f)$, it is clear that $\overline{K}(f)$ is countable in $B$. Hence it follows on applying the first result to $f$ and $-f$ that $f$ is $GBV$ on $A$ and $B$, and so it is $GBV$ on $E$.

Re (b). Suppose the hypothesis of (b) holds. Let $A$ denote the set of points in $E$ where $D_+ f > -\infty$, and set $B = E \sim A$. Then $D_- f > -\infty$ n.e. in $B$. Hence on applying Theorem 14.5.2(a) to $f$ and $-f_-$ it follows that $f$ is $GLAC$ on $A$ and $B$, and so it is clearly so on $E$ as well. Further, on applying this result to $-f$, it follows under the present hypothesis that $f$ is $GUAC$ on $E$. Hence it is clear that $f$ is $GAC$ on $E$.

Next, if the hypothesis of (b) holds for some fixed lower and upper derivates of $f$, then it follows from the above arguments that $f$ is $G_\delta\text{-}GLAC$ and $G_\delta\text{-}GUAC$ on $E$, and so in this case $f$ is $G_\delta\text{-}GAC$ on $E$.

Finally, to obtain the last part, suppose $f$ is continuous relative to $E$. Then if $f$ is $BV$, $L\text{-}BV_*$, $BV_*$, or AC on a set $A \subset E$, it is easy to see that $f$ also has that property on $E \cap \bar{A}$. Hence the last part follows clearly from the results obtained. □

Finally, we deduce from the Second Derivate Theorem the following version of another theorem of Denjoy on the property $GAC_*$ (see [S8, p. 235]).

11.5.5. Theorem. *Let $f: X \to \mathbb{R}$ and $E \subset X$.* (a) *If the two right (or left) derivates of $f$ are finite at each point of $E$, then $f$ is $F\text{-}GAC_*$ on $E$.*

(b) *If at nearly every point of $E$ at least on one side the two unilateral derivates of $f$ are finite, then $f$ is $GAC_*$ on $E$, and in case $f$ is continuous relative to $E$, then it is $F\text{-}GAC_*$ on $E$.*

Proof. First, to obtain (a), suppose its hypothesis holds for the right derivates of $f$. Then it follows from the Second Derivate Theorem 2.2.7 that $f$ is $F\text{-}GRL$ on $E$. Hence it follows from Lemma 11.5.1(c) that $f$ is $F\text{-}GAC_*$ on $E$. A similar argument holds in the other case.

Next, to obtain (b), suppose its hypothesis holds. Let $A$ and $B$ denote the sets of points in $E$ where the two right or left derivates, respectively, of $f$ are finite, and set $C = E \sim (A \cup B)$. Then $C$ is clearly countable, and it follows from (a) that $f$ is $GAC_*$ on both $A$ and $B$. Hence it is clear that $f$ is $GAC_*$ on $E$. The last part of (b) is on the other hand obvious as before. □

## §11.6 Descriptive definitions of lower and upper Perron integrals.

In this section we present descriptive definitions of lower and upper Perron integrals and establish their equivalence with the latter integrals (see Theorem 11.6.8). The descriptive definition of weak Perron integral will be considered later in §11.8.

Further, in Theorem 11.6.9, we prove that a function is Perron integrable iff it is simultaneously lower and upper Perron integrable; and in Theorem 11.6.10 we obtain a reduction of hypotheses for lower and upper Perron integrabilities in the case of a measurable function. A similar reduction theorem has been obtained by Marcinkiewicz [Mar] for Perron integrability. Finally, with the help of their descriptive definitions, we show in Example 11.6.11 that the lower and upper Perron integrals are more general than the Perron integral.

It should be pointed out here that the notions of $LD_*$ and $UD_*$ integrals defined below have been considered earlier by Ionescu Tulcea in his brief note [I-T] under different nomenclature. He made a few observations on these integrals without proof or examples (see Remark 11.6.12).

Given a function $f: I \to \overline{\mathbb{R}}$, we will call $f$ *$LD_*$-integrable* if there exists a continuous function $F: I \to \mathbb{R}$ which is $GL\text{-}AC^*$ such that $f(x) = LF'(x)$ for a.e. $x$. The function $F$ is then called an *indefinite* $LD_*$*-integral* of $f$, and the difference $F(b) - F(a)$ is called the *definite* $LD_*$*-integral* of $f$ over $I$ and is denoted by $(LD_*)\int_a^b f\,dx$.

Similarly, we will call a function $f$ $UD_*$*-integrable* if there exists a continuous function $F: I \to \mathbb{R}$ which is $GU$-$AC_*$ such that $f(x) = UF'(x)$ for a.e. $x$. The function $F$ is then called an *indefinite* $UD_*$*-integral* of $f$, and the difference $F(b) - F(a)$ is called the *definite* $UD_*$*-integral* of $f$ over $I$ and is denoted by $(UD_*)\int_a^b f\,dx$.

Because the integrals $LD_*$ and $UD_*$ are clearly generalizations of the Denjoy-Perron integral $D_*$ (see Saks [S8, 241]), they may be called *lower* and *upper Denjoy-Perron integrals*, respectively.

In the remainder of this chapter we will use the term "interval" only for closed intervals. Further, for each function $F: I \to \mathbb{R}$, we will use $F(J)$ to denote the interval function induced by $F$ (see [S8, p. 96]); that is, if $J = [c, d]$, then $F(J) = F(d) - F(c)$.

We begin with the following relation between $UD_*$ and $LD_*$ integrals.

11.6.1. THEOREM. *A function* $f: I \to \overline{\mathbb{R}}$ *is* $UD_*$*-integrable iff* $-f$ *is* $LD_*$*-integrable; and if so, then* $(UD_*)\int_a^b f\,dx = -(LD_*)\int_a^b - f\,dx$.

PROOF. Given $F: I \to \mathbb{R}$, it is easy to see that for each (closed) interval $J \subset I$, $\overline{0}(F; J) = \underline{0}(-F; J)$. Hence it is clear that $F$ is $GU$-$AC_*$ iff $-F$ is $GL$-$AC_*$. The result follows now clearly from the fact that $F$ is upper derivable at any point $x \in I$ iff $-F$ is lower derivable at $x$; and if so, then $UF'(x) = -L(-F)'(x)$. □

Thus it is clear from the above theorem that it is enough to consider the integral $LD_*$. Hence while referring to the results of §§11.2 and 11.3 which are stated in terms of $WP$-integral, it will be understood in this section that we are referring to their versions in terms of $LP$-integral.

To obtain the equivalence of $LD_*$-integral with $LP$-integral, we will prove here two theorems similar to the classical theorems of Hake [Hak] and Alexandroff [A1, A2] and Looman [Lo] dealing with the equivalence of the $D_*$-integral with the ordinary Perron integral (or see Saks [S8, pp. 250, 251]). The main ideas behind the proofs of these two theorems are adopted from the proofs of the two classical theorems (see Saks [S8, pp. 247–252]). The equivalence of $UD_*$-integral with $UP$-integral will be seen later in Theorem 11.6.8 to follow directly from the obtained equivalence.

We begin with the theorem similar to the Alexandroff-Looman theorem (see [S8, p. 251]).

11.6.2. THEOREM. *Suppose* $f: I \to \overline{\mathbb{R}}$ *is* $LP$*-integrable. Then the indefinite* $LP$*-integral of* $f$ *is continuous and* $GL$-$AC_*$. *Consequently,* $f$ *is* $LD_*$*-integrable and*

$$(LD_*)\int_a^b f\,dx = (LP)\int_a^b f\,dx.$$

PROOF. Let $F$ denote the indefinite $LP$-integral of $f$. Then $F$ is continuous by Theorem 11.3.2. We will first prove that $F$ is $GLL\text{-}AC_*$. First, let $U$ be any $L$-major function of $f$. Then according to Corollary 11.5.3, $U$ is $ILL\text{-}AC_*$. Hence it follows from Theorem 11.4.3 that $U$ is $IL\text{-}BV_*$. Further, the function $U - F$ is nondecreasing by Lemma 11.3.1, and so it is clear that $F = U - (U - F)$ is $IL\text{-}BV_*$. Hence, given any closed set $K$ in $I, F$ is $L\text{-}BV_*$ on some portion $K_0$ of $K$. Let $P = \overline{K}_0$. Now since $F$ is continuous, it follows from Theorem 5.2.1 that $F$ is also $L\text{-}BV_*$ on $P$. We claim that $F$ is $LL\text{-}AC_*$ on $P$. To prove this, it is enough to show by Theorem 11.4.6 that $F$ is $LAC$ on $P$.

Let $J_0 = [a_0, b_0] = \text{co}(P)$, and let $F_1$ and $U_1$ denote the functions on $J_0$ which coincide with $F$ and $U$, respectively, on $P$ and are linear on the contiguous intervals of $P$ in $J_0$. Then since $F$ is $BV$ on $P$, it is clear that $F_1$ is $BV$ on $J_0$. Further, since $U$ is an $L$-major function of $f$, it is clear that $D_+U_1 > -\infty$ n.e. Also, since $U = F + (U - F)$, where $U - F$ is nondecreasing, it is easy to see that $D_+U_1 \geq F_1'$ a.e. in $J_0$. Now define $f_1(x) = F_1'(x)$ at the points where the last inequality holds, and $f_1(x) = -\infty$ elsewhere. Then it is clear that $f_1$ is summable on $J_0$ and that $U_1$ is an $L$-major function of $f_1$. Let $\mathcal{J}$ denote the family of all closed subintervals of $J_0$, and let $F_{1a}$ and $F_{1s}$ denote the $AC$ and singular parts of $F_1$. Then since $U_1 - F_1$ is nondecreasing, for each $J \in \mathcal{J}$ we clearly have

$$(1) \qquad U_1(J) \geq \int_J f_1(x)\,dx = F_{1a}(J).$$

Now, given $\varepsilon > 0$, suppose $U$ is chosen such that $U(b) - F(b) < \varepsilon$. Then since $U - F$ is nondecreasing, so is the interval function induced by it. Hence we obtain

$$(2) \qquad U_1(J_0) - F_1(J_0) = U(J_0) - F(J_0) \leq U(I) - F(I) < \varepsilon.$$

Further, since $U_1 - F_1$ is nondecreasing on $J_0$, for each interval $J \in \mathcal{J}$ we obtain from (2)

$$(3) \qquad U_1(J) - F_1(J) \leq U_1(J_0) - F_1(J_0) < \varepsilon.$$

Hence for each $J \in \mathcal{J}$ we obtain from (1) and (3)

$$F_1(J) > U_1(J) - \varepsilon \geq F_{1a}(J) - \varepsilon,$$

that is, $F_{1s}(J) > -\varepsilon$. Thus we have proved that $F_{1s}(J) > -\varepsilon$ for every $J \in \mathcal{J}$. Because this holds for every $\varepsilon > 0$, this proves indeed that $F_{1s}(J) \geq 0$ for every $J \in \mathcal{J}$. Hence $F_{1s}$ is nondecreasing, and so it is clearly $LAC$. Consequently, the function $F_1$ is $LAC$ on $J_0$ (see [G16, p. 35]), and the function $F$ is therefore $LAC$ on $P$.

We have thus proved that $F$ is $ILL\text{-}AC_*$, and hence it follows from Theorem 1.3.1 that $F$ is $GLL\text{-}AC_*$. Using an $L$-minor function in the place of $U$, it is proved similarly that $F$ is $GUL\text{-}AC_*$. Hence, it is clear that $F$ is $GL\text{-}AC_*$. Further, by the Fundamental Theorem 11.3.3, $f(x) = LF'(x)$ for a.e. $x$. This proves that $f$ is $LD_*$-integrable and that $F$ is an indefinite $LD_*$-integral of $f$. Also, it is clear that

$$(LP)\int_a^b f\,dx = F(b) = F(b) - F(a) = (LD_*)\int_a^b f\,dx. \quad \square$$

In the following theorem we deduce from the above theorem a somewhat stronger version of the Alexandroff-Looman theorem which will be used later.

11.6.3. THEOREM (ALEXANDROFF-LOOMAN). *If a function $f: I \to \overline{\mathbb{R}}$ is lower and upper Perron integrable, or, in particular, Perron integrable, then it is $D_*$-integrable.*

PROOF. It is enough to consider the case when $f$ is lower and upper Perron integrable; for in case $f$ is $P$-integrable, then it is so by Theorem 11.1.4. Let $F$ denote the indefinite $LP$-integral of $f$. Then according to Theorem 11.1.2, $F$ is also the indefinite $UP$-integral of $f$. Hence it follows from Theorem 11.6.2 that $F$ is continuous and $GL\text{-}AC_*$ and $GU\text{-}AC_*$. So $F$ is $GAC_*$ by Lemma 11.4.1. Further, it follows from the Fundamental Theorem 11.3.3 that $f(x) = LF'(x) = UF'(x)$ for a.e. $x$. Hence $f(x) = F'(x)$ for a.e. $x$, which proves that $f$ is $D_*$-integrable. $\square$

The proof of the converse of Theorem 11.6.2 is much more involved. We will need the following three lemmas to obtain this result.

11.6.4. LEMMA. *Suppose $f$ is $LP$-integrable on each interval in the interior of $I = [a, b]$. Then if the $LP$-integral of $f$ on $[a + \varepsilon, b - \eta]$ has a finite limit as $\varepsilon \to 0+$ and $\eta \to 0+$, then $f$ is $LP$-integrable on $I$.*

PROOF. By dividing the interval $I$ into two subintervals, it is clear that it is enough to consider the case when $f$ is $LP$-integrable on each interval of the form $[a, b - \eta]$, where $0 < \eta < b - a$. Define $F(x) = (LP)\int_a^x f(t)\,dt$ for $a \le x < b$, and let $c = F(b-)$.

Let $\{a_n\}$ be an increasing sequence of points in $I$ such that $a_1 = a$ and $a_n \to b$ as $n \to \infty$. Then according to Theorem 11.2.3, $f$ is $LP$-integrable on the interval $I_n = [a_n, a_{n+1}]$ for each $n$. Hence given $\varepsilon > 0$, it is easy to define on account of Lemma 11.3.1 a continuous function $U$ on $[a, b)$ such that the following three properties hold: (i) $U(a) = 0$, $D_+U \ge f$ a.e. and

$D_+U > -\infty$ n.e., (ii) $U - F$ is nondecreasing on $[a, b)$, and (iii) for each $n$,

$$(4) \qquad \{U(x) - U(a_n)\} - \{F(x) - F(a_n)\} < \varepsilon/2^n \quad \text{for} \quad x \in I_n.$$

Then it is clear that $U - F$ is bounded on $[a, b)$. Hence $U - F$ has a finite limit at $b$, and so does $F$ by hypothesis. Consequently, $U$ also has a finite limit at $b$. Define $U(b) = U(b-)$. Then $U$ is continuous on $I$, and so it is an $L$-major function of $f$ on $I$. Further, from (4) we obtain

$$\begin{aligned} U(b) = U(b-) - U(a) &= \sum_n \{U(a_{n+1}) - U(a_n)\} \\ &< \sum_n \{F(a_{n+1}) - F(a_n)\} + \sum_n \varepsilon/2^n \\ &= F(b-) - F(a) + \varepsilon = c + \varepsilon. \end{aligned}$$

Using a similar argument we define an $L$-minor function $V$ of $f$ on $I$ such that $V(b) > c - \varepsilon$. Hence it follows from Lemma 11.2.2 that $f$ is $LP$-integrable on $I$. □

11.6.5. Lemma. *Suppose $F$ is the indefinite LP-integral of a function $f: I \to \overline{\mathbb{R}}$. Then for each $\varepsilon > 0$ there are L-major and L-minor functions $U$ and $V$ of $f$ such that for each closed subinterval $J$ of $I$ we have*

$$(5) \qquad \underline{0}(U; J) < \underline{0}(F; J) + \varepsilon, \qquad \underline{0}(V; J) > \underline{0}(F; J) - \varepsilon.$$

Proof. Given $\varepsilon > 0$, choose $L$-major and $L$-minor functions $U$ and $V$ of $f$ such that $U(b) - F(b) < \varepsilon$ and $F(b) - V(b) < \varepsilon$. Then for each $x \in I$ we have, by Lemma 11.3.1, $U(x) \geq F(x)$. Hence, given any closed subinterval $J = [c, d]$ of $I$, we have $\inf\{U(x): x \in J\} \geq \inf\{F(x): x \in J\}$. Further, since $U - F$ is nondecreasing by the same lemma, $U(c) - F(c) \leq U(d) - F(d) \leq U(b) - F(b) < \varepsilon$. Hence we obtain

$$0_-(U; J) = U(c) - \inf_{x \in J} U(x) < F(c) + \varepsilon - \inf_{x \in J} F(x) = 0_-(F; J) + \varepsilon,$$

$$0_+(U; J) = U(d) - \inf_{x \in J} U(x) < F(d) + \varepsilon - \inf_{x \in J} F(x) = 0_+(F; J) + \varepsilon.$$

Now the first inequality of (5) follows clearly from these two inequalities. The second inequality is deduced by a similar argument from the fact that $F - V$ is nondecreasing. □

11.6.6. Lemma. *Let $K$ be a closed subset of $I = [a, b]$ such that $I = \operatorname{co}(K)$, and let $\{I_n = [a_n, b_n]\}$ be the sequence of contiguous intervals of $K$ in $I$. Suppose $f: I \to \overline{\mathbb{R}}$ is LP-integrable on each $I_n$ and summable on $K$. Let $0_n$ be for each $n$ the lower oscillation of the indefinite LP-integral of $f$ on $I_n$. Then if the series*

$\Sigma_n 0_n$ *is convergent, then* $f$ *is LP-integrable on* $I$ *and*

(6) $$(LP)\int_a^b f\,dx = \int_K f\,dx + \sum_n (LP)\int_{a_n}^{b_n} f\,dx.$$

PROOF. Given $\varepsilon > 0$, choose a positive integer $p$ such that

(7) $$\sum_{n>p} 0_n < \varepsilon.$$

Set $K_1 = K \cup \bigcup_{n \le p} I_n$, and define $f_1(x) = f(x)$ for $x \in K_1$ and $f_1(x) = 0$ for $x \in I \sim K_1$. Then $f_1$ is clearly summable on $K \cup \bigcup_{n>p} I_n$, and so it is $LP$-integrable on $I$ by Theorems 11.1.4 and 11.2.3. Hence, with the help of a known theorem on summable functions (see Saks [S8, p. 191]), we can choose $L$-major and $L$-minor functions $U_1$ and $V_1$ of $f_1$ such that

(8) $$U_1(b) - \varepsilon \le \int_K f\,dx + \sum_{n\le p} (LP)\int_{I_n} f\,dx \le V_1(b) + \varepsilon.$$

Next, we will define $L$-major and $L$-minor functions of the function $f_2 = f - f_1$. First, with the help of the first inequality of Lemma 11.6.5 we choose for each $n$ an $L$-major function $G_n$ of $f$ on $I_n$ such that

(9) $$\underline{0}(G_n;I_n) \le 20_n.$$

Then $G_n(a_n) = 0$ for each $n$. Next, we define for each $n$ two functions $P_n$ and $Q_n$ on $I_n$ as follows:

(10) $$P_n(x) = 0_-(G_n;[a_n,x]), \qquad Q_n(x) = 0_+(G_n;[x,b_n]), \qquad x \in I_n.$$

Further, we define

(11) $$G(x) = G_n(x) + P_n(x) - Q_n(x) + Q_n(a_n) \quad \text{for} \quad x \in I_n^0, \quad n > p,$$

and $G(x) = 0$ elsewhere.

We first obtain the following properties of the function $G$:

(a) $G(x) \ge 0$ for each $x \in I$;

(b) $G(a_n+) = G(a_n) = 0$ for each $n$;

(c) $P_n - Q_n$ is nondecreasing on $I_n$ for each $n$;

(d) $0_-(G;I_n) \le 20_n$ for each $n$; and

(e) $0 \le G(b_n-) \le 40_n$ for each $n$.

To prove (a), first suppose $x \in I_n^0$ for some $n > p$. Then since $G_n(a_n) = 0$, it is easy to see from (10) and (11) that

$$G(x) = G_n(x) - \inf_{a_n \le t \le x} G_n(t) + \inf_{x \le t \le b_n} G_n(t) - \inf_{a_n \le t \le b_n} G_n(t),$$

and so it is clear that $G(x) \geq 0$. Hence it follows from the definition of $G$ that this relation holds for every $x \in I$.

Next, since $G(x) = 0$ for $x \in I_n$ for each $n \leq p$, the properties (b) to (e) hold trivially for $n \leq p$. Hence suppose $n > p$. Then since $G_n(a_n) = 0$, (b) follows clearly from (10) and (11) due to the continuity of $G_n$ on $I_n$. Property (c) follows on the other hand from (10) since $P_n$ is nondecreasing and $Q_n$ is nonincreasing on $I_n$. Further, as for (d), we obtain the following from (b), (c), (9), and (11):

$$\begin{aligned} 0_-(G;I_n) &= -\inf_{x\in I_n} G(x) \leq -\inf_{x\in I_n} G_n(x) - \inf_{x\in I_n}\{P_n(x) - Q_n(x)\} - Q_n(a_n) \\ &= -\inf_{x\in I_n} G_n(x) - \{P_n(a_n) - Q_n(a_n)\} - Q_n(a_n) \\ &= 0_-(G_n;I_n) \leq 20_n. \end{aligned}$$

Finally, the first inequality of (e) follows clearly from (a). As for the second inequality, it is clear from the definition of lower oscillation that

$$|G_n(b_n)| = |G_n(b_n) - G_n(a_n)| \leq \underline{0}(G_n;I_n) \leq 20_n.$$

Further, since $G_n$ is continuous, from (10) and (11) we obtain

$$G(b_n-) = G_n(b_n) + 0_-(G_n;I_n) + 0_+(G_n;I_n).$$

Hence it follows from (9) that

$$G(b_n-) \leq |G_n(b_n)| + 20_n \leq 40_n.$$

Finally, with the help of $G$ we define

$$U_2(x) = G(x) + \sum\nolimits_n^{(x)} G(b_n-), \qquad x \in I, \tag{12}$$

where the summation $\Sigma_n^{(x)}$ is taken over those $n$ for which $b_n \leq x$. It is clear from (e) and (7) that the function $U_2$ is well-defined.

We claim that $U_2$ is an $L$-major function of $f_2$. It is clear from the continuity of the functions $G_n$, (b), (10), (11), and (12) that $U_2$ is continuous. Also, it is clear that $U_2(a) = G(a) = 0$. Further, it is clear from the definitions of $f_1$ and $f_2$ that $f_2(x) = 0$ for $x \in K_1$, and $f_2(x) = f(x)$ for $x \in I_n^0$ for each $n > p$.

First, let $x \in K_1$. Then $G(x) = 0$, and so $U_2(x) = \Sigma_n^{(x)} G(b_n-)$. Let $y \in I$ and $y > x$. Then it follows from (a) and (e) that

$$U_2(y) = G(y) + \sum\nolimits_n^{(y)} G(b_n-) \geq U_2(x).$$

Consequently, $D_+U_2(x) \geq 0 = f_2(x)$. Next, suppose $x \in I_n^0$ for some $n > p$.

Then since $P_n - Q_n$ is nondecreasing on $I_n$ by (c), it follows from (11) and (12) that

$$D_+U_2(x) = D_+G(x) \geq D_+G_n(x).$$

But since $G_n$ is an $L$-major function of $f$ on $I_n$, it thus follows that $D_+U_2 > -\infty$ n.e. in $I_n$ and $D_+U_2 \geq f = f_2$ a.e. in $I_n$. This proves that $U_2$ is an $L$-major function of $f_2$ on $I$. Further, since $G(b) = 0$ and $G$ is continuous from the left at $b_n$ for $n \leq p$, we obtain from (7), (12), and (e),

$$U_2(b) = \sum_n G(b_n-) \leq \sum_n 4 0_n < 4\varepsilon. \tag{13}$$

Using similar arguments we define with the help of the second inequality of (5) an $L$-minor function $V_2$ of $f_2$ such that $V_2(b) > -4\varepsilon$. Now let $U = U_1 + U_2$ and $V = V_1 + V_2$. Then it is easy to see that $U$ and $V$ are $L$-major and $L$-minor functions of $f = f_1 + f_2$. Also, if $s$ denotes the sum on the right side of (6), it follows from (7) and (8) that $U(b) - 6\varepsilon < s < V(b) + 6\varepsilon$. This proves that $f$ is $LP$-integrable on $I$ and that (6) holds. □

Now we obtain the following theorem similar to the Hake's theorem (see [S8, p. 250]). If $F$ is the indefinite $LP$-integral of some function $f$, then we will call here a function of the form $F + c$, where $c \in \mathbb{R}$, "an" indefinite $LP$-integral of $f$.

11.6.7. THEOREM. *If a function $f : I \to \overline{\mathbb{R}}$ is $LD_*$-integrable, then it is $LP$-integrable.*

PROOF. Suppose $f$ is $LD_*$-integrable, and let $F$ be an indefinite $LD_*$-integral of $f$ on $I$. Let $\mathcal{J}$ denote this time the family of all closed subintervals of $I$. We will call an interval $J \in \mathcal{J}$ "regular" if $f$ is $LP$-integrable on $J$ and $F$ is an indefinite $LP$-integral of $f$ on $J$. Further, we will call a point $x \in I$ regular if there is a regular interval $J$ such that $x \in J^0$. Now, let $P$ denote the set of points in $I$ which are not regular. Then $P$ is clearly closed, and if $J \in \mathcal{J}$ is disjoint with $P$, then $J$ is regular. Hence it is enough to show that $P = \varnothing$.

Suppose, if possible, that $P \neq \varnothing$. Then it follows from Lemma 11.6.4 that every contiguous interval of $P$ in $I$ is regular and that $P$ is dense in itself. Further, since $F$ is continuous and $GL$-$AC_*$, it is clearly $F$-$GL$-$AC_*$. Hence it follows from Theorem 1.3.1 that $F$ is $L$-$AC_*$ on some portion $P_0$ of $P$. Let $J_0 = \overline{\text{co}}\,(P_0)$. Then since $P$ is dense in itself, it is clear that $P \cap J_0^0 \neq \varnothing$. Hence it is enough to prove that $J_0$ is regular.

Now, let $J \in \mathcal{J}$ be any given subinterval of $J_0$. Let $K$ be the set obtained on adjoining the endpoints of $J$ to $P \cap J$. Then $K$ is a closed subset of $J$. Let $\{I_n\}$ denote the sequence of contiguous intervals of $K$ in $J$. Now let $G$ be the function which coincides with $F$ on $K$ and is linear on the intervals $I_n$. Then since $F$ is $L$-$AC_*$ on $K$, it is clearly $AC$ on $K$. Hence $G$ is $AC$ on $J$, and so its derivative is summable on $J$. Further, it is clear that $G'(x) = LF'(x) = f(x)$ for a.e. $x$ in $K$, and $G(I_n) = F(I_n)$ for each $n$. Hence we obtain

$$F(J) = G(J) = \int_J G'\,dx = \sum_n F(I_n) + \int_K f\,dx. \tag{14}$$

The function $f$ is thus $LP$-integrable on each $I_n$ and summable on $K$, and $F$ is an indefinite $LP$-integral of $f$ on each $I_n$. Further, since $F$ is $L$-$AC_*$ on $P_0$, by Theorem 11.4.3 it is $L$-$BV_*$ on $P_0$. Hence the series $\Sigma_n \underline{O}(F;I_n)$ is convergent, and so it follows from Lemma 11.6.6 and (14) that $f$ is $LP$-integrable on $J$ and $F(J) = (LP)\int_J f\,dx$. Since this holds for every subinterval $J$ of $J_0$, this proves that $J_0$ is regular. □

Now with the help of Theorems 11.6.2 and 11.6.7 we obtain the following:

11.6.8. THEOREM. *A function $f: I \to \overline{\mathbb{R}}$ is $LP$-integrable iff it is $LD_*$-integrable; and if so, then $(LD_*)\int_a^b f\,dx = (LP)\int_a^b f\,dx$.*

*Similarly, $f$ is $UP$-integrable iff it is $UD_*$-integrable; and if so, then $(UD_*)\int_a^b f\,dx = (UP)\int_a^b f\,dx$.*

PROOF. The first part is obtained directly on combining Theorems 11.6.2 and 11.6.7. Further, it is clear that the second part follows from the first with the help of Theorems 11.1.1 and 11.6.1. □

Next, we obtain Perron integrability in terms of lower and upper Perron integrabilities. It is somewhat surprising that this seemingly apparent result turned out to be not so easy to prove.

11.6.9. THEOREM. *A function $f: I \to \overline{\mathbb{R}}$ is $P$-integrable iff it is simultaneously $LP$- and $UP$-integrable. Consequently, $f$ is $D_*$-integrable iff it is simultaneously $LD_*$- and $UD_*$-integrable.*

PROOF. The necessity part of the first result follows directly from Theorem 11.1.4. To prove its sufficiency, suppose $f$ is simultaneously $LP$- and $UP$-integrable. Then according to Theorem 11.6.3, $f$ is $D_*$-integrable, and so it follows from the theorem of Hake (see [S8, p. 250]) that $f$ is $P$-integrable.

The second part follows on the other hand from the first with the help of Theorem 11.6.8 and the well-known equivalence of $D_*$-integrability with $P$-integrability (see [S8, pp. 250, 251]). □

Next, in the case of measurable functions we obtain in the following theorem a reduction of hypotheses for $LP$- and $UP$-integrabilities. This theorem is similar to a theorem obtained by Marcinkiewicz [Mar] (or see [S8, p. 253]) on the reduction of hypothesis for *P-integrability*.

11.6.10. THEOREM (REDUCTION). *A function $f : I \to \overline{\mathbb{R}}$ is LP-integrable iff it is measurable and it has at least one L-major function and at least one L-minor function. A similar statement holds for UP-integrability.*

PROOF. We give the proof here for the first result. The necessity part of this result holds clearly due to Corollary 11.3.5. To prove its sufficiency, suppose $f$ is measurable and it has some $L$-major and $L$-*minor* functions $U$ and $V$. We will call here a point $x \in I$ "regular" if $f$ is $LP$-integrable on some interval $J \subset I$ such that $x \in J^0$. Now let $K$ be the set of points in $I$ which are not regular. Then $K$ is clearly closed, and $f$ is $LP$-integrable on each interval in $I$ which is disjoint with $K$. Hence it is enough to show that $K = \varnothing$.

Suppose, if possible, that $K \neq \varnothing$. If $f$ is $LP$-integrable on any interval $J \subset I$, it is clear that $U(J) \geq (LP)\int_J f\,dx \geq V(J)$. Hence if $[c, d]$ is any contiguous interval of $K$, then the definite $LP$-integral of $f$ on the interval $[c + \varepsilon, d - \eta]$ has a finite limit as $\varepsilon \to 0+$ and $\eta \to 0+$. Hence it follows from Lemma 11.6.4 that $f$ is $LP$-integrable on each contiguous interval of $K$ in $I$. Consequently, $K$ is dense in itself.

Now, it is clear from Theorems 11.5.2 and 11.4.3 that the functions $U$ and $V$ are $GL$-$BV_*$. Further, since these functions are continuous, they are indeed $F$-$GL$-$BV_*$ by Theorem 5.2.1, and hence they are $IL$-$BV_*$ by Theorem 1.3.1. Hence there is a portion $K_0$ of $K$ on which $U$ and $V$ are $L$-$BV_*$. Each of these functions is thus $BV$ on $P = \overline{K_0}$. Let $J_0 = \mathrm{co}(P)$, and let $U_1$ and $V_1$ be the functions on $J_0$ which coincide with $U$ and $V$, respectively, on $P$ and are linear on the contiguous intervals of $P$ in $J_0$. Then $U_1$ and $V_1$ are clearly $BV$ on $J_0$, and so there derivatives are summable on $J_0$. Hence it is clear that the lower derivatives of $U$ and $V$ are summable on $P$. Further, since $f$ is measurable and $D_+U \geq f \geq D^-V$ a.e., it thus follows that $f$ is summable on $P$.

Now let $\{I_n\}$ be the sequence of contiguous intervals of $P$ in $J_0$, and let, for each $n$, $F_n$ denote the indefinite $LP$-integral of $f$ on $I_n$ and $0_n = \underline{0}(F_n; I_n)$. Then since $F_n$ is continuous by Theorem 11.3.2, and $U - F_n$ and $F_n - V$ are by Lemma 11.3.1 nondecreasing on $I_n$, it is easy to see that for each $n$, $0_+(F_n; I_n) \leq 0_+(U; I_n)$ and $0_-(F_n; I_n) \leq 0_-(V; I_n)$. Hence for each $n$ we obtain

$$0_n \leq 0_+(F_n; I_n) + 0_-(F_n; I_n) \leq \underline{0}(U; I_n) + \underline{0}(V; I_n).$$

Now since $U$ and $V$ are $L$-$BV_*$ on $K_0$, it is thus clear that $\Sigma_n 0_n < +\infty$. Hence it follows from Lemma 11.6.6 that $f$ is $LP$-integrable on $J_0$. But this contradicts the fact that $J_0 \cap K \neq \varnothing$. □

Finally, we present two examples of functions to show that the $LP$- and $UP$-integrals are strictly more general than the ordinary Perron integral; or, equivalently, the $LD_*$ and $UD_*$ integrals are more general than the $D_*$ integral (see Theorem 11.6.8). These examples show further due to Theorem 11.6.9 that the $LP$- and $UP$-integrals are not comparable in general.

Given any two generalized integrals $G_1$- and $G_2$-, we will call the $G_1$-integral (strictly) *more general than* the $G_2$-integral if it is weaker than the $G_2$-integral but not equivalent to it—that is, if there exists a function $f: I \to \overline{\mathbb{R}}$ on some compact interval $I$ which is $G_1$-integrable but not $G_2$-integrable.

11.6.11. EXAMPLE. The $LD_*$-integral is clearly weaker than the $D_*$-integral. Hence to prove that it is more general than the $D_*$-integral, it is enough to show that there exists a continuous function $F: I \to \mathbb{R}$ which is $GL\text{-}AC_*$ but not $GAC_*$. For, such a function $F$ is clearly an indefinite $LD_*$-integral of the function $f = LF'$ which may be defined arbitrarily at the points where $F$ is not lower derivable (see Theorems 11.4.3 and 5.3.1), but is not an indefinite $D_*$-integral of $f$ (see Theorems 11.6.8 and 11.1.4). Also, it is easy to see from the functions $-F$ and $-f$ that the $UD_*$-integral is more general than the $D_*$-integral.

The following example was considered earlier by Saks [S8, p. 224] to prove the existence of a continuous $GAC$ function which is not derivable in the ordinary sense at the points of a set whose measure is positive.

Let $P$ be any nowhere dense perfect set in $I$ such that $I = \mathrm{co}(P)$ and $|P| > 0$. Let $\{I_n = [a_n, b_n]\}$ be the sequence of contiguous intervals of $P$ in $I$, and let $c_n = (a_n + b_n)/2$ for each $n$. Further, let $r_n$ denote, for each $n$, the length of the largest interval in $I$ which does not overlap with any of the intervals $I_1, I_2, \ldots, I_n$. Then $\{r_n\}$ is a nonincreasing sequence which converges to zero.

Now, define $F(x) = 0$ for $x \in P$, and let, for each $n$, $F(c_n) = |I_n| + r_n$, and let $F$ be defined linearly on $[a_n, c_n]$ and $[c_n, b_n]$. Then since $|I_n|$ and $r_n$ both converge to zero as $n \to \infty$, it is clear that $F$ is continuous. Further, $F$ is clearly $AC_*$ on each $I_n$, and since $\underline{O}(F; I_n) = 0$ for each $n$, $F$ is trivially $L\text{-}AC_*$ on $P$. Hence $F$ is $GL\text{-}AC_*$ on $I$. We claim that $F$ is not $GU\text{-}AC_*$ on $I$.

Given any point $x \in P^b$, let for every sufficiently large $n$, $i_n$ (or $k_n$) denote the subscript of the interval among $I_1, I_2, \ldots, I_n$ which is on the left (or right) of $x$ and is nearest to $x$. Then $\lim_n i_n = \lim_n k_n = \infty$ and $c_{i_n}$ and $c_{k_n}$ converge to $x$ from the left or right, respectively, as $n \to \infty$. Now, for every large $n$ we have

$$0 < c_{k_n} - x < |I_{k_n}| + r_n \le |I_{k_n}| + r_{k_n} = F(c_{k_n}) - F(x).$$

Hence $QF(x, c_k) > 1$ for every large $n$, and so $D^+F(x) \ge 1$. Similarly, for every large $n$ we have

$$0 < x - c_{i_n} < |I_{i_n}| + r_n \le |I_{i_n}| + r_{i_n} = F(c_{i_n}) - F(x),$$

so that $\{F(c_{i_n}) - F(x)\}/(x - c_{i_n}) > 1$, or $QF(x, c_{i_n}) < -1$. Hence $D_-F(x) \le -1$. This proves that $F$ is not upper derivable at any point of $P^b$. Now since $|P^b| > 0$, it follows from Theorem 5.3.1 that $F$ is not $GU\text{-}BV_*$, and so it is not $GU\text{-}AC_*$ by Theorem 11.4.3.

11.6.12. REMARK. As mentioned earlier, the notions of $LP$- and $UP$-integrals have been considered earlier in terms of somewhat different definitions by Ridder [Ri1] and Ionescu Tulcea [I-T]. Ridder called them $L_2$ and $L_2^*$ integrals, respectively. Also, using different definitions and nomenclature, they defined $LD_*$ and $UD_*$ integrals.

They did not obtain the properties of these integrals, nor did they prove the equivalence of $LP$- and $UP$-integrals with $LD_*$ and $UD_*$ integrals, respectively. Ridder obtained, however, the relations between $LD_*$, $UD_*$, and $D_*$ integrals considered in Example 11.6.11. See further Example 11.8.6 for the relation between $LD_*$ and $UD_*$ integrals with the general $D$-integral.

**§11.7 Semistrong properties $S\text{-}BV_*$ and $S\text{-}AC_*$.** To obtain a descriptive definition of weak Perron integral, we define in this section semistrong properties $S\text{-}BV_*$ and $S\text{-}AC_*$ by combining the lower and upper versions of these two properties defined earlier. Further, we obtain here results on these two properties similar to the ones obtained in Chapter 5 and §§11.4 and 11.5, and also on the lower and upper versions of the property $S\text{-}AC_*$.

Suppose $f : X \to \mathbb{R}$, where $X \subset \mathbb{R}$, and let $E \subset X$. Then we will call $f$ $S\text{-}BV_*$ (or *semistrongly of bounded variation*) on $E$ if it is either $L\text{-}BV_*$ or $U\text{-}BV_*$ on $E$. Similarly, we will call $f$ $S\text{-}AC_*$ (or *semistrongly AC*) on $E$ if it is either $L\text{-}AC_*$ or $U\text{-}AC_*$ on $E$.

Further, we will call $f$ $LS\text{-}AC_*$ on $E$ if it is either $LL\text{-}AC_*$ or $LU\text{-}AC_*$ on $E$; and $f$ will be called similarly $US\text{-}AC_*$ on $E$ if it is either $UL\text{-}AC_*$ or $UU\text{-}AC_*$ on $E$. The properties $LS\text{-}AC_*$ and $US\text{-}AC_*$ may be called *lower* and *upper semistrong absolute continuities*.

Now the generalized properties $GS\text{-}BV_*$, $GS\text{-}AC_*$, $GLS\text{-}AC_*$, and $GUS\text{-}AC_*$ of $f$ on $E$ are defined as usual as in §1.3. We begin with two lemmas which will be used frequently. The following lemma on these generalized properties is obtained directly from the above definitions.

11.7.1. LEMMA. *Suppose $f : X \to \mathbb{R}$ and $E \subset X$. Then $f$ is $GS\text{-}BV_*$ (or $GS\text{-}AC_*$) on $E$ iff $E = A \cup B$, where $f$ is $GL\text{-}BV_*$ (or $GL\text{-}AC_*$) on $A$ and $GU\text{-}BV_*$ (or $GU\text{-}AC_*$) on $B$. Similarly, $f$ is $GLS\text{-}AC_*$ (or $GUS\text{-}AC_*$) on $E$ iff $E = A \cup B$, where $f$ is $GLL\text{-}AC_*$ (or $GUL\text{-}AC_*$) on $A$ and $GLU\text{-}AC_*$ (or $GUU\text{-}AC_*$) on $B$.*

Next, in the following lemma we deal with a relation between the above properties.

11.7.2. LEMMA. *Suppose $f : X \to \mathbb{R}$ and $E \subset X$. Then $f$ is $S\text{-}AC_*$ on $E$ iff it is $LS\text{-}AC_*$ and $US\text{-}AC_*$ on $E$. Consequently, $f$ is $GS\text{-}AC_*$ on $E$ iff it is $GLS\text{-}AC_*$ and $GUS\text{-}AC_*$ on $E$.*

PROOF. The necessity part of the first result follows directly from the definitions. Its sufficiency is easily seen on the other hand to follow from Lemma 11.4.1. The second part is of course a direct consequence of the first. □

The following extension theorem on the property $S\text{-}BV_*$ is obtained on applying Theorem 5.2.1 to $f$ or $-f$ as necessary.

11.7.3. THEOREM. *Suppose $f : X \to \mathbb{R}$ and $E \subset X$ is a bounded set such that the extreme points of $\text{co}(E)$ are in $X$. Then if $f$ is $S\text{-}BV_*$ on $E$ and it is continuous at the points of $\overline{E} \sim E$, then $f$ is $S\text{-}BV_*$ on $\overline{E}$. Consequently, if $f$ is continuous and $GS\text{-}BV_*$ on $E$, then it is $F\text{-}GS\text{-}BV_*$ on $E$.*

Next, in the following theorem we deal with differentiability of $GS\text{-}BV_*$ functions.

11.7.4. THEOREM. *Let $f : X \to \mathbb{R}$ and $E \subset X$. If $f$ is $GS\text{-}BV_*$ on $E$, then it is normally semidifferentiable* a.e. *in $E$; and further if $N = E \sim \Delta_s(f)$, then $|f(N)| = \Lambda\{Gf(N)\} = 0$.*

PROOF. Suppose $f$ is $GS\text{-}BV_*$ on $E$. Then according to Lemma 11.7.1, $E = A \cup B$, where $f$ is $GL\text{-}BV_*$ on $A$ and $GU\text{-}BV_*$ on $B$. Now on applying Theorem 5.3.1 to $f$ and $-f$ it follows that $f$ is lower differentiable a.e. in $A$ and upper differentiable a.e. in $B$. Hence it follows from the Dejoy-Young-Saks Theorem 5.5.2 that $f$ is normally semidifferentiable a.e. in $E$. Also, the second part follows similarly from the last part of Theorem 5.3.1. □

The following relation between the properties $S\text{-}AC_*$ and $S\text{-}BV_*$ is obtained similarly from Theorem 11.4.3 with the help of Lemma 11.7.1.

11.7.5. THEOREM. *Let $f : X \to \mathbb{R}$ and $E \subset X$. Suppose $E$ is bounded and $f$ is bounded on $X \cap (\text{co } E)$. Then if $f$ is either $LS\text{-}AC_*$ or $US\text{-}AC_*$ on $E$, then it is $S\text{-}BV_*$ on $E$. Consequently, if $f$ is either $GLS\text{-}AC_*$ or $GUS\text{-}AC_*$ on $E$, then it is $GS\text{-}BV_*$ on $E$.*

In the following theorem we deal with the converse of the above theorem.

11.7.6. THEOREM. *Let $f : X \to \mathbb{R}$ and suppose $E \subset X$ is compact and $f$ is $S\text{-}BV_*$ on $E$. Then if $f$ is LAC, UAC, or AC on $E$, then it is $LS\text{-}AC_*$, $US\text{-}AC_*$, or $S\text{-}AC_*$, respectively, on $E$.*

PROOF. This result is obtained directly from Theorem 11.4.6 when $f$ is $L\text{-}BV_*$ on $E$; and when $f$ is $U\text{-}BV_*$ on $E$, the result is obtained on applying this theorem to $-f$. □

The following characterization of the properties $LS\text{-}AC_*$, $US\text{-}AC_*$, and $S\text{-}AC_*$ is obtained in a similar manner from Theorem 11.4.7 with the help of Lemma 11.7.1.

11.7.7. THEOREM. *Let $I = [a, b]$, and suppose $f : I \to \mathbb{R}$ is continuous and $E \subset I$ is closed. Then $f$ is $LS\text{-}AC_*$, $US\text{-}AC_*$, or $S\text{-}AC_*$ on $E$ iff it is simultaneously $S\text{-}BV_*$ and LAC, UAC, or AC, respectively, on $E$.*

*Consequently, $f$ is $F\text{-}GLS\text{-}AC_*$, $F\text{-}GUS\text{-}AC_*$, or $F\text{-}GS\text{-}AC_*$ on $E$ iff it is simultaneously $F\text{-}GS\text{-}BV_*$ and F-GLAC, F-GUAC, or F-GAC, respectively, on $E$.*

Now, in the case of a continuous function, the following result in terms of inherent properties follows clearly from the above theorem with the help of Theorem 1.3.1.

11.7.8. Corollary. *Let $I = [a, b]$, and suppose $f : I \to \mathbb{R}$ is continuous and $E \subset I$ is closed. Then f is GLS-$AC_*$, GUS-$AC_*$, or GS-$AC_*$ on E iff it is simultaneously IS-$BV_*$ and ILAC, IUAC, or IAC, respectively, on E.*

Next, in the following theorem we deal with criteria for these three generalized properties in terms of weak derivates.

11.7.9. Theorem. *Let $f : X \to \mathbb{R}$ and $E \subset X$. Then if $\underline{D}_w f > -\infty$ or $\overline{D}_w f < +\infty$* n.e. *in E, then f is GLS-$AC_*$ or GUS-$AC_*$, respectively, on E; and if both of these inequalities hold* n.e. *in E, then f is GS-$AC_*$ on E. Further, if f is continuous relative to E, then each of these results holds with "F-G" in the place of "G".*

Proof. First, suppose $\underline{D}_w f > -\infty$ n.e. in $E$. Let $A$ denote the set of points in $E$ where $D_+ f > -\infty$, and set $B = E \sim A$. Then $D_- f > -\infty$ n.e. in $B$. Hence it follows on applying Theorem 11.5.2(a) to $f$ and $-f_-$ (see Lemma 1.1.2) that $f$ is $GLL$-$AC_*$ on $A$ and $GLU$-$AC_*$ on $B$. Consequently, $f$ is $GLS$-$AC_*$ on $E$. Now on applying this result to $-f$, the result is obtained when $\overline{D}_w f < +\infty$ n.e. in $E$. Further, when the two inequalities hold simultaneously n.e. in $E$, the result is obtained on combining these two results with the help of Lemma 11.7.2. The last part is obtained by similar arguments from the last part of Theorem 11.5.2(a). □

Now, since weak major and minor functions are continuous according to their definitions, we obtain their following inherent properties from the above theorem with the help of Theorem 1.3.1.

11.7.10. Corollary. *Let $f : I \to \overline{\mathbb{R}}$, where $I = [a, b]$. Then every W-major or W-minor function of f is ILS-$AC_*$ or IUS-$AC_*$, respectively.*

**§11.8 Descriptive definition of weak Perron integral.** In this final section of Part III we present the descriptive definition of weak Perron integral and establish its equivalence with the latter integral (see Theorems 11.8.1 and 11.8.5). Also, in Example 11.8.6, we show by two examples that the weak Perron integral is more general than the lower and upper Perron integrals, and that it is not as general as the $D$-integral of Denjoy and Khintchine.

Given a function $f : I \to \overline{\mathbb{R}}$, we will call $f$ *$SD_*$-integrable* if there exists a continuous function $F : I \to \mathbb{R}$ which is $GS$-$AC_*$ such that $f(x) = SF'(x)$ for a.e. $x$. The function $F$ is then called an *indefinite $SD_*$-integral* of $f$, and the difference $F(b) - F(a)$ is called the *definite $SD_*$-integral* of $f$ over $I$ and is denoted by $(SD_*)\int_a^b f dx$.

The integral $SD_*$ is clearly weaker than $LD_*$ and $UD_*$ integrals of §11.6, and may be called the *semi-Denjoy-Perron integral.*

The results of this section are obtained by arguments similar to the ones used in §11.6, and so we will indicate here only the changes needed in the proofs of §11.6 to obtain the results of this section.

We begin with the result similar to Alexandroff-Looman theorem for the weak Perron integral.

11.8.1. THEOREM. *Suppose $f: I \to \overline{\mathbb{R}}$ is WP-integrable. Then f is $SD_*$-integrable and*

$$(SD_*)\int_a^b f\,dx = (WP)\int_a^b f\,dx.$$

PROOF. This result is obtained by an argument similar to the one used for Theorem 11.6.2, on replacing $L$-major by $W$-major, $L$-minor by $W$-minor, $LP$ by $WP$, $L\text{-}AC_*$ by $S\text{-}AC_*$, $L\text{-}BV_*$ by $S\text{-}BV_*$, $D_+$ by $\underline{D}_w$, $LF'$ by $SF'$, and $LD_*$ by $SD_*$; and using Corollary 11.7.10 in the place of Corollary 11.5.3, and using Theorems 11.7.5, 11.7.3, and 11.7.6 in the place of Theorems 11.4.3, 5.2.1, and 11.4.6, respectively. □

The next two lemmas are obtained by similar modifications in the proofs of Lemmas 11.6.4 and 11.6.5.

11.8.2. LEMMA. *Suppose f is WP-integrable on each interval in the interior of $I = [a, b]$. Then if the WP-integral of f on $[a + \varepsilon, b - \eta]$ has a finite limit as $\varepsilon \to 0+$ and $\eta \to 0+$, then f is WP-integrable on I.*

11.8.3. LEMMA. *Suppose F is the indefinite WP-integral of a function $f: I \to \overline{\mathbb{R}}$. Then for each $\varepsilon > 0$ there are W-major and W-minor functions U and V of f such that for each closed subinterval J of I we have*

$$\underline{O}(U;J) < \underline{O}(F;J) + \varepsilon, \qquad \underline{O}(V;J) > \underline{O}(F;J) - \varepsilon.$$

11.8.4. LEMMA. *Let K be a closed subset of $I = [a, b]$ such that $I = \operatorname{co}(K)$, and let $\{I_n = [a_n, b_n]\}$ be the sequence of contiguous intervals of K in I. Suppose $f: I \to \overline{\mathbb{R}}$ is WP-integrable on each $I_n$ and summable on K. Let $O_n$ be for each n the lower (or upper) oscillation of the indefinite WP-integral of f on $I_n$. Then if the series $\Sigma_n O_n$ is convergent, then f is WP-integrable on I and*

$$(WP)\int_a^b f\,dx = \int_K f\,dx + \sum_n (WP)\int_{a_n}^{b_n} f\,dx.$$

PROOF. Let $F_n$ denote for each $n$ the indefinite $WP$-integral of $f$ on $I_n$. It is enough to prove the result here in the case when $O_n = \underline{O}(F;I_n)$ for each $n$. When $O_n = \overline{O}(F;I_n)$ for each $n$, the result follows with the help of Theorem 11.2.6 on applying this result to $-f$.

The result in this case is obtained on the other hand by an argument similar to the one used for Lemma 11.6.6, on replacing *LP* by *WP*, *L*-major by *W*-major and *L*-minor by *W*-minor; and using Lemma 11.8.3 in the place of Lemma 11.6.5. Some modifications are needed however in the proof of Lemma 11.6.6 to prove in this case that the function $U_2$ defined there is now a *W*-major function of $f_2$.

Suppose $p$, $K_1$, $f_2$, $G_n$, $P_n$, $Q_n$, $G$, and $U_2$ are defined as in the proof of Lemma 11.6.6 with the above-mentioned modifications. Then as seen in that proof, $U_2$ is continuous, $U_2(a) = 0$, $f_2(x) = 0$ for $x \in K_1$, and $f_2(x) = f(x)$ for $x \in I_n^0$ for each $n > p$. Now, when $x \in K_1$, it was seen in that proof that $D_+U_2(x) \geq 0 = f_2(x)$, and so we clearly have $\underline{D}_w U_2(x) \geq D_+U_2(x) \geq 0 = f_2(x)$. Hence, suppose $x \in I_n^0$ for some $n > p$. Then since $P_n - Q_n$ is nondecreasing on $I_n$, it follows clearly from the definitions of $U_2$ and $G$ that $D_\pm U_2(x) = D_\pm G(x) \geq D_\pm G_n(x)$, and hence $\underline{D}_w U_2(x) \geq \underline{D}_w G_n(x)$. Now since $G_n$ is a *W*-major function of $f$ on $I_n$, it is thus clear that $U_2$ is a *W*-major function of $f = f_2$ on $I_n$. It is thus proved that $U_2$ is a *W*-major function of $f_2$ on $I$. □

Now we obtain the following theorem similar to the Hake's theorem.

11.8.5. THEOREM. *If a function $f: I \to \overline{\mathbb{R}}$ is $SD_*$-integrable, then it is WP-integrable.*

PROOF. This result is obtained by an argument similar to the one used for Theorem 11.6.7, on replacing $LD_*$ by $SD_*$, *LP* by *WP*, $L\text{-}AC_*$ by $S\text{-}AC_*$, and $LF'$ by $SF'$; and using Lemmas 11.8.2 and 11.8.4 in the place of Lemmas 11.6.4 and 11.6.6, respectively, and using Theorem 11.7.5 in the place of Theorem 11.4.3. For, in this case $F$ is found to be $S\text{-}BV_*$ on $K$, and so it is either $L\text{-}BV_*$ or $U\text{-}BV_*$ on $K$. Hence one of the series $\Sigma_n \underline{O}(F;I_n)$ and $\Sigma_n \overline{O}(F;I_n)$ is convergent, and so the result follows from Lemma 11.8.4 as in the proof of Theorem 11.6.7. □

It is somewhat surprising to note here that the Reduction Theorem 11.6.10 does not seem to hold for *WP*-integral. For, on making the necessary modifications in its proof, what follows is that $U$ and $V$ are $S\text{-}BV_*$ on $K_0$, and the argument breaks down when $U$ is $L\text{-}BV_*$ and $V$ is $U\text{-}BV_*$ on $K_0$.

Next, we present two examples to prove that (a) the $SD_*$-integral is more general than the $LD_*$ and $UD_*$ integrals, or, equivalently, the *WP*-integral is more general than the *LP*- and *UP*-integrals, and (b) the $SD_*$-integral is less general than the general *D*-integral of Denjoy and Khintchine (see Saks [S8, p. 241]).

11.8.6. EXAMPLE. First, to prove (a), it follows by an argument similar to the one used at the beginning of Example 11.6.11 that it is enough to show that there exists a continuous function $G: I \to \mathbb{R}$ which is $GS\text{-}AC_*$ but not $GL\text{-}AC_*$ or $GU\text{-}AC_*$.

Given $I = [a, b]$, let $c = (a + b)/2$, and set $J_1 = [a, c]$ and $J_2 = [c, b]$. Now, in terms of the function $F$ defined in Example 11.6.11, we define $G(x) = F(2x - a)$ for $a \leq x \leq c$, and $G(x) = -F(2x - b)$ for $c < x \leq b$. Then since $F$ is continuous and $F(a) = F(b) = 0$, it is easy to see that $G$ is continuous on $I$. Further, since $F$ is $GL\text{-}AC_*$, it is clear that $G$ is $GL\text{-}AC_*$ on $J_1$ and $GU\text{-}AC_*$ on $J_2$, and so it is $GS\text{-}AC_*$ on $I$. Also, since $F$ is not $GU\text{-}AC_*$, it is easy to see that $G$ is neither $GL\text{-}AC_*$ nor $GU\text{-}AC_*$ on $I$.

Next, to prove (b), it is similarly enough to show that there exists a continuous function $H: I \to \mathbb{R}$ which is $GAC$ but not $GS\text{-}AC_*$. We define such a function $H$ by modifying the function $F$ of Example 11.6.11 suitably.

Suppose $P$, $\{I_n = [a_n, b_n]\}$, $\{c_n\}$, and $\{r_n\}$ are defined as in Example 11.6.11. Let $d_n = (c_n + b_n)/2$ for each $n$. Now define $H(x) = 0$ for $x \in P$, and for each $n$ let $H(c_n) = |I_n| + r_n$ and $H(d_n) = -|I_n| - r_n$. Further, let $H$ be defined linearly on the intervals $[a_n, c_n]$, $[c_n, d_n]$, and $[d_n, b_n]$ for each $n$. Then $H$ is seen as before to be continuous on $I$. Further, $H$ is clearly $AC$ on $P$ and on each $I_n$, and so it is $GAC$ on $I$.

Now, given any point $x \in P^b$, define the sequences $\{i_n\}$ and $\{k_n\}$ as before. Then it follows by the arguments used earlier that $D^+H(x) \geq 1$ and $D_-H(x) \leq -1$. Further, for every large $n$ we obtain, as before,

$$0 < d_{k_n} - x < |I_{k_n}| + r_n \leq |I_{k_n}| + r_{k_n} = -H(d_{k_n}) + H(x),$$

$$0 < x - d_i < |I_{i_n}| + r_n \leq |I_{i_n}| + r_i = -H(d_{i_n}) + H(x).$$

Thus it is clear that for every large $n$ we obtain $QH(x, d_{k_n}) < -1$ and $QH(d_{i_n}, x) > 1$, and hence $D_+H(x) \leq -1$ and $D^-H(x) \geq 1$. This proves that $H$ is not weakly derivable at any point of $P^b$. Now since $|P^b| > 0$, it follows from Theorem 11.7.4 that $H$ is not $GS\text{-}BV_*$, and so it is not $GS\text{-}AC_*$ by Theorem 11.7.5.

11.8.7. Remark. In connection with the new Perron integrals considered in this chapter, Ridder [Ri2] also considered the right and left Perron integrals that are defined in terms of right and left derivates separately. He proved that each of these two integrals is equivalent to the Perron integral, from which it is clear that the McShane's Perron integral [Mc2] is equivalent to the Perron integral. This was proved earlier by McShane in [Mc1], but his proof was much more involved due to the use of 16 major and minor functions in his definition.

# *PART IV*

# SOME DIRECT APPLICATIONS OF THE THEORY OF NEW DERIVATIVES

**Introduction to Part IV.** In this part we present some direct applications of the results of the theory of new derivatives obtained in Parts II and III. A few of the results of this part are also deduced from the results of Part I. The applications of the results of the theory of new derivatives presented in Parts V and VI are, on the other hand, more of a theoretical nature since they deal with unified theories of various aspects of generalized derivatives.

First, in Chapter 12, we deal with symmetric, smooth, and quasismooth functions (i.e., functions which satisfy the Zygmund's condition $\Lambda$). We introduce here decompositions of these three properties into their lower and upper versions. Then we deduce from a result of Part I two theorems on the relations between unilateral and strong derivates for lower symmetric and lower smooth functions. Also, we deduce from the derivability theorems of Part II several results dealing with the existence of points where a lower quasismooth or lower smooth function is derivable or differentiable in the ordinary sense; the results obtained here on lower quasismooth functions are altogether new.

Further, we deduce from the results of Part III monotonicity and other properties of lower quasismooth and lower smooth functions in terms of their derivative. The theorem of Saks for functions with property $(T_2)$ is found here to hold also for lower quasismooth functions. Also, we deduce from the results of Part III the mean value and Darboux properties of lower and ordinary derivatives for lower smooth and smooth functions, respectively. The results obtained in this chapter strengthen or generalize several known results in their direction.

Next, in Chapter 13, we investigate the differential structure (i.e., the nature of derivates, medians and various derivatives) of any function $f$ at the points of its nonmonotonicity set $N(f)$ (i.e., points in no neighborhood of which $f$ is nondecreasing).

First we deduce from the first three monotonicity theorems of Parts I and III some results on the unilateral derivates and lower derivative of $f$ at the points of $N(f)$. In the cases when $f$ has property $(T_2)$ or $(N)$, or $f$ is lower quasismooth or lower smooth, we deduce on the other hand from the Sixth and Ninth Monotonicity Theorems the properties of ordinary derivative of $f$ at the points of $N(f)$. Also, when $f$ is lower singular (i.e., when the derivative of $f$ is $\geq 0$ at almost all the points where it exists), we deduce from the results obtained and the Second Median Theorem of Part III the differential properties of $f$ at the points of $N(f)$.

Further, we deduce from the results obtained the differential properties of nowhere monotone and strongly nowhere monotone functions—that is, functions $f$ for which $f_{+c}$ is nowhere monotone for every $c \in \mathbb{R}$. Also, we deduce from these results and the Third and Fourth Derivability Theorems of Part II some results on the nature of new derivatives of continuous nonderivable functions. These results seem to clarify some of the mystery of nonderivable functions and reveal some order in almost all paths of one-dimensional Brownian motion. The results obtained in this chapter generalize several known results on these classes of functions.

Finally, in Chapter 14, we investigate the notions of strong derivates and strong derivative as suggested by Peano. The properties of strong derivates, strong derivative, strong derivability sets, and so on, are found to be considerably stronger than the ones of the corresponding entities related to ordinary and other notions of derivates. Also, the values of strong derivates at any single point are found to determine the behavior of the function in a neighborhood of that point in terms of monotonicity and Lipschitz properties.

Further, with the help of the monotonicity and median theorems of Part III, we obtain relations between unilateral and strong derivates which in turn lead to strengthened versions of the two classical theorems of Dini on unilateral derivates. Also, characterizations of strong derivability are obtained in terms of local properties, unilateral derivates, and ordinary derivative. Further, the mean value property and the Darboux and Denjoy properties of strong derivative are obtained for strongly derivable functions without any continuity hypothesis.

# CHAPTER 12

# Derivates and Derivability of Symmetric, Quasismooth and Smooth Functions

In this chapter we deal with applications of the results of the first three parts to symmetric, quasismooth, and smooth functions and their lower and upper versions.

First, in §12.1, we introduce decompositions of the properties of symmetric, quasismooth, and smooth functions into their lower and upper versions. Then, in §12.2, we obtain some results on the continuity of lower symmetric and quasismooth functions. Further, in §12.3, we obtain two theorems on the relations between unilateral and strong derivates of lower symmetric and lower smooth functions.

Next, in §12.4, we deduce from the Third Derivability Theorem of Part II the existence of points of ordinary derivability for lower quasismooth functions. Then, in §12.5, we obtain monotonicity and other properties of lower quasismooth functions in terms of their derivates and ordinary derivative. Also, the Saks' theorem on functions with property $(T_2)$ is obtained here in the case of lower quasismooth functions (see Theorem 12.5.5).

Next, in §12.6, we deduce from the First Derivability Theorem the existence of points of ordinary differentiability for lower smooth functions. Also, two monotonicity theorems are obtained for lower smooth functions in terms of ordinary derivative. Finally, in §12.7, we obtain the mean value and Darboux properties of lower and ordinary derivatives in the case of lower smooth and smooth functions, respectively.

The results obtained in this chapter on lower smooth functions are known in the case of smooth functions, but the results obtained on quasismooth functions are altogether new. Also, the results of this chapter strengthen or generalize several known results of Mazurkiewicz, Marcinkiewicz and Zygmund, Stein and Zygmund, and Neugebauer on symmetric, quasismooth and smooth functions.

We will assume throughout this chapter that $f: I \to \mathbb{R}$, where $I$ is an "open" subinterval of $\mathbb{R}$.

**§12.1 Lower and upper symmetric, quasismooth and smooth functions.** In this section we introduce decompositions of the properties of symmetric, quasismooth and smooth functions into lower and upper properties which will be used throughout this chapter.

Let us recall here that we assume throughout this chapter that $f: I \to \mathbb{R}$, where $I$ is open. Given $x \in I$, let $\Delta^2 f(x, h)$ denote the *second symmetric difference* of $f$ at $x$, namely,

$$\Delta^2 f(x,h) = f(x+h) + f(x-h) - 2f(x), \quad \text{provided that} \quad x \pm h \in I.$$

Then $f$ is said to be *symmetric* or *smooth* at $x$ if $\Delta^2 f(x, h)$ is $o(1)$ or $o(h)$, respectively, as $h \to 0$. Further, when $\Delta^2 f(x, h) = O(h)$ as $h \to 0$, $f$ is said to be *quasismooth* (or $QS$) at $x$ (see, e.g., Thomson [T2, p. 112]).

The function $f$ will be said as usual to be symmetric, quasismooth or smooth if it has that property at each point of its domain.

All these three properties are known to play an important role in the theory of trigonometric series (see, e.g., [Zy3]). The notion of smoothness goes back to Riemann [Rie]; and the quasismoothness was introduced by Zygmund, who called it the condition $\Lambda$ [Zy1].

It is clear here that quasismoothness is weaker than smoothness, but stronger than the property of being symmetric.

We will further use here the following canonical decompositions of the above three properties into lower and upper properties. These lower and upper properties play a natural role in the present context and provide generalizations of some of the related known results. The notions of lower and upper symmetric and smooth functions have also been considered by Manna [Man3] using different terminology.

Given $f: I \to \mathbb{R}$ and $x \in I$, we will call $f$ (i) *lower* or *upper symmetric* at $x$ if

$$\liminf_{h \to 0+} \Delta^2 f(x,h) \geq 0 \quad \text{or} \quad \limsup_{h \to 0+} \Delta^2 f(x,h) \leq 0,$$

respectively, (ii) *lower* or *upper smooth* at $x$ if

$$\liminf_{h \to 0+} \frac{\Delta^2 f(x,h)}{h} \geq 0 \quad \text{or} \quad \limsup_{h \to 0+} \frac{\Delta^2 f(x,h)}{h} \leq 0,$$

respectively, and (iii) *lower* or *upper quasismooth* (or $LQS$ or $UQS$) at $x$ if

$$\liminf_{h \to 0+} \frac{\Delta^2 f(x,h)}{h} > -\infty \quad \text{or} \quad \limsup_{h \to 0+} \frac{\Delta^2 f(x,h)}{h} < \infty,$$

respectively.

Further, $f$ will be said as before to be lower or upper symmetric, quasismooth, or smooth if it has that property at each point of its domain.

It should be noted here that $f$ is symmetric at a point $x \in I$ iff it is lower and upper symmetric at $x$; and the same holds for the other two properties. Also, the lower quasismoothness is weaker (as before) than lower smoothness, but stronger than the property of being lower symmetric; the same holds for the upper properties.

Further, since $f$ is clearly upper symmetric, $UQS$, or upper smooth at a point $x \in I$ iff $-f$ is lower symmetric, $LQS$, or lower smooth, respectively, at $x$, it is enough to consider these lower properties.

**§12.2 Continuity of symmetric and quasismooth functions.** We first obtain in this section some results on the continuity of symmetric and $QS$ functions which will be needed in the other sections of this chapter. The results of this section provide strengthening and extensions of some of the known results due to Mazurkiewicz and others.

Given $f: I \to \mathbb{R}$, it is clear that $f$ is lower or upper symmetric at every point of $I$ where it is lower or upper semicontinuous, respectively. The converse, however, does not hold in general. For, it is well known that there exist nonmeasurable functions $f$ such that $f(x+y) = f(x) + f(y)$ for each $x, y \in \mathbb{R}$, and it is easy to see that such a function $f$ is even smooth.

We begin with a lemma. We will use here $m$ and $M$ to denote the lower and upper Baire functions [Ba] of $f$, namely,

$$m(x) = \sup_{\delta>0;|t-x|<\delta} \inf f(t), \qquad M(x) = \inf_{\delta>0;|t-x|<\delta} \sup f(t), \qquad x \in I.$$

12.2.1. LEMMA. *If $f$ is lower symmetric at $x \in I$, then*

$$(1) \qquad f(x) \le \tfrac{1}{2}\min\Big\{\underline{f}(x-0) + \bar{f}(x+0),$$
$$\bar{f}(x-0) + \underline{f}(x+0),\ m(x) + M(x)\Big\},$$

*where the inequality holds for the sums on the right that are defined.*

PROOF. Suppose $f$ is lower symmetric at $x$. Then it is clear that

$$(2) \qquad f(x) \le \tfrac{1}{2}\liminf_{h\to 0+}\{f(x+h) + f(x-h)\}.$$

The inequality (1) for the first two sums on its right follows clearly from this inequality. The inequality for the last sum holds clearly when $f(x) = m(x)$. Otherwise, $m(x) < f(x)$, and so there exists a sequence of numbers $\{h_n\}$ which converges to zero such that $f(x+h_n) \to m(x)$ as $n \to \infty$. Now, since

$\limsup_{n\to\infty} f(x - h_n) \le M(x)$, from (2) we obtain

$$f(x) \le \tfrac{1}{2} \liminf_{n\to\infty} \{f(x + h_n) + f(x - h_n)\} \le \tfrac{1}{2}\{m(x) + M(x)\}. \quad \square$$

In the following theorem we obtain some results on the reflectivity of some continuities and local boundedness of a function at a point where it is lower symmetric. When the function is upper symmetric at a point, a similar result is obtained on applying this theorem to $-f$.

12.2.2. THEOREM. *Suppose f is lower symmetric at $x \in I$. Then*

(a) *f is LC or LI at x iff it is so from the left; and f is UC or UI at x iff it is so from the right;*

(b) *f is continuous at x iff it is USC at x; and*

(c) *if f is bounded from above in some neighborhood of x, then it is bounded in some neighborhood of x.*

PROOF. To prove (a), first suppose $f$ is $LC_-$ at $x$. Then $f(x) \ge \bar{f}(x - 0)$, and so it follows clearly from the above lemma that $f(x) \le \underline{f}(x + 0)$. Consequently, $f$ is $LC$ at $x$. When $f$ is $LI_-$ at $x$, we have $f(x) \ge \underline{f}(x - 0)$, and so in this case it follows from the above lemma that $f(x) \le \bar{f}(x + 0)$, which proves $f$ to be $LI$ at $x$. Similar arguments hold for the other two parts of (a).

Next, to prove (b), suppose $f$ is $USC$ at $x$. Then $f$ is $LC_-$ and $UC_+$ at $x$. Hence it follows from (a) that $f$ is $LC$ and $UC$ at $x$, and so it is continuous at $x$.

Finally, to prove (c), suppose $f$ is bounded from above in some neighborhood of $x$, but is not bounded from below in any neighborhood of $x$. Then $M(x) < \infty$ and $m(x) = -\infty$. Hence from the above lemma we obtain

$$f(x) \le \tfrac{1}{2}\{m(x) + M(x)\} = -\infty,$$

which is not possible. $\square$

In the following theorem we obtain an extended version of a result of Mazurkiewicz [Maz1], who proved that every bounded symmetric function is of $B_1$.

12.2.3. THEOREM (MAZURKIEWICZ). *Suppose f is bounded from above or below. Then f is of $B_1$ on the set of points where it is symmetric.*

*Consequently, if f is symmetric* a.e., *then it is measurable.*

PROOF. Let $E$ denote the set of points where $f$ is symmetric. Then, given $x \in E$, it follows from Theorem 12.2.2(c) under the present hypothesis that $m(x)$ and $M(x)$ are finite. Hence it follows from Lemma 12.2.1 on applying it to $f$ and $-f$ that $f(x) = \frac{1}{2}\{m(x) + M(x)\}$. Now since $m$ and $M$ are $LSC$ and

*USC*, respectively, it is clear that $f$ is of $B_1$ on $E$. This proves the first part, and the second part follows from it on taking into account the completeness of the Lebesgue measure. □

Next, we will obtain an extended version of a theorem of Stein and Zygmund [SZ, p. 266] by a suitable modification of their proof. The term "point of density" is usually applied only to measurable sets (see, e.g., Saks [S8, p. 129]); we will use it here in the following more general sense.

Given an arbitrary set $E \subset \mathbb{R}$, we will call a point $x \in \mathbb{R}$ a *point of density* of $E$ if for each positive number $\alpha < 1$ there is a $\delta > 0$ such that $m_*(E \cap (x-h, x+h)) > 2\alpha h$ whenever $0 < h < \delta$. It is easy to see that this is equivalent to saying that $E$ contains a measurable subset of which $x$ is a point of density in the usual sense.

First, we obtain a strengthened version of a lemma given by Zygmund [Zy3, p. 78]. The full strength of this lemma is used later in §12.4 (see the proof of Theorem 12.4.5).

12.2.4. LEMMA. *Suppose $P$ and $Q$ are two subsets of $\mathbb{R}$ such that zero is a point of density of $P$ and a point of outer density of $Q$. Then, given $\varepsilon \in \mathbb{R}_+$, $\varepsilon < 2$, there is a $\delta > 0$ such that for every point $x$ for which $0 < |x| < \delta$, there exists a point $y \in P$ such that $|y| < |x|$, $|x - y| < \varepsilon|x|$, and $\frac{1}{2}(x + y) \in P \cap Q$.*

PROOF. Let $(0, x)$ also denote the interval $(x, 0)$ when $x < 0$. Then on choosing $\alpha = 1 - \varepsilon/2$ it follows easily from the hypothesis that there is a $\delta > 0$ such that for each $x$ with $0 < |x| < \delta$, we obtain

$$m_*(P \cap (0, x)) > (1 - \varepsilon)|x| \quad \text{and} \quad |Q \cap (0, x)| > (1 - \varepsilon)|x|.$$

Let $x$ be any such point, and for the sake of definiteness suppose $x > 0$. Let $P_0$ be a measurable subset of $P \cap (0, x)$ such that $|P_0| = m_*(P \cap (0, x))$, and set $Q_0 = Q \cap (0, x)$. Then $|P_0| > (1 - \varepsilon)x$ and

$$|P_0 \cap Q_0| = |P_0| + |Q_0| - |P_0 \cup Q_0| > 2(1 - \varepsilon)x - x = (1 - 2\varepsilon)x.$$

Now, set

$$A = \{y \in (0, x) : \tfrac{1}{2}(x + y) \in P_0 \cap Q_0\}, \qquad B = P_0 \cap Q_0 \cap (\tfrac{1}{2}x, x).$$

Then $A = 2B - x$, so that

$$|A| = 2|B| \geq 2\{|P_0 \cap Q_0| - \tfrac{1}{2}x\} > 2\{(1 - 2\varepsilon)x - \tfrac{1}{2}x\} = (1 - 4\varepsilon)x.$$

Hence,

$$|P_0 \cap A| = |P_0| + |A| - |P_0 \cup A| > (1 - \varepsilon)x + (1 - 4\varepsilon)x - x = (1 - 5\varepsilon)x.$$

This proves that there exists a point $y \in P \cap A$ such that $(1 - 5\varepsilon)x < y < x$, and then $\frac{1}{2}(x + y) \in P \cap Q$. Since this holds for arbitrary $\varepsilon$, the lemma is thus established. □

In the second part of the following theorem we obtain an extended version of a theorem of Stein and Zygmund [SZ], who proved it for a measurable function $f$ when the set of points where $f$ is symmetric is measurable.

12.2.5. THEOREM. *Suppose $f$ is measurable on a measurable set $E \subset I$. Then $f$ is LSC at almost all the points of $E$ where it is lower symmetric. Consequently, $f$ is continuous at almost all the points of $E$ where it is symmetric.*

PROOF. It is enough to prove here the first part, for the second part follows clearly on combining the first part with the result obtained on applying it to $-f$.

Let $A$ and $E_0$ be the sets of points in $E$ where $f$ is lower symmetric or approximately continuous, respectively. Then, according to a theorem of Denjoy [S8, p. 132], $|E \sim E_0| = 0$.

Further, for each pair of positive integers $n, k$, let $A_{n,k}$ denote the set of points $x$ in $A$ for which $\Delta^2 f(x, h) > -1/n$ whenever $0 < h < 1/k$, and let $B_{n,k}$ be the set of points in $A_{n,k}$ that are not points of outer density of $A_{n,k}$. Then, by the Lebesgue's density theorem, the set $B = \bigcup_n \bigcup_k B_{n,k}$ is of measure zero. Hence, it is enough to show that $f$ is $LSC$ at each point $x_0 \in A \cap E_0 \sim B$.

Given such a point $x_0$, we can clearly assume without loss of generality that $x_0 = 0$ and $f(x_0) = 0$. Now since $f$ is approximately continuous at 0, it is indeed continuous at 0 relative to a set $P \subset I$ of which 0 is a point of density. Hence, given $\varepsilon > 0$, there is a $\delta_1 > 0$ such that $|f(x)| < \varepsilon$ for $x \in P \cap (-\delta_1, \delta_1)$. Let $n_0$ be any integer greater than $1/\varepsilon$. Then since $A \subset \bigcup_k A_{n_0,k}$, there is another integer $k_0$ such that $0 \in A_{n_0,k_0}$, and then 0 is a point of outer density of $A_{n_0,k_0}$. Now let $\delta_2$ be the positive number determined by the above lemma for $P, Q \equiv A_{n_0,k_0}$, and $\varepsilon$. Set $\delta = \min\{\delta_1, \delta_2, 1/k_0\}$.

Now, given any point $x$ for which $0 < |x| < \delta$, there is a point $y \in P$ such that $|y| < |x|$ and $z \equiv \frac{1}{2}(x + y) \in P \cap A_{n_0,k_0}$. Set $h = \frac{1}{2}|x - y|$. Then $0 < h < \delta \le 1/k_0$, and hence it is clear that

$$
\begin{aligned}
f(x) &= \{f(x) + f(y) - 2f(z)\} - f(y) + 2f(z) \\
&\ge \Delta^2 f(z, h) - |f(y)| - 2|f(z)| > -\frac{1}{n_0} - 3\varepsilon > -4\varepsilon.
\end{aligned}
$$

This proves that $f$ is $LSC$ at $x_0 = 0$. □

From the above theorem we obtain, with the help of Theorem 12.2.3, the following extended version of another theorem of Mazurkiewicz [Maz1] in which he proved every bounded symmetric function to be continuous a.e.

12.2.6. COROLLARY. *Suppose $f$ is bounded from above or below. If $E \subset I$ is measurable and $f$ is symmetric* a.e. *in $E$, then $f$ is continuous* a.e. *in $E$.*

Finally, in the following theorem we obtain a stronger result on the continuity of measurable $QS$ functions which will be needed in §12.4. This result was obtained originally by Mazurkiewicz [Maz1] for a bounded smooth function, and he obtained it later for measurable smooth functions [Maz2] (see also Neugebauer [Ne2]). Mazurkiewicz's first result is included in the following theorem due to Theorem 12.2.3, and the following proof is based on his proof.

12.2.7. THEOREM. *If $f$ is a measurable $QS$ function, then the set of points where $f$ is discontinuous is nowhere dense.*

PROOF. Since $f$ is clearly symmetric, it follows from Theorem 12.2.5 that $f$ is continuous a.e. Hence, given any open subinterval $J$ of $I$, there exists an open subinterval $J_0$ of $J$ on which $f$ is bounded. Now it follows from the proof of the above-mentioned theorem of Mazurkiewicz [Maz1, p. 236], which extends without difficulty to bounded $QS$ functions, that $f$ is continuous on an open subinterval of $J_0$. □

12.2.8. REMARK. We include here some comments on symmetrical continuity, which also plays an important role in the theory of trigonometric series, but is not included in the main text since it is not needed here. We present here a decomposition of this property as well, and discuss results on these notions related with the results of the present section.

Given $f: I \to \mathbb{R}$, $f$ is said to be *symmetrically continuous* (see e.g. [Zy2, p. 48]) at a point $x \in I$ if $f(x+h) - f(x-h) = o(1)$ as $h \to 0$. Decomposing this property, let $f$ be called *symmetrically* $LC$ or $UC$ at $x$ if $\liminf_{h \to 0+}\{f(x+h) - f(x-h)\} \geq 0$ or $\limsup_{h \to 0+}\{f(x+h) - f(x-h)\} \leq 0$, respectively. Then it is clear that $f$ is symmetrically continuous at $x$ iff it is symmetrically lower and upper continuous at $x$. Also, when $f$ is $LC$ or $UC$ at $x \in I$, it is easy to see that it is symmetrically so.

It is interesting to note here, as can be easily verified, that the indefinite integral of an integrable function $g$ is lower or upper smooth at every point where $g$ is symmetrically $LC$ or $UC$, respectively.

Analogous to Theorem 12.2.2, when $f$ is symmetrically $LC$ at $x$, it is easy to see that $\underline{f}(x-0) \leq \underline{f}(x+0)$ and $\bar{f}(x-0) \leq \bar{f}(x+0)$. Further, using arguments similar to the ones used for Theorem 12.2.5, it can be proved under its hypothesis that $f$ is continuous at almost all the points of $E$ where it is symmetrically lower or upper continuous. This in turn provides an extension of another result of Stein and Zygmund [SZ, p. 266].

## §12.3 Symmetry of derivates of symmetric and lower smooth functions.

In this section we obtain two theorems on the symmetry of unilateral derivates of symmetric and lower, or upper, smooth functions.

As was pointed out earlier in Remark 2.5.7, Neugebauer [Ne1] proved in connection with Theorem 2.5.6 that every measurable symmetric function $f$ on an interval $I$ has symmetrical derivates r.e. Generalizing this result it was proved later by Manna [Man3] that if $f$ is lower symmetric and of $B_1$, then $D_+f \le D_-f$ and $D^+f \le D^-f$ r.e.

We show here, as mentioned earlier in Remark 2.5.7, that these results are contained in Theorem 2.5.4, and deduce from this theorem strengthened versions of the two results.

It should be recalled here that a function $f: I \to \mathbb{R}$ which is continuous at a dense set of points in $I$ is called *pointwise discontinuous* (see, e.g., [K1, p. 105]).

We begin with the following lemma, which plays an essential role here.

12.3.1. LEMMA. *If $f$ is pointwise discontinuous and lower symmetric, then it is QLSC. Conversely, if $f$ is QLSC, then it is pointwise discontinuous.*

PROOF. Suppose $f$ is pointwise discontinuous and lower symmetric, but it is not *QLSC* at some point $x \in I$. Then there exist $\varepsilon, \delta \in \mathbb{R}_+$ such that $J \equiv (x - \delta, x + \delta) \subset I$ and the set $A = \{t \in J : f(t) \le f(x) - \varepsilon\}$ is dense in $J$. Let $C = J \cap C(f)$. Then it is clear that $C \subset A$; and since $C$ is a $G_\delta$-set, it follows that $A$ is residual in $J$. Now, since $B \equiv 2x - A$ is also residual in $J$, so is the set $A \cap B$. Hence we can find a sequence of positive numbers $\{h_n\}$ which converges to zero such that $x + h_n$, $x - h_n \in I$ for each $n$. Then, for each $n$, we have

$$\Delta^2 f(x, h_n) = f(x + h_n) + f(x - h_n) - 2f(x) \le -2\varepsilon.$$

But this clearly contradicts the fact that $f$ is lower symmetric at $x$. The first part is thus established.

Next, to prove the converse, suppose $f$ is *QLSC*. Let $\omega$ denote the oscillation function of $f$; and for each positive integer $n$, set

$$E_n = \{x \in I : \omega(x) \ge 1/n, f(x) \le n\}.$$

Then $I \sim C(f) = \bigcup_n E_n$. Hence to prove $C(f)$ to be dense in $I$, it is enough to show that $E_n$ is nowhere dense for each $n$.

Suppose there is an $n$ for which $E_n$ is dense in some open subinterval $J$ of $I$. Now, since $f$ is *QLSC*, it is clear that $f(x) \le n$ for each $x \in J$, and since the set $\{x \in I : \omega(x) \ge 1/n\}$ is closed in $I$, it follows that $J \subset E_n$. Let $c$ be the supremum of $f$ on $J$. Then $c$ is finite, and there exists a point $x_0 \in J$ such that $f(x_0) > c - 1/(2n)$. Hence there exists an open subinterval $U$ of $J$ such that $f(x) > c - 1/(2n)$ for each $x \in U$. Now since $f(x) \le c$ for each $x \in U$, it follows that $\omega(x) \le 1/(2n) < 1/n$ for each $x \in U$. Hence $U \cap E_n = \varnothing$, which is a contradiction. □

Now, with the help of the above lemma, we deduce from Theorem 2.5.4 the following strengthened versions of the above-mentioned results of Neugebauer [Ne2] and Manna [Man3]. For, a function of $B_1$ is well known to be pointwise discontinuous, and a measurable symmetric function is so by Theorem 12.2.5.

12.3.2. THEOREM. (a) *If $f$ is pointwise discontinuous and lower symmetric, then $D_+ f = \underline{D}_* f$ and $D^- f = \overline{D}_* f$* r.e.

(b) *If $f$ is either* (i) *pointwise discontinuous and symmetric, or* (ii) *QUSC and lower symmetric, then $D_- f = D_+ f = \underline{D}_* f$ and $D^- f = D^+ f = \overline{D}_* f$* r.e.

PROOF. Part (a) follows directly from the above lemma and Theorem 2.5.4. On combining (a) with the result obtained on applying it to $-f$, we obtain on the other hand (b) in case (i). Further, in case (ii), it follows from the second part of Lemma 12.3.1 that $f$ is pointwise discontinuous, and hence from the first part of that lemma that $f$ is *QBSC*; consequently the result is obtained in this case on applying Theorem 2.5.4 to $f$ and $-f$. □

The unilateral derivates of a smooth function are known to be symmetrical everywhere. In the next theorem we obtain the symmetry of derivates of pointwise discontinuous lower and upper smooth functions r.e.

First we prove a general result on the derivates of lower smooth and smooth functions in the following lemma which is due to Manna [Man3].

12.3.3. LEMMA. *If $f$ is lower smooth at $x \in I$, then $D_- f(x) \le D_+ f(x)$ and $D^- f(x) \le D^+ f(x)$. Consequently, if $f$ is smooth at $x$, then its derivates are symmetrical at $x$.*

PROOF. First, suppose $f$ is lower smooth at $x$. Then it is clear that

$$0 \le \liminf_{h \to 0+} \left\{ \frac{f(x+h) - f(x)}{h} - \frac{f(x-h) - f(x)}{-h} \right\}$$
$$\le \min\{D_+ f(x) - D_- f(x), D^+ f(x) - D^- f(x)\},$$

where the last inequality holds for the differences that are defined. This proves each of the two required inequalities when the difference between the two derivates involved in it is defined; otherwise it holds trivially. The first part is thus established.

The second part follows on the other hand on combining the first with the result obtained on applying it to $-f$. □

It is now easy to see that on combining the first part of the above lemma with Theorem 12.3.2(a), the following result is obtained for lower smooth functions. The result on upper smooth functions is obtained of course on applying this result to $-f$, which in turn strengthens another result of Manna [Man3, p. 298].

12.3.4. THEOREM. *Suppose $f$ is pointwise discontinuous and lower or upper smooth. Then $D_-f = D_+f = \underline{D}_*f$ and $D^-f = D^+f = \overline{D}_*f$* r.e.

**§12.4 Ordinary derivability of quasismooth functions.** In this section we obtain the existence of points of derivability of $LQS$ functions, and we also obtain an extended version of a theorem of Marcinkiewicz and Zygmund regarding differentiability of a $QS$ function at the points of approximate derivability.

First, in Theorem 12.4.2, we deduce from the Third Derivability Theorem the existence of points of ordinary derivability of $LQS$ and quasismooth functions. This altogether new result came somewhat as a surprise. Then, in Theorem 12.4.5, we obtain an extended version of the Marcinkiewicz-Zygmund theorem establishing normal lower differentiability of an $LQS$ function at almost all the points where it is approximately derivable.

First, to deal with ordinary derivability of $LQS$ and $QS$ functions, we need the following lemma.

12.4.1. LEMMA. *Suppose $f$ is $LQS$ at $x \in I$. Then $f$ is lower knotted at $x$ iff it is biknotted at $x$. Consequently, $f$ is normally upper derivable at $x$ iff it is derivable at $x$.*

PROOF. It is clearly enough to prove here the first part. Hence suppose $f$ is lower knotted at $x$. Since $f$ is $LQS$ at $x$, there exists a positive integer $k$ such that

$$\liminf_{h\to 0+} \frac{\Delta^2 f(x,h)}{h} > -k.$$

Further, it is clear that

$$\frac{\Delta^2 f(x,h)}{h} = \frac{f(x+h)-f(x)}{h} - \frac{f(x-h)-f(x)}{-h}.$$

Now since $D^-f(x) = +\infty$, we obtain

$$D^+f(x) = \limsup_{h\to 0+}\left\{\frac{f(x-h)-f(x)}{-h} + \frac{\Delta^2 f(x,h)}{h}\right\} \geq D^-f(x) - k = +\infty;$$

and since $D_+f(x) = -\infty$, we obtain similarly the following:

$$D_-f(x) = \liminf_{h\to 0+} \left\{ \frac{f(x+h) - f(x)}{h} - \frac{\Delta^2 f(x,h)}{h} \right\} \le D_+f(x) + k = -\infty.$$

Consequently, $f$ is biknotted at $x$. □

Now, with the help of the above lemma, we deduce in the following theorem from the Third Derivability Theorem the existence of points of ordinary derivability for *LQS* and *QS* functions which seems to be altogether new. Also, in the case of functions with property $(S_2)$ we deduce here a stronger result from the Fifth Derivability Theorem.

12.4.2. THEOREM (EIGHTH DERIVABILITY). (a) *Suppose $f$ is either* (i) *USC from left or right and LQS,* (ii) *$L^*LI$ or $R^*UI$ and LQS* n.e., *or* (iii) *a measurable QS function. Then $f$ is derivable at a dense set of points in almost all of its level sets, and it is derivable at a c-dense set of points in $I$.*

(b) *If $f$ is a weakly Darboux, QS function with property $(S_2)$, then it is everywhere derivable in almost all of its level sets.*

PROOF. To prove (a), first suppose (i) holds. Then $f$ is either $LC_-$ or $UC_+$, and hence it follows from Lemma 12.2.2 that $f$ is either $LC$ or $UC$. Consequently, the same holds for $-f$, and hence the result is obtained in terms of normal upper derivability from the Third Derivability Theorem 6.2.2 on applying it to $-f$. Consequently, the result holds in terms of ordinary derivability due to Lemma 12.4.1.

Next, suppose (ii) holds. Then the result is obtained directly from Theorem 6.2.2 on applying it to $-f$ in terms of normal upper derivability. Now, because the set $E$ of points where $f$ is not $LQS$ is countable, so is $f(E)$, and so the result follows from Lemma 12.4.1 on restricting to level sets $f^{-1}(c)$ for which $c \notin f(E)$.

Finally, in case (iii), $f$ is by Theorem 12.2.7 continuous on a dense set of intervals in $I$, and so the result is obtained from the result in case (i).

Part (b) follows on the other hand directly from the Fifth Derivability Theorem 6.5.2 with the help of Lemma 12.4.1. □

Now, with the help of Corollaries 6.1.3 and 6.1.4 and Lemma 6.2.1, the following result is obtained from the above theorem as in §6.2.

12.4.3. COROLLARY. *Suppose $f$ satisfies one of the three hypotheses of the above theorem.*

(a) *If $f$ is regular, then it is differentiable at a dense set of points in almost all of its level sets, and it is differentiable at a metrically dense set of points in $I$.*

(b) *If $f$ is singular, then it has an infinite derivative at a dense set of points in almost all of its level sets, and it has an infinite derivative at a c-dense set of points in $I$ provided that $f$ is a nowhere constant Darboux function.*

Next, in the following theorem we obtain an extension of one part Lusin-Denjoy-Saks Theorem 5.3.2 on $GBV_*$ functions to $GU\text{-}BV_*$ functions that are $LQS$ n.e.

12.4.4. THEOREM. *Suppose $f$ is $GU\text{-}BV_*$. Then $f$ is differentiable at almost all the points where it is $LQS$. Further, if $f$ is $LQS$ n.e. and $N$ is the set of points where $f$ is not derivable, then $|N| = |f(N)| = \Lambda\{Gf(N)\} = 0$.*

PROOF. Since $f$ is $GU\text{-}BV_*$, it follows from the Second Derivability Theorem 5.3.1 with the help of the Denjoy-Young-Saks Theorem 5.5.2 that $f$ is normally upper differentiable a.e. Hence the first part follows from Lemma 12.4.1.

Next, to obtain the second part, there is clearly no loss of generality in assuming $f$ to be $LQS$ everywhere. Then it follows from Lemma 12.4.1 that $\Delta(f) = \overline{\Delta}_n(f)$, and so $N = I \sim \overline{\Delta}_n(f)$. Hence the result follows clearly from Theorem 6.7.2 on applying it to $-f$. □

Next, in the following theorem we obtain normal lower differentiability of a $LQS$ function at almost all the points where it is approximately derivable. The second part of this theorem dealing with differentiability of a $QS$ function was obtained by Marcinkiewicz and Zygmund [MZ] for a measurable function for which the set of points where it is approximately derivable is also measurable. We obtain this result here without these measurability hypotheses. The following proof is an adaptation of the proof given by Marcinkiewicz [Mar, p. 67].

12.4.5. THEOREM. *Every function $f : I \to \mathbb{R}$ is normally lower differentiable at almost all the points where it is approximately derivable and $LQS$.*

*Consequently, $f$ is differentiable at almost all the points where it is approximately derivable and $QS$.*

PROOF. It is clearly enough to prove here the first part.

Hence, let $E$ denote the set of points in $I$ where $f$ is approximately derivable and $LQS$. We may assume here that $E$ does not contain any point where $f$ has an infinite approximate derivative since such points form a set of measure zero (see [S8, p. 297]), and similarly no point where $f$ is lower derivable but not normally lower differentiable (see Theorem 5.5.2).

Now, let $E_k$ denote for each positive integer $k$ the set of points $x$ in $E$ for which $\Delta^2 f(x, h) > -kh$ for $0 < h < 1/k$. Then $E = \bigcup_k E_k$. Hence, given any $k$, it is enough to show that $f$ is lower derivable at each point $x_0$ of $E_k$ that is a point of outer density of the set.

We may clearly assume here for the sake of simplicity that $x_0$, $f(x_0)$, and the approximate derivative of $f$ at $x_0$ are all zero. Then there is a set $P \subset I$ such that 0 is a point of density of $P$ and $f$ has a zero derivative at 0 relative to $P$. Hence, given $\varepsilon > 0$, there is a $\delta_1 > 0$ such that $|f(x)| \le \varepsilon|x|$ for $x \in P \cap (-\delta_1, \delta_1)$. Let $\delta_2$ be the positive number determined by Lemma 12.2.4 for $P, Q = E_k$ and $\varepsilon$, and set $\delta = \min\{\delta_1, \delta_2\}$.

Now, let $x$ be any point of $I$ such that $0 < |x| < \delta$. Then there exists a point $y \in P$ such that $|y| < |x|, |x - y| < \varepsilon|x|$ and $z \equiv \frac{1}{2}(x + y) \in P \cap E_k$. Let $h = \frac{1}{2}|x - y|$. Then we obtain

$$\begin{aligned} f(x) &= \{f(x) + f(y) - 2f(z)\} - f(y) + 2f(z) \\ &> -kh - |f(y)| - 2|f(z)| \\ &> -k\varepsilon|x| - 3\varepsilon|x| = -(k+3)\varepsilon|x|. \end{aligned}$$

This proves that $f$ has a zero lower derivative at 0, which establishes the result. □

Finally, on taking into account the theorem of Denjoy and Khintchine on approximate derivability of measurable *GBV* functions a.e. (see [S8, p. 222]), we obtain the following from the above theorem.

12.4.6. Corollary. *If $f$ is measurable and GBV, then it is normally lower differentiable at almost all the points where it is LQS.*

12.4.7. Remark. In connection with Theorem 12.4.2, the question arises whether every continuous *QS* function is also differentiable at some points. As we see below, the answer to this question is negative, and we discuss here a few related questions on *QS* functions.

The Swiss mathematician Cellérier [Ce, pp. 152–155] is claimed to have proved before Weierstrass, although his works were published only posthumously in 1890, that for any sufficiently large even integer $a$ the continuous function $f(x) = \Sigma_n a^{-n} \sin a^n x$ is nondifferentiable. This function has been shown by Zygmund [Zy2, p. 47] to be *QS* for any number $a > 1$.

Since $f$ is clearly singular, it follows from Corollary 12.4.3(b) that it has an infinite derivative at a $c$-dense set of points. This function was somehow believed for some time to be nonderivable (see, e.g., G. C. Young [Y.G2, pp. 171–173] and [H2, p. 402]). However, it was proved later by Hardy [Har] that for any number $a > 1$ this function $f$ is nondifferentiable but has a derivative $+\infty$ at $x = 0$. It was discovered later from Hardy's proof that $f$ has derivatives $+\infty$ and $-\infty$ at two (countable) dense sets of points (see Hobson [H2, p. 407]). In fact, since $f$ is singular, it follows easily from Lemma 12.4.1 and Theorem 9.1.5 that $f$ has derivatives $+\infty$ and $-\infty$ at two $c$-dense sets of points. Also, it follows clearly from Lemma 12.4.1 and Theorem 5.5.2 that $f$ is biknotted a.e. like Weierstrass' function.

Next, given an arbitrary *QS* function $f: I \to \mathbb{R}$, it is easy to see by arguments similar to the ones used for Lemma 12.4.1 that (a) each of the four derivates of $f$ is finite at every point where it is semidifferentiable, and (b) $f$ is derivable at every point where it has an infinite weak gradient. Thus such a function cannot be cusped anywhere, although it is not nonangular in general like smooth functions; for the Cellérier's function is known to be angular at a dense set of points (see Hardy [Har, p. 310]). It is not clear, however, whether a continuous *QS* function can be semidifferentiable only at a countable set of points like the Weierstrass' function (see Theorem 5.1.2 and Remark 5.1.4).

**§12.5 Properties of quasismooth functions in terms of derivates and ordinary derivative.** In this section we deduce from the results of Chapters 2, 5 and 11 various generalized properties of *LQS* functions in terms of unilateral derivates. Also, several monotonicity and related results are deduced from the monotonicity theorems of §9.1 for this class of functions in terms of ordinary derivative.

In the first two theorems we obtain generalized $BV_*$, $LAC_*$, $AC_*$ and other properties of *LQS* functions in terms of unilateral derivates. In Theorem 12.5.3 these properties are obtained on a portion of a given set. All these results are deduced from the theorems established earlier in §§2.2, 5.4, and 11.5 in terms of unilateral derivates.

Next, in Theorem 12.5.5, we obtain a result on *LQS* functions in terms of ordinary derivative which is similar to the Saks' theorem on functions with property $(T_2)$. We include here some interesting consequences of this theorem including a monotonicity theorem (see Theorem 12.5.7) for such functions in terms of ordinary derivative.

It indeed came as a surprise that the various properties of lower and upper *QS* functions can be investigated in terms of their ordinary derivative. Each of these two classes of functions thus resembles in a sense the class of functions that have property $(T_2)$ (see Problem 12.5.10). Part of the reason behind this mysterious phenomenon may be found in Theorem 12.5.1, where every *LQS* function is proved to be differentiable at almost all the points where it is not upper knotted.

The full explanation of this phenomenon seems to lie, however, in Lemma 12.5.4, where the points where a *LQS* function is upper derivable but not derivable are shown to form a set which is inconsequential in a sense. And, of course, as was seen in Part III, the properties of every continuous function can be investigated in terms of its upper (or lower) derivative in general.

We begin in the following two theorems with some generalized properties of *LQS* functions in terms of unilateral derivates (see §§2.1, 5.2, 5.4, and 11.4 for definitions of these properties).

12.5.1. THEOREM. *Suppose $f$ is LQS. Then $f$ is GBMT (hence $GBV_*$) on $I \sim \overline{K}(f)$. Consequently, $f$ is differentiable* a.e. *in $I \sim \overline{K}(f)$. Further, when $f$ is QS, the same holds on $I \sim K_b(f)$.*

*Moreover, if $f$ is continuous, then these generalized properties are strengthened to inherent properties.*

PROOF. Let $E = I \sim \overline{K}(f)$. Then since $\underline{K}(f) \subset \overline{K}(f)$ by Lemma 12.4.1, $E \cap \underline{K}(f) = \varnothing$. Hence it follows from Theorem 5.4.1 on applying it to $f$ and $-f$ that $f$ is $GL\text{-}MT_*$ and $GU\text{-}MT_*$ on $E$. This proves that $f$ is *GBMT* on $E$. Hence it follows from Lemma 5.5.1 that $f$ is $GL\text{-}BV_*$ and $GU\text{-}BV_*$ on $E$, and so it is $GBV_*$ on $E$. Consequently, it follows from Corollary 5.3.2 that $f$ is differentiable a.e. in $E$.

The second part follows clearly from the first since $K_b(f) = \underline{K}(f) \cap \overline{K}(f)$. Finally, when $f$ is continuous, according to the last part of Theorem 5.4.1 and Theorem 1.3.1 the above generalized properties strengthen to inherent properties. □

12.5.2. THEOREM. *Suppose $f$ is LQS and $E \subset I$.*

(a) *If $D_-f > -\infty$ n.e. in $E$, then $f$ is $GLAC_*$ on $E$.*

(b) *If $D_-f > -\infty$ and $D^+f < \infty$ at the points of $E$, then $f$ is $F$-$GL_*$ (hence $F$-$GAC_*$) on $E$.*

*Moreover, if $f$ is QS, then the conclusion of* (b) *holds provided that $\underline{D}_w f > -\infty$ and $\overline{D}_w f < +\infty$ at the points of $E$.*

PROOF. First, to prove (a), suppose $D_-f > -\infty$ n.e. in $E$. Then it follows from Lemma 12.4.1 that $D_+f > -\infty$ n.e. in $E$. Hence $\underline{D}f > -\infty$ n.e. in $E$, and so it follows from Theorem 11.5.2 on applying it to $f$ and $-f_-$ that $f$ is $GLAC_*$ on $E$.

Next, to prove (b), suppose $D_-f > -\infty$ and $D^+f < \infty$ at the points of $E$. Then by Lemma 12.4.1, $D_+f > -\infty$ and $D^-f < \infty$ on $E$. Thus $\underline{D}f > -\infty$ and $\overline{D}f < \infty$ on $E$. Hence it follows from Theorem 2.2.7 that $f$ is $F$-$GRL$ and $F$-$GLL$ on $E$. Consequently, $f$ is $F$-$GL_*$ on $E$, which clearly implies that $f$ is $F$-$GAC_*$ on $E$ (see §11.4).

The last part follows from Lemma 12.4.1 by a similar argument. □

The following theorem deals with the above properties on a portion of a set in the case when the requisite conditions hold at a nonmeager set of points in the set.

12.5.3. THEOREM. *Suppose $f$ is LQS and it is QBSC relative to $E \subset I$.*

(a) *If $E \sim \overline{K}(f)$ is nonmeager in $E$, then $f$ is BMT (hence $BV_*$) on some portion of $E$.*

(b) *If $D_-f > -\infty$ and $D^+f < \infty$ at a nonmeager set of points in $E$, then $f$ is $L_*$ (hence $AC_*$) on some portion of $E$.*

PROOF. First, to prove (a), suppose $A = E \sim \overline{K}(f)$ is nonmeager in $E$. Then it follows from Theorem 5.4.3 on applying it to $-f_-$ that $f$ is $U$-$MT_*$ on some portion $P$ of $E$. Thus $f$ is either $RUL$ or $LLL$ on $P$, and so it is clear that $P \cap \overline{K}(f) = \varnothing$. Hence it follows from Lemma 12.4.1 that $P \cap \underline{K}(f) = \varnothing$. Now, it follows again from Theorem 5.4.3 that $f$ is $L$-$MT_*$ on some portion $Q$ of $P$. Consequently, $f$ is $BMT$ on $Q$, which is of course a portion of $E$. This in turn implies by Lemma 5.5.1 that $f$ is also $BV_*$ on $Q$.

Next, to prove (b), suppose its hypothesis holds. Then on applying Theorem 5.4.4 to $-f_-$ it follows that $f$ is $U$-$L_*$ on some portion $P$ of $E$. Thus $f$ is $RUL$ and $LLL$ on $P$. Now, at each $x \in P$, it is clear that $D_-f(x) > -\infty$ and $D^+f(x) < \infty$, and so by Lemma 12.4.1 $D_+f(x) > -\infty$ and $D^-f(x) < \infty$.

Hence it follows again from Theorem 5.4.4 that $f$ is $L$-$L_*$ on some portion $Q$ of $P$. Consequently, $f$ is $L_*$ on the portion $Q$ of $E$, which of course implies that $f$ is $AC_*$ on $Q$. □

**A. Monotonicity of lower quasismooth functions in terms of ordinary derivative.** We next obtain some monotonicity and related results on $LQS$ functions in terms of ordinary derivative.

First, we deal in the following lemma with the relationship between the ordinary and upper derivatives of a $LQS$ function.

12.5.4. LEMMA. *Suppose $f$ is $LQS$. If $E$ denotes the set of points in $I$ where $f$ is upper derivable but not derivable, then*

$$|E| = |f(E)| = \Lambda\{Gf(E)\} = 0.$$

PROOF. According to Lemma 12.4.1, $\Delta(f) = \overline{\Delta}_n(f)$. Hence it is clear that $E = \overline{\Delta}(f) \sim \overline{\Delta}_n(f)$. Consequently, the result is obtained from Theorems 5.5.2, 6.1.2 and 6.7.1. □

We begin in the following theorem with the analogue of Saks' Theorem 9.3.3 on functions with property $(T_2)$ in the case of $LQS$ functions.

12.5.5. THEOREM. *Suppose $f$ is $LQS$. Given $a, b \in I$, $a < b$, let $E_-$ and $E_+$ denote the sets of points in $(a, b)$ where $f$ has a negative or positive derivative, respectively. Then if $f$ is $LC_-$, then $f(a) - f(b) \le |f(E_-)|$; and if $f$ is $UC_+$, then $f(b) - f(a) \le |f(E_+)|$.*

PROOF. According to Theorem 12.2.2, $f$ is $LC$ when it is $LC_-$, and $f$ is $UC$ when it is $UC_+$. Hence the two inequalities follow from the two parts of Theorem 9.1.5 within brackets with the help of the above lemma. □

On combining the above theorem with Theorem 6.7.3, the following inequalities are obtained in terms of integrals which may be compared with Corollary 9.3.4.

12.5.6. COROLLARY. *Suppose $f$ is $USC$, $LQS$ and regular. Given $a, b \in I$, $a < b$, let $E_-$ and $E_+$ denote the sets of points in $(a, b)$ where $f$ has a finite negative or positive derivative, respectively. Then*

$$\int_{E_-} f'\,dm \le f(b) - f(a) \le \int_{E_+} f'\,dm.$$

With the help of Corollary 6.1.4 of Saks we obtain on the other hand the following monotonicity theorem from the above theorem for $LQS$ functions in terms of ordinary derivative. This result should in turn be compared with the Sixth Monotonicity Theorem 9.3.5 on functions with property $(T_2)$. The necessity part of this result is of course obvious.

12.5.7. THEOREM (NINTH MONOTONICITY). *Suppose $f$ is $LC_-$ and $LQS$. Then $f$ is nondecreasing iff it is lower regular and lower singular.*

We include here the following consequence of this monotonicity result on the Lipschitz property of a $LQS$ function.

12.5.8. COROLLARY. *Suppose $f$ is USC, LQS, and regular. Then, given $c \in \mathbb{R}_+$, $f$ is c-Lipschitz iff $f'$ is essentially bounded on $\Delta(f)$ by $c$, and $f$ is constant iff it is singular.*

12.5.9. REMARK. It should be noted here that some of the results of this chapter on quasismooth functions were announced earlier in [G13]. Noticing their significance, some of the results of §12.4 and the present section have been included by B. Thomson in his recent monograph on symmetric properties of real functions (see [T2, pp. 154–156]).

12.5.10. PROBLEM. In connection with the relation between $QS$ functions and functions with property $(T_2)$, it has been proved by Anderson and Pitt [AP] that every uniformly $QS$ function has property $(T_2)$. However, it is not known if it holds for a general $QS$ function. It will thus be interesting to investigate if every lower or upper $QS$ function has property $(T_2)$.

## §12.6 Differentiability of smooth functions and their properties in terms of derivates and ordinary derivative.

In this section we first deal in Theorem 12.6.1 with the relation between lower differentiability and lower smoothness. In the rest of the section we obtain results on smooth functions which do not hold for $QS$ functions.

In Theorem 12.6.3 we obtain two generalized versions of the well known result on the differentiability of continuous smooth functions. Then, in Theorems 12.6.4 and 12.6.5, we obtain some generalized properties of lower smooth functions in terms of derivates. Finally, in the last two theorems we obtain two monotonicity theorems for lower smooth functions.

First, a function is known to be smooth at every point where it is differentiable. In the following theorem we obtain its analogue for lower and upper smoothness.

12.6.1. THEOREM. *If $f$ is lower differentiable at $x \in I$, then it is lower smooth at $x$.*

PROOF. Suppose $f$ is lower differentiable at $x$. Then $f$ has some finite lower gradient $c$ at $x$. Hence $D^{-}f(x) \le c \le D_{+}f(x)$, and so we obtain

$$\liminf_{h\to 0+} \frac{\Delta^2 f(x,h)}{h} \ge \liminf_{h\to 0+} \frac{f(x+h)-f(x)}{h} + \liminf_{h\to 0+} \frac{f(x-h)-f(x)}{h}$$
$$= D_{+}f(x) - \limsup_{h\to 0+} \frac{f(x-h)-f(x)}{-h}$$
$$= D_{+}f(x) - D^{-}f(x) \ge c - c = 0. \qquad \square$$

Next, to obtain the derivability theorem for lower smooth and smooth functions, we need the following lemma.

12.6.2. LEMMA. *Suppose $f$ is lower smooth at $x \in I$. Then $f$ is upper nonangular at $x$, and $f$ is derivable at $x$ iff it is upper derivable at $x$.*

PROOF. Under the present hypothesis it is clear from Lemma 12.3.3 that $D_{-}f(x) \le D_{+}f(x) \le D^{+}f(x)$, and so $f$ is upper nonangular at $x$. Further, if $f$ is upper derivable at $x$, then it follows from that lemma that $D^{-}f(x) \le D^{+}f(x) \le D_{-}f(x) \le D_{+}f(x)$, from which it is clear that $f$ is derivable at $x$. $\square$

Now, with the help of the above lemma we deduce from the First Derivability Theorem the following derivability theorem for lower smooth and smooth functions. This theorem was obtained originally by Zalcwasser for continuous smooth functions (see [Zy2, pp. 43, 375]), and it was extended later by Neugebauer [Ne2] to measurable smooth functions. The result on lower smooth functions has been obtained on the other hand by Manna [Man3] for continuous functions.

12.6.3. THEOREM (NINTH DERIVABILITY). *Suppose $f$ is either* (i) *USC and lower smooth, or* (ii) *measurable and smooth. Then $f$ is differentiable at a $c$-dense set of points in $I$.*

PROOF. First, suppose (i) holds. Then $f$ is upper nonangular by Lemma 12.6.2. Hence it follows from the First Derivability Theorem 5.1.2, on applying it to $-f$, that $f$ is uniquely upper differentiable at a $c$-dense set of points $E$ in $I$. It follows now from the second part of Lemma 12.6.2 that $f$ is differentiable at each point of $E$.

In case (ii), $f$ is by Theorem 12.2.7 continuous on a dense set of intervals in $I$, and so the result follows clearly from the above result. $\square$

In the next two theorems we obtain some other generalized properties in the case of lower smooth functions.

12.6.4. THEOREM. *Suppose $f$ is lower smooth and $E \subset I$.*

(a) *If $D_-f > 0$ n.e. in $E$, then $f$ is GSBI on $E$, and if $f$ is continuous on $E$, then it is F-GSBI on $E$.*

(b) *If $f$ is QBSC relative to $E$ and $D_-f > 0$ at a nonmeager set of points in $E$, then $f$ is SBI on some portion of $E$.*

PROOF. First, to prove (a), suppose $D_-f > 0$ n.e. in $E$. Then it follows from Lemma 12.3.3 that $D_+f$ also is positive n.e. in $E$. Hence it follows from the First Derivate Theorem 2.2.2 on applying it to $f$ and $-f_-$ that $f$ is *GSBI* on $E$. The other part of (a) also follows from that theorem.

Next, to prove (b), suppose its hypothesis holds. Then on applying the Third Derivate Theorem 2.4.1 to $-f_-$ it follows that $f$ is *SLI* on some portion $P$ of $E$. Now, for each $x \in P$, it is clear that $D_-f(x) > 0$, and so by Lemma 12.3.3, $D_+f(x) > 0$. Hence it follows again from Theorem 2.4.1 that $f$ is *SRI* on some portion $Q$ of $P$. Consequently, $f$ is *SBI* on the portion $Q$ of $E$. □

12.6.5. THEOREM. *Suppose $f$ is lower smooth, $E \subset I$, and $c \in \mathbb{R}_+$.*

(a) *If $D_-f > -c$ and $D^+f < c$ n.e. in $E$, then $f$ is F-$GL_*(c)$ on $E$.*

(b) *If $f$ is QBSC relative to $E$, and $D_-f > -c$ and $D^+f < c$ at a nonmeager set of points in $E$, then $f$ is $L_*(c)$ on some portion of $E$.*

PROOF. To prove (a), suppose its hypothesis holds. Then it follows from Lemma 12.3.3 that $\underline{D}f > -c$ and $\overline{D}f < c$ n.e. in $E$. Hence the result follows from the Second Derivate Theorem 2.2.7 on applying it to $f$ and $f_-$.

Next, to prove (b), suppose its hypothesis holds. Then it is easy to see that there exists a positive number $d < c$ such that $D_-f > -d$ and $D^+f < d$ at a nonmeager set of points in $E$. Now it follows from the second part of Theorem 5.4.4 on applying it to $-f_-$ that $f$ is $U\,L_*(d)$ on some portion $P$ of $E$. Hence it is clear that $D_-f \geq -d > -c$ and $D^+f \leq d < c$ at each point of $P$. Now it follows from Lemma 12.3.3 that $D_+f > -c$ and $D^-f < c$ at each point of $P$, and so it follows from Theorem 5.4.4 that $f$ is $L$-$L_*(c)$ on some portion $Q$ of $P$. Consequently, $f$ is $L_*(c)$ on the portion $Q$ of $E$. □

Finally, in the next two theorems we deduce from the Second and Fifth Monotonicity Theorems of §9.1 two monotonicity results on lower smooth functions which do not hold in general for *LQS* functions. The unbracketed part of the following theorem was obtained by Manna [Man3] in the case when $f$ is continuous and $E_-$ is empty.

12.6.6. THEOREM (TENTH MONOTONICITY). *Suppose $f$ is $I^*USC$ and lower smooth. Let $E_-$ and $E_+$ denote the sets of points where $f$ has a finite negative or positive derivative, respectively. Then $f$ is nondecreasing (or nonincreasing) iff $E_-$ (or $E_+$) is countable.*

PROOF. Since $f$ is lower smooth, according to Lemma 12.6.2 it is upper nonangular, and its ordinary derivative coincides with the upper derivative. Hence the result is obtained from the Second Monotonicity Theorem 9.1.3 on applying it to $-f_{-}$. □

12.6.7. THEOREM (ELEVENTH MONOTONICITY). *Suppose $f$ is $I^*USC$ and lower smooth.*

(a) *If $f$ is nowhere decreasing and has nowhere a zero derivative, then $f$ is increasing.*

(b) *If $f$ is lower singular and there is a $c \in \mathbb{R}_+$ such that $f$ has nowhere a derivative equal to $-c$, then $f$ is nondecreasing.*

PROOF. Part (a) is obtained as above with the help of Lemma 12.6.2 on applying the Fifth Monotonicity Theorem 9.1.10 to $-f_{-}$. Part (b) is obtained similarly from Corollary 9.1.11(b). □

**§12.7 Properties of ordinary derivative of smooth functions.** In this final section of this chapter we obtain the mean value and Darboux properties of lower and ordinary derivatives of lower smooth and smooth functions, respectively.

We begin with the mean value property of these derivatives. Because the following theorem holds without any derivability hypothesis, what we mean by the *mean value property* of lower derivative of $f$ is that for every $a, b \in I$, $a < b$, there exists a point $x$ between $a$ and $b$ where $f$ is lower derivable such that $Qf(a, b) \in Lf'(x)$. The mean value property of the ordinary derivative of $f$ is defined similarly when it is derivable at some point $x$ between $a$ and $b$ such that $Qf(a, b) = f'(x)$.

12.7.1. THEOREM (FOURTH MEAN VALUE). *Suppose $f$ is $I^*C$. If $f$ is lower smooth, then $Lf'$ possesses the mean value property, and if $f$ is smooth, then $f'$ also possesses the mean value property.*

PROOF. First, suppose $f$ is lower smooth. Then $f$ is upper nonangular by Lemma 12.6.2, and so the result is obtained from the Second Mean Value Theorem 8.1.4. When $f$ is smooth, it has symmetrical derivates by Lemma 12.3.3, and so the result follows from the General Mean Value Theorem 8.1.2. □

The following theorem dealing with the Darboux property of these derivatives is obtained similarly from Theorem 10.4.1 with the help of Lemmas 12.6.2 and 12.3.3; see §10.4 for the definition of this property. In the case of a continuous smooth function $f$, the Darboux property of $f'$ has been known for some time (see, e.g., Zygmund [Zy2, p. 44]).

12.7.2. THEOREM. *Suppose $f$ is $I^*C$. If $f$ is lower smooth, then $Lf'$ possesses the Darboux property; and if $f$ is smooth, then $f'$ also possesses the Darboux property.*

*CHAPTER 13*

# Differential Structure at Nonmonotonicity Points and Nowhere Monotone and Nonderivable Functions

In this chapter we investigate the differential structure of a function $f$ at the points of its nonmonotonicity set $N(f)$. Also, the differential properties of nowhere monotone, strongly nowhere monotone, and nonderivable functions are investigated here.

First, in §13.1, we obtain some results on the unilateral derivates and lower derivative of $f$ at the points of $N(f)$. Also, in the case when $f$ has property $(T_2)$ or $(N)$, or $f$ is lower quasismooth or lower smooth, we obtain here the properties of ordinary derivative of $f$ at the points of $N(f)$.

Next, in §13.2, we deduce from the results of §13.1 the differential properties of nowhere monotone functions, and in §13.3, the differential properties of strongly nowhere monotone functions. The results of these two sections generalize the results obtained earlier in [G2, G3, G4] on these two classes of functions.

Next, in §13.4, we deduce from the results of §13.1 and the Second Median Theorem the differential properties of a lower singular function $f$ at the points of $N(f)$. Then, in §13.5, we deduce from the results of §§13.2 and 13.3, and from the Third and Fourth Derivability Theorems, results on the nature of lower and upper derivatives of nonderivable functions. The results of this section generalize some of the results obtained earlier in [G8] on nonderivable functions. Finally, in §13.6, we obtain some results dealing with the nature of lower and upper derivatives of almost all paths of one-dimensional Brownian motion. The existence of these derivatives is assured due to the continuity of the paths.

It may be pointed out here that the results of §§13.1 and 13.4 at the points of the nonmonotonicity set $N(f)$ are investigated here for the first time—except for [G11], where the set $N(f)$ was introduced and the nature of bilateral derivates of $f$ was investigated at the points of this set. The results of these two sections are somewhat deeper and more general, for they apply to all the functions that are not constant.

We revert here to our original convention; that is, unless stated otherwise, $f: I \to \mathbb{R}$ where $I$ is an arbitrary subinterval of $\mathbb{R}$.

## §13.1 Differential structure at nonmonotonicity points.

We first investigate in this section the differential structure of a function $f: I \to \mathbb{R}$ at the points of its nonmonotonicity set $N(f)$. More specifically, properties dealing with unilateral derivates and lower and ordinary derivatives of $f$ are investigated at the points of this set. The results of this section are obtained from the various monotonicity theorems proved earlier in Chapter 9, and they play a basic role in the present chapter; the results of the other sections are deduced from the results of this section.

First, in Theorem 13.1.1, we obtain a result on unilateral derivates similar to Morse's Theorem 3.5.5. In the next two theorems the lower derivative of $f$ is investigated at the points of $N(f)$. Then in Theorems 13.1.4 and 13.1.5 we investigate the properties of ordinary derivative of a function $f$ which has property $(T_2)$ or $(N)$, or is $LQS$ or lower smooth, at the points of $N(f)$.

It may be recalled here that the set $N(f)$ denotes, as defined in §9.1, the set of points $x$ in $I$ for which $f$ is not nondecreasing in any neighborhood of $x$. The set $N(-f)$ is also in a sense a nonmonotonicity set of $f$, but it is clearly enough to consider $N(f)$.

The significance of these two nonmonotonicity sets should be clear from the roles they played in the proofs of Theorems 10.4.6 and 10.4.8 which deal with two extensions of the Denjoy property of ordinary derivative. We refer here further to the Third Median Theorem 9.4.6 for a result obtained earlier on the differential structure of $f$ at the points of $N(f)$. The first two Goldowski-Tonelli Theorems 9.6.3 and 9.6.7 are deduced in fact from this median theorem.

We begin with a result in the direction of the extended version of Morse's Theorem 3.5.5 on the set of points where $D^+f = 0$. On applying the following theorem to $-f_-$, a similar result can be obtained on $D^-f$; and then on applying these two results to $-f$, similar results follow on the two lower derivates of $f$ at the points of $N(-f)$.

13.1.1. THEOREM. *Suppose $f$ is nowhere strongly decreasing. Then*

(a) *if $f$ is $I^*LC_-$ and $LI_+$, then $D^+f = 0$ at a c-dense set of points in $N(f)$; and*

(b) *if $f$ is $I^*LC$, then $D^+f = 0 \in M_-f$ and $D^-f = 0 \in M_+f$ at two c-dense sets of points in $N(f)$.*

PROOF. Suppose the set $N \equiv N(f)$ is nonempty, for there is nothing to prove otherwise. Let $U$ be any subinterval of $I$ which is open in $I$ and meets $N$.

Then there exist $a, b \in U$, $a < b$, such that $f(a) > f(b)$. Now, it is easy to see in the proof of the First Monotonicity Theorem 3.5.3 that the set $E$ defined there corresponding to $a, b$ is a subset of $N$. Hence it is easy to verify that in each of the two cases the result is obtained by the arguments used in the proof of that theorem. □

Next, we investigate the lower derivative of $f$ at the points of $N(f)$.

13.1.2. THEOREM. *Suppose $I$ is open and $f$ is $I^*LSC$. Then $f$ has a finite negative lower gradient at a dense set of points in $N(f)$. Further,*

(a) *if $f$ is lower nonangular, then it has a finite negative unique lower derivative at a c-dense set of points in $N(f)$; and*

(b) *if $f$ is nowhere nonincreasing, then it also has a zero lower gradient at a dense set of points in $N(f)$.*

PROOF. Suppose as before that $N \equiv N(f)$ is nonempty, and let $U$ be any open subinterval of $I$ which meets $N$. Now let $E$ denote the set of points in $U$ where $f$ has a finite negative lower gradient. Then it is clear that $E \subset N$.

Now, since $f$ is not nondecreasing on $U$, it follows from the Second Monotonicity Theorem 9.1.3 that $E$ is nonempty, which proves the first part. Also, when $f$ is lower nonangular, according to that theorem $E$ has cardinality $c$, which proves (a).

Finally, to prove (b), suppose $f$ is nowhere nonincreasing. Then it follows from Lemma 9.1.9 that $f$ has a relative minimum at some point $x$ in $U$. Now since $f$ is nowhere constant, $x \in N$, and of course $f$ has a zero lower gradient at $x$. Part (b) is thus also established. □

In the following theorem we deal with normal lower derivative of $f$ at the points of $N(f)$.

13.1.3. THEOREM. *Suppose $f$ is either* (i) *$LI_-$ and $I^*LC_+$, or* (ii) *$I^*LC$. Then $f$ has a negative normal lower derivative at a c-dense set of points in $N(f)$. Further,*

(a) *if $f$ is lower $L$-regular, then it has a finite negative normal lower derivative at a metrically dense set of points in $N(f)$; and*

(b) *if $f$ is lower $L$-singular, then it has a normal lower derivative $-\infty$, and thus a left derivative $-\infty$, at a c-dense set of points in $N(f)$.*

PROOF. Suppose $N \equiv N(f)$ is nonempty, and let $U$ be any open subinterval of $I$ which is open relative to $I$ and meets $N$. Let $E$ be the set of points in $U$ where $f$ has a negative normal lower derivative, and let $A$ and $B$ denote the sets of points in $E$ where $Lf'$ is finite or $-\infty$, respectively. Now since $f$ is not nondecreasing on $U$, it follows from the Third Monotonicity Theorem 9.1.7 that $f$ is not simultaneously lower $L$-regular and lower $L$-singular on $U$. If $f$

is not lower $L$-singular on $U$, then $|A| > 0$; and if $f$ is not lower $L$-regular on $U$, then clearly $|f(B)| > 0$, so that $B$ has cardinality $c$. Hence, in any case, card $(E) = c$. It is now clear that all the three results follow from these facts. □

On applying part (b) of the above theorem to $f$ and $f_-$ we obtain the following with the help of Lemma 1.1.2(b).

13.1.4. Corollary. *Suppose $f : I \to \mathbb{R}$ is $L$-singular and nowhere monotone and that it is either* (i) *$I^*C$, or* (ii) *$I^*LSC$ and $USI$. Then $f$ has normal lower derivatives $-\infty$ and $+\infty$ at two $c$-dense sets of points in $I$. Consequently, $f$ has left and right derivatives $-\infty$ and $+\infty$, respectively, at two $c$-dense sets of points in $I$.*

It should be noted here that in the case of a function with symmetrical derivates (e.g., a smooth function) the last two theorems hold for ordinary derivative in the place of lower derivative.

In the case of functions with property $(T_2)$ and $LQS$ functions we still have the following theorem on ordinary derivative. This theorem is obtained from the Sixth and Ninth Monotonicity Theorems 9.3.5 and 12.5.7 by an argument similar to the one used for the last theorem.

13.1.5. Theorem. *Suppose either* (i) *$f$ is lower internal, of $B_1$, and has property $(T_2)$, or* (ii) *$I$ is open and $f$ is $LC_-$ and $LQS$. Then $f$ has a negative derivative at a $c$-dense set of points in $N(f)$. Further,*

(a) *if $f$ is lower regular, or it has property $(N)$ in the place of $(T_2)$, then $f$ has a finite negative derivative at a metrically dense set of points in $N(f)$; and*

(b) *if $f$ is lower singular, then it has a derivative $-\infty$ at a $c$-dense set of points in $N(f)$.*

In the case of lower smooth functions we have on the other hand the following result on finite ordinary derivative.

13.1.6. Theorem. *Suppose $I$ is open and $f$ is lower smooth.*

(a) *If $f$ is $I^*USC$, then it has a finite negative derivative at a $c$-dense set of points in $N(f)$; and if $f$ is further nowhere nonincreasing, then it also has a zero derivative at a dense set of points in $N(f)$.*

(b) *If $f$ is $I^*LC$ and nowhere strongly decreasing, then $D^+f = D^-f = 0$ at a $c$-dense set of points in $N(f)$.*

Proof. First, to prove (a), suppose $f$ is $I^*USC$. Then the first part of (a) is obtained from the Ninth Monotonicity Theorem 12.5.7 by an argument similar to the one used for Theorem 13.1.3. For, since $f'$ is of $B_1$ on $\Delta(f)$, which in turn is an $F_{\sigma\delta}$-set in $I$ (see, e.g., [G11, pp. 315, 322]), the set of points where $f$ has a finite negative derivative is a Borel set in $\mathbb{R}$, and so its cardinality is $c$ when it is uncountable.

Now, to prove the other part of (a), suppose $f$ is further nowhere nonincreasing, and let $U$ be any open subinterval of $I$ which meets $N$. Then the function $-f_-$ is also nowhere nonincreasing, and it is not increasing on $-U$. Hence on applying Lemma 9.1.9 to this function it follows that $f$ has a relative maximum at some point $x \in U$. Now since $f$ is nowhere constant, it is clear that $x \in N$. Also, since $f$ is then upper derivable at $x$, it follows from Lemma 12.6.2 that it is derivable at $x$. Consequently, $f$ has a zero derivative at $x$, which proves the result.

Next, to prove (b), suppose its hypothesis holds. Then according to Theorem 13.1.1(b), $D^+f = 0 \in M_-f$ at a $c$-dense set of points $E$ in $N \equiv N(f)$. Now, given $x \in E$, it follows from Lemma 12.3.3 that $0 \leq D^-f(x) \leq D^+f(x) = 0$, which proves this part. □

## §13.2 Nowhere monotone functions.

In this section we deduce differential properties of nowhere monotone functions from the results of the last section.

First, in Theorem 13.2.1, we obtain a result on various medians of a nowhere monotone function. Theorem 13.2.2 deals with unilateral derivates, and in the next two theorems we obtain properties of lower derivative of a nowhere monotone function. Finally, in Theorems 13.2.5 and 13.2.6 we deal with the ordinary derivative of nowhere monotone functions that have property $(T_2)$ or $(N)$, or that are $LQS$ or lower smooth.

Given any function theoretic property $P$, it should be recalled here that a function $f : X \to \mathbb{R}$ is said to be *nowhere* $P$ if it does not have $P$ on any portion of $X$ (see §9.4).

Some of the results of this section and of the next section were obtained earlier in [G2, G3, G4] in the case of continuous functions. The present results provide generalizations of those results.

We begin with the following median theorem similar to the First Median Theorem 9.4.1. This theorem deals with right and ordinary medians and is obtained from the Third Derivate Theorem 2.4.1 on applying it to $f$ and $-f$; the last part of this theorem is obtained on the other hand with the help of Theorem 2.5.6.

13.2.1. THEOREM (SIXTH MEDIAN). *Suppose $f : X \to \mathbb{R}$, where $X \subset \mathbb{R}$, and let $E \subset X$. If $f$ is QBSC relative to $E$, and it is nowhere strongly right monotone (or bimonotone) on $E$, then $0 \in M_+f$ (or $Mf$)* r.e. *in $E$.*

*Consequently, if $f$ is QBSC and nowhere monotone on $X$, then $D_+f = D_-f \leq 0 \leq D^+f = D^-f$* r.e. *in $X$.*

From now on, $f$ denotes, as mentioned in the introduction of this chapter, a function defined on an interval $I \subset \mathbb{R}$. When $f$ is nowhere monotone, it is clear that $N(f) = I$ and $N(f_-) = -I$. The following theorem on unilateral derivates of nowhere monotone functions is thus obtained directly from Theorem 13.1.1 on applying it to $f$ and $f_-$.

13.2.2. THEOREM. *Suppose f is nowhere monotone. Then*
(a) *if f is $I^*LC_-$ and $LI_+$, then $D^+f = 0$ at a c-dense set of points in I; and*
(b) *if f is $I^*LC$, then $D^+f = 0 \in M_-f$ and $D^-f = 0 \in M_+f$ at two c-dense sets of points in I.*

The following two theorems on the lower derivative of nowhere monotone functions are obtained similarly from Theorems 13.1.2 and 13.1.3 on applying them to $f$ and $f_-$.

13.2.3. THEOREM. *Suppose f is $I^*LSC$ and nowhere monotone. Then f has finite positive, negative, and zero lower gradients at dense sets of points in I. Further, in case f is lower nonangular, then it has finite positive, negative and zero unique lower derivatives at c-dense sets of points in I.*

13.2.4. THEOREM. *Suppose f is nowhere monotone and either* (i) *$I^*C$, or* (ii) *$I^*LSC$ and USI. Then f has positive and negative normal lower derivatives at two c-dense sets of points in I. Further, if f is L-regular, then it has finite positive and negative normal lower derivatives at two metrically dense sets of points in I.*

The next two theorems on the ordinary derivative of nowhere monotone functions with property $(T_2)$ or $(N)$, or which are $LQS$ or lower smooth, are obtained similarly on applying Theorems 13.1.5 and 13.1.6 to $f$ and $f_-$.

13.2.5. THEOREM. *Suppose f is nowhere monotone and that either* (i) *f is internal, of $B_1$, and has property $(T_2)$, or* (ii) *I is open and f is USC and LQS. Then f has positive and negative derivatives at two c-dense sets of points in I. Further,*
(a) *if f is regular, or it has property $(N)$ in the place of $(T_2)$, then f has finite positive and negative derivatives at two metrically dense sets of points in I; and*
(b) *if f is singular, then it has derivatives $+\infty$ and $-\infty$ at two c-dense sets of points in I.*

13.2.6. THEOREM. *Suppose I is open and f is nowhere monotone and lower smooth. If f is $I^*USC$, then it has finite positive and negative derivatives at two c-dense sets of points in I, and has a zero derivative at a dense set of points in I. Further, if f is $I^*C$, then $D^+f = D^-f = 0$ and $D_+f = D_-f = 0$ at two c-dense sets of points in I.*

**§13.3 Strongly nowhere monotone functions.** In this section we deduce from the results of the last section the differential properties of strongly nowhere monotone functions.

First, in Theorem 13.3.1, we obtain a median theorem on functions which are nowhere $L\text{-}MT_*$, $MT$, $RMT$, or $BMT$. The next theorem deals with unilateral derivates of strongly nowhere monotone functions, and then in Theorems 13.3.3 and 13.3.4 we obtain properties of lower derivative of such functions. Finally, in the last two theorems we deal with the ordinary derivative of strongly nowhere monotone functions that have property $(T_2)$ or $(N)$, or that are *LQS* or lower smooth.

It may be recalled here that a function $f: X \to \mathbb{R}$ was defined in §9.4 to be *strongly nowhere monotone* if $f_{+c}$ is nowhere monotone for every $c \in \mathbb{R}$, or, equivalently, if $f$ is nowhere of monotonic type.

Some of the results of this section were obtained earlier in [G4] in the case of continuous functions. The present results provide generalizations of those results.

To begin with, let us recall here that if a function $f: X \to \mathbb{R}$ is strongly nowhere monotone on $E \subset X$ and it is *QLSC* or *QBSC* relative to $E$, then as proved in the First Median Theorem 9.4.1, $f$ is lower knotted or biknotted, respectively, r.e. in $E$. In the following median theorem we state some further results in this direction which follow easily from Corollaries 2.4.2 and 2.4.3 of the Third Derivate Theorem.

13.3.1. THEOREM (SEVENTH MEDIAN). *Suppose $f: X \to \mathbb{R}$, where $X \subset \mathbb{R}$, and let $E \subset X$. If $f$ is QLSC relative to E and it is nowhere $L\text{-}MT_*$ on E, then $f$ is lower knotted* r.e. *in E. Further, if $f$ is QBSC relative to E and it is nowhere MT, RMT, or BMT on E, then $f$ is biknotted, right knotted, or knotted, respectively,* r.e. *in E.*

The following theorem on unilateral derivates of strongly nowhere monotone functions is obtained on applying Theorem 13.2.2 to $f_{-r}$ for different real values of $r$.

13.3.2. THEOREM. *Suppose $f$ is strongly nowhere monotone. Then*

(a) *if $f$ is $I^*LC_-$ and $LI_+$, then $D^+f$ assumes each real value at a c-dense set of points in I; and*

(b) *if $f$ is $I^*LC$, then for each $r \in \mathbb{R}$, $D^+f = r \in M_-f$ and $D^-f = r \in M_+f$ at two c-dense sets of points in I.*

The following two theorems on the lower derivative of strongly nowhere monotone functions are obtained similarly from Theorems 13.2.3 and 13.2.4 on applying them to $f_{-r}$ for different $r \in \mathbb{R}$.

13.3.3. THEOREM. *Suppose $f$ is $I^*LSC$ and strongly nowhere monotone. Then every real number is a lower gradient of $f$ at a dense set of points in I. Further, if $f$ is lower nonangular, then for each $r \in \mathbb{R}_+$, $f$ has finite unique lower derivatives $> r$ and $< -r$ at two c-dense sets of points in I.*

13.3.4. THEOREM. *Suppose $f$ is strongly nowhere monotone and is either* (i) *$I^*C$, or* (ii) *$I^*LSC$ and USI. Then for each $r \in \mathbb{R}_+$, $f$ has normal lower derivatives $> r$ and $< -r$ at two c-dense sets of points in $I$. Further, if $f$ is L-regular, then it has finite normal lower derivatives $> r$ and $< -r$ at two metrically dense sets of points in $I$.*

The next two theorems on the ordinary derivative of strongly nowhere monotone functions that have property $(T_2)$ or $(N)$, or that are $LQS$ or lower smooth, are obtained similarly on applying Theorems 13.2.5 and 13.2.6 to $f_{-r}$ for different $r \in \mathbb{R}$.

13.3.5. THEOREM. *Suppose $f$ is strongly nowhere monotone and that either* (i) *$f$ is internal, of $B_1$, and has property $(T_2)$, or* (ii) *$I$ is open and $f$ is USC and LQS. Then for each $r \in \mathbb{R}_+$, $f$ has derivatives $> r$ and $< -r$ at two c-dense sets of points in $I$. Further, if $f$ is regular, or has property $(N)$ in the place of $(T_2)$, then $f$ has finite derivatives $> r$ and $< -r$ at two metrically dense sets of points in $I$.*

13.3.6. THEOREM. *Suppose $I$ is open and $f$ is strongly nowhere monotone and lower smooth. If $f$ is $I^*USC$, then each real number is a derivative of $f$ at a dense set of points in $I$. Further, if $f$ is $I^*C$, then for each $r \in \mathbb{R}$, $D^+f = D^-f = r$ and $D_+f = D_-f = r$ at two c-dense sets of points in $I$.*

13.3.7. REMARK. In connection with our earlier investigation of nowhere monotone functions, practically all the results of §13.2 and the present section, except for the ones involving $LQS$ and lower smooth properties, were obtained in [G2, G3, G4] in the case of continuous functions with $E = I$. (See Manna [Man1, Man2] and Manna and Mukhopadhyay [MM] for extensions of the results obtained in [G2, G4] in connection with Theorems 13.2.1 and 13.3.1 from continuous functions to $LSC$ functions.)

As for the earlier historical background of these results, it may be added here that in connection with Theorem 13.2.2 it was proved by Sierpiński [Si2] in 1914 that each derivate of a continuous nowhere monotone function vanishes at a dense set of points. Further, in connection with Theorem 13.3.1, it was proved by W. H. Young [Y4] in 1908 that every continuous nonderivable function is biknotted r.e.; this in turn was extended by Denjoy [D4, p. 180] in 1916 to continuous functions that are nowhere of bounded variation.

## §13.4 Differential structure of lower singular functions at nonmonotonicity points.

In this section we investigate the differential structure of lower singular functions at the points of nonmonotonicity.

First, in Theorem 13.4.2, we obtain a result on the medians of nowhere monotone singular functions. Theorem 13.4.3 deals with unilateral derivates of a lower singular function $f$ at the points of $N(f)$. Then in Theorem 13.4.4 we obtain a result on the lower derivative of a lower singular function $f$ at the points of $N(f)$, and in Theorem 13.4.5 a result is obtained on the ordinary derivative of $f$ when $f$ is lower smooth. Finally, in Theorem 13.4.6, a result is obtained on unilateral derivates of a singular increasing function.

It should be recalled here that we do not assume a lower singular or singular function to be of bounded variation. A function $f: X \to \mathbb{R}$ was defined in §6.2 to be lower singular (or singular) if its derivative is $\geq 0$ (or $= 0$) a.e. in $\Delta(f)$, or, equivalently, if $\underline{D}f \geq 0$ (or $0 \in Mf$) a.e. in $X$. Thus every nondifferentiable function is automatically singular (see Theorem 5.5.2).

Similar to the notation $\mathbb{R}_+$, it will be convenient to use here $\mathbb{R}_-$, $\mathbb{R}_+^*$ and $\mathbb{R}_-^*$ to denote the intervals $(-\infty, 0)$, $[0, \infty)$ and $(-\infty, 0]$, respectively.

The following lemma on nowhere monotone singular functions follows directly from Lemma 9.4.5.

13.4.1. Lemma. *If $f$ is singular and nowhere monotone, then it is strongly nowhere monotone.*

Thus all the results obtained in §13.3 on strongly nowhere monotone functions, except for the first one, hold for every nowhere monotone singular function. Some of these results were obtained earlier in [G9] in the case of continuous singular functions of bounded variation.

The following median theorem on nowhere monotone singular functions follows from the Second Median Theorem 9.4.2 on applying it to $f$ and $f_-$.

13.4.2. Theorem (Eighth Median). *Suppose $f$ is singular and nowhere monotone. If $f$ is QLSC (or QBSC), then it is lower knotted (or biknotted)* r.e. *in $I$.*

Next, we deduce from Theorem 13.1.1 the following theorem on unilateral derivates of lower singular functions.

13.4.3. Theorem. *Suppose $f$ is lower singular, and let $r \in \mathbb{R}_-^*$.*

(a) *If $f$ is $I^*LC_-$ and $LI_+$, then $D^+f = r$ at a c-dense set of points in $N(f)$.*

(b) *If $f$ is $I^*LC$, then $D^+f = r \in M_-f$ and $D^-f = r \in M_+f$ at two c-dense sets of points in $N(f)$.*

Proof. Let $r \in \mathbb{R}_-^*$. Then since $\underline{D}f_{-r} \geq -r \geq 0$ a.e., it is clear that $f_{-r}$ is nowhere strongly decreasing. Now since $N(f_{-r}) = N(f)$ by Lemma 9.4.5, each part is obtained on applying Theorem 13.1.1 to $f_{-r}$. □

In the following theorem we deal with the lower derivative of a lower singular function at its points of nonmonotonicity.

13.4.4. Theorem. *Suppose $I$ is open and $f$ is $I^*LSC$ and lower singular. Then every $r \in \mathbb{R}_-$ is a lower gradient of $f$ at a dense set of points in $N(f)$. Consequently, if $f$ is lower nonangular, then for every $r \in \mathbb{R}_-$, $f$ has a finite unique lower derivative $< r$ at a c-dense set of points in $N(f)$.*

PROOF. Given $r \in \mathbb{R}_-$, we have $N(f_{-r}) = N(f)$ by Lemma 9.4.5. Further, since $\overline{D}f_{-r} \geq -r > 0$ a.e., $f_{-r}$ is nowhere nonincreasing. Hence the first part is obtained on applying Theorem 13.1.2(b) to $f_{-r}$. The second part follows of course from the first since the lower derivative of a lower nonangular function is unique at every point where it exists. □

Next, the following theorem on the ordinary derivative and unilateral derivates of lower smooth, lower singular functions is obtained directly from Theorem 13.1.6 with the help of Lemma 9.4.5.

13.4.5. THEOREM. *Suppose $I$ is open and $f$ is lower singular and lower smooth. Then if $f$ is $I^*USC$, then each $r \in \mathbb{R}_-$ is a derivative of $f$ at a dense set of points in $N(f)$; and if $f$ is $I^*LC$, then for each $r \in \mathbb{R}_-^*$, $D^+f = D^-f = r$ at a c-dense set of points in $N(f)$.*

Finally, in the following theorem we obtain a result on increasing singular functions.

13.4.6. THEOREM. *Suppose $f$ is singular and increasing. Then for each $r \in \mathbb{R}_+^*$, $D^+f = r \in M_-f$ and $D^-f = r \in M_+f$ at two c-dense sets of points in $I$. Further, if $f$ is $I^*UC$, then $D_+f = r \in M_-f$ and $D_-f = r \in M_+f$ at two c-dense sets of points in $I$, and also $f$ has a derivative $+\infty$ at a c-dense set of points in $I$.*

PROOF. Since $f$ is in this case derivable a.e., $f' = 0$ a.e. Hence both the results hold obviously for $r = 0$. Hence suppose $r > 0$. Then the function $g = f_{-r}$ is nowhere decreasing. For, if $g$ is decreasing on some subinterval $J$ of $I$, then $f$ is clearly upper Lipschitz on $J$, and so it follows from Proposition 2.24 of [G11] that $f$ is nonincreasing on $J$, which is contrary to the hypothesis.

Now, since $g' = -r < 0$ a.e., $g$ is further nowhere nondecreasing. Also, since $f$ is clearly LC, so is $g$. The first part follows now on applying Theorem 13.1.1(b) to $g$.

Next, to prove the second part, suppose $f$ is $I^*UC$. Then the first result in the second part is obtained on applying Theorem 13.1.1(b) to $-f$. Further, since $f$ is monotone, it is clearly derivable at every point where it is normally semiderivable. Hence the second result of this part is obtained on applying Theorem 13.1.3(b) to $-f$. □

13.4.7. REMARK. It should be noted here in connection with Theorem 13.1.1, and its consequences obtained in Theorems 13.2.2, 13.3.2, and 13.4.3, that the two sets considered in parts (b) of these theorems may on the one hand be identical, and on the other mutually disjoint. As was seen in Theorems 13.1.6, 13.2.6, 13.3.6, and 13.4.5, they do become identical in the case of a lower smooth function. However, in the case of a knotted function it is easy to see that the two sets under consideration are mutually disjoint.

**§13.5 The nature of new derivatives of nonderivable functions.** In this section we deduce from the Third and Fourth Derivability Theorems results on the nature of lower and upper derivatives of nonderivable functions. These results seem to clarify some of the mystery of nonderivable functions. We will state most of the results here in terms of lower derivative; the results on upper derivative follow on applying these results to $-f$.

First, in Theorem 13.5.1, we obtain the existence of points of knotted lower derivability for a nonderivable function in almost all of its level sets, and we deduce from it some results in two particular cases. In Theorems 13.5.4 and 13.5.6 we deal on the other hand with the values of these knotted lower derivatives. Further, in Theorem 13.5.7, we establish in the case of a *LQS* function which is nondifferentiable the existence of infinite ordinary derivative at a *c*-dense set of points.

Finally, in Remark 13.5.8 we deal with the historical background of continuous nonderivable functions, and we present in Problem 13.5.9 a problem posed by Zahorski on unilateral derivates as mentioned earlier.

We refer here further to Remarks 4.4.4, 5.1.4, 6.1.5, and 12.4.7 for some earlier comments on nonderivable functions.

Let us recall here that we call a function $f : I \to \mathbb{R}$ nonderivable or nondifferentiable if it is not derivable or differentiable, respectively, at any point of $I$. Also, we call $f$ unilaterally nonderivable if it is not unilaterally derivable from any side at any point of $I$. Further, $f$ is called knotted if it is knotted at each point of $I$—that is, if $\underline{D}f = -\infty$ and $\overline{D}f = +\infty$ everywhere.

Let us observe here first that a nonderivable function is normally lower derivable at any point iff it is knottedly lower derivable there. Using this fact, the following theorem on the existence of points of knotted lower derivability is obtained directly from the Third and Fourth Derivability Theorems 6.2.2 and 6.3.5. The results of this theorem were obtained earlier in some particular cases in [G8] in the case of continuous functions.

13.5.1. THEOREM. *Suppose $f$ is nonderivable, and either $R^*LI$ or $L^*UI$. Then $f$ is knottedly lower derivable at a dense set of points in almost all of its level sets, and at a c-dense set of points in $I$. Further,*

(a) *if $f$ is L-regular, then it is knottedly lower differentiable at a dense set of points in almost all of its level sets, and at a metrically dense set of points in $I$; and*

(b) *if $f$ is L-singular, then it has an infinite knotted lower derivative at a dense set of points in almost all of its level sets, and at a c-dense set of points in $I$ provided that $f$ is a nowhere constant Darboux function.*

Now, let us recall that we call a function $S$-regular or $S$-singular if it is regular or singular, respectively, relative to semiderivative. Thus a function is $S$-regular iff it is simultaneously regular relative to lower and upper derivatives, and a similar statement holds for $S$-singularity.

It is easy to see that at each point where a function has an infinite knotted lower or upper derivative, it must be unilaterally derivable from some side. Hence in the case of a continuous unilaterally nonderivable function (e.g., the Besicovitch's function [Be]), we obtain the following on applying part (a) of the above theorem to $f$ and $-f$.

13.5.2. COROLLARY. *Suppose $f$ is continuous and unilaterally nonderivable. Then $f$ is S-regular, and it has finite knotted lower and upper derivatives at two dense sets of points in almost all of its level sets, and at two metrically dense sets of points in $I$.*

In the particular case of the Besicovitch's function it was proved in [G8] that its bilateral knot set is of measure zero, and so it follows from Theorem 5.5.2 that this function has finite knotted semiderivative a.e.

On applying part (b) of the above theorem to $f$ and $-f$, we obtain similarly the following result on continuous nonderivable functions which are biknotted a.e. The Weierstrass' nonderivable function constitutes an example of such functions as shown by G. C. Young [Y.G2, p. 171].

13.5.3. COROLLARY. *Suppose $f$ is a continuous nonderivable function which is biknotted* a.e. *Then $f$ is S-singular, and it has infinite knotted lower and upper derivatives at two dense sets of points in almost all of its level sets, and at two c-dense sets of points in $I$.*

The following theorem on the values of knotted lower derivatives of nonderivable functions is obtained similarly from Theorem 13.3.4 and Corollary 13.1.4.

13.5.4. THEOREM. *Suppose $f$ is nonderivable, and either* (i) *$I^*C$, or* (ii) *$I^*LSC$ and USI. Then for each $r \in \mathbb{R}_+$, $f$ has knotted lower derivatives $> r$ and $< -r$ at two c-dense sets of points in $I$. Further,*

(a) *if $f$ is L-regular, then it has finite knotted lower derivatives $> r$ and $< -r$ at two metrically dense sets of points in $I$; and*

(b) *if $f$ is L-singular, then it has knotted lower derivatives $+\infty$ and $-\infty$ at two c-dense sets of points in $I$.*

Now, if $f$ has knotted lower derivative $+\infty$ or $-\infty$ at a point $x$, it is clear that it has right derivative $+\infty$ or left derivative $-\infty$, respectively, at $x$. Similarly, if $f$ has knotted upper derivative $+\infty$ or $-\infty$ at $x$, then it has left derivative $+\infty$ or right derivative $-\infty$, respectively, at $x$. Hence from part (b) of the above theorem we obtain the following.

13.5.5. COROLLARY. *Suppose $f$ is a continuous, nonderivable function which is S-singular* (*e.g., biknotted* a.e.). *Then $f$ has right* (*and left*) *derivatives $+\infty$ and $-\infty$ at two c-dense sets of points in $I$.*

To provide some insight into the nature of infinite knotted lower and upper derivatives, we include here in Figures 1(a) and 1(b) on p. 310 the graph of a continuous function $f$ in the neighborhood of a point $x$ where $f$ has a knotted lower derivative $-\infty$ or $+\infty$, respectively; and, similarly, in Figures 2(a) and 2(b) on page 311 the graph of $f$ in the neighborhood of a point $x$ where $f$ has a knotted upper derivative $-\infty$ or $+\infty$, respectively. Here we use $KLf'$ and $KUf'$ to denote knotted lower and upper derivatives, respectively, and use $z$ to denote the point $(x, f(x))$.

In the case of a lower nonangular, knotted function it is easy to see that it is knottedly lower derivable at every point where it is lower derivable. Hence for such a function we obtain the following result from Theorem 13.3.3 on finite knotted lower derivatives.

13.5.6. THEOREM. *If $f$ is an $I^*LSC$, lower nonangular, knotted function, then every real number is a knotted lower derivative of $f$ at a dense set of points in $I$.*

Finally, in the case of a $LQS$ function which is nondifferentiable, we obtain the following result from Theorem 13.2.5(b) on the existence of infinite ordinary derivatives.

13.5.7. THEOREM. *Suppose $I$ is open and $f$ is USC and LQS. Then if $f$ is nondifferentiable, then it has derivatives $+\infty$ and $-\infty$ at two c-dense sets of points in $I$.*

Instead of obtaining the differential properties of typical continuous functions here, which are of course nonderivable, we will obtain them later in Chapter 19 where we present a unified theory of generalized derivates of typical continuous functions. We refer to Theorem 19.5.5 or [G10] for various properties of ordinary derivates of such functions.

13.5.8. REMARK. As for the historical background of nonderivable functions, we refer to the outstanding monograph of A. N. Singh [Sing1] on the theory and construction of nonderivable functions which includes exhaustive references on the subject. This monograph was republished later as a part of [H3]. However, it should be pointed out here that the statement made in this monograph on pages 76 and 96 that it can be shown that every continuous nonderivable function is biknotted a.e. is not true in general. This statement was based on a lemma of Singh which was later found to be not valid. In the case of the Besicovitch's unilaterally nonderivable function the set of its bilateral knot points was seen in [G8, p. 137] to be only of measure zero.

Further, as mentioned earlier in Remark 4.4.4, a complete description of the derivates of Weierstrass' nonderivable function has been obtained in [G8] with the help of some results obtained earlier by G. C. Young [Y.G2] on this function. Also, in connection with a problem posed by W. H. Young [Y4, p. 69], it was shown in [G8, p. 139] that the Besicovitch's function has asymmetrical derivates a.e.

A typical continuous function is also known to be biknotted a.e. (see Jarník [J1]). Hence on taking into account Remark 5.1.4 and Corollary 13.5.3, it follows that a typical continuous function has both finite and infinite knotted lower derivatives at two $c$-dense sets of points, and the same holds for its knotted upper derivative. The points of

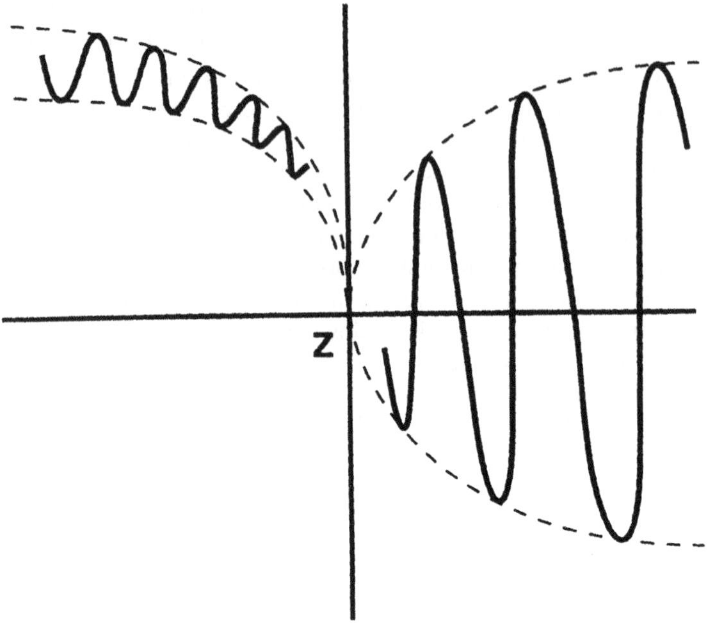


Figure 1(a). $KLf'(x) = -\infty$.

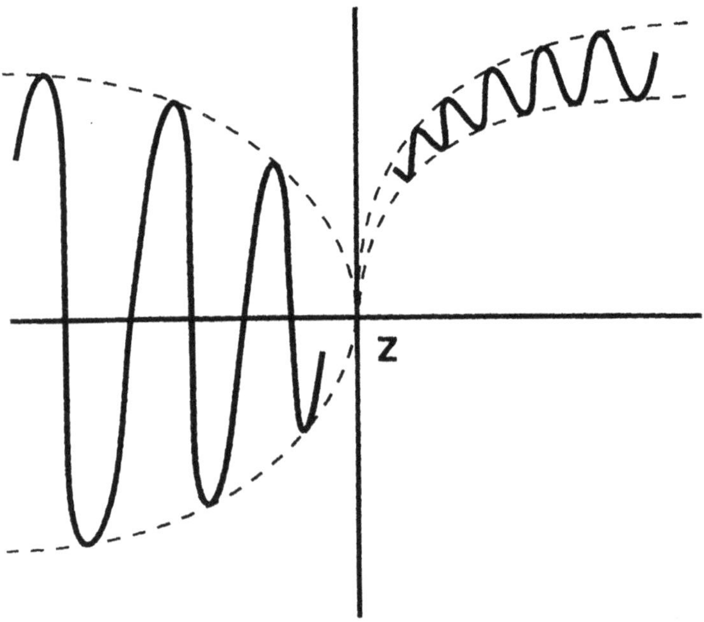


Figure 1(b). $KLf'(x) = +\infty$.

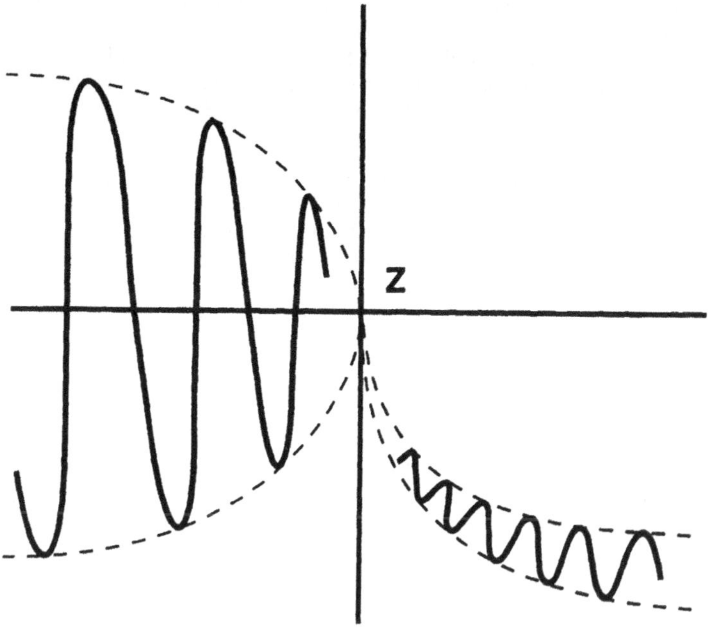


Figure 2(a). $KUf'(x) = -\infty$.

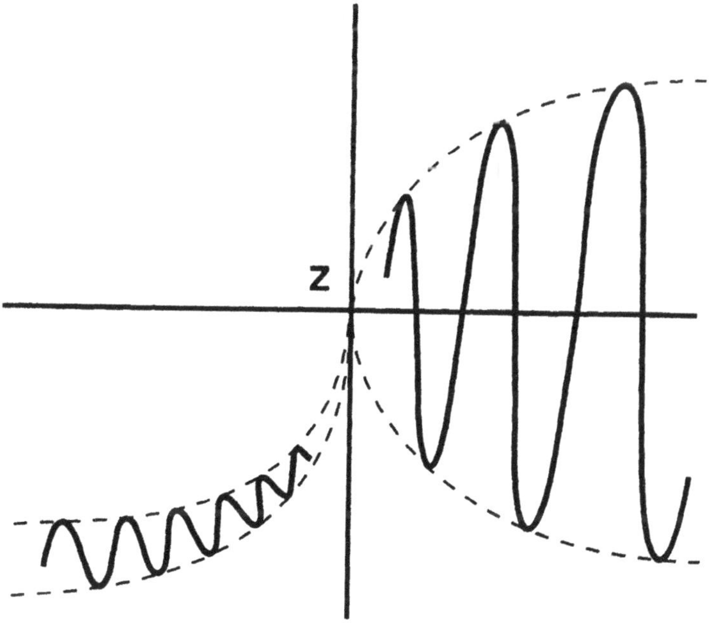


Figure 2(b). $KUf'(x) = +\infty$.

semiderivability of such a function form, of course, a set of measure zero, and this set is further meager (see [G3] or Theorem 13.3.1).

It is further clear from Corollary 13.5.5 and the Jarník's result cited above that unilaterally nonderivable functions form a meager set in $\mathbb{C}$; this was proved initially by Saks [S5] in 1932 using the sophisticated analysis of analytic sets. See §19.5 for further results on the differential structure of a typical continuous function.

13.5.9. PROBLEM. We include here as mentioned earlier a problem on unilateral derivates that was posed by Zahorski [Z3] in 1947: Does there exist a continuous function $f$ on $I$ whose unilateral derivates have exactly two, or four, distinct values at each point of $I$? As is clear from Corollary 3.4.2, a continuous function cannot have more than two values everywhere. However, for two values this problem remains unresolved.

To resolve this problem an example of such a nonderivable function will have to be constructed. The Weierstrass' nonderivable function came somewhat close to such a function, but this function was found in [G8, p. 142] to have three distinct derivates at a $c$-dense set of points. We indicate below some of the properties that such a function would have to have.

Suppose, if possible, that there is a continuous function $f$ on $I$ which has exactly two distinct derivates at each point of $I$. Then, according to the Denjoy-Young-Saks Theorem 5.5.2, $f$ must be biknotted a.e., and so it is $S$-singular. Hence, according to the Fourth Derivability Theorem 6.3.5(b), $f$ has infinite knotted lower and upper derivatives at two dense sets of points in almost all of its level sets. See Theorem 3.5.5(b) regarding the finite derivates of $f$. According to this theorem, $f$ cannot be, on the other hand, a knotted function like the Weierstrass' function. Further, if $f$ is not nonangular, then $f$ must be unilaterally derivable from both sides at each point where it is angular.

## §13.6 New derivatives of Brownian paths.

In this section we investigate the nature of lower and upper derivatives of Brownian paths. We deal here only with one-dimensional Brownian motion. It has been known for some time that almost all Brownian paths are nonderivable. This has been referred to sometimes as "chaos" in Brownian motion.

We obtain here some properties of lower and upper derivatives of almost all Brownian paths which in turn indicate some order in Brownian motion (see Theorems 13.6.3 and 13.6.4).

The one-dimensional Brownian motion is a stochastic process which is defined in terms of a probability measure space $(\Omega, \mathscr{B}, P)$ whose elements $\omega \in \Omega$ consist of all continuous real-valued functions $x_\omega(t)$, $t \in \mathbb{R}$, with $x_\omega(0) = 0$. We refer here to the paper of Dvoretzky, Erdös, and Kakutani [DEK] for the definition and other details of this probability measure space.

Further, for each $\omega \in \Omega$, $x_\omega$ is called a *Brownian path*; and any proposition is said to hold *for almost all Brownian paths* if it holds for $x_\omega$ corresponding to a set $E$ of $\omega$ in $\Omega$ such that $P(E) = 1$. The term "almost surely" is also used sometimes for "almost all" here.

In the place of the notation $x_\omega$ used in [DEK] for a Brownian path, we will use here $B_\omega$, or $B$ when $\omega$ is fixed, which is more commonly used now (see, e.g., Freedman [Fr] and Knight [Kni]).

We deduce the results of this section in fact from some of the known results on Brownian paths with the help of some of the earlier results of this chapter on nowhere monotone and nonderivable functions.

We first state in the following theorem two results obtained by Dvoretzky, Erdös, and Kakutani [DEK] on the nonderivability of almost all Brownian paths. (See Knight [Kni, p. 150] for another proof of part (a) of this theorem, and see Freedman [Fr, p. 40] for another proof of its part (b).)

13.6.1. THEOREM (DVORETZKY-ERDÖS-KAKUTANI). *The following holds for almost all Brownian paths $B$:*

(a) *$B$ is a knotted function, and hence it is nonderivable; and*

(b) *at each point $t \in \mathbb{R}$, and from each side, at least one derivate of $B$ is infinite.*

13.6.2. THEOREM. *Almost all Brownian paths $B$ are biknotted* a.e. *and* r.e.

PROOF. Since almost all paths $B$ are nonderivable by the last theorem, they are clearly strongly nowhere monotone. Hence the result re "r.e." follows directly from Theorem 13.3.1 due to the continuity of $B$.

We have not been able to find any reference to the result re "a.e.". However, this result does hold, as was pointed out by Professor Kakutani when asked during his talk at our University. □

Since each Brownian path $B$ is continuous, according to the derivability theorems of Part II, $B$ is lower and upper derivable at some points. In the next two theorems we deal with the nature of these new derivatives for almost all Brownian paths.

The following theorem follows directly from Corollary 13.5.3 on taking into account the above Theorems 13.6.1(a) and 13.6.2.

13.6.3. THEOREM. *Almost all Brownian paths $B$ have infinite knotted lower and upper derivatives at two dense sets of points in almost all of their level sets, and at two c-dense sets of points in $\mathbb{R}$.*

Next, in the following theorem we deal with the values of knotted lower and upper derivatives. Here we will call an $r \in \mathbb{R}$ a *knotted lower* (or *upper*) *gradient* of $B$ at $t \in \mathbb{R}$ if it is a lower (or upper) gradient of $B$ at $t$ and if $B$ is further upper (or lower) knotted at $t$.

13.6.4. THEOREM. *The following holds for almost all Brownian paths $B$:*

(a) *every real number $r$ is a knotted lower and upper gradient of $B$ at two dense sets of points in $\mathbb{R}$; and*

(b) *$B$ has knotted lower (and upper) derivatives $+\infty$ and $-\infty$ at two c-dense sets of points in $\mathbb{R}$.*

PROOF. Part (a) follows directly from the first part of Theorem 13.3.3 on taking into account Theorems 13.6.1(a) and 13.6.2. Part (b) is obtained on the other hand from Theorem 13.5.4(b) with the help of the same theorems. □

Using an argument similar to the one used earlier for Corollary 13.5.5, we obtain from part (b) of the above theorem the following result on the existence of infinite unilateral derivatives for almost all Brownian paths.

13.6.5. COROLLARY. *Almost all Brownian paths $B$ have right derivatives $+\infty$ and $-\infty$ at two c-dense sets of points; similarly, they have left derivatives $+\infty$ and $-\infty$ at two c-dense sets of points.*

We include here a result on the derivates of almost all Brownian paths. This result follows easily from Theorem 13.3.2(b) with the help of the two parts of Theorem 13.6.1.

13.6.6. THEOREM. *For almost all Brownian paths $B$ and for each real number $r$, the following four sets are c-dense in $\mathbb{R}$: the sets $E_{r,1}$ and $E_{r,2}$ of points where $B$ is upper knotted and* (i) $D_+B = r \leq D^-B$ *or* (ii) $D_+B \leq r = D^-B$, *respectively; and the sets $E_{r,3}$ and $E_{r,4}$ of points where $B$ is lower knotted and* (iii) $D^+B = r \geq D_-B$ *or* (iv) $D^+B \geq r = D_-B$, *respectively.*

13.6.7. PROBLEM. It would be worthwhile to investigate whether almost all Brownian paths are nonangular. If so, then the term "gradient" in Theorem 13.6.4(a) can be replaced by "derivative".

# CHAPTER 14

# Strong Derivates and Strong Derivative

In this chapter we obtain various results related with strong derivates and strong derivative defined earlier in §2.5. The results on strong derivates will be found to put the results on the ordinary and generalized derivates considered in the next part in a somewhat better perspective.

As mentioned in §2.5, the notions of strong derivates and strong derivative were introduced by Peano [P] in 1892 without any nomenclature, and he suggested the investigation of these notions. No systematic study of these derivates seems to have been conducted till now, and the present chapter is in a sense a response to Peano's suggestion. Some results were obtained earlier in this direction in Theorem 2.5.4 and in §§3.4 and 8.3. To motivate the results of this chapter, we first discuss their nature briefly.

Although the notion of strong derivability is found to be too strong for even a monotone or a Lipschitz function to be strongly derivable at any point (see Example 14.1.4), the strong derivates and other entities related with them like strong median, strong derivative, and strong derivability sets are found to have considerably stronger properties than the ones of the corresponding entities related with the ordinary and other notions of derivates. Thus at the points where any other derivate coincides with the corresponding strong derivate (see, e.g., Theorem 2.5.4), the stronger property of the strong derivate is shared in a sense by the other derivate.

Further, the values of strong derivates of a function at any point are found to determine the behavior of the function in a neighborhood of that point in terms of the monotonicity and Lipschitz properties considered in Part I. This phenomenon is of course typical of strong derivates since it does not hold for any other known derivates.

Due to the above two reasons, a systematic investigation of strong derivates and related entities appears to be worthwhile. Some of the results of this chapter are obtained directly from definitions, but the others are deduced from the monotonicity and median theorems of Part III.

First, in §14.1, we establish the stronger properties of strong derivates, strong derivative, and related entities like strong median and strong derivability and strong knot sets. Then, in §14.2, we obtain local and global properties of functions in terms of strong derivates.

Next, in §14.3, relations between unilateral and strong derivates are investigated, which lead to strengthened versions of the two classical theorems of Dini on unilateral derivates. Then, in §14.4, we obtain some characterizations of strong derivability in terms of local properties, unilateral derivates, and ordinary derivative.

Finally, in §14.5, the mean value property and the Darboux and Denjoy properties of strong derivative are obtained for strongly derivable functions without any continuity hypothesis.

**§14.1 Properties of strong derivates, strong median, and strong derivative.** In this section we obtain the Baire class of strong derivates, strong median, and strong derivative (see Theorem 14.1.1), and investigate the Borel class of strong derivability and strong knot sets (Theorem 14.1.3).

Also, we include here an example to show that a monotone or Lipschitz function need not be strongly derivable at any point (see Example 14.1.4), and we compare the results of this section on strong derivates and related entities with the corresponding results on bilateral and unilateral derivates (see Remark 14.1.5).

We begin by recalling the nomenclature related with strong derivates and introduce here some notations.

Let $f: X \to \mathbb{R}$, where $X \subset \mathbb{R}$. The *lower* and *upper strong derivates* of $f$ at a point $x \in X^d$ were defined in §2.5 to be the lower and upper limits, respectively, of $Qf(y, z)$ as $y, z$ approach $x$ through $X$. We denote these derivates by $\underline{D}_*f(x)$ and $\overline{D}_*f(x)$, respectively. When these two derivates are equal, then $f$ is said to be *strongly derivable* at $x$, and the common value of these derivates is called the *strong derivative* of $f$ at $x$ and is denoted by $f'_*(x)$. Also, when $f'_*(x)$ is finite, $f$ will be said to be *strongly differentiable* at $x$.

Further, the interval $[\underline{D}_*f(x), \overline{D}_*f(x)]$ was defined in §3.3 to be the *strong median* of $f$ at $x$, which we denote by $M_*f(x)$; as was seen in Remark 3.3.7, this median is closely related with the notion of paratingent introduced by Bouligand [Bou].

Next, we will use $\Delta_*(f)$ and $\Delta_*^*(f)$ to denote the sets of points in $X^d$ where $f$ is strongly derivable or strongly differentiable, respectively. These two sets will be called the *strong derivability sets* of $f$. Also, $f$ will be said to be *strongly knotted* at $x$ if $\underline{D}_*f(x) = -\infty$ and $\overline{D}_*f(x) = +\infty$, and the set of all strong knot points of $f$ will be called the *strong knot set* of $f$ and denoted by $K_*(f)$.

In the case when $x \in X'$, it is clear that $\underline{D}_*f(x) \leq \underline{D}f(x) \leq \overline{D}f(x) \leq \overline{D}_*f(x)$. Hence if $f$ is strongly derivable at such a point $x$, then it is also derivable at $x$ and $f'(x) = f'_*(x)$.

In connection with the following theorem it was proved by Peano [P] that the strong derivative is continuous.

14.1.1. THEOREM. *Let $f: X \to \mathbb{R}$. Then $\underline{D}_* f$ is LSC and $\overline{D}_* f$ is USC on $X^d$. Consequently, $M_* f$ is USC on $X^d$, and $f'_*$ is continuous on $\Delta_*(f)$.*

PROOF. The semicontinuities of $\underline{D}_* f$ and $\overline{D}_* f$ follow directly from their definitions. Hence it follows from Theorem 1.5.1(b) that $M_* f$ is *USC* as a multifunction. Further, since $f'_*$ coincides with $\underline{D}_* f$ and $\overline{D}_* f$ on $\Delta_*(f)$, it is clear that $f'_*$ is continuous on this set. □

From the *USC* of $M_* f$ we easily obtain the following result.

14.1.2. COROLLARY. *Suppose $f: X \to \mathbb{R}$ and $c \in \overline{\mathbb{R}}$. Then $K_*(f)$ and the set $\{x \in X^d : c \in M_* f(x)\}$ are closed in $X$.*

In the following theorem we deal with the Borel class of strong derivability sets.

14.1.3. THEOREM. *Let $f: X \to \mathbb{R}$. Then each of the strong derivability sets $\Delta_*(f)$ and $\Delta^*_*(f)$ is a $G_\delta$-set in $\mathbb{R}$.*

PROOF. Let $Q$ denote the set of all rational numbers. Then

$$X^d \sim \Delta_*(f) = \{x \in X^d : \underline{D}_* f(x) < \overline{D}_* f(x)\}$$
$$= \bigcup_{r,s \in Q;\, r<s} \left[\{x \in X^d : \underline{D}_* f(x) \le r\} \cap \{x \in X^d : \overline{D}_* f(x) \ge s\}\right].$$

Hence it follows clearly from the first part of Theorem 14.1.1 that this set is an $F_\sigma$-set in $X^d$. Now since $X^d$ is closed in $\mathbb{R}$, this proves that $\Delta_*(f)$ is a $G_\delta$-set in $\mathbb{R}$.

Further, since

$$\Delta^*_*(f) = \Delta_*(f) \sim \left[\{x \in X^d : \underline{D}_* f(x) = -\infty\} \cup \{x \in X^d : \overline{D}_* f(x) = \infty\}\right],$$

where the two sets within chain brackets on the right side are closed by Theorem 14.1.1, it is clear that $\Delta^*{}_*(f)$ is also a $G_\delta$-set in $\mathbb{R}$. □

Next, we present an example to show that a monotone or Lipschitz function need not be strongly derivable at any point. It may be pointed out here in this connection, as we see later in Corollary 14.2.3, that every strongly nowhere monotone function is strongly knotted everywhere. Hence for such a function strong derivates become somewhat redundant. The same holds, however, for the bilateral derivates of a knotted function.

14.1.4. EXAMPLE. Let $I = [0, 1]$, and let $A$ be any measurable subset of $I$ such that both $A$ and $I \sim A$ are metrically dense in $I$. Now let $g$ be the characteristic function of $A$ on $I$, and define

$$f(x) = \int_0^x g(t)\, dt, \qquad x \in I.$$

Then for each $x, y \in I$, $x < y$, it is clear that $0 \leq Qf(x, y) \leq 1$. Hence $f$ is a nondecreasing, Lipschitz function, and it follows from a theorem of Lebesgue [N1, p. 253] that $f' = g$ a.e. Thus $f$ has derivatives 0 and 1 at two dense sets of points in $I$. Hence it is clear that $\underline{D}_* f = 0$ and $\overline{D}_* f = 1$ at two dense sets of points in $I$, and so it follows from Theorem 14.1.1 that the same holds at each point of $I$. Consequently, $f$ is not strongly derivable at any point.

14.1.5. REMARK. To see the strength of the properties of various entities related with strong derivates, we include here a comparison of the results of this section with the corresponding results on similar entities related with bilateral and unilateral derivates.

For the sake of simplicity, we assume here that $f : I \to \mathbb{R}$; the following comments hold also for a function $f : X \to \mathbb{R}$, but then only relative to $X$. Then

(a) $\underline{D}_* f$ is always $LSC$ (see Theorem 14.1.1), whereas $\underline{D} f$ is only of $B_2$ in general, and of $LB_1$ when $f$ is $WUC$ (see [G11, Theorem 3.4]); $D_+ f$ is, on the other hand, not even measurable in general (see Example 3.2.8(c) and Theorems 3.2.2 and 3.2.4 for the Baire class of $D_+ f$);

(b) $M_* f$ is always $USC$ (see Theorem 14.1.1), whereas $Mf$ is only of $B_2$ in general, and of $UB_1$ when $f$ is $WC$ (see Theorem 3.3.1); $M_+ f$ is, on the other hand, not even measurable in general (see Theorem 3.3.2 for the Borel class of $M_+ f$);

(c) $f'_*$ is always continuous (see Theorem 14.1.1), whereas $f'$ is only of $B_1$ in general (see [G11, Theorem 5.1]); $f'_+$ is, on the other hand, $L$-measurable but not $B$-measurable in general (see Theorem 10.3.4);

(d) $K_*(f)$ is always closed (see Corollary 14.1.2), whereas $K(f)$ is only a $G_{\delta\sigma}$-set in general, and a $G_\delta$-set when $f$ is $WC$ (see [G11, Corollary 3.7]); $K_+(f)$ is, on the other hand, not even measurable in general (see Example 10.1.7(a) and Theorem 10.1.2 for the Borel class of $K_+(f)$); and

(e) $\Delta_*(f)$ and $\Delta^*_*(f)$ are always $G_\delta$-sets (see Theorem 14.1.3), whereas $\Delta(f)$ and $\Delta^*(f)$ are only $F_{\sigma\delta}$-sets in general (see [G11, Theorem 3.12]); $\Delta_+(f)$ is, on the other hand, measurable but not $B$-measurable in general, and $\Delta^*_+(f)$ is an $F_{\sigma\delta}$-set in general (see Theorem 10.1.4).

## §14.2 Local and global properties of functions in terms of strong derivates.

In this section we obtain local and global versions of the monotonicity and Lipschitz properties defined in §2.1 in terms of strong derivates. By local properties we mean properties determined in some neighborhood of a point by the values of strong derivates at that point. This phenomenon is, of course, typical of strong derivates, for it does not hold for any other known derivates.

First, in Theorems 14.2.1 and 14.2.4 and their corollaries, we obtain local properties in terms of strong derivates. Then, in Theorems 14.2.6 and 14.2.10, several (global) monotonicity conditions are obtained in terms of strong derivates. Some consequences of these two theorems in terms of other global properties are included as corollaries.

We begin with local properties determined by strong derivates at a point.

14.2.1. THEOREM. *Suppose $f : X \to \mathbb{R}$ and $x \in X^d$. Then*

(a) $\underline{D}_* f(x) > 0$ *iff $f$ is strongly increasing in some neighborhood of $x$; and*

(b) $\underline{D}_* f(x) > -\infty$ *iff $f$ is lower Lipschitz in some neighborhood of $x$.*

PROOF. The sufficiency part holds obviously in each of the two results.

To prove the necessity part of (a), suppose $\underline{D}_* f(x) > 0$. Let $0 < c < \underline{D}_* f(x)$. Then there exists a neighborhood $U$ of $x$ such that for each $y, z \in U \cap X$, $y < z$, $Qf(y, z) > c$. Clearly, for such $y, z$, $f_{-c}(y) < f_{-c}(z)$, so that $f_{-c}$ is increasing on $U \cap X$. Consequently, $f$ is strongly increasing on $U \cap X$.

Next, to prove the necessity part of (b), suppose $\underline{D}_* f(x) > -\infty$. Choose $c \in \mathbb{R}$ such that $\underline{D}_* f(x) > c$. Then $\underline{D}_* f_{-c}(x) > 0$. Hence, by (a), $f_{-c}$ is increasing in some neighborhood of $x$, and $f$ is consequently lower Lipschitz in that neighborhood of $x$. □

Following are some easy consequences of the above theorem.

14.2.2. COROLLARY. *Suppose* $f: X \to \mathbb{R}$ *and* $x \in X^d$. *Then*

(a) $0 \in M_* f(x)$ *iff* $f$ *is not strongly monotone in any neighborhood of* $x$;

(b) $x \in K_*(f)$ *iff* $f$ *is not* MT *in any neighborhood of* $x$; *and*

(c) *if* $\underline{D}_* f(x) > -c$ *and* $\overline{D}_* f(x) < c$, *where* $c \in \mathbb{R}_+$, *then* $f$ *is* $c$-*Lipschitz in some neighborhood of* $x$.

The following result is obtained, on the other hand, from the first two parts of the above corollary with the help of Corollary 14.1.2.

14.2.3. COROLLARY. *Let* $f: X \to \mathbb{R}$. *Then*

(a) $f$ *is nowhere strongly monotone iff* $0 \in M_* f$ *everywhere, or, equivalently, at a dense set of points in* $X$; *and*

(b) $f$ *is strongly nowhere monotone iff it is strongly knotted everywhere, or, equivalently, at a dense set of points in* $X$.

Next, in the following theorem we obtain some local properties of lower singular and singular functions from Theorem 14.2.1.

14.2.4. THEOREM. *Suppose* $f: I \to \mathbb{R}$ *and* $x \in \bar{I}$.

(a) *If* $f$ *is lower singular and* $\underline{D}_* f(x) > -\infty$, *then* $f$ *is nondecreasing in some neighborhood of* $x$.

(b) *If* $f$ *is singular and* $\underline{D}_* f(x)$ *and* $\overline{D}_* f(x)$ *are finite, then* $f$ *is constant in some neighborhood of* $x$.

(c) *If* $f$ *is lower singular (or singular) and it is strongly derivable at* $x$, *then* $f$ *is nondecreasing (or constant) in some neighborhood of* $x$.

PROOF. Here it is enough to prove parts (a) and (c) for lower singular functions; for the results on singular functions follow clearly on applying these results to $f$ and $-f$.

Hence suppose $f$ is lower singular. First, to prove (a), suppose $\underline{D}_* f(x) > -\infty$. Then by Theorem 14.2.1(b) there is a subinterval $U$ of $\bar{I}$, open relative to $\bar{I}$, such that $x \in U$ and $f$ is lower Lipschitz on $U$. Hence it follows from Proposition 2.24 of [G11] that $f$ is nondecreasing on $U \cap I$, which proves (a).

Next, to prove (c), suppose $f$ is strongly derivable at $x$. Since $f$ is lower singular, we have $\overline{D}_* f \geq \overline{D} f \geq 0$ a.e. by Theorem 5.5.2. Hence it follows from Theorem 14.1.1 that $\overline{D}_* f \geq 0$ everywhere in $\overline{I}$. Now since $f$ is strongly derivable at $x$, we thus have $f'_*(x) \geq 0$, and so it follows from (a) that $f$ is nondecreasing in some neighborhood of $x$. □

Next, we deal with global properties in terms of strong derivates. We begin with the following lemma which seems to be of some interest on its own.

14.2.5. LEMMA. *Suppose $f : X \to \mathbb{R}$ and $E \subset X^d$. If $\underline{D}_* f(x) > 0$ for each $x \in E$, then $E$ has a countable open cover $\{U_n\}$ such that $f$ is strongly increasing on $X \cap U_n$ for each $n$.*

PROOF. Suppose $\underline{D}_* f(x) > 0$ for each $x \in E$. Then for each $x \in E$, there is, by Theorem 14.2.1(a), a neighborhood $U_x$ of $x$ such that $f$ is strongly increasing on $X \cap U_x$. Now the result follows clearly from the fact that the topology of $\mathbb{R}$ is second countable. □

It is interesting to note here that results similar to the earlier results of this section can be deduced from this lemma.

In the rest of this chapter it will be understood according to our convention, unless stated otherwise, that $f : I \to \mathbb{R}$, where $I$ is an interval.

14.2.6. THEOREM. (a) *If $I$ is bounded and $\underline{D}_* f > 0$ at each point of $\overline{I}$, then $f$ is strongly increasing.*

(b) *If $I$ is bounded and $\underline{D}_* f > -\infty$ at each point of $\overline{I}$, then $f$ is lower Lipschitz.*

(c) *If $\underline{D}_* f \geq 0$ at each point of $\overline{I}$, then $f$ is nondecreasing.*

PROOF. First, to prove (a), suppose its hypothesis holds. Then since $\overline{I}$ is compact, it follows from the above lemma that $I$ has a finite open cover $\{U_i : i = 1, 2, \ldots, n\}$ such that $f$ is strongly increasing on $I \cap U_i$ for each $i$. Hence, for each $i \leq n$, there is a $c_i \in \mathbb{R}_+$ such that $f_{-c_i}$ is increasing on $I \cap U_i$. Let $c = \min\{c_i : i \leq n\}$. Then $c > 0$, and it is clear that $f_{-c}$ is increasing on $I$. Consequently, $f$ is strongly increasing.

Next, to prove (b), suppose its hypothesis holds. Then since $\underline{D}_* f$ is $LSC$ on $\overline{I}$ by Theorem 14.1.1, and $\overline{I}$ is compact, $\underline{D}_* f$ is bounded from below and attains its infimum (say $a$) on $\overline{I}$. Choose $c \in \mathbb{R}$ such that $c < a$. Then $\underline{D}_* f_{-c} > 0$ at each point of $\overline{I}$, and so, by (a), $f_{-c}$ is (strongly) increasing. Consequently, $f$ is lower Lipschitz.

Finally, to prove (c), suppose $\underline{D}_* f \geq 0$ at each point of $\overline{I}$, but $f$ is not nondecreasing on $I$. Then there exist $a, b \in I$, $a < b$, such that $f(a) > f(b)$.

Then $c = \frac{1}{2}Qf(a,b) < 0$, and hence $Qf(a,b) = 2c < c$. However, since $\underline{D}_* f_{-c}(x) \geq -c > 0$ for each $x \in \bar{I}$, and $J \equiv [a,b]$ is compact, it follows from (a) that $f_{-c}$ is (strongly) increasing on $J$. Thus $Qf_{-c}(a,b) > 0$, so that $Qf(a,b) > c$, which is a contradiction. □

Following are some further global properties that follow easily from the last two parts of the above theorem.

14.2.7. COROLLARY. (a) *If $I$ is bounded and $\underline{D}_* f$ and $\overline{D}_* f$ are finite at each point of $\bar{I}$, then $f$ is Lipschitz.*

(b) *If $\underline{D}_* f \geq -c$ and $\overline{D}_* f \leq c$ at each point of $\bar{I}$, where $c \geq 0$, then $f$ is $c$-Lipschitz.*

In the case of a strongly derivable function $f$, since $f'_*$ is continuous by Theorem 14.1.1, we further obtain the following result from Theorem 14.2.6(c).

14.2.8. COROLLARY. *Suppose $f$ is strongly derivable on $\bar{I}$. If $f'_*$ is nonnegative, bounded from below, bounded, or zero on a dense set of points in $\bar{I}$, then $f$ is nondecreasing, lower Lipschitz, Lipschitz, or constant, respectively.*

In the next theorem we obtain some monotonicity conditions in terms of strong derivates. First, in the following lemma we deal with the strong median of a function at its nonmonotonicity points.

14.2.9. LEMMA. *If $\overline{D}_* f \geq 0$ at a dense set of points in $I$, then $0 \in M_* f$ at each point of $N(f)$; and if $f$ is lower singular, then $[-\infty, 0] \subset M_* f$ at each point of $N(f)$.*

PROOF. First, suppose $\overline{D}_* f \geq 0$ at a dense set of points in $I$. Then since $\overline{D}_* f$ is *USC* (see Theorem 14.1.1), the same holds at each point of $\bar{I}$. Further, given $x \in N(f)$, it follows clearly from Theorem 14.2.1(a) that $\underline{D}_*(x) \leq 0$, and hence $0 \in M_* f(x)$.

Next, suppose $f$ is lower singular. Then $\overline{D}_* f \geq \overline{D} f \geq 0$ a.e. in $I$, and so, as just seen, $\overline{D}_* f \geq 0$ at each point of $\bar{I}$. Further, it is clear from Theorem 14.2.4(a) that $\underline{D}_* f = -\infty$ at each point of $N(f)$. Consequently, $[-\infty, 0] \subset M_* f$ at each point of $N(f)$. □

14.2.10. THEOREM. *Suppose $f$ is WLC. Then $f$ is nondecreasing whenever one of the following holds:*

(a) *$\overline{D}_* f \geq 0$ at a dense set of points in $I$ and* $\operatorname{card}\{x : 0 \in M_* f(x)\} < c$.

(b) *$f$ is lower singular and* $\operatorname{card}\{x : \underline{D}_* f(x) = -\infty, \overline{D}_* f \geq 0\} < c$.

PROOF. Since $f$ is *WLC*, the set $N \equiv N(f)$ is, by Lemma 9.1.2, perfect in $I$. Hence in each of the two cases the result follows from the two parts of the above lemma. □

Finally, in the case of functions that are strongly derivable n.e. we obtain the following from the above theorem.

14.2.11. COROLLARY. *Suppose f is strongly derivable* n.e. *in I.*

(a) *If f is WLC and lower singular, then it is nondecreasing.*

(b) *If f is WC and singular, then it is constant.*

14.2.12. REMARK. We include here some examples to show that Theorem 14.2.6(a) does not hold on replacing $\underline{I}$ by $I$, and that this theorem and Theorem 14.2.6(b) and Corollary 14.2.7(a) do not hold for unilateral or bilateral derivates even in the case of a differentiable function.

First, let $f(x) = x^2$, $0 < x < 1$. Then it is easy to see that $f$ is strongly differentiable everywhere and $f'_*(x) > 0$ for each $x$ in $(0, 1)$, but $f$ is not strongly increasing.

Next, to see that Theorem 14.2.6(a) does not hold for bilateral derivates even for a function defined on a compact interval, define

$$f(x) = 24x + x^4 e^{-x^2/4} \sin(8x^{-3}) \quad \text{for} \quad 0 < x \leq 1, \quad \text{and} \quad f(0) = 0.$$

Then using arguments similar to the ones used by Gelbaum and Olmsted in [GO, pp. 37–38], it can be verified that $f$ is derivable and it has a positive derivative everywhere, but it is not strongly increasing since $f'$ takes arbitrarily small values in the neighborhood of zero.

As for Theorem 14.2.6(b) and Corollary 14.2.7(a), define $f(x) = x^2 \sin(x^{-2})$ for $0 < x \leq 1$, and $f(0) = 0$. Then it is easy to see that $f$ is everywhere differentiable but is not of bounded variation, and so $f$ is not even lower Lipschitz.

## §14.3 Relations between unilateral and strong derivates; Dini's theorems.

In this section we deduce from the monotonicity theorems of Chapter 9 two results on the relations between unilateral and strong derivates. These results provide stronger versions of two theorems of Dini on unilateral derivates ([Di], or see [S8, p. 204]). Dini's theorems constitute indeed the first two known results on unilateral derivates which were introduced by him, and thus they are commonly known as Dini derivates.

Dini proved in 1878 that each unilateral derivate of a continuous function $f$ on $I \equiv [a, b]$ has the same lower and upper bounds as the ones of $Qf$, and he deduced from this theorem that if a derivate of $f$ is continuous at some point $x \in I$, then so are the other three derivates of $f$ and $f$ is derivable at $x$.

It should be noted here that these two theorems of Dini do not hold in general for bilateral or strong derivates, which in turn is clear from any continuous knotted function.

Following is a strengthened version of the first theorem of Dini. We state the result in terms of $D_+f$; similar results hold of course for the other three derivates.

14.3.1. THEOREM (DINI). (a) *If f is WLC$_-$, then $D_+f$ and $Qf$ have the same lower bounds, and $D_+f = \underline{D}_*f$ at the points where $D_+f$ is LSC.*

(b) *If f is either* (i) *I*LSC, or* (ii) *UI$_+$ and I*UC$_-$, then $D_+f$ and $Qf$ have the same upper bounds, and $D_+f = \overline{D}_*f$ at the points where $D_+f$ is USC.*

PROOF. First, to prove (a), suppose $f$ is $WLC_-$. Let $a$ and $b$ denote the infima of $Qf$ and $D_+f$, respectively, on $I$. Then clearly $a \le b$. Suppose $a < b$, and choose $c$ such that $a < c < b$. Then it is clear that $D_+f_{-c} > 0$ at each point of $I$. Hence it follows from the Fourth Median Theorem 9.6.1(a) that $f_{-c}$ is increasing on $I^0$. But since $f_{-c}$ is then $WLC$, this implies that $f_{-c}$ is increasing on $I$, which contradicts the fact that $a < c$. Consequently, $a = b$.

Next, suppose $D_+f$ is $LSC$ at $x \in I$. Clearly, $\underline{D}_*f(x) \le D_+f(x)$. To obtain the reverse inequality, suppose $D_+f(x) > c > -\infty$. Then there is a subinterval $U$ of $I$, open relative to $I$, such that $x \in U$ and $D_+f > c$ on $U$. Now, since $D_+f_{-c} > 0$ on $U$, it follows again from Theorem 9.6.1(a) as before that $f_{-c}$ is increasing on $U$. Consequently, $\underline{D}_*f_{-c}(x) \ge 0$, so that $\underline{D}_*f(x) \ge c$, which proves that $\underline{D}_*f(x) = D_+f(x)$.

Part (b) is obtained by a similar argument from the Second Monotonicity Theorem 9.1.3(a) in case (i), and on applying the Third Monotonicity Theorem 9.1.7 to $f_-$ in case (ii). □

Following is the strengthened version of the second theorem of Dini. Its necessity part follows from Theorem 14.1.1 since $\underline{D}_*f \le D_+f \le \overline{D}_*f$ everywhere, and the sufficiency part is obtained on combining the two parts of the above theorem.

14.3.2. COROLLARY (DINI). *If $f$ is $WLC_-$ and either* (i) *$I^*LSC$, or* (ii) *$UI_+$ and $I^*UC_-$, then $f$ is strongly derivable at $x \in I^+$ iff $D_+f$ is continuous at $x$.*

Finally, on applying this corollary twice, we obtain the following part of Dini's second theorem.

14.3.3. COROLLARY (DINI). *If $f$ is continuous and one of the derivates of $f$ is continuous at some point $x \subset I$, then so are the other derivates of $f$ that are defined at $x$.*

## §14.4 Characterizations of strong derivability.

In this section we obtain some characterizations of strong derivability in terms of local properties, unilateral derivates, and ordinary derivative. Some similar characterizations of strong derivability were obtained earlier in §8.3 in terms of limits of weak and lower derivatives.

First, in Theorem 14.4.1, we obtain a characterization of strong derivability in terms of continuity of ordinary derivative. Then, in Theorem 14.4.2, a characterization is obtained in terms of local monotonicity and Lipschitz property.

Theorems 14.4.4 and 14.4.5 deal, on the other hand, with characterizations of strong derivability in terms of unilateral derivates and limits of ordinary derivative. Finally, in Theorem 14.4.6, a characterization of strong differentiability is obtained in terms of uniform derivability.

We begin with the characterization of strong derivability in terms of continuity of ordinary derivative. The following theorem was obtained by Peano [P] in the case of continuous functions with the help of the mean value theorem.

14.4.1. THEOREM (PEANO). *If $f$ is strongly derivable on $I$, then it has a continuous derivative on $I$. Conversely, if $f$ is WC and either $I^*LSC$ or $I^*USC$, then if $f$ has a continuous derivative on $I$, then it is strongly derivable on $I$.*

PROOF. The first part follows clearly from Theorem 14.1.1. The second part is obtained, in the case when $f$ is $I^*LSC$, on applying Corollary 14.3.2 to $f$ and $f_-$, and in the other case on applying this result to $-f$. □

Next, in the following theorem we obtain a characterization of strong derivability in terms of local monotonicity and local Lipschitz property.

14.4.2. THEOREM. *Let $f: X \to \mathbb{R}$ and $x \in X^d$. Then*
(a) *$f$ has a finite strong derivative $c$ at $x$ iff for every $\varepsilon > 0$, $f_{-c}$ is $\varepsilon$-Lipschitz in some neighborhood of $x$; and*
(b) *$f$ has a strong derivative $+\infty$ (or $-\infty$) at $x$ iff for each positive integer $k$, $f_{-k}$ is increasing (or $f_{+k}$ is decreasing) in some neighborhood of $x$.*

PROOF. First, to prove the necessity part of (a), suppose $f$ has a finite strong derivative $c$ at $x$, and let $\varepsilon > 0$. Then $f_{-c+\varepsilon}$ and $f_{-c-\varepsilon}$ have strong derivatives $> 0$ and $< 0$, respectively, at $x$, and so it follows from Theorem 14.2.1(a) that there is a neighborhood $U$ of $x$ such that $f_{-c+\varepsilon}$ is increasing and $f_{-c-\varepsilon}$ is decreasing on $U \cap X$. Consequently, $f_{-c}$ is $\varepsilon$-Lipschitz on $U \cap X$.

Now, to prove the converse, suppose the condition in (a) holds. Then, given $\varepsilon > 0$, there is a neighborhood $U$ of $x$ such that $f_{-c}$ is $\varepsilon$-Lipschitz on $U \cap X$. Hence, if $y, z \in U \cap X$, $y \neq z$, it is clear that

$$|Qf(y,z) - c| = |Qf_{-c}(y,z)| \leq \varepsilon.$$

This proves that $f$ has a strong derivative $c$ at $x$.

Next, to prove (b), it is enough to prove the result for $+\infty$, for the result on $-\infty$ follows on applying this result to $-f$.

First, suppose $f$ has a strong derivative $+\infty$ at $x$, and let $k$ be any positive integer. Then $f_{-k}$ also has a strong derivative $+\infty$ at $x$, and so it is, by Theorem 14.2.1(a), increasing in some neighborhood of $x$. Next, to prove the converse, suppose the condition holds. Then for each positive integer $k$, $f_{-k}$ is increasing in some neighborhood of $x$, so that $\underline{D}_* f_{-k}(x) \geq 0$. Thus $\underline{D}_* f(x) \geq k$ for each positive integer $k$, which proves that $\underline{D}_* f(x) = +\infty$. □

Following is an easy consequence of the above theorem on the behavior of various derivates and derivatives in the neighborhood of a point of strong derivability.

14.4.3. Corollary. *Suppose* $f: X \to \mathbb{R}$ *is strongly derivable at* $x \in X^d$. *If* $x$ *is a limit point of the set of points where a given unilateral, bilateral, symmetric, approximate, or strong derivate (or derivative) of* $f$, *say* $Df$, *is defined, then* $f'_*(x)$ *is the limit of* $Df$ *at* $x$; *and if* $Df$ *is also defined at* $x$, *then it is continuous at* $x$.

In the following theorem we obtain a characterization of strong derivability in terms of a unilateral derivate with the help of the Second and Third Monotonicity Theorems of §9.1 and the Fourth Median Theorem of §9.6.

14.4.4. Theorem. *Let* $x \in \bar{I}$. *Then*

(a) $f$ *has a finite strong derivative* $c$ *at* $x$ *iff it is WC and* $I^*LSC$ *in some neighborhood of* $x$ *and* $D_+f$ *has limit* $c$ *at* $x$, *and*

(b) $f$ *has a strong derivative* $+\infty$ *at* $x$ *iff it is* $I^*LC$ *in some neighborhood of* $x$ *and* $D_+f$ *has limit* $+\infty$ *at* $x$.

Proof. In each of the two cases the necessity of the continuity condition is obvious (see Theorem 14.4.2(a)), and the rest of the necessity part follows from the last corollary. Hence we need to prove here only the sufficiency parts.

First, to prove (a), suppose its conditions hold. Then, given $\varepsilon > 0$, there is a neighborhood $U$ of $x$, open relative to $I$, such that $f$ is $WC$ and $I^*LSC$ on $U$ and, for each $y \in U \sim \{x\}$, $|D_+f(y) - c| < \varepsilon$. Hence, for such a point $y$, $D_+f_{-c+\varepsilon}(y) > 0$ and $D_+f_{-c-\varepsilon}(y) < 0$. Now, since $f_{-c+\varepsilon}$ and $f_{-c-\varepsilon}$ are also $WC$ and $I^*LSC$ on $U$, it follows from the Fourth Median Theorem 9.6.1(a) that $f_{-c+\varepsilon}$ is increasing on $U$, and it follows from the Second Monotonicity Theorem 9.1.3(a) that $f_{-c-\varepsilon}$ is nonincreasing on $U$. Consequently, $f_{-c}$ is $\varepsilon$-Lipschitz on $U$, and so the result follows from Theorem 14.4.2(a).

The sufficiency part of (b) is proved by a similar argument with the help of the Third Monotonicity Theorem 9.1.7 and Theorem 14.4.2(b). □

Next, in the following theorem we obtain a characterization of strong derivative in terms of the ordinary derivative. Similar characterizations were obtained earlier in §8.3 in terms of weak and lower derivatives; see also Esser and Shisha [ES] for the part (a) of the following theorem.

14.4.5. Theorem. *Let* $x \in \bar{I}$. *Then*

(a) $f$ *has a finite strong derivative* $c$ *at* $x$ *iff it is* AC *in some neighborhood of* $x$ *and* $c$ *is the limit of* $f'(t)$ *as* $t$ *approaches* $x$ *through* $\Delta^*(f)$; *and*

(b) $f$ *has strong derivative* $+\infty$ *at* $x$ *iff it is* LAC *in some neighborhood of* $x$ *and* $+\infty$ *is the limit of* $f'(t)$ *as* $t$ *approaches* $x$ *through* $\Delta^*(f)$.

Proof. In each of the two cases we need to prove here only the sufficiency part, for the necessity part follows from Theorem 14.4.2 and its Corollary 14.4.3.

Hence, to prove (a), suppose its conditions hold, and let $\varepsilon > 0$. Then there is a neighborhood $U$ of $x$, open relative to $I$, such that $f$ is AC on $U$, and for each $y \in U \cap \Delta^*(f)$, $c - \varepsilon < f'(y) < c + \varepsilon$. Thus $f_{-c+\varepsilon}$ and $f_{-c-\varepsilon}$ are lower and upper singular, respectively, on $U$, and hence they are, by Proposition 2.24 of [G11], nondecreasing and nonincreasing, respectively, on $U$. Thus $f_{-c}$ is $\varepsilon$-Lipschitz on $U$, and so it follows from Theorem 14.4.2(a) that $f$ has strong derivative $c$ at $x$.

The sufficiency part of (b) is proved similarly with the help of the part (b) of Theorem 14.4.2. □

Finally, in the following theorem we obtain a characterization of strong differentiability on an interval in terms of uniform derivability considered by Ostrowski [O]. He called a differentiable function $f: I \to \mathbb{R}$ *uniformly derivable* (on $I$) if for every $\varepsilon > 0$ there is $\delta > 0$ such that $|Qf(x, y) - f'(x)| < \varepsilon$ whenever $x, y \in I$ and $0 < |x - y| < \delta$.

14.4.6. THEOREM. *Suppose $I = [a, b]$. Then $f$ is strongly differentiable on $I$ iff it is uniformly derivable on $I$.*

PROOF. First, suppose $f$ is strongly differentiable on $I$. Then $f$ is clearly differentiable on $I$, and by Theorem 14.1.1, $f' = f'_*$ is continuous on $I$. Hence $f'$ is uniformly continuous on $I$, and so, given $\varepsilon > 0$, there is a $\delta > 0$ such that $|f'(y) - f'(x)| < \varepsilon$ whenever $|y - x| < \delta$.

Now, suppose $x, y \in I$ and $0 < |y - x| < \delta$. Then according to the mean value theorem there is a point $z$ between $x$ and $y$ such that $Qf(x, y) = f'(z)$. Thus $|Qf(x, y) - f'(x)| = |f'(z) - f'(x)| < \varepsilon$, which proves that $f$ is uniformly derivable.

Next, to prove the converse, suppose $f$ is uniformly derivable. Then $f$ is differentiable on $I$, and given $\varepsilon > 0$, there is a $\delta > 0$ such that $|Qf(x, y) - f'(x)| < \varepsilon$ whenever $x, y \in I$ and $0 < |y - x| < \delta$. Hence for such $x, y$ we have

$$|f'(y) - f'(x)| \leq |f'(y) - Qf(x, y)| + |Qf(x, y) - f'(x)| < 2\varepsilon.$$

This proves that $f'$ is (uniformly) continuous on $I$, and so it follows from Theorem 14.4.1 that $f$ is strongly derivable on $I$. But since $f$ is differentiable on $I$, this implies that it is strongly differentiable on $I$. □

**§14.5 Mean value property and the Darboux and Denjoy properties of strong derivative.** In this section we obtain the mean value property and the Darboux and Denjoy properties of strong derivative. These properties are found to hold without any continuity hypothesis, whereas in the case of ordinary derivative they hold only under some continuity hypothesis (see Remark 14.5.2).

14.5.1. THEOREM. *Suppose $f$ is strongly derivable everywhere. Then the mean value theorem holds for $f$ in terms of $f'_*$, and $f'_*$ possesses the Darboux and Denjoy properties.*

PROOF. The Darboux and Denjoy properties of $f'_*$ follow directly from its continuity established earlier in Theorem 14.1.1.

To prove the mean value property of $f'_*$, suppose $J \equiv [a, b] \subset I$. We may assume here without loss of generality that $f(a) = f(b)$; for if $c \equiv Qf(a, b) \neq 0$, then the result follows on applying the result in this particular case to $f_{-c}$.

Let $E_+$, $E_-$, and $E_0$ denote the sets of points of $J^0$ where $f'_*$ is positive, negative, or zero, respectively. We need to show here that $E_0 \neq \varnothing$. When both $E_+$ and $E_-$ are nonempty, the result follows again from the continuity of $f'_*$. Hence, suppose that one of them, say $E_-$, is empty and that $E_0$ also is empty. Then $J^0 \subset E_+$, and so it follows easily from Theorem 14.2.6(a) that $f$ is increasing on $J^0$. Then $f'_* \geq 0$ on $J^0$, and so it follows from the continuity of $f'_*$ that $f'_* \geq 0$ on $J$. Hence $f$ is clearly increasing on $J$, which is a contradiction.

A similar argument holds when $E_+$ is empty. □

14.5.2. REMARK. It is somewhat surprising that the above theorem holds without any continuity hypothesis. When $f$ is derivable in the ordinary sense, the mean value, Darboux, and Denjoy properties of $f'$ were found in [G11, p. 325] to hold when $f$ is *WC*. To see the necessity of weak continuity in the case of ordinary derivative, let $f(x) = -1, 0, 1$ according to whether $x < 0, = 0$, or $> 0$, respectively. Then it is easy to see that $f$ is derivable everywhere, but $f'$ does not possess the Darboux or mean value property; this function is, of course, not strongly derivable at $x = 0$.

# PART V

# UNIFIED AXIOMATIC THEORIES OF GENERALIZED DERIVATIVES

**Introduction to Part V.** In this part we present unified axiomatic theories of three kinds of generalized derivatives. Most of the results of these three theories are deduced from the results of the theory of new derivatives obtained in Parts II and III, and a few of them are obtained by methods used in those parts. It is somewhat surprising that the various known generalized derivatives (except for ordinary derivative) are not comparable with new derivatives, but still their results can be deduced from the results of new derivatives due to the compatibility of the two derivatives.

In a recent edition of A. M. Bruchner's well-known monograph on differentiation of real functions, he posed the following two problems on generalized derivatives in his final chapter [Br2, p. 158]:

> What is that causes some generalized derivatives to enjoy so many desirable properties, while others possess so few? When a new generalized derivative is invented, can one determine easily which, if any, of the desirable properties it possesses?

The three axiomatic theories of generalized derivatives presented here address both of these problems. The results of each of these three theories are obtained in terms of some admissibility and regularity hypotheses on the corresponding generalized derivates, and the status of admissibility and regularity is determined here in the case of several known generalized derivates and generalized symmetric derivates.

Further, while considering any new generalized derivative, it will be possible to determine the nature of its admissibility easily. It will also be possible to determine the nature of its regularity quickly using the methods presented here. We illustrate this phenomenon here in the case of several new generalized derivatives obtained on combining (the notions of) different pairs of known generalized derivatives. Similarly, we consider several new generalized symmetric derivatives to illustrate the same phenomenon for them.

First, in Chapter 15, we present two axiomatic models of generalized limits and generalized derivates. The unified theories of generalized derivatives are developed in terms of these models. Here we use the term "generalized limits" for generalized lower and upper unilateral limits, and we use the term "generalized derivates" for generalized unilateral derivates. We define here in terms of several axioms the notions of admissible, $s$-admissible, and $s^*$-admissible generalized limits. Similarly, we define in terms of several axioms the notions of admissible, $s$-admissible, and $s^*$-admissible generalized quotients and generalized derivates.

Further, among the admissible generalized derivates, we define here in terms of the lower and upper Baire classes $LB_1$ and $UB_1$ the notions of regular and totally regular generalized derivates, and we define similarly the notions of subregular and totally subregular generalized derivates. Also, when any generalized derivates are admissible or regular in any of the above senses, the generalized derivative induced by them is also said to be so. We investigate here which of these admissibilities and regularities hold for various known generalized derivates. All the known generalized derivates considered here are found to be admissible, and they are either regular or subregular (see the Status Theorems 15.3.3, 15.4.2, and 15.4.4 for their actual status).

Next, in Chapter 16, we present in terms of the above axiomatic model a unified theory of generalized derivatives, except for generalized symmetric derivatives. The various results on generalized derivatives are obtained here in terms of various admissibility and regularity hypotheses considered in Chapter 15, which in turn explain why some of the results do not hold in some particular cases. All admissible generalized derivatives are found to be compatible with new derivatives, and it is due to this compatibility that the various results of this chapter on generalized derivatives are obtained from the results of the theory of new derivatives.

To provide some insight into the nature of results obtained in the three unified theories of this part, we describe here briefly the main results obtained in Chapter 16. Let $G$-derivates denote any admissible generalized derivates, and suppose for the sake of simplicity that $f: I \to \mathbb{R}$ is continuous.

As for the Calculus of generalized derivatives, (i) the additivity and linearity are found to hold for all $s$-admissible generalized derivatives; and (ii) the product and quotient rules hold for $s^*$-admissible generalized derivatives. (iii) The mean value theorem is found to hold for $f$ in terms of $G$-derivative provided that $f$ is $G$-derivable at the points where it is semi-differentiable; and (iv) the Cauchy's mean value theorem and l'Hôpital's rule hold in terms of any $s$-admissible generalized derivative.

Further, we deduce from the Fourth Monotonicity Theorem (v) a similar monotonicity theorem in terms of $G$-derivates; and from the other monotonicity theorems of Part III we deduce (vi) similar monotonicity theorems for $f$ in terms of $G$-derivative provided that $f$ is $G$-derivable at nearly all the points where it is lower derivable. (vii) The Goldowski-Tonelli theorem is found on the other hand to hold for every subregular generalized derivative. As for the relations between $G$-derivates and strong derivates, (viii) the Dini's theorem (see Theorem 14.3.1) is found to hold for (any admissible) $G$-derivates; and (ix) Denjoy's Theorem 2.5.4 holds for $G$-derivates provided that they are subregular.

Further, (x) when $G$-derivates are regular, the two $G$-derivability sets are found to be $F_{\sigma\delta}$-sets; and (xi) every subregular generalized derivative is found to be of $B_1$. (xii) The $G$-derivative of $f$ possesses the Darboux property provided that $f$ is $G$-derivable at the points where it is semidifferentiable; and (xiii) when $G$-derivates are subregular and $f$ is nonangular and $G$-derivable n.e., the $G$-derivative of $f$ is found to possess the Denjoy property.

As the ordinary derivates are found to be $s^*$-admissible and regular, all the above results hold in particular for ordinary derivates and ordinary derivative, which in turn strengthen some of the known results on them. Further, in the case of various known generalized derivatives, several new results follow from the results of this chapter (as mentioned above), whereas the known results are strengthened.

Next, in Chapter 17, we present a similar unified axiomatic theory of generalized symmetric derivatives. We define here similar notions of admissibility and regularity for generalized symmetric derivates. Then in terms of these notions we deduce from the results of Parts II and III results similar to above for generalized symmetric derivatives, but for one major difference. Namely, the results on these derivatives hold only for functions that are nonangular in the corresponding generalized sense. This restriction is, however, not entirely unexpected, because in the case of ordinary symmetric derivative the mean value theorem is found to hold only for nonangular functions. On the other hand, as we see later in Chapter 19, a typical continuous function is nonangular relative to all the known generalized derivates considered here. The results of this chapter yield several new results on ordinary symmetric derivates and derivative, and strengthen the known results. Further, in the case of various known generalized symmetric derivates and derivatives, several new results follow from the results of this chapter whereas the known results are strengthened.*

A natural question arises here: What happens if we combine a notion of new derivative with that of a generalized derivative. In Chapter 18 we present a unified axiomatic theory of "generalized new derivatives" that are obtained

*The axiomatic theories of Chapters 15–17 were presented in brief at a symposium at Chattanooga (Tennessee). A summary of this talk appeared in the proceedings of this symposium in [G17]. Due to the complex nature of these axiomatic theories, this summary may be found to provide a good introduction and motivation to the developments of these axiomatic theories.

on combining the notions of new derivatives with the ones of various known generalized derivatives. This theory is based on the axiomatic model of Chapter 15, and its results are to some extent more similar to the results of Parts II and III from which these results are deduced. In particular, from the First and Third Derivability Theorems of Part II we obtain here similar derivability theorems in terms of generalized new derivatives.

# *CHAPTER 15*

# An Axiomatic Model of Generalized Limits and Derivates

In this chapter we introduce the axiomatic models of generalized limits and generalized derivates in terms of which the various unified theories of generalized derivatives are developed in the later chapters.

First, in §15.1, we define in terms of some axioms the notion of admissible generalized limits—that is, generalized lower and upper unilateral limits. Also, we define here two stronger versions of admissible generalized limits which we call $s$-admissible and $s^*$-admissible generalized limits. In §15.2, we define similarly admissible, $s$-admissible, and $s^*$-admissible generalized quotients and generalized derivates—that is, generalized unilateral derivates. In each of these sections we include some preliminary results dealing with these notions which are used throughout the following chapters.

Next, in §15.3, we present several known examples of admissible generalized limits and derivates, and we investigate which of them are $s$- or $s^*$-admissible.

Further, among the admissible generalized derivates, we define in §15.4 the notions of regular and totally regular generalized derivates, and similarly we define the notions of subregular and totally subregular generalized derivates. We investigate here which of these regularities holds for various known generalized derivates. Finally, to deal with a symmetry of generalized limits, we define in §15.5 the notion of $s_1$-admissible generalized limits in terms of another axiom.

**§15.1 Admissible and $s$- and $s^*$-admissible generalized limits.** In this section we present an axiomatic model of generalized limits—that is, generalized lower and upper unilateral limits. We define in fact their admissibility in terms of necessary axioms, and two stronger versions of their admissibility are considered which are called $s$-admissible and $s^*$-admissible.

Another axiom will be added later to this model in §15.5 to define $s_1$-admissible generalized limits.

Examples of admissible generalized limits are considered later in §15.3. Most of the known generalized derivates are obtained in that section in terms of these generalized limits. Also, this model of generalized limits is used again in Chapters 17 and 20 to define various generalized symmetric derivatives and generalized smooth functions.

Given $x \in \mathbb{R}$, we will use throughout this chapter $\mathscr{F}_x$ to denote the set of all real-valued functions defined in some neighborhood of $x$, except possibly at $x$. Also, we will assume $I$ to be an "open" subinterval of $\mathbb{R}$, and we will use $\mathscr{F}(I)$ to denote the space of all real-valued functions on $I$.

Let $G$-limits denote any notion of lower and upper generalized unilateral limits. Corresponding to these limits, we will assume throughout that the lower and upper bilateral $G$-limits of a function $f \in \mathscr{F}_x$, $x \in \mathbb{R}$, are defined at $x$ in the usual manner as follows:

(A0)
$$G\text{-}\liminf_{h \to 0} f(x+h) = \wedge G\text{-}\liminf_{h \to 0\pm} f(x+h),$$
$$G\text{-}\limsup_{h \to 0} f(x+h) = \vee G\text{-}\limsup_{h \to 0\pm} f(x+h).$$

We will call the $G$-limits *admissible* if the folowing three axioms hold for every $f \in \mathscr{F}_x$, $x \in \mathbb{R}$, and $c \in \mathbb{R}$:

(A1)
$$G\text{-}\limsup_{h \to 0\pm} f(x+h) = -G\text{-}\liminf_{h \to 0\pm} (-f(x+h));$$

(A2)
$$\liminf_{h \to 0\pm} f(x+h) \le G\text{-}\liminf_{h \to 0\pm} f(x+h) \le G\text{-}\limsup_{h \to 0\pm} f(x+h);$$

(A3)
$$G\text{-}\liminf_{h \to 0\pm} \{f(x+h) + c\} = G\text{-}\liminf_{h \to 0\pm} f(x+h) + c.$$

Now, when the lower and upper $G$-limits of $f$ at $x$ coincide, their common value will be called, as usual, a *unique $G$-limit of $f$ at $x$* and will be denoted by $G\text{-}\lim_{h \to 0} f(x+h)$. The right and left unique $G$-limits of $f$ at $x$ are defined similarly.

When the $G$-limits are admissible, it is clear from the axioms (A1) and (A3) that a relation similar to (A3) holds also for the left and right upper $G$-limits. Further, with the help of the axiom (A1), we easily obtain the following from the first inequality of (A2).

15.1.1. Lemma. *If $G$-limits are admissible, then for each $f \in \mathscr{F}_x$, $x \in \mathbb{R}$, we obtain*

$$G\text{-}\limsup_{h \to 0\pm} f(x+h) \le \limsup_{h \to 0\pm} f(x+h).$$

Now with the help of this lemma we obtain from the axiom (A2) the following relation between ordinary and $G$-limits, which explains the reason behind the use of the term "admissible" in the context of generalization of ordinary limit.

15.1.2. THEOREM. *Suppose $G$-limits are admissible, $f \in \mathscr{F}_x$, $x \in \mathbb{R}$. Then if $f$ has an ordinary right, left, or bilateral limit at $x$, then it also has a right, left, or bilateral $G$-limit, respectively, at $x$, and the two limits are equal.*

PROOF. The result on the right and left limits follows directly from the axiom (A2) with the help of the above lemma, and then the result on bilateral limit follows with the help of (A0). □

In the next two subsections we introduce some further axioms to define $s$-admissible and $s^*$-admissible generalized limits. We include in these subsections a series of lemmas on these two stronger versions of admissibility which will be needed in the following sections.

**A. $s$-admissible generalized limits.** Given any admissible generalized $G$-limits, we will call them *$s$-admissible*, or *strongly admissible*, if the following two additional axioms hold for every $f, g \in \mathscr{F}_x$, $x \in \mathbb{R}$, and $c \in \mathbb{R}$:

(A4)
$$G\text{-}\liminf_{h\to 0\pm} f(x+h) + G\text{-}\liminf_{h\to 0\pm} g(x+h) \le G\text{-}\liminf_{h\to 0\pm}\{f(x+h)+g(x+h)\}$$

provided that the sum on the left side is defined; and

(A5)
$$G\text{-}\liminf_{h\to 0\pm}\{cf(x+h)\} = c\,G\text{-}\liminf_{h\to 0\pm} f(x+h) \quad \text{if} \quad c>0.$$

We begin with some consequences of $s$-admissibility. Each relation in the following lemmas will be assumed to hold only when the sum involved in it is defined.

15.1.3. LEMMA. *Suppose $G$-limits are $s$-admissible and $f, g \in \mathscr{F}_x$, $x \in \mathbb{R}$. Then*

$$\begin{aligned} G\text{-}\liminf_{h\to 0\pm}&\{f(x+h)+g(x+h)\} \\ &\le G\text{-}\liminf_{h\to 0\pm} f(x+h) + G\text{-}\limsup_{h\to 0\pm} g(x+h) \\ &\le G\text{-}\limsup_{h\to 0\pm}\{f(x+h)+g(x+h)\} \\ &\le G\text{-}\limsup_{h\to 0\pm} f(x+h) + G\text{-}\limsup_{h\to 0\pm} g(x+h). \end{aligned}$$

PROOF. We give the proof here for the right limits; in the case of left limits the result is obtained by a similar argument. Thus to prove the first inequality we need to show that

$$(1)\qquad G\text{-}\liminf_{h\to 0+}\{f(x+h)+g(x+h)\} \le G\text{-}\liminf_{h\to 0+} f(x+h) + G\text{-}\limsup_{h\to 0+} g(x+h).$$

It is clear from (A4) that

$$G\text{-}\liminf_{h\to 0+}\{f(x+h)+g(x+h)\} + G\text{-}\liminf_{h\to 0+}\{-g(x+h)\} \le G\text{-}\liminf_{h\to 0+} f(x+h).$$

Hence with the help of (A1) we obtain

$$(2)\qquad G\text{-}\liminf_{h\to 0+}\{f(x+h)+g(x+h)\} - G\text{-}\limsup_{h\to 0+} g(x+h) \le G\text{-}\liminf_{h\to 0+} f(x+h).$$

Now suppose the right side of (1) is defined. Further, suppose that the limit on the left side of (1) is $> -\infty$ and that the two limits on the right side of (1) are $< +\infty$; for (1) holds trivially otherwise. Then the left side of (2) is defined. When $c \equiv G\text{-}\limsup_{h\to 0+} g(x+h)$ is finite, the inequality (1) follows directly from (2). Further, when $c = -\infty$, it follows from (2) that $G\text{-}\liminf_{h\to 0+} f(x+h) = +\infty$, which is a contradiction.

The second inequality of the lemma is obtained similarly from the first with the help of (A1), and the third inequality is obtained from (A4) with the help of (A1). □

15.1.4. LEMMA. *Suppose G-limits are s-admissible and $f, g \in \mathcal{F}_x$, $x \in \mathbb{R}$. Then if $f$ has a right or left G-limit at $x$, then*

$$G\text{-}\liminf_{h\to 0\pm}\{f(x+h)+g(x+h)\} = G\text{-}\lim_{h\to 0\pm} f(x+h) + G\text{-}\liminf_{h\to 0\pm} g(x+h),$$

$$G\text{-}\limsup_{h\to 0\pm}\{f(x+h)+g(x+h)\} = G\text{-}\lim_{h\to 0\pm} f(x+h) + G\text{-}\limsup_{h\to 0\pm} g(x+h).$$

PROOF. The first relation follows from (A4) and the first inequality of Lemma 15.1.3 on inverting the roles of $f$ and $g$, and the second relation follows directly from the other two inequalities of that lemma. □

15.1.5. LEMMA. *Suppose G-limits are s-admissible,* $f \in \mathscr{F}_x$, $x \in \mathbb{R}$, *and* $c \in \mathbb{R}$. *Then if* $c > 0$, *then* $G\text{-}\limsup_{h \to 0\pm}\{cf(x+h)\} = c\, G\text{-}\limsup_{h \to 0\pm} f(x+h)$, *and if* $c < 0$, *then* $G\text{-}\liminf_{h \to 0\pm}\{cf(x+h)\} = c\, G\text{-}\limsup_{h \to 0\pm} f(x+h)$ *and* $G\text{-}\limsup_{h \to 0\pm}\{cf(x+h)\} = c\, G\text{-}\liminf_{h \to 0\pm} f(x+h)$.

*Further, each of these relations including the one in the axiom* (A5) *holds also when* $c = 0$.

PROOF. First, suppose $c > 0$. Then from axioms (A1) and (A5) we obtain

$$G\text{-}\limsup_{h\to 0\pm}(cf)(x+h) = -G\text{-}\liminf_{h\to 0\pm}\{c(-f)(x+h)\}$$

$$= -c\, G\text{-}\liminf_{h\to 0\pm}(-f)(x+h) = c\, G\text{-}\limsup_{h\to 0\pm} f(x+h).$$

Further, when $c < 0$, the two desired equations follow similarly from (A1), (A5), and the above equation.

Next, when $c = 0$, the function $cf$ is identically zero. Hence in this case it follows clearly from (A2) and Lemma 15.1.1 that $cf$ has a $G$-limit equal to zero at $x$. The product on the right side of each of the equations in question is, on the other hand, zero in this case by definition even when it is indeterminate. □

15.1.6. THEOREM. *Suppose G-limits are s-admissible and* $c \in \mathbb{R}$. *If* $f, g \in \mathscr{F}_x$, $x \in \mathbb{R}$, *have right or left G-limits at* $x$, *then so do* $f + g$ *and* $cf$, *and*

$$G\text{-}\lim_{h\to 0\pm}\{f(x+h) + g(x+h)\} = G\text{-}\lim_{h\to 0\pm} f(x+h) + G\text{-}\lim_{h\to 0\pm} g(x+h);$$

$$G\text{-}\lim_{h\to 0\pm}\{cf(x+h)\} = c\, G\text{-}\lim_{h\to 0\pm} f(x+h).$$

PROOF. The first part follows clearly on combining the two relations of Lemma 15.1.4, and the second part is obtained similarly from (A5) and Lemma 15.1.5. □

Let a function $f \in \mathscr{F}(I)$ be called *G-continuous* at $x \in I$ if it has a unique $G$-limit at $x$ equal to $f(x)$. Then from the last theorem we obtain immediately the following result on the linearity of $G$-continuity in the case when the $G$-limits are $s$-admissible.

15.1.7. THEOREM. *Suppose G-limits are s-admissible. Then if* $f, g \in \mathscr{F}(I)$ *are G-continuous at* $x \in I$, *then so are* $f + g$ *and* $cf$ $(c \in \mathbb{R})$.

**B. $s^*$-admissible generalized limits.** We next define $s^*$-admissible generalized limits which are stronger than $s$-admissible limits. Given any

$s$-admissible $G$-limits, we will call them *$s^*$-admissible* if the following two axioms hold for every $f \in \mathscr{F}(I)$ and $g \in \mathscr{F}_x$, $x \in I$:

(A6) If $f$ is $G$-continuous at $x$ and $f(x) > 0$, then

$$G\text{-}\liminf_{h \to 0\pm} \{f(x+h)g(x+h)\} = f(x) G\text{-}\liminf_{h \to 0\pm} g(x+h).$$

(A7) If $f$ is $G$-continuous at $x$, $f(x) = 0$ and $g$ has finite unilateral $G$-limits at $x$, then $fg$ has a unique $G$-limit at $x$ equal to zero.

It may be noted here that in the case of $s^*$-admissibility the axiom (A5) becomes redundant since it is implied by (A6).

We include here some consequences of $s^*$-admissibility. The following lemma follows easily from the axiom (A6) with the help of (A1).

15.1.8. Lemma. *Suppose $G$-limits are $s^*$-admissible, $f \in \mathscr{F}(I)$, $g \in \mathscr{F}_x$, $x \in I$, and that $f$ is $G$-continuous at $x$. Then if $f(x) > 0$, then*

$$G\text{-}\limsup_{h \to 0\pm} \{f(x+h)g(x+h)\} = f(x) G\text{-}\limsup_{h \to 0\pm} g(x+h);$$

*and if $f(x) < 0$, then*

$$G\text{-}\liminf_{h \to 0\pm} \{f(x+h)g(x+h)\} = f(x) G\text{-}\limsup_{h \to 0\pm} g(x+h)$$

*and*

$$G\text{-}\limsup_{h \to 0\pm} \{f(x+h)g(x+h)\} = f(x) G\text{-}\liminf_{h \to 0\pm} g(x+h).$$

Finally, on combining this lemma with the axioms (A6) and (A7) we obtain the following theorem.

15.1.9. Theorem. *Suppose $G$-limits are $s^*$-admissible, $f \in \mathscr{F}(I)$ and $g \in \mathscr{F}_x$, $x \in I$. Then if $f$ is $G$-continuous at $x$, $f(x) \neq 0$, and $g$ has a $G$-limit at $x$, then so does $fg$ and*

$$G\text{-}\lim_{h \to 0} \{f(x+h)g(x+h)\} = f(x) G\text{-}\lim_{h \to 0} g(x+h).$$

*Further, the same holds when $f(x) = 0$ provided that the limit on the right is finite.*

## §15.2 Admissible and $s$- and $s^*$-admissible generalized quotients and derivates.

In this section we present an axiomatic model of generalized (unilateral) derivates. Analogous to the axioms of the last section, we present here axioms in terms of generalized derivates to define admissible,

$s$-admissible, and $s^*$-admissible generalized derivates. Also, to deal with some known generalized derivates, we define here similar admissibilities of generalized quotients.

We will consider in the next section examples of generalized derivates with these admissibilities, to which the results of the remaining chapters will apply.

Given any generalized $G$-derivates, we will use $GD_+$, $GD_-$, $GD^+$, and $GD^-$ to denote these derivates. Corresponding to these derivates, we will assume throughout that the *lower* and *upper bilateral G-derivates* of a function $f \in \mathscr{F}(I)$ are also defined as usual at any point of $I$ as follows:

(D0) $$G\underline{D}f = \wedge GD_{\pm}f, \qquad G\overline{D}f = \vee GD^{\pm}f.$$

Further, we will call the $G$-derivates *admissible* if the following three axioms hold for each $f \in \mathscr{F}(I)$ at each point of $I$:

(D1) $$GD^+f = -GD_+(-f), \qquad GD^-f = -GD_-(-f);$$

(D2) $$D_+f \leq GD_+f \leq GD^+f, \quad D_-f \leq GD_-f \leq GD^-f;$$

and for each $c \in \mathbb{R}$ and $x \in I$ we obtain

(D3) $$GD_+(f_{+c})(x) = GD_+f(x) + c, \qquad GD_-(f_{+c})(x) = GD_-f(x) + c.$$

Now, when the two bilateral $G$-derivates of $f$ in (D0) coincide at some point $x$, $f$ will be called as usual *G-derivable* at $x$, and the common value of the two derivates will be called the *G-derivative* of $f$ at $x$ and denoted by $Gf'(x)$. The right and left $G$-derivatives of $f$ at $x$ are defined similarly, and they will be denoted by $Gf'_+(x)$ and $Gf'_-(x)$, respectively. Also, when $Gf'(x)$ is finite, $f$ will be called as before *G-differentiable* at $x$.

When the $G$-derivates are admissible, it is clear from the axiom (D1) that the two relations of (D3) hold also for the upper $G$-derivates. Further, with the help of (D1) we obtain the following lemma directly from the axiom (D2).

15.2.1. Lemma. *If G-derivates are admissible, then for each $f \in \mathscr{F}(I)$, the following holds at each point of $I$:*

$$GD^+f \leq D^+f, \qquad GD^-f \leq D^-f.$$

Now with the help of this lemma we obtain from (D2) the following relation between ordinary and $G$-derivatives.

15.2.2. Theorem. *Suppose G-derivates are admissible, $f \in \mathscr{F}(I)$ and $x \in I$. Then if $f$ is derivable at $x$, then it is also G-derivable at $x$ and the two derivatives are equal. The same holds for the left and right derivabilities of $f$ at $x$.*

Next, in analogy with $s$-admissible generalized limits, we define $s$-admissible generalized derivates. Each of the following relations will again be assumed to hold only if the sum involved in it is defined.

Given any admissible generalized $G$-derivates, we will call them *s-admissible* if the following two axioms hold for every $f, g \in \mathcal{F}(I)$, $x \in I$, and $c \in \mathbb{R}$:

$$\text{(D4)} \qquad GD_{\pm}f(x) + GD_{\pm}g(x) \le GD_{\pm}(f+g)(x);$$

$$\text{(D5)} \qquad GD_{\pm}(cf)(x) = c\,GD_{\pm}f(x) \quad \text{if} \quad c > 0.$$

When the $G$-derivates are admissible or $s$-admissible, the $G$-derivative induced by them will also sometimes be called *admissible* or *s-admissible*, respectively. Further, in analogy with the ordinary median $Mf$, we will use $GMf(x)$ to denote the (*ordinary*) *G-median* of $f$ at $x$, namely, the interval $[G\underline{D}f(x), G\overline{D}f(x)]$. The other $G$-medians of $f$ can also be defined similarly.

15.2.3. Lemma. *Suppose G-derivates are s-admissible, $f, g \in \mathcal{F}(I)$, $x \in I$, and $c \in \mathbb{R}$. Then*

(a) $GD_{\pm}(f + g)(x) \le GD_{\pm}f(x) + GD^{\pm}g(x) \le GD^{\pm}(f + g)(x) \le GD^{\pm}f(x) + GD^{\pm}g(x)$.

(b) *If $c > 0$, then $GD^{\pm}(cf)(x) = c\,GD^{\pm}f(x)$, and if $c < 0$, then $GD_{\pm}(cf)(x) = c\,GD^{\pm}f(x)$ and $GD^{\pm}(cf)(x) = c\,GD_{\pm}f(x)$. Further, each of these equations, and the one in the axiom* (D5), *holds also when $c = 0$.*

Proof. We give the proof here for the right derivates; in the case of left derivates the result is obtained by similar arguments. Thus to prove the first inequality of (a) we need to show that

$$(1) \qquad GD_{+}(f+g)(x) \le GD_{+}f(x) + GD^{+}g(x).$$

In this direction we obtain clearly from the axiom (D4) the inequality

$$GD_{+}(f+g)(x) + GD_{+}(-g)(x) \le GD_{+}f(x),$$

and so with the help of (D1) we obtain

$$(2) \qquad GD_{+}(f+g)(x) - GD^{+}g(x) \le GD_{+}f(x).$$

Now suppose the right side of (1) is defined. Further, suppose $GD_{+}(f+g)(x) > -\infty$, $GD_{+}f(x) < +\infty$ and $GD^{+}g(x) < +\infty$; for (1) holds trivially otherwise. Then the left side of (2) is defined. When $GD^{+}g(x)$ is finite, the inequality (1) follows directly from (2). Further, when $GD^{+}g(x) = -\infty$, it follows from (2) that $GD_{+}f(x) = +\infty$, which is a contradiction.

The second inequality of (a) is obtained similarly from the first with the help of the axiom (D1), and the last inequality of (a) is obtained from (D4) with the help of (D1).

Next, to obtain (b), first suppose $c > 0$. Then from (D1) and (D5) we obtain

$$GD^+(cf)(x) = -GD_+(-cf)(x) = -c\,GD_+(-f)(x) = c\,GD^+f(x).$$

Further, when $c < 0$, the two desired equations follow similarly from the axioms (D1) and (D5) and the above equation. Finally, when $c = 0$, the function $cf$ is identically zero, and so it follows from Theorem 15.2.2 that $G(cf)'(x) = 0$. The product on the right side of each of the equations in question is, on the other hand, zero in this case by definition even when it is indeterminate. □

Now, when $f$ is $G$-differentiable at $x$, we obtain the following theorem needed later on combining part (a) of the above lemma with axiom (D4).

15.2.4. THEOREM. *Suppose $G$-derivates are $s$-admissible, $f$, $g \in \mathscr{F}(I)$, and let $h = f + g$. Then if $f$ is $G$-differentiable at $x \in I$, then*

$$GD_{\pm}h(x) = Gf'(x) + GD_{\pm}g(x) \quad \textit{and} \quad GD^{\pm}h(x) = Gf'(x) + GD^{\pm}g(x).$$

Since the axiom (D3) is not used in proving the above theorem, it is clear from this theorem that this axiom becomes redundant for the $s$-admissibility of $G$-derivates.

**A. Admissibility of generalized quotients.** To deal with some of the generalized derivates, we need to consider some generalizations of the ordinary quotient $Qf(x, x + h)$. By a *generalized $G$-quotient $GQ$* we will mean an operator which assigns a real number $GQf(x, x + h)$ to each $f \in \mathscr{F}(I)$, $x \in I$, and $h \neq 0$ provided that $x + h \in I$.

Given such a $G$-quotient, we will call it *admissible* if the following three axioms hold for each $f \in \mathscr{F}(I)$, $x \in I$, $h \neq 0$, and $c \in \mathbb{R}$ provided that $x + h \in I$:

(Q1) $$GQ(-f)(x, x + h) = -GQf(x, x + h);$$

(Q2) $$\liminf_{h \to 0\pm} Qf(x, x + h) \leq \liminf_{h \to 0\pm} GQf(x, x + h);$$

(Q3) $$GQ(f_{+c})(x, x + h) = GQf(x, x + h) + c.$$

Next, given any admissible $G$-quotient, we will call it *$s$-admissible* if the following two axioms hold for each $f, g \in \mathscr{F}(I)$, $x \in I$, $h \neq 0$, and $c \in \mathbb{R}$ provided that $x + h \in I$:

(Q4) $$GQ(f + g)(x, x + h) = GQf(x, x + h) + GQg(x, x + h);$$

(Q5) $$GQ(cf)(x, x + h) = c\,GQf(x, x + h).$$

Finally, given any $s$-admissible $G$-quotient, we will call it *$s^*$-admissible relative to* some generalized *$G$-limits* if to each $f \in \mathscr{F}(I)$, $x \in I$, and $h \neq 0$ there corresponds a function $f_G(x, h)$ such that the following three axioms hold for each $f, g \in \mathscr{F}(I)$, $x \in I$, and $h \neq 0$ provided that $x + h \in I$:

(Q6) if $f$ is $G$-continuous at $x$, then $G\text{-}\lim_{h\to 0} f_G(x,h) = f(x)$;

(Q7) $GQ(fg)(x, x+h) = f_G(x,h)GQg(x, x+h) + g(x)GQf(x, x+h)$;

(Q8) if $f(x) \neq 0$, then $GQ(1/f)(x, x+h)$

$$= -GQf(x, x+h)/\{f(x)f_G(x,h)\}.$$

Further, when the $G$-quotient is $s^*$-admissible relative to every admissible generalized limits, it will be called simply *$s^*$-admissible.*

**B. Induced generalized derivates.** We now define generalized derivates induced by generalized limits and generalized quotients. Given any generalized $G$-limits and a generalized $G$-quotient, we will call the following *$G$-derivates* the *derivates induced by the $G$-limits and $G$-quotient*; that is, for each $f \in \mathscr{F}(I)$ and $x \in I$,

$$GD_{\pm}f(x) = G\text{-}\liminf_{h\to 0\pm} GQf(x, x+h), \tag{3}$$

$$GD^{\pm}f(x) = G\text{-}\limsup_{h\to 0\pm} GQf(x, x+h).$$

We have used here the same prefix $G$- in $G$-limits and $G$-quotient only for notational simplicity, but they need not be related. The derivates induced by $G$-limits and $G_1$-quotient, obtained on replacing $GQ$ by $G_1Q$ in (3), will be called *$GG_1$-derivates.*

Further, the $G$-derivates induced by $G$-limits and the ordinary quotient $Qf$ will be called simply the *derivates induced by $G$-limits*; and similarly the $G$-derivates induced by the ordinary limits and $G$-quotient will be called the *derivates induced by $G$-quotient.* All the known generalized derivates are of one of these two kinds.

Next, we define any generalized $G$-derivates to be $s^*$-admissible if they are induced by some $G$-limits and $G$-quotient such that the $G$-limits are $s^*$-admissible and the $G$-quotient is $s^*$-admissible relative to $G$-limits.

Now we obtain the following induction theorem on the admissibility of induced derivates.

15.2.5. THEOREM (FIRST INDUCTION). *If G-limits and G-quotient are simultaneously admissible, or s- or s*-admissible, then so are the G-derivates induced by them.*

PROOF. First, suppose the $G$-limits and $G$-quotient are admissible. Then, for the $G$-derivates induced by them, it is clear from the definition (3) that (D0) follows from (A0). Further, axiom (D1) follows similarly from axioms (Q1) and (A1); (D2) follows from (Q2) and (A2); and (D3) is obtained in turn from (Q3) and (A3). Hence the $G$-derivates are admissible.

Next, suppose the $G$-limits and $G$-quotient are $s$-admissible. Then axiom (D4) follows clearly from axioms (Q4) and (A4), and (D5) is obtained similarly from (Q5) and (A5); which proves the $G$-derivates to be $s$-admissible.

As for $s^*$-admissibility, the result follows directly from the above definition of $s^*$-admissible $G$-derivates. □

**C. Comparison of generalized derivates.** We finally introduce here some terminology for the comparison of two different admissible generalized derivates which will be used frequently.

Given any two admissible generalized derivates $G$- and $G_1$-, we will call $G$-derivates *stronger than* $G_1$-derivates, or $G_1$-derivates *weaker than* $G$-derivates, if for each $f \in \mathscr{F}(I)$, $GD_+f \le G_1D_+f$ and $GD_-f \le G_1D_-f$ at every point of $I$. In such a case, as it follows from the axiom (D1), $GD^+f \ge G_1D^+f$ and $GD^-f \ge G_1D^-f$ everywhere.

In the case when $G$-derivates are stronger than $G_1$-derivates, it should be noted here that in terms of their induced derivatives the $G$-derivative is stronger than $G_1$-derivative; that is, when any function $f \in \mathscr{F}(I)$ is $G$-derivable at some point $x \in I$, then it is also $G_1$-derivable at $x$ and the $G$- and $G_1$-derivatives of $f$ at $x$ are equal.

Further, it should be noted here that all the admissible generalized derivates are due to the axiom (D2) weaker than ordinary derivates, and it should also be noted that the approximate derivates are weaker than ordinary derivates but stronger than preponderant derivates (see the next section for their definitions).

## §15.3 Examples of admissible generalized limits and derivates.

In this section we present several examples of admissible generalized limits and derivates, and we investigate their stronger forms of admissibilities.

We begin with examples of generalized limits. Some of these limits are known, whereas the others are introduced here since they are associated with corresponding notions of generalized derivates or derivatives in a natural manner. Also, these generalized limits will be used later in Chapter 17 to define corresponding notions of generalized symmetric derivates, and they will be used in Chapter 20 to define corresponding notions of generalized smooth functions.

Given $f \in \mathscr{F}_x$, $x \in \mathbb{R}$, in each of the following cases it is enough to define the lower unilateral limits of $f$ at $x$; for the definitions of upper unilateral limits will then follow from the axiom (A1). We present here the definitions of lower right limits; the definitions of lower left limits are similar as indicated at the end of the definitions. We also introduce here convenient notations for these limits which will be used throughout this part.

Hence let $f \in \mathscr{F}_x$, $x \in \mathbb{R}$. We will use here, for each $c \in \mathbb{R}$, $E_c$ to denote the set of values of $h$ for which $f(x + h) < c$, and we will assume that the sup of the empty set is $-\infty$.

(a) **Approximate limits:** The following approximate (lower right) limit of $f$ at $x$ corresponds to the approximate derivative introduced by Denjoy [D4] (see Saks [S8] for definition):

$$A\text{-}\liminf_{h \to 0+} f(x+h) = \sup\{c \in \mathbb{R} : x \text{ is a point of dispersion of } E_c \cap \mathbb{R}_+\}.$$

(b) **Preponderant limits:** The following preponderant limit corresponds in turn to the preponderant derivative introduced by Denjoy [D4]:

$$P\text{-}\liminf_{h \to 0+} f(x+h) = \sup\left\{c \in \mathbb{R} : \text{the upper right density of } E_c \text{ is } < \frac{1}{2} \text{ at } x\right\}.$$

(c) **Sequentially congruent limits:** Let $K$ denote any sequence of positive numbers $\{h_n\}$ which converges to zero. The following sequentially congruent limit, or $S_K$-*limit*, corresponds to Sindalovskii's sequentially congruent derivative [Sin1]:

$$S_K\text{-}\liminf_{h \to 0+} f(x+h) = \liminf_{n \to \infty} f(x+h_n).$$

(d) **Congruent limits:** Let $K$ be any subset of $\mathbb{R}$ such that $0 \in K^b$; we will refer to this set $K$ as a *congruence*. Generalizing the notion of sequentially congruent derivative, Sindalovskii [Sin1, Sin2] introduced further a more general notion of congruent derivative. The following congruent limit corresponds to that derivative:

$$K\text{-}\liminf_{h \to 0+} f(x+h) = \liminf_{h \in K;\, h \to 0+} f(x+h).$$

(e) **Borel limits:** When $f$ is $D_*$-integrable, the following Borel's mean limit corresponds to the mean derivative introduced by Borel [Bo2], which is also called the Borel derivative. The corresponding Borel derivates were introduced later by Sargent [Sa1]:

$$B\text{-}\liminf_{h\to 0+} f(x+h) = \liminf_{h\to 0+} \;\; \liminf_{\delta\to 0+} (1/h)\int_\delta^h f(x+t)\,dt.$$

(f) **Cesàro limits:** We define these limits corresponding to Cesàro derivates introduced by Burkill [Bur2] and Sargent [Sa2]. Suppose $f \in \mathscr{F}(I)$ is $D_*$-integrable. Given $x \in I$, we first associate with $f$ and $x$ the following *average function* $f_x^a$ of $f$ *relative to* $x$ defined as follows:

$$f_x^a(x \pm h) = \frac{1}{2h}\int_0^{2h} f(x \pm t)\,dt \quad \text{for} \quad h > 0, \quad f_x^a(x) = f(x).$$

Thus if $h > 0$, then $f_x^a(x+h)$ is the average of $f$ on $[x, x+2h]$, and $f_x^a(x-h)$ is the average of $f$ on $[x-2h, x]$.

Now, we define the various Cesàro limits, or *C-limits*, of $f$ at $x$ simply as the corresponding ordinary limits of $f_x^a$ at $x$. Thus, for example,

$$C\text{-}\liminf_{h\to 0+} f(x+h) = \liminf_{h\to 0+} f_x^a(x+h) = \liminf_{h\to 0+} \frac{1}{2h}\int_0^{2h} f(x+t)\,dt.$$

As for the lower left $C$-limit of $f$ at $x$, it is thus clear that

$$C\text{-}\liminf_{h\to 0-} f(x+h) = \liminf_{h\to 0+} f_x^a(x-h) = C\text{-}\liminf_{h\to 0+} f(x-h).$$

(g) **Qualitative limits:** The following qualitative limit corresponds to the qualitative derivates introduced by Marcus [M1]:

$$Q\text{-}\liminf_{h\to 0+} f(x+h) = \lim_{\delta\to 0+} \sup\{c \in \mathbb{R} : E_c \cap (x, x+\delta) \text{ is meager}\}.$$

(h) **Negligent limits:** Let $\mathscr{N}$ be any $\sigma$-ideal of subsets of $\mathbb{R}$—that is, a family of sets which includes subsets of its elements and is stable under countable union. Such a family of sets has been called by Császár [Cs] a family of *negligible sets*. We will assume here further that "$\mathscr{N}$ does not include any nonempty open set," for if it includes some open set $U$, then as recognized by Császár [Cs, p. 139], the second inequality of (A2) does not hold for the following $\mathscr{N}$-limits at the points of $U$. Császár defined the lower right *$\mathscr{N}$-limit* as follows (Császár called these limits as limits under negligence; the term "negligent limits" was introduced later by Thomson [T1, p. 21]):

$$\mathscr{N}\text{-}\liminf_{h\to 0+} f(x+h) = \lim_{\delta\to 0+} \sup\{c \in \mathbb{R} : E_c \cap (x, x+\delta) \in \mathscr{N}\}.$$

(i) **Limits relative to a set:** Given $E \subset \mathbb{R}$, the *limit relative* to $E$ is defined as follows:

$$E\text{-}\liminf_{h\to 0+} f(x+h) = \liminf_{\substack{x+h\in E\\ h\to 0+}} f(x+h).$$

(j) **Selective limits:** Let $s$ be a "selection" which associates with every closed interval $I$ a point $s_I \in I^0$. The following selective limit relative to $s$ corresponds to the selective derivative introduced by O'Malley [O-M]:

$$s\text{-}\liminf_{h\to 0+} f(x+h) = \liminf_{h\to 0+} f(s_{[x,x+h]}).$$

(k) **Path limits:** These limits correspond to the notion of path derivative introduced by Bruckner, O'Malley, and Thomson [BOT]. Given $x \in \mathbb{R}$, they called a set $P_x \subset \mathbb{R}$ a *path leading to* $x$ if $x \in (P_x)'$, and they called a collection $\mathcal{P} = \{P_x : x \in \mathbb{R}\}$ a *system of paths* if each $P_x$ is a path leading to $x$. We will assume here that $x \in (P_x)^b$ for each $x$; that is, $\mathcal{P}$ is a "bilateral" system of paths. The need of unilateral paths considered by them is avoided here by taking unilateral limits through these paths. Now, given such a system of paths $\mathcal{P}$, the *path limit* relative to $\mathcal{P}$ is defined as follows:

$$\mathcal{P}\text{-}\liminf_{h\to 0+} f(x+h) = \liminf_{\substack{x+h\in P_x\\ h\to 0+}} f(x+h).$$

Let us indicate here how the lower left limits are defined in the above cases. In case (a) it is defined simply by replacing $\mathbb{R}_+$ by $\mathbb{R}_-$; in case (b) by replacing the right density by left density; in case (c) by replacing $h_n$ by $-h_n$; in cases (d), (i), (j), and (k) by taking the lower limit as $h \to 0$ from the left; in case (e) by replacing $f(x+t)$ by $f(x-t)$; in case (f) as mentioned above by replacing $x+h$ by $x-h$; and in cases (g) and (h) by replacing the interval $(x, x+\delta)$ by $(x-\delta, x)$.

Next, we define generalized derivates corresponding to each of the above generalized limits. In each of these cases except for (f) and (j), the corresponding generalized derivates are induced by the generalized limits as defined in the last section; that is, they are obtained in terms of the generalized limits of the ordinary quotient $Qf(x, x+h)$. But in cases (f) and (j), the Cesàro and selective derivates are induced by generalized quotients as defined below.

(i) **Cesàro quotient:** Given a $D_*$-integrable function $f \in \mathcal{F}(I)$ and $x \in I$, we define the *Cesàro quotient* $CQf$ of $f$ on $[x, x+h]$ to be the ordinary quotient of $f_x^a$ on this interval. Thus

$$CQf(x, x+h) = Qf_x^a(x, x+h) = \frac{1}{h}\{f_x^a(x+h) - f_x^a(x)\}.$$

The $C$-derivates of $f$ at $x$ are thus simply the corresponding ordinary derivates of $f_x^a$ at $x$. It is now easy to see that the two lower $C$-derivates of $f$ at $x$ are indeed

$$CD_+f(x) = \liminf_{h\to 0+} \frac{1}{h}\left\{\frac{1}{2h}\int_0^{2h} f(x+t)\,dt - f(x)\right\},$$

$$CD_-f(x) = \liminf_{h\to 0+} \frac{1}{h}\left\{f(x) - \frac{1}{2h}\int_0^{2h} f(x-t)\,dt\right\}.$$

(ii) **Selective quotient:** Given any selection $s$, we define the *selective quotient* $sQf$ of $f$ relative to $s$ in terms of the ordinary quotient as follows:

$$sQf(x, x+h) = Qf(x, s_{[x,x+h]}), \qquad sQf(x-h, x) = Qf(s_{[x-h,x]}, x).$$

We now deal in the following theorem with the status of admissibility of the various quotients considered here.

15.3.1. THEOREM (FIRST STATUS). *The ordinary and selective quotients are $s^*$-admissible, and the Cesàro quotient is $s$-admissible.*

PROOF. In the cases of ordinary and selective quotients, it is easy to verify directly from their definitions that axioms (Q1) to (Q5) hold for them. Hence these two quotients are $s$-admissible.

Further, in the case of Cesàro quotient, all of these axioms except for (Q2) follow directly from its definition. We give the proof of (Q2) for $h \to 0+$; a similar argument holds for $h \to 0-$. Suppose $\liminf_{h\to 0+} Qf(x, x+h) > -\infty$, for there is nothing to prove otherwise. Then, given any real number $c < \liminf_{h\to 0+} Qf(x, x+h)$, there is a $\delta > 0$ such that for each $t \in (x, x+\delta)$, we obtain $Qf(x, x+t) > c$, or $f(x+t) > f(x) + ct$. Hence if $x < h < x + \delta$, then

$$f_x^a(x+h) = \frac{1}{2h}\int_0^{2h} f(x+t)\,dt \geq \frac{1}{2h}\int_0^{2h} \{f(x) + ct\}\,dt = f(x) + ch.$$

Thus $CQf(x, x+h) = Qf_x^a(x, x+h) = \frac{1}{h}\{f_x^a(x, x+h) - f_x^a(x)\} \geq c$. This clearly proves the desired inequality, and hence the $C$-quotient is $s$-admissible.

Next, to prove the $s^*$-admissibility of the ordinary quotient, let $G$-limits be any given admissible generalized limits, and let $f,\ g \in \mathscr{S}(I)$. Choose $f_G(x, h)$ to be $f(x+h)$. Then if $f$ is $G$-continuous at $x$, it is clear that $G\text{-}\lim_{h\to 0} f_G(x, h) = G\text{-}\lim_{h\to 0} f(x+h) = f(x)$. Further, it is easy to verify that

$$(1)\quad Q(fg)(x, x+h) = f(x+h)Qg(x, x+h) + g(x)Qf(x, x+h),$$

and that if $f(x) \neq 0$, then

$$(2)\qquad Q(1/f)(x, x+h) = -Qf(x, x+h)/\{f(x)f(x+h)\}.$$

It is thus clear that axioms (Q6) to (Q8) hold for the ordinary quotient, and hence it is $s^*$-admissible.

Next, to consider selective quotient, let $s$ be any selection. Given any admissible $G$-limits, choose this time $f_G(x, h)$ to be $f(s_{[x,x+h]})$. Then if $f$ is $G$-continuous at $x$, then since $s_{[x,x+h]} \to x$ as $h \to 0$, it is clear that

$G\text{-}\lim_{h \to 0} f_G(x, h) = G\text{-}\lim_{h \to 0} f(s_{[x, x+h]}) = f(x)$. Further, with the help of (1) we obtain

$$\begin{aligned} sQ(fg)(x, x+h) &= Q(fg)(x, s_{[x, x+h]}) \\ &= f(s_{[x, x+h]})Qg(x, s_{[x, x+h]}) + g(x)Qf(x, s_{[x, x+h]}) \\ &= f(s_{[x, x+h]})sQg(x, x+h) + g(x)sQf(x, x+h); \end{aligned}$$

and if $f(x) \neq 0$, we obtain similarly from (2) the following:

$$sQ(1/f)(x, x+h) = -sQf(x, x+h)/\{f(x)f(s_{[x, x+h]})\}.$$

The $s^*$-admissibility of the $s$-quotient is thus established. □

Now, with the help of the above status theorem, we obtain the following version of the First Induction Theorem for generalized derivatives that are induced either by some generalized limits or by some generalized quotient, as defined in §15.2B.

15.3.2. THEOREM (SECOND INDUCTION).
(a) *If some generalized G-limits are admissible, or s- or $s^*$-admissible, then so are the G-derivates induced by them.*
(b) *If some generalized G-quotient is admissible, or s- or $s^*$-admissible, then so are the G-derivatives induced by it.*

PROOF. Since the ordinary quotient is $s^*$-admissible by the above theorem, part (a) follows clearly from the First Induction Theorem 12.2.5.

Further, it is easy to see that the ordinary limits are $s^*$-admissible, and so part (b) also follows from the same induction theorem. □

To include the ordinary unilateral limits and derivates in the axiomatic model developed here, we will consider these limits and derivates as trivial (or extreme) cases of admissible generalized limits and admissible generalized derivates, respectively.

Finally, in the following theorem we obtain the status of admissibility of all the generalized limits and derivates considered above.

15.3.3. THEOREM (SECOND STATUS). *The following lower and upper unilateral limits and derivates are $s^*$-admissible*:
(a) *ordinary limits and derivates;*
(b) *approximate limits and derivates (Denjoy);*
(c) *sequentially congruent limits and derivates (Sindalovskii);*
(d) *congruent limits and derivates (Sindalovskii);*

(e) *qualitative limits and derivates (Marcus);*
(f) *negligent limits and derivates (Császár);*
(g) *limits and derivates relative to a set;*
(h) *selective limits and derivates (O'Malley); and*
(i) *path limits and derivates (Bruckner-O'Malley-Thomson).*

*Further, the following limits and derivates are s-admissible:*
(j) *Borel limits and derivates; and*
(k) *Cesàro limits and derivates.*

*Furthermore,* (l) *the preponderant limits and derivates are admissible.*

PROOF. We will first prove the results for the various limits. In the case of ordinary limits, axioms (A1) to (A7) are known to hold, and so they are $s^*$-admissible.

In each of the cases (b) to (j), it is easy to verify directly from the definitions of these limits, using usual arguments, that axioms (A1) to (A5) hold for these limits. Hence these generalized limits are $s$-admissible. Further, in case (k), the first inequality of (A2) is proved for $C$-limits by an argument similar to the one used for (Q2) in the proof of Theorem 15.3.1 in the case of $C$-quotient, and the other axioms are obtained directly from the definition of $C$-limits. Hence, $C$-limits are also $s$-admissible. Similarly, in case (l), it is easy to verify that the preponderant limits are admissible.

As for $s^*$-admissibility, it is easy to verify using usual arguments that axioms (A6) and (A7) hold in the cases (b), (c), (d), (g), (h), and (i). Hence these generalized limits are $s^*$-admissible. We will obtain here these axioms in case (f); the limits in (e) constitute on the other hand a particular case of (f).

Let $\mathcal{N}$ be any given family of negligible sets, $f \in \mathcal{F}(I)$ and $g \in \mathcal{F}_x$, $x \in I$. We will give the proof here for the right $\mathcal{N}$-limit; a similar argument holds for the left $\mathcal{N}$-limit. Let $\mathcal{N}_x$ denote the family of all sets $E \subset I$ for which there is a $\delta > 0$ such that $E \cap (x, x + \delta) \in \mathcal{N}$.

To obtain the axiom (A6) for $\mathcal{N}$-limits, suppose $f$ is $\mathcal{N}$-continuous at $x$, $f(x) > 0$, and set $a = f(x)$, $b = \mathcal{N}\text{-}\liminf_{h \to 0+} g(x + h)$, and $L = \mathcal{N}\text{-}\liminf_{h \to 0+}\{f(x + h)g(x + h)\}$. Then we need to show that $L = ab$. Define for each $c \in \mathbb{R}$,

$$A_c = \{y \in I : f(y)g(y) < c\}.$$

Now, since $f$ is $\mathcal{N}$-continuous at $x$ and $f(x) > 0$, for every $\varepsilon > 0$ there clearly exists a set $E_\varepsilon \subset I$ such that $I \sim E_\varepsilon \in \mathcal{N}_x$ and for each $y \in E_\varepsilon$, $f(y) > 0$ and $a - \varepsilon < f(y) < a + \varepsilon$.

First, let $c < ab$. Then we need to show that $A_c \in \mathcal{N}_x$. First, suppose $b > 0$. Then we can clearly choose an $\varepsilon > 0$ such that $\varepsilon < a \wedge b$ and $c < (a - \varepsilon)(b - \varepsilon)$. Hence if $y \in A_c \cap E_\varepsilon$, then

$$g(y) < c/f(y) < (a - \varepsilon)(b - \varepsilon)/f(y) < b - \varepsilon.$$

Consequently, $A_c \cap E_\varepsilon \subset \{y : g(y) < b - \varepsilon\} \in \mathscr{N}_x$, from which it is clear that $A_c \in \mathscr{N}_x$. Further, if $b \le 0$, then $c < 0$, and so we can choose an $\varepsilon > 0$ such that $c < (a + \varepsilon)(b - \varepsilon)$. If $y \in A_c \cap E_\varepsilon$, then since $b - \varepsilon < 0$, we obtain

$$g(y) < c/f(y) < (a + \varepsilon)(b - \varepsilon)/f(y) < b - \varepsilon.$$

Hence in this case it follows as before that $A_c \in \mathscr{N}_x$.

This proves that $L \ge ab$. To prove the reverse inequality, let $c > ab$. Then we need to show that $A_c \notin \mathscr{N}_x$. First, suppose $b \ge 0$. Then $c > 0$, and so we can choose an $\varepsilon > 0$ such that $c > (a + \varepsilon)(b + \varepsilon)$. Hence if $y \in E_\varepsilon$ and $g(y) < b + \varepsilon$, then $f(y)g(y) < (a + \varepsilon)(b + \varepsilon) < c$, so that $A_c$ includes the set $E_\varepsilon \cap \{y : g(y) < b + \varepsilon\}$, which is not in $\mathscr{N}_x$. Consequently, $A_c \notin \mathscr{N}_x$. Further, when $b < 0$, we can choose $\varepsilon > 0$ such that $\varepsilon < a \wedge (-b)$ and $c > (a - \varepsilon)(b + \varepsilon)$. If $y \in E_\varepsilon$ and $g(y) < b + \varepsilon$, then $f(y)g(y) < (a - \varepsilon)(b + \varepsilon) < c$, and so it follows as before that $A_c \notin \mathscr{N}_x$. Thus it has been proved that $L = ab$.

Next, to obtain (A7) in the present case, suppose $f$ is $\mathscr{N}$-continuous at $x$, $f(x) = 0$, and $g$ has finite unilateral $\mathscr{N}$-limits at $x$. Then there is an $M \in \mathbb{R}_+$ such that $\{y : |g(y)| > M\} \in \mathscr{N}_x$. Hence there is a set $A \subset I$ such that $I \sim A \in \mathscr{N}_x$ and $|g(y)| \le M$ on $A$. Further, given $\varepsilon > 0$, there is a set $B \subset I$ such that $I \sim B \in \mathscr{N}_x$ and $|f(y)| < \varepsilon/M$ for $y \in B$. It is thus clear that $|f(y)g(y)| < \varepsilon$ for $y \in A \cap B$, which proves that $fg$ has an $\mathscr{N}$-limit equal to zero at $x$.

The result has thus been established for all the generalized limits.

Next we consider various generalized derivates. In each of the above cases except for (h) and (k), the derivates are induced by the corresponding generalized limits, and so the result for these derivates follows from the above proved result on these limits with the help of part (a) of the Second Induction Theorem 15.3.2. Further, since the derivates in (h) and (k) are induced by selective and Cesàro quotients, respectively, the results for these derivates follow from the First Status Theorem 15.3.1 with the help of part (b) of Theorem 15.3.2. □

15.3.4. EXAMPLE. We include here an example to show that the preponderant limits are not $s$-admissible. Let $A$ and $B$ be two disjoint sets with right density $1/3$ at 0. Define $f$ and $g$ to be the characteristic functions of the complements of $A$ and $B$, respectively. Then it is clear that

$$P\text{-}\liminf_{x \to 0+} f(x) + P\text{-}\liminf_{x \to 0+} g(x) = 1 + 1 = 2.$$

However, since $f + g$ is 1 on $A \cup B$, whose right density is $2/3$ at 0, it is further clear that $P\text{-}\liminf_{x \to 0+} \{f(x) + g(x)\} = 1$. Hence (A4) does not hold for $P$-limits in general.

15.3.5. PROBLEM. Due to the integrals involved in the definitions of Borel and Cesàro limits and derivates, they do not seem to be $s^*$-admissible in general. It will be interesting to find suitable hypotheses under which these derivates are $s^*$-admissible.

15.3.6. REMARK. Given any admissible generalized unilateral $G$-limits, it is interesting to note here that the various weaker forms of continuity and internal properties introduced in §1.2 can be generalized in a natural manner by replacing the ordinary limits by $G$-limits.

Given $f \in \mathscr{F}(I)$ and $x \in I$, the function $f$ is defined to be (i) $G$-$LSC$ at $x$ if $f(x) \leq G\text{-}\liminf_{h \to 0} f(x+h)$, (ii) $G$-$LC$ at $x$ if $G\text{-}\limsup_{h \to 0-} f(x+h) \leq f(x) \leq G\text{-}\liminf_{h \to 0+} f(x+h)$, and (iii) $G$-$LI$ at $x$ if $G\text{-}\liminf_{h \to 0-} f(x+h) \leq f(x) \leq G\text{-}\limsup_{h \to 0+} f(x+h)$. The notions of $G$-$USC$, $G$-$UC$, and $G$-$UI$ are defined analogously.

It is reasonable to expect that in some of the results of this part the ordinary continuity can be weakened to various weaker $G$-continuities. However, we will not pursue these generalized continuities in the present work; they will be used subsequently in [G20].

**§15.4 Regular, subregular, and totally regular generalized derivates.** In this section we define the notions of regular and totally regular generalized derivates, and similarly the notions of subregular and totally subregular generalized derivates. Further, we investigate here which of these regularities holds for various known generalized derivates.

The subregularity of generalized derivates, which is weaker than regularity, is found to be adequate for the validity of many results, and so it will be used frequently in the following chapters.

From now on we will use $\mathscr{F}$ to denote the space of all real-valued functions defined on open subintervals of $\mathbb{R}$, and we will use $\mathscr{C}$ to denote the space of continuous functions in $\mathscr{F}$.

Now, given any admissible generalized $G$-derivates, we will call them *regular* if for every function $f \in \mathscr{C}$, its lower $G$-derivates $GD_+f$ and $GD_-f$ are of $LB_1$; and we will call them *totally regular* if the same holds for every $f \in \mathscr{F}$. Further, the derivative induced by any regular or totally regular $G$-derivates will also be called regular or totally regular, respectively.

In connection with the above definitions it should be noted here that the ordinary unilateral derivates are regular but not totally regular (see Theorem 3.2.2 and Example 3.2.8(b)). Further, as we will see in Theorem 15.4.2, the preponderant derivates are totally regular.

The following result on regular $G$-derivates follows directly from the axioms (D1) and (D0).

15.4.1. THEOREM. *If G-derivates are regular, then for every $f \in \mathscr{C}$, $G\underline{D}f$ is of $LB_1$, and the upper G-derivates $GD^+f$, $GD^-f$, and $G\overline{D}f$ are of $UB_1$; and if G-derivates are totally regular, the same holds for every $f \in \mathscr{F}$.*

Before coming to examples of regular generalized derivates, we introduce here some further nomenclature to include in the present model the approximate derivates which are more commonly used among the various generalized derivates.

Given any admissible generalized $G$-derivates, we will call them *subregular* if there exist some regular $G_1$-derivates which are weaker than the $G$-derivates. Also, when the $G_1$-derivates can be chosen to be totally regular, the $G$-derivates will be called *totally subregular*. It should be noted here that regular $G$-derivates are also subregular, and totally regular $G$-derivates are similarly also totally subregular.

We first establish in the following theorem the total regularity of preponderant derivates, and hence the total subregularity of ordinary and approximate derivates. The total regularity of preponderant derivates came indeed as a surprise; for these derivates are on the one hand the weakest in terms of admissibility as seen in the Second Status Theorem 15.3.3, but they turn out to be the strongest in terms of regularity. Also, we do not know if any other generalized derivates considered in Theorem 15.3.3 are totally regular.

15.4.2. THEOREM (THIRD STATUS). *The preponderant derivates are totally regular. Consequently, the ordinary and approximate derivates are totally subregular.*

PROOF. Since the ordinary and approximate derivates are stronger than $P$-derivates, it is enough to prove the first part here.

Since $P$-derivates are admissible by the Second Status Theorem 15.3.3, we need to show here that for every $f \in \mathscr{F}$, the lower unilateral $P$-derivates of $f$ are of $LB_1$. We assume here for the sake of simplicity that $I = \mathbb{R}$, which does not involve any loss of generality. Hence let $f : \mathbb{R} \to \mathbb{R}$. We will give the proof for $PD_+ f$, and the result on $PD_- f$ follows on applying this result to $-f_-$.

Given $c \in \mathbb{R}$, we need to show that the set

$$E = \{x \in \mathbb{R} : PD_+ f(x) > c\}$$

is an $F_\sigma$-set. Let $\{r_n\}$ be any increasing sequence of positive numbers which converges to $1/2$, and let $E_{n,k}$ denote, for each pair of positive integers $n, k$, the set of points $x$ in $\mathbb{R}$ for which the following holds whenever $0 < h < 1/k$:

$$|\{y \in (x, x+h) : Qf(x,y) < c + 1/n\}| \le r_n h.$$

Then it is easy to see that $E = \bigcup_n \bigcup_k E_{n,k}$. Hence it is enough to show that $E_{n,k}$ is closed for each $n$ and $k$.

Given $n$ and $k$, suppose $x$ is a limit point of $E_{n,k}$. First, suppose $x$ is a left limit point of the set. Then there is an increasing sequence of points $\{x_i\}$ in $E_{n,k}$ which converges to $x$. Now, given $0 < h < 1/k$, there is a positive integer $j$ such that $x - x_j < (1/k) - h$. Hence, for each $i > j$ we obtain

$$\begin{aligned}
&|\{y \in (x, x+h) : Qf(x,y) < c + 1/n\}| \\
&\quad \le |\{y \in (x_i, x+h) : Qf(x,y) < c + 1/n\}| \\
&\quad \le r_n |(x_i, x+h)| < r_n h + (x - x_i).
\end{aligned}$$

Since this holds for each $i > j$, it is thus clear that $x \in E_{n,k}$.

Next, suppose $x$ is a right limit point of $E_{n,k}$. Then there is a decreasing sequence of points $\{x_i\}$ in $E_{n,k}$ which converges to $x$. Given $0 < h < 1/k$, there is this time an integer $j$ such that $x_j < x + h$. Hence

$$\begin{aligned}|\{y \in (x, x+h) : Qf(x,y) < c + 1/n\}| &\le \sum_{i \ge j} |\{y \in (x_{i+1}, x_i) : Qf(x,y) < c + 1/n\}| \\ &\quad + |\{y \in (x_j, x+h) : Qf(x,y) < c + 1/n\}| \\ &\le \sum_{i \ge j} r_n |(x_{i+1}, x_i)| + r_n |(x_j, x+h)| = r_n h.\end{aligned}$$

Hence $x \in E_{n,k}$ again, which proves that $E_{n,k}$ is closed. □

In the next theorem we establish the regularity of other generalized derivates considered in §15.3. The following lemma is needed for this purpose.

15.4.3. Lemma. (a) *If $f$ is the pointwise limit of a nondecreasing sequence of functions $\{f_n\}$ on $I$ where $f_n$ is of $LB_1$ for each $n$, then $f$ also is of $LB_1$.*

(b) *If $f(x) = \liminf_{h \to 0+} g(x,h)$, $x \in I$, where $g(x,h)$ is continuous in $x$ for each $h$, then $f$ is of $LB_1$.*

Proof. Given $c \in \mathbb{R}$, let $E = \{x \in I : f(x) > c\}$. Then if the hypothesis of (a) holds, it is clear that

$$E = \bigcup_n \{x \in I : f_n(x) > c\}.$$

Now since each $f_n$ is of $LB_1$, it is clear that $E$ is an $F_\sigma$-set, and hence $f$ is of $LB_1$.

Next, to prove (b), define the following for each positive integer $n$:

$$f_n(x) = \inf\{g(x,h) : 0 < h < 1/n\}, \qquad x \in I.$$

Then it is clear that $\{f_n\}$ is a nondecreasing sequence of functions which converges pointwise to $f$. Further, since each $f_n$ is the infimum of a family of continuous functions of $x$, it is *USC*, and thus it is of $B_1$. Hence it follows from (a) that $f$ is of $LB_1$. □

Among the other generalized derivates, selective and path derivates are not regular in general, and so we need some restrictions on these derivates.

We will call a selection $s$ *continuous* if the function $s_{[x,y]}$ is continuous in $x$ and $y$. The derivates defined by a continuous selection $s$ will in turn be called *continuously selective derivates*.

Further, given a (bilateral) system of paths $\mathscr{P} = \{P_x : x \in \mathbb{R}\}$, we will call it *subcontinuous* if $P_x \cap [x, x+h]$ and $P_x \cap [x-h, x]$, $x \in \mathbb{R}$, $h \in \mathbb{R}_+$, have selections which are continuous in $x$ and $h$. The derivates defined by a subcontinuous system of paths will be called, on the other hand, *subcontinuous path derivates*.

15.4.4. THEOREM (FOURTH STATUS). *The following derivates are regular:*
(a) *ordinary derivates,*
(b) *sequentially congruent derivates,*
(c) *congruent derivates,*
(d) *Borel derivates,*
(e) *Cesàro derivates,*
(f) *qualitative derivates,*
(g) *negligent derivates,*
(h) *derivates relative to a set E at the points of E, and*
(i) *continuously selective derivates.*
*Further,* (j) *subcontinuous path derivates are subregular.*

PROOF. Since each of these derivates are admissible by the Second Status Theorem 15.3.3, we need to show here that in each of these cases, except for (j), the lower unilateral derivates of a function $f \in \mathscr{C}$ are of $LB_1$. We will give the proof, in each case, for the lower right derivate; the proof for the lower left derivate is similar. Further, there is no loss of generality in assuming $I = \mathbb{R}$. Hence, suppose $f: \mathbb{R} \to \mathbb{R}$ is continuous.

To avoid repetition, we will adopt the following approach in most of the cases to be considered. On account of part (b) of the above lemma, it is enough to define in each case a function $g(x, h)$, $x \in \mathbb{R}$, $h > 0$, such that the following two properties hold: (i) the given lower right derivate is the lower right limit of $g(x, h)$ as $h \to 0$ from the right, and (ii) $g(x, h)$ is continuous in $x$ for each $h$. We will give below, in each case, the definition of $g(x, h)$ and prove (ii); the proof of (i) will be omitted since it is straightforward.

**Re** (a). *Ordinary derivates.* In this case the result follows directly from Theorem 3.2.2 on choosing $E = X = \mathbb{R}$.

Next we consider case (c), for the derivates in (b) constitute a particular case of (c).

**Re** (c). *Congruent derivates.* Let $K$ be any congruence, i.e. a subset of $\mathbb{R}$ for which $0 \in K^b$. Define for each $n$,

$$f_n(x) = \inf\{Qf(x, x+h) : h \in K, 0 < h < 1/n\}, \qquad x \in I.$$

Then $KD_+f$ is clearly the pointwise limit of the nondecreasing sequence of functions $\{f_n\}$. Further, since $Qf(x, x+h)$ is continuous in $x$ for each $h$, each $f_n$ is $USC$ as seen earlier in the proof of part (b) of Lemma 15.4.3. Hence it follows from part (a) of that lemma that $KD_+f$ is of $LB_1$.

**Re** (d). *Borel derivates.* To deal with the Borel derivate $BD_+f$, define the following for each $x \in \mathbb{R}$ and $h > 0$:

$$g(x, h) = \liminf_{\delta \to 0+} \frac{1}{h} \int_\delta^h Qf(x, x+t)\, dt.$$

Then since $Qf(x, x+t)$ is continuous in $x$ and $t$, and it is $D_*$-integrable by hypothesis, it can be easily verified that $g(x, h)$ is continuous in $x$.

**Re** (e). *Cesàro derivates.* To deal with the Cesàro derivate $CD_+f$, define

$$g(x,h) = CQf(x, x+h) = \frac{1}{h}\left\{\frac{1}{2h}\int_0^{2h} f(x+t)\,dt - f(x)\right\}.$$

Once again, since $f$ is continuous, it is clear that $g$ is continuous in $x$.

Next we consider case (g); for the derivates in (f) constitute a particular case of (g).

**Re** (g). *Negligent derivates.* Let $\mathscr{N}$ be any given family of negligible sets which does not include any nonempty open set. Then since $f$ is continuous, it is easy to see that $\mathscr{N}$-$D_+f$ coincides with $D_+f$ at every point. Hence it follows from Theorem 3.2.2 that this derivate is of $LB_1$.

**Re** (h). *Derivates relative to a set.* Given a subset $E$ of $\mathbb{R}$, the result on the derivates relative to $E$, at the points of $E$, follows from Theorem 3.2.2 on choosing $X = E$.

**Re** (i). *Continuously selective derivates.* Given a continuous selection $s$, define $g(x, h) = sQf(x, x+h) = Qf(x, s_{[x, x+h]})$. Then since $f$ is continuous, it is clear that $g(x, h)$ is continuous in $x$.

**Re** (j). *Subcontinuous path derivates.* Given a subcontinuous system of paths $\mathscr{P} = \{P_x : x \in \mathbb{R}\}$, let $s$ be a continuous selection of $P_x \cap [x, x+h]$, $x \in \mathbb{R}$, $h \in \mathbb{R}_+$. Then for each $f \in \mathscr{F}(I)$, it is clear that $\mathscr{P}D_+f \leq sD_+f$ at each point of $I$. Now since the $s$-derivates are regular as seen above, this proves that the derivate $\mathscr{P}D_+f$ is subregular. The derivate $\mathscr{P}D_-f$ is proved similarly to be subregular. □

15.4.5. PROBLEM. As proved in Theorem 15.4.2, the approximate derivates are totally subregular. It will be interesting to investigate whether these derivates are totally regular, or even regular. The proof given in that theorem for preponderant derivates does not extend to approximate derivates. What is known in this direction is that the approximate derivative is always of $B_1$ (see Preiss [Pr1]), but this follows easily from Theorem 15.4.2 as seen in the next chapter (see Corollary 16.7.4).

## §15.5 $s_1$-admissible generalized limits.

In this section we introduce another axiom to define $s_1$-admissible generalized limits. This axiom will not be needed in the next chapter, but will play a critical role in Chapters 17 and 20 while dealing with generalized symmetric derivates and generalized smooth functions.

Given any $s$-admissible generalized $G$-limits, we will call them *$s_1$-admissible* if the following axiom holds for every $f \in \mathscr{F}_x$, $x \in \mathbb{R}$:

(A8) $$G\text{-}\liminf_{h \to 0\pm} f(x-h) = G\text{-}\liminf_{h \to 0\mp} f(x+h).$$

Further, when the $G$-limits are simultaneously $s_1$- and $s^*$-admissible, they will be called simply *$s_1^*$-admissible*.

The following lemma on the upper limit follows directly from the axioms (A1) and (A8).

15.5.1. LEMMA. *If G-limits are $s_1$-admissible, then for each $f \in \mathcal{F}_x$, $x \in \mathbb{R}$ we obtain*

$$G\text{-}\limsup_{h \to 0\pm} f(x-h) = G\text{-}\limsup_{h \to 0\mp} f(x+h).$$

Axiom (A8) holds clearly for the ordinary and sequentially congruent limits, but it does not hold in general for congruent, negligent, selective, and path limits. To obtain this property in these cases, we define (i) a congruence $K$ to be *symmetric* if it is symmetric about zero, (ii) a family $\mathcal{N}$ of negligent sets to be *symmetric* if for each set $E \in \mathcal{N}$ and $x \in \mathbb{R}$, $2x - E \in \mathcal{N}$, (iii) a selection $s$ to be *symmetric* if for each $x \in \mathbb{R}$ and $h > 0$,

$$s_{[x, x+h]} - x = x - s_{[x-h, x]},$$

and (iv) a system of paths $\mathcal{P} = \{P_x : x \in \mathbb{R}\}$ to be *symmetric* if for each $x \in \mathbb{R}$, $P_x$ is symmetric about $x$.

Now, if $K$, $\mathcal{N}$, $s$, or $\mathcal{P}$ is symmetric, we will call the $K$-, $\mathcal{N}$-, $s$- or $\mathcal{P}$-limits *symmetrically congruent*, *symmetrically negligent*, *symmetrically selective*, or *symmetric path limits*, respectively.

In the following theorem we now state the status of $s_1$-admissibility of various generalized limits.

15.5.2. THEOREM (FIFTH STATUS). *The following generalized limits are $s_1^*$-admissible*:

(a) *ordinary limits*,

(b) *approximate limits*,

(c) *sequentially congruent limits*,

(d) *symmetrically congruent limits*,

(e) *qualitative limits*,

(f) *symmetrically negligent limits*,

(g) *symmetrically selective limits, and*

(h) *symmetric path limits*.

*Further, the following limits are $s_1$-admissible*:

(i) *Borel limits, and*

(j) *Cesàro limits*.

PROOF. Each of these generalized limits was seen earlier in Second Status Theorem 15.3.3 to be $s$-admissible. The fact that they satisfy axiom (A8) can be verified directly from their definitions. Now, the $s_1^*$-admissibility of the limits in (a) to (h) also follows from Theorem 15.3.3. □

# CHAPTER 16

# A Unified Theory of Generalized Derivatives

In this chapter we present a unified theory of generalized derivatives in terms of the axiomatic model developed in the last chapter. The symmetric and generalized symmetric derivatives will be considered later in the next chapter since they involve a different quotient. As the various results on generalized derivatives are obtained here under various admissibility and regularity hypotheses, it thus becomes clear why some of the results do not hold in some particular cases.

First, in §16.1, we demonstrate the noncomparability of new derivatives with some of the more commonly used generalized derivatives. Then, in §16.2, we establish the compatibility of admissible generalized derivatives with new derivatives. It is due to this compatibility that most of the results of this chapter follow from the results of the theory of new derivatives obtained in Parts II and III.

Next, in §16.3, we deal with the Calculus of generalized derivatives. Also, in the case of Borel and Cesàro derivatives which are not $s^*$-admissible, we obtain here the product and quotient rules under a mild hypothesis on one of the two functions at the point in question. Next, in §16.4, we obtain the mean value theorem, Cauchy's mean value theorem, and l'Hôpital's rule in terms of generalized derivatives. In §16.5 we obtain some monotonicity theorems in terms of generalized derivates and derivatives, and the Goldowski-Tonelli theorem for such derivatives.

Next, in §16.6, we deal with the relations between generalized derivates and strong derivates, and in §16.7 we obtain the Darboux and Denjoy properties, and the Baire class, of generalized derivatives. Finally, to demonstrate the applicability of the theory presented in this chapter to new generalized derivates and derivatives, we present in §16.8 several new generalized derivates and obtain their admissibility and regularity.

In the case of various known generalized derivatives to which this unified theory applies, several new results follow from this theory whereas the existing results are strengthened.

We assume throughout this chapter, unless stated otherwise, that $f : I \to \mathbb{R}$ where $I$ is an "open" subinterval of $\mathbb{R}$. This assumption is made for the sake of simplicity of presentation; many of the results obtained here hold also when $I$ is closed (see Remark 16.7.6 for details).

**§16.1 Comparison of new derivatives with other generalized derivatives.** We begin in this section with a comparison of new derivatives (i.e., lower, upper, semi-, and weak derivatives) with various generalized derivatives considered in §15.3. First, we make some general comparisons in the case of continuous functions, and then we look into some specific questions that remain in the case of some of the more commonly used generalized derivatives, namely, unilateral, symmetric, and approximate derivatives (see §17.1 for the definition of symmetric derivative).

Given a continuous function $f : I \to \mathbb{R}$, according to the First Derivability Theorem 5.1.2, $f$ is lower and upper differentiable at two dense sets of points in $I$, whereas these two sets are $c$-dense in $I$ in case $f$ is nonangular. Further, according to the Third Derivability Theorem 6.2.2, $f$ is normally lower and upper derivable at two dense sets of points in almost all of its level sets.

This does not hold, however, for any generalized derivative considered in §15.3. To set the issue in perspective, we summarize in the following theorem some of the results obtained in this direction in Chapter 19 on typical continuous functions (see §1.1 for definition).

16.1.1. THEOREM. *A typical continuous function is not differentiable from any side, at any point, in terms of any of the following derivatives:*

(a) *ordinary derivative,*

(b) *approximate derivative,*

(c) *preponderant derivative,*

(d) *sequentially congruent derivative,*

(e) *congruent derivative,*

(f) *Borel derivative,*

(g) *Cesàro derivative,*

(h) *continuously selective derivative, and*

(i) *subcontinuous path derivative.*

*Further, in terms of the derivatives in* (a) *to* (e), *and* (j) *symmetric derivative, a typical continuous function is not derivable at any point.*

We refer to Corollary 19.2.2 for the proof of the first part, and we refer to Theorems 19.4.1 and 19.4.2 for the second part.

The above theorem holds also for qualitative and negligent derivatives; but we did not include them here since they coincide with ordinary derivative in the case of continuous functions. Also, this theorem will be found to hold further for various generalized symmetric derivatives defined later in §17.1 (see Corollary 19.3.3 and Theorem 19.4.1).

It is thus clear from the initial observation and the above theorem that the semi- and weak derivabilities of a continuous function at a point do not imply derivability in any of the senses considered in Theorem 16.1.1 at that point.

Before coming to other issues regarding this comparison, we first put together the following consequences of two generalizations of Denjoy's theorem obtained in §§16.6 and 17.6 regarding the equivalence of various derivatives at a residual set of points. Given any two notions of derivatives $G_1$- and $G_2$-, we will call them *equivalent for a function* $f : I \to \mathbb{R}$ *at a point* $x \in I$ if the following holds: $f$ is $G_1$-derivable at $x$ iff it is $G_2$-derivable there; and if so, then the two derivatives at $x$ are equal.

16.1.2. THEOREM. *Each of the following derivatives is equivalent to strong derivative for a continuous function* $f : I \to \mathbb{R}$ *at a residual set of points in* $I$:

(a) *weak derivative,*
(b) *lower (or upper, or semi-) derivative,*
(c) *ordinary derivative,*
(d) *unilateral derivative,*
(e) *symmetric derivative,*
(f) *approximate derivative,*
(g) *preponderant derivative,*
(h) *sequentially congruent derivative,*
(i) *congruent derivative,*
(j) *Borel derivative,*
(k) *Cesàro derivative,*
(l) *continuously selective derivative, and*
(m) *subcontinuous path derivative.*

We refer to Theorem 16.6.3 (along with Theorems 15.4.2 and 15.4.4) for the proof in all the cases except for (e), and we refer to Corollary 17.6.4 for (e).

Once again, the above theorem will be seen later in Theorem 17.6.3 to hold also for various generalized symmetric derivatives defined in §17.1 under certain hypotheses.

Thus to compare any generalized derivative with new derivatives, it remains to investigate the following two questions:

**Question 1:** Given any generalized $G$-derivative and a function $f : I \to \mathbb{R}$, does the $G$-derivability of $f$ at a point $x \in I$ imply the semi- or weak derivability of $f$ at $x$?

**Question 2:** Given any generalized $G$-derivative and a function $f: I \to \mathbb{R}$, how thick can the following two sets be in terms of measure: the set $A_{G,f}$ of points where $f$ is $G$-derivable but not semiderivable, and the set $B_{G,f}$ of points where $f$ is semiderivable but not $G$-derivable?

We will look into these two questions in the particular cases of unilateral, symmetric, and approximate derivatives which are more commonly used. As for the ordinary derivative, it is clear from the characteristic function of the set of rational numbers that there exist measurable functions which are normally semidifferentiable everywhere without being derivable at any point.

**A. Unilateral derivative.** The unilateral derivability does imply weak derivability since it is clear from the definition of the latter. But it does not imply semiderivability in general. Let $f(x) = x \sin(1/x)$ if $x > 0$, and $f(x) = 0$ otherwise. Then it is clear that $f$ is unilaterally derivable at 0 but not semiderivable there.

As for the converse, there exist continuous functions that are normally semidifferentiable almost everywhere but not unilaterally derivable anywhere. For example, the Besicovitch's unilaterally nonderivable function [Be] is such a function; for as shown in [G8, p. 137], its bilateral knot points form a set of measure zero, and so the claim follows from Theorem 5.5.2. Indeed, it is clear from the Fourth Derivability Theorem 6.3.5 that every continuous unilaterally nonderivable function is normally lower and upper differentiable at two metrically dense sets of points.

The points where a function is unilaterally derivable but not semiderivable form on the other hand a set of measure zero, which is clear from the following consequence of the Denjoy-Young-Saks Theorem 5.5.2.

16.1.3. THEOREM. *Every function is differentiable at almost all the points where it is unilaterally derivable.*

**B. Symmetric derivative.** First, to see that symmetric derivability does not imply weak derivability, let $f(x) = x \sin(1/x)$ if $x \neq 0$, and $f(0) = 0$. Then it is easy to see that $f$ is symmetrically derivable at 0, but not weakly derivable there.

To investigate the converse, let $f$ be the characteristic function of the set $Q$ of rational numbers. Then $f$ is clearly semiderivable everywhere. We claim that $f$ is not symmetrically derivable at any irrational point. Let $x \in \mathbb{R} \sim Q$ and $n$ be any positive integer. First, choose $h > 0$ such that $h < 1/(2n)$ and $x + h \in Q$. Then $x - h \notin Q$; otherwise $x = \{(x+h) + (x-h)\}/2 \in Q$, which is a contradiction. Hence $\{f(x+h) - f(x-h)\}/(2h) = 1/(2h) > n$. Similarly, on choosing $h$ such that $x - h \in Q$, it follows that $\{f(x+h) - f(x-h)\}/(2h) < -n$, which proves the claim.

The points where a measurable function is symmetrically derivable but not weakly derivable form on the other hand a set of measure zero which follows from the following theorem of Khintchine [Kh]:

16.1.4. THEOREM (KHINTCHINE). *Every measurable function is differentiable at almost all the points where it is symmetrically derivable.*

**C. Approximate derivative.** We begin here with the following theorem on the measure of the set of points where a function is weakly derivable but not approximately derivable.

16.1.5. THEOREM. *Every measurable function is approximately derivable at almost all the points where it is weakly derivable.*

PROOF. Suppose $f: I \to \mathbb{R}$ is measurable, and let $E$ be the set of points where $f$ is weakly derivable. Then it follows from Theorem 9.5.1 that $f$ is $GMT$ on $E$. Hence $f$ is $GBV$ on $E$, and so according to a theorem of Denjoy and Khintchine (see [S8, p. 222]), $f$ is approximately derivable a.e. in $E$. □

Next, we present some examples in connection with the above theorem.

16.1.6. EXAMPLE. (a) First, we show that the hypothesis of measurability is necessary in Theorem 16.1.5. Let $I = (0, 1)$, and let $E$ be a set of points in $I$ such that both $E$ and $I \sim E$ have outer density 1 at each point of $I$ (see, e.g., [S8, p. 187]). Then the characteristic function $f$ of $E$ on $I$ is clearly semiderivable everywhere, but not approximately derivable anywhere.

(b) We next show that the converse of Theorem 16.1.5 does not hold in general. Let $A$ and $B$ be two countable, mutually disjoint, dense sets in $\mathbb{R} \sim \{0\}$, and set $C = \mathbb{R} \sim (A \cup B)$. Now define $f(x) = x$, $-x$, or 0 according to whether $x \in A$, $B$, or $C$, respectively. Then $A \cup B$ is the set of points where $f$ is weakly (or semi-) derivable, but $C$ is the set of points where $f$ is approximately derivable. Curiously enough, for this measurable function, weak and approximate derivabilities occur at two complementary sets of points.

(c) We add here an example of any everywhere semiderivable, measurable function which is approximately derivable a.e. to show that it may still assume almost all of its values only at the points where it is not approximately derivable.

Let $g$ be a continuous, increasing, singular function which maps $I = [0, 1]$ onto itself (see [S8, p. 101] for an example). Then $E \equiv \{x : g'(x) = +\infty\}$ is a $c$-dense set of measure zero on which $g$ assumes almost all of its values (see, e.g., [G9, Theorem 1]). Now, define $f(x) = g(x)$ for $x \in E$ and $f(x) = 0$ for $x \in I \sim E$. Then $f$ is easily seen to have all the desired properties.

16.1.7. REMARK. We include here two observations on approximate derivative without details since they are not needed in the sequel.

(a) Firstly, in connection with Example 16.1.6(b), it may be pointed out that there also exist continuous functions which are approximately derivable a.e. but weakly derivable only at a set of points of measure zero.

(b) Secondly, it has been proved by Khintchine [Kh, p. 242] that any monotone function is derivable at every point where it is approximately derivable. This is not true for weak and semi- derivabilities.

For, it is easy to see that any nondecreasing function is semiderivable at every point where it is discontinuous, and, of course, it is not always derivable at such a point. Further, let $f$ be the Cantor's continuous, nondecreasing, singular function defined in terms of his nowhere dense perfect set $P$ in $[0,1]$. Then it is easy to see that $f$ is lower and upper derivable at two $c$-dense sets of points in $P$ without being derivable there.

It may be added here, on the other hand, that a monotone function is derivable at every point where it is normally semiderivable; for such a function is not knotted at any point.

16.1.8. PROBLEM. As seen above, the symmetric and approximate derivatives are not comparable with weak or semi- derivatives in general. The same seems to be true for all the other known generalized derivatives. However, we leave it as an open problem to look into the Questions 1 and 2 for the other generalized derivatives considered in Theorem 15.3.1.

## §16.2 Compatibility of admissible generalized derivatives with new derivatives.

As we saw in the last section, the new derivatives are not comparable in general with other generalized derivatives. However, as we see in the later sections of this chapter, most of the known results on other generalized derivatives can still be deduced, and sometimes improved, from the results on new derivatives obtained in Parts II and III. This is indeed due to the compatibility of admissible generalized derivatives with weak derivative as obtained here. This compatibility will be used repeatedly in the following sections.

We will call any generalized $G$-derivative *compatible with weak derivative* if for every function $f: I \to \mathbb{R}$, if $f$ is simultaneously $G$-derivable and weakly derivable at $x \in I$, then $Gf'(x) = Wf'(x)$.

16.2.1. LEMMA. *Suppose $G$-derivates are admissible, $f \in \mathscr{F}(I)$ and $x \in I$. Then if $f$ is $G$-derivable at $x$, then it is nonangular at $x$.*

*Moreover, if $f$ is weakly derivable at $x$, then $Wf'(x) \subset GMf(x)$.*

PROOF. The first part follows clearly from the axiom (D2) and Lemma 15.2.1. To prove the second part, suppose $f$ is weakly derivable at $x \in I$. Then it follows clearly from (D2) and (D3) that

$$G\underline{D}f(x) \le \wedge GD^{\pm}f(x) \le \wedge D^{\pm}f(x) \le \vee D_{\pm}f(x) \le \vee GD_{\pm}f(x) \le G\overline{D}f(x).$$

Hence it is clear that $Wf'(x) \subset GMf(x)$. □

Now the following theorem follows directly from the above lemma. We call this theorem the "First Compatibility Theorem" since another such theorem will be obtained in the next chapter. Both of these compatibility theorems play a basic role throughout this part.

16.2.2. THEOREM (FIRST COMPATIBILITY). *If $G$-derivates are admissible, then the $G$-derivative is compatible with weak derivative.*

**§16.3 Calculus of generalized derivatives.** In this section we deal with the Calculus of generalized derivatives under stronger versions of admissibility.

First, in Theorem 16.3.1, we deal with generalized continuity under generalized differentiability. Then, in Theorem 16.3.2, we obtain the additivity and linearity of $s$-admissible generalized derivatives.

Next, in the part A of this section we obtain the product and quotient rules for $s^*$-admissible generalized derivatives (see Theorems 16.3.3 and 16.3.6). Further, in the cases of Borel and Cesàro derivatives which are not $s^*$-admissible, we obtain these rules in part B under certain hypotheses (see Theorems 16.3.8 and 16.3.11).

We assume throughout this section that $f, g \in \mathscr{F}(I)$, $x \in I$, and $c \in \mathbb{R}$.

We begin with a generalized version of the well-known result regarding continuity under differentiability.

16.3.1. THEOREM. (a) *If G-derivates are induced by $s^*$-admissible G-limits and f is G-differentiable at x, then f is G-continuous at x.*

(b) *If f is C-differentiable at x, then it is C-continuous at x.*

(c) *If f is s-differentiable at x, where s is any selection, then f is s-continuous at x.*

PROOF. To prove (a), suppose $G$-limits are $s^*$-admissible and $f$ is $G$-differentiable at $x$. It is clear from the definition of $Qf(x, x+h)$ that

$$f(x+h) = f(x) + hQf(x,x+h). \tag{1}$$

Now since $Qf(x, x+h)$ has a unique finite $G$-limit as $h \to 0$, it is clear from the axiom (A7) that

$$G\text{-}\lim_{h\to 0} hQf(x, x+h) = 0.$$

Hence it follows from (1) with the help of axiom (A3) that $G\text{-}\lim_{h\to 0} f(x+h) = f(x)$; that is, $f$ is $G$-continuous at $x$.

Next, to prove (b), suppose $f$ is $C$-differentiable at $x$. According to the definition of the Cesàro quotient, we have

$$CQf(x, x+h) = \{f_x^a(x+h) - f_x^a(x)\}/h,$$

and so

$$f_x^a(x+h) = f(x) + hCQf(x, x+h).$$

Now since $CQf(x, x+h)$ has a finite limit as $h \to 0$, the second term on the right side converges to zero as $h \to 0$. Hence we obtain

$$C\text{-}\lim_{h\to 0} f(x+h) = \lim_{h\to 0} f_x^a(x+h) = f(x).$$

This proves that $f$ is $C$-continuous at $x$.

Finally, to prove (c), let $s$ be any selection and suppose $f$ is $s$-differentiable at $x$. It is clear from the definition of $sQf(x, x+h)$ that

$$sQf(x, x+h) = \{f(s_{[x,x+h]}) - f(x)\}\{s_{[x,x+h]} - x\},$$

and so

$$f(s_{[x,x+h]}) = f(x) + \{s_{[x,x+h]} - x\} sQf(x, x+h).$$

Now since $s_{[x,x+h]} \to x$ as $h \to 0$, the second term on the right converges to zero as $h \to 0$. Hence we obtain

$$s\text{-}\lim_{h\to 0} f(x+h) = \lim_{h\to 0} f(s_{[x,x+h]}) = f(x).$$

This in turn proves that $f$ is $s$-continuous at $x$. □

In the following theorem we deal with the additivity and linearity of $s$-admissible generalized derivatives.

16.3.2. THEOREM. *Suppose G-derivates are s-admissible. Then if f is G-derivable at x and g is G-differentiable there, then f + g and cf are G-derivable at x and*

$$G(f+g)'(x) = Gf'(x) + Gg'(x), \qquad G(cf)'(x) = c\,Gf'(x).$$

PROOF. The first relation follows directly from Theorem 15.2.4, and the second relation is obtained on combining Lemma 15.2.3(b) with axiom (D5). □

Next, we deal with the product and quotient rules for generalized derivatives. We need to consider here two cases separately.

**A. Product and quotient rules for $s^*$-admissible generalized derivatives.** We first obtain the product rule for such derivatives.

16.3.3. THEOREM. *Suppose G-derivates are $s^*$-admissible and f is G-continuous at x. Then if f and g are G-differentiable at x, then so is fg and*

$$G(fg)'(x) = f(x)Gg'(x) + g(x)Gf'(x). \tag{2}$$

*Moreover, when the G-derivates are induced by G-limits, the same holds without the continuity hypothesis.*

PROOF. Since the $G$-derivates are $s^*$-admissible, they are induced by some $G$-limits and $G$-quotient, where the $G$-limits are $s^*$-admissible and the $G$-quotient is $s^*$-admissible relative to $G$-limits. Now, since $f$ is $G$-continuous at $x$, there is a function $f_G(x, h)$, $h \neq 0$, such that $G\text{-}\lim_{h\to 0} f_G(x, h) = f(x)$ and the following holds for $h \neq 0$ provided that $x + h \in I$:

$$GQ(fg)(x, x+h) = f_G(x, h)GQg(x, x+h) + g(x)GQf(x, x+h).$$

Now since the $G$-limits are $s^*$-admissible, we obtain the following with the help of Theorems 15.1.6 and 15.1.9:

$$\begin{aligned} G(fg)'(x) &= G\text{-}\lim_{h\to 0}\{f_G(x, h)GQg(x, x+h)\} + g(x)G\text{-}\lim_{h\to 0} GQf(x, x+h) \\ &= f(x)Gg'(x) + g(x)Gf'(x). \end{aligned}$$

The second part follows, on the other hand, from the first due to Theorem 16.3.1. □

In the case of derivatives induced by generalized quotients, we obtain the following result from the above theorem.

16.3.4. COROLLARY. *Suppose G-derivates are induced by an $s^*$-admissible G-quotient and f is continuous at x. Then if f and g are G-differentiable at x, then so is fg and the relation* (2) *holds.*

To obtain the quotient rule we need the following lemma.

16.3.5. LEMMA. *Suppose G-derivates are $s^*$-admissible, f is G-continuous at x and $f(x) \neq 0$. Then if f is G-differentiable at x, then so is $1/f$ and*

$$G(1/f)'(x) = -\{1/f(x)\}^2 Gf'(x). \tag{3}$$

*Moreover, when the G-derivates are induced by G-limits, the same holds without the continuity hypothesis.*

PROOF. This result is obtained from axiom (Q8) and Theorem 15.1.9 by an argument similar to the one used for the last theorem. □

Now on combining the above lemma with Theorem 16.3.3 we obtain the following quotient rule for $s^*$-admissible generalized derivatives.

16.3.6. THEOREM. *Suppose G-derivates are $s^*$-admissible, g is G-continuous at x, and $g(x) \neq 0$. Then if f and g are G-differentiable at x, then so is $f/g$ and*

$$G(f/g)'(x) = \{1/g(x)\}^2\{g(x)Gf'(x) - f(x)Gg'(x)\}. \tag{4}$$

*Moreover, when the G-derivates are induced by G-limits, the same holds without the continuity hypothesis.*

**B. Product and quotient rules for Borel and Cesàro derivatives.** Since the $B$-limits and $C$-quotient are defined in terms of integrals, it is not possible to obtain their $s^*$-admissibility. Hence the Borel and Cesàro derivates are not $s^*$-admissible. Khintchine [Kh, p. 222] has already shown, by constructing an example, that the product rule does not hold for Borel derivative in general. The same seems to hold for the Cesàro derivative.

In the rest of this section we will obtain some hypotheses under which the product and quotient rules hold for Borel and Cesàro derivatives.

Given $f \in \mathcal{F}(I)$ and $x \in I$, we will call $f$ *B-finite at* $x$ if the following two limits are finite:

$$\limsup_{h \to 0\pm} \limsup_{\delta \to 0+} \frac{1}{h} \int_\delta^h |Qf(x, x \pm t)|\, dt. \tag{5}$$

Similarly, we will call $f$ *C-finite at* $x$ if the following two limits are finite:

$$\limsup_{h \to 0\pm} \frac{1}{2h^2} \int_0^{2h} |f(x \pm t) - f(x)|\, dt. \tag{6}$$

Due to the similar nature of the results obtained on Borel and Cesàro derivatives, we combine these results in the following theorems.

We begin with the following lemma on $B$- and $C$-finiteness. Generalizing the notion of $D$-finiteness, we will call a function $f \in \mathcal{F}(I)$ *BD-finite* or *CD-finite at* a point $x \in I$ if its Borel or Cesàro derivates, respectively, are finite at $x$.

16.3.7. Lemma. (a) *If $f$ is D-finite at $x$, then it is B- and C-finite at $x$.*

(b) *If $f$ is B- or C-finite at $x$, then it is BD- or CD-finite, respectively, at $x$.*

Proof. To prove (a), suppose $f$ is $D$-finite at $x$. Then there is an $M \in \mathbb{R}_+$ and $\eta > 0$ such that $|Qf(x, x+t)| < M$ for $0 < |t| < \eta$. Hence if $0 < \delta < h < \eta$, then

$$\limsup_{\delta \to 0+} \frac{1}{h} \int_\delta^h |Qf(x, x \pm t)|\, dt \le \limsup_{\delta \to 0+} \frac{M}{h}(h - \delta) = M.$$

Hence the two limits in (5) are bounded by $M$.

Further, if $0 < |t| < \eta$, then $|f(x+t) - f(x)| < M|t|$. Hence if $0 < t < \eta$, then

$$\frac{1}{2h^2} \int_0^{2h} |f(x \pm t) - f(x)|\, dt \le \frac{1}{2h^2} \int_0^{2h} Mt\, dt = M,$$

which proves the limits in (6) to be bounded by $M$.

Next, to prove (b), first suppose $f$ is $B$-finite at $x$. Then the two limits in (5) are bounded by some $M \in R_+$. Hence,

$$BD^+f(x) = \limsup_{h\to0+} \limsup_{\delta\to0+} \frac{1}{h}\int_\delta^h Qf(x, x+t)\,dt$$

$$\leq \limsup_{h\to0+} \limsup_{\delta\to0+} \frac{1}{h}\int_\delta^h |Qf(x, x+t)|\,dt \leq M,$$

$$BD_+f(x) = \liminf_{h\to0+} \liminf_{\delta\to0+} \frac{1}{h}\int_\delta^h Qf(x, x+t)\,dt$$

$$\geq -\limsup_{h\to0+} \limsup_{\delta\to0+} \frac{1}{h}\int_\delta^h |Qf(x, x+t)|\,dt \geq -M.$$

The result on left $B$-derivates is obtained by a similar argument.

The other part of (b) is also proved by a similar argument. □

Next, in the following theorem we obtain the product rule for Borel and Cesàro derivatives.

16.3.8. THEOREM. *Suppose $G = B$ or $C$, $f$ is $G$-finite at $x$ and $g$ is continuous at $x$. Then if $f$ and $g$ are $G$-differentiable at $x$, then so is $fg$ and the relation* (2) *holds.*

PROOF. We begin with the following relation which is easy to verify:

$$(7)\qquad \begin{aligned} &f(x+t)g(x+t) - f(x)g(x) \\ &\quad= f(x)\{g(x+t) - g(x)\} + g(x)\{f(x+t) - f(x)\} \\ &\qquad+ \{f(x+t) - f(x)\}\{g(x+t) - g(x)\}. \end{aligned}$$

First, suppose $f$ is $B$-finite at $x$ and $f$ and $g$ are $B$-differentiable at $x$. We present the proof of (2) here for the right $B$-derivative; a similar argument holds for the left $B$-derivative. It is clear from (7) that

$$\begin{aligned} Q(fg)(x, x+t) &= f(x)Qg(x, x+t) + g(x)Qf(x, x+t) \\ &\quad+ \{g(x+t) - g(x)\}Qf(x, x+t). \end{aligned}$$

Hence if $0 < \delta < h$, then

$$\frac{1}{h}\int_\delta^h Q(fg)(x, x+t)\,dt$$

$$= f(x)\frac{1}{h}\int_\delta^h Qg(x, x+t)\,dt + g(x)\frac{1}{h}\int_\delta^h Qf(x, x+t)\,dt + I,$$

where

$$I = \frac{1}{h}\int_{\delta}^{h}\{g(x+t) - g(x)\}Qf(x, x+t)\,dt.$$

Thus to obtain (2) for the right $B$-derivative, it is enough to show that $I$ converges to zero as $\delta \to 0$ and $h \to 0$.

Now, since $f$ is $B$-finite at $x$, the two limits in (5) are bounded by some $M \in \mathbb{R}_+$. Further, given $\varepsilon > 0$, since $g$ is continuous at $x$, there is an $\eta > 0$ such that $|g(x+t) - g(x)| < \varepsilon/M$ for $|t| < \eta$. Hence on choosing $h < \eta$ we obtain

$$\begin{aligned}
\limsup_{h\to 0+}\ \limsup_{\delta\to 0+}|I| &\leq \limsup_{h\to 0+}\ \limsup_{\delta\to 0+}\frac{1}{h}\int_{\delta}^{h}|g(x+t) - g(x)|\,|Qf(x, x+t)|\,dt \\
&\leq \frac{\varepsilon}{M}\limsup_{h\to 0+}\ \limsup_{\delta\to 0+}\frac{1}{h}\int_{\delta}^{h}|Qf(x, x+t)|\,dt \\
&\leq \frac{\varepsilon}{M}M = \varepsilon.
\end{aligned}$$

The result is thus proved for $B$-derivative.

Next, suppose $f$ is $C$-finite at $x$ and $f$ and $g$ are $C$-differentiable at $x$. We will prove here (2) again for the right $C$-derivative; a similar argument holds for the left $C$-derivative. It is clear from (7) that

$$\begin{aligned}
Q(fg)_x^a(x, x+h) &= \frac{1}{2h^2}\int_0^{2h}\{f(x+t)g(x+t) - f(x)g(x)\}\,dt \\
&= f(x)Qg_x^a(x, x+h) + g(x)Qf_x^a(x, x+h) + J,
\end{aligned}$$

where

$$J = \frac{1}{2h^2}\int_0^{2h}\{f(x+t) - f(x)\}\{g(x+t) - g(x)\}\,dt.$$

Hence, to obtain (2) for the right $C$-derivative, it is enough to show that $J$ converges to zero as $h \to 0$. This, however, is obtained by an argument similar to above. □

In the case when one of the functions is differentiable in the ordinary sense at $x$, we obtain the following from the above theorem with the help of Lemma 16.3.7(a).

16.3.9. Corollary. *Suppose $G = B$ or $C$, $f$ is differentiable at $x$ and $g$ is continuous at $x$. Then if $g$ is $G$-differentiable at $x$, then so is $fg$ and the relation (2) holds.*

Next, to obtain the quotient rule we need the following lemma.

16.3.10. LEMMA. *Suppose $G = B$ or $C$, $f$ is $G$-finite and continuous at $x$, and $f(x) \neq 0$. Then $1/f$ also is $G$-finite at $x$; and if $f$ is further $G$-differentiable at $x$, then so is $1/f$ and the relation* (3) *holds.*

PROOF. We first observe here that

$$(8)\qquad \frac{1}{f(x+t)} - \frac{1}{f(x)} = -\frac{f(x+t)-f(x)}{f(x+t)f(x)}$$

$$= -\frac{f(x+t)-f(x)}{\{f(x)\}^2} + \frac{\{f(x+t)-f(x)\}^2}{f(x+t)\{f(x)\}^2}.$$

First, suppose $f$ is $B$-finite at $x$. It is clear from (8) that

$$(9)\qquad Q(1/f)(x, x+t) = -\frac{Qf(x, x+t)}{\{f(x)\}^2} + T,$$

$$\text{where}\quad T = \frac{f(x+t)-f(x)}{f(x+t)\{f(x)\}^2} Qf(x, x+t).$$

We claim that

$$(10)\qquad \limsup_{h\to 0+} \limsup_{\delta\to 0+} \frac{1}{h}\int_\delta^h |T|\,dt = 0.$$

Since $f$ is $B$-finite at $x$, the limits in (5) are bounded by some $M \in \mathbb{R}_+$. Further, given $\varepsilon > 0$, since $f$ is continuous at $x$ and $|f(x)| > 0$, we can choose an $\eta > 0$ such that if $|t| < \eta$, then $|f(x+t)| > \frac{1}{2}|f(x)|$ and $|f(x+t) - f(x)| < \varepsilon|f(x)|^3/(2M)$. Hence

$$\limsup_{h\to 0+} \limsup_{\delta\to 0+} \frac{1}{h}\int_\delta^h |T|\,dt$$

$$= \limsup_{h\to 0+} \limsup_{\delta\to 0+} \frac{1}{h}\int_\delta^h \frac{|f(x+t)-f(x)|}{|f(x+t)|\,|f(x)|^2} |Qf(x, x+t)|\,dt$$

$$\le \frac{\varepsilon}{M} \limsup_{h\to 0+} \limsup_{\delta\to 0+} \frac{1}{h}\int_\delta^h |Qf(x, x+t)|\,dt \le \varepsilon.$$

This proves (10). It is now clear from (9) and (10) that for the function $1/f$ the right limit in (5) is bounded by $M$. The same is obtained for the left limit on replacing $x + t$ in the above argument by $x - t$. Hence $1/f$ is $B$-finite at $x$.

Now, suppose $f$ is further $B$-differentiable at $x$. Then it follows clearly from (9) and (10) that $1/f$ is right $B$-differentiable at $x$ and that the relation (3) holds for the right $B$-derivative at $x$. On replacing $x + t$ by $x - t$ in the above argument, the same is obtained for the left $B$-derivative of $1/f$ at $x$.

Next, suppose $f$ is $C$-finite at $x$. With the help of (8) we obtain in this case

$$(11)\quad (1/f)_x^a(x, x+h)$$
$$= \frac{1}{2h^2}\int_0^{2h}\left\{\frac{1}{f(x+t)} - \frac{1}{f(x)}\right\} dt$$
$$= -\frac{1}{\{f(x)\}^2} Qf_x^a(x, x+h) + \frac{1}{2h^2}\int_0^{2h}\frac{\{f(x+t)-f(x)\}^2}{f(x+t)\{f(x)\}^2}\, dt.$$

It is proved in this case by an argument similar to above that

$$(12)\qquad \limsup_{h\to 0+}\frac{1}{2h^2}\int_0^{2h}\frac{\{f(x+t)-f(x)\}^2}{|f(x+t)|\{f(x)\}^2}\, dt = 0.$$

Now, it follows from (11) and (12) that for $1/f$ the right limit in (6) is bounded by $M$, and the same is obtained for the left limit as before. Hence $1/f$ is $C$-finite at $x$.

Further, in case $f$ is further $C$-differentiable at $x$, the result on the $C$-derivative of $1/f$ is obtained from (11) and (12) by an argument similar to the one above. □

Now, on combining the above lemma with Theorem 16.3.8, we obtain the following quotient rule for Borel and Cesàro derivatives.

16.3.11. THEOREM. *Suppose $G = B$ or $C$, $f$ and $g$ are continuous at $x$, $g$ is $G$-finite at $x$, and $g(x) \neq 0$. Then if $f$ and $g$ are $G$-differentiable at $x$, then so is $f/g$ and the relation* (4) *holds.*

Finally, with the help of Lemma 16.3.7 we obtain the following from the above theorem.

16.3.12. COROLLARY. *Suppose $G = B$ or $C$, $f$ is continuous at $x$, $g$ is differentiable at $x$, and $g(x) \neq 0$. Then if $f$ is $G$-differentiable at $x$, then so is $f/g$ and the relation* (4) *holds.*

16.3.13. REMARK. In the case of Cesàro derivative it is interesting to note here that the product rule obtained in Corollary 16.3.9 was obtained by Burkill [Bur2] under the hypothesis that $f$ has a bounded derivative in some neighborhood of $x$ but without the hypothesis of continuity on $g$.

16.3.14. PROBLEM. We have not investigated here the chain rule for various generalized derivatives. It will be worthwhile to investigate hypotheses under which the chain rule holds for each admissible generalized derivative.

**§16.4 Mean value theorems and l'Hôpital's rule in terms of generalized derivatives.** In this section we obtain a mean value theorem and a version of Cauchy's mean value theorem in terms of generalized derivatives. Further, from the latter theorem we deduce l'Hôpital's rule in terms of such derivatives.

First, in Theorem 16.4.1, we obtain the mean value theorem for admissible generalized derivatives. Then, in Theorem 16.4.2, we obtain Cauchy's mean value theorem for $s$-admissible generalized derivatives. Finally, in Theorem 16.4.3, from the latter theorem we deduce l'Hôpital's rule in terms of such derivatives.

We begin with the mean value theorem for admissible generalized derivatives. This theorem is known for some of the known generalized derivatives, but only under the hypothesis of everywhere derivability in the considered sense. In the following theorem this derivability hypothesis is considerably reduced.

16.4.1. THEOREM (FIFTH MEAN VALUE). *Suppose $f: I \to \mathbb{R}$ is $I^*C$, where $I = [a, b]$. If G-derivates are admissible and $f$ is G-derivable at the points where it is semidifferentiable, then there is a point $x \in I^0$ such that $f$ is G-differentiable at $x$ and $Qf(a, b) = Gf'(x)$.*

PROOF. This result follows directly from the General Mean Value Theorem 8.1.2 with the help of Theorem 16.2.2. □

Next, we deduce from the General Mean Value Theorem the following version of Cauchy's mean value theorem in the case of $s$-admissible generalized derivatives.

16.4.2. THEOREM (SECOND CAUCHY'S MEAN VALUE). *Let $f, g: I \to \mathbb{R}$, where $I = [a, b]$. Suppose* (i) *G-derivates are s-admissible,* (ii) *$f$ is $I^*C$ and $g$ is continuous on $I$,* (iii) *$f$ is G-derivable and $g$ is G-differentiable on $I^0$, and* (iv) *$Gg'(x) \neq 0$ for $x \in I^0$. Then there is a point $x \in I^0$ such that*

$$\frac{f(b) - f(a)}{g(b) - g(a)} = \frac{Gf'(x)}{Gg'(x)}. \tag{1}$$

PROOF. It is clear from the above mean value theorem that $g(a) \neq g(b)$. Hence the ratio on the left side of (1) is defined. Let $c$ denote this ratio, and define

$$h(x) = f(x) - f(a) - c\{g(x) - g(a)\}, \qquad x \in I. \tag{2}$$

Then $h(a) = 0 = h(b)$; and since $h$ is $I^*C$, it follows from the General Mean Value Theorem 8.1.2 that there is a point $x \in I^0$ such that $h$ is semidifferentiable at $x$ and $0 \in Sh'(x)$. Further, due to (i) and (2), it follows from Theorems 16.2.2 and 16.3.2 that $Sh'(x) = Gh'(x) = Gf'(x) - cGg'(x)$. Hence $Gf'(x) = cGg'(x)$, which proves (1). □

Next, we deduce from the above Cauchy's mean value theorem the following extended version of l'Hôpital's rule in terms of generalized derivatives. We state the result as before in terms of the right limit.

16.4.3. THEOREM (SECOND L'HÔPITAL'S RULE). *Let $f, g : I \to \mathbb{R}$, where $I = (a, b)$, $-\infty \le a < b < \infty$. Suppose* (i) *G-derivates are s-admissible,* (ii) *$f$ is $I^*C$ and $g$ is continuous on $I$,* (iii) *$f$ is G-derivable and $g$ is G-differentiable on $I$, and* (iv) *$Gg'(x) \neq 0$ for $x \in I$. Then if the following limit on the left side is of indeterminate form $0/0$ or $\infty/\infty$, then*

$$\lim_{x \to a+} \frac{f(x)}{g(x)} = \lim_{x \to a+} \frac{Gf'(x)}{Gg'(x)},$$

*provided that the limit on the right exists (finite or infinite).*

PROOF. We omit the proof here since this result is obtained by an argument similar to the one used for Theorem 8.2.3, using this time the above Cauchy's Mean Value Theorem 16.4.2 in the place of Theorem 8.2.1, and on replacing $Sf'$ by $Gf'$, $g'$ by $Gg'$, $\varepsilon$ by $=$, and $\subset$ by $\in$. □

## §16.5 Monotonicity in terms of generalized derivates and derivatives; Goldowski-Tonelli theorem.

In this section we deduce from the monotonicity theorems of §9.1 similar monotonicity results in terms of admissible generalized derivates and derivatives. Also, we obtain here a general version of the Goldowski-Tonelli theorem for subregular generalized derivatives (see Theorem 16.5.7). Some of the results obtained here are known in some particular cases (see Remark 16.5.10).

Further, in terms of subregular generalized derivates, we obtain here weaker versions of the Third and Fourth Derivate Theorems.

We begin with a monotonicity theorem in terms of generalized derivates.

16.5.1. THEOREM. *Suppose $I$ is closed and $f$ is either* (i) *$I^*LC$, or* (ii) *$LI_+$ and $I^*LC_-$. If G-derivates are admissible and $GD^+f \ge 0$* a.e. *and* $> -\infty$ n.e., *then $f$ is nondecreasing.*

PROOF. This result follows directly from the Fourth Monotonicity Theorem 9.1.8 on taking into account Lemma 15.2.1. □

Next, due to the compatibility of admissible generalized derivatives with weak derivative as obtained in Theorem 16.2.2, with the help of the above theorem we obtain the following monotonicity result from the Second Monotonicity Theorem 9.1.3. It may be recalled here that in this chapter $I$ is assumed to be open.

16.5.2. THEOREM. *Suppose G-derivates are admissible, $f$ is $I^*LSC$ and $f$ is G-derivable at the points where it is lower differentiable. If $E$ is the set of points where $f$ has a finite negative (or positive) G-derivative, then $f$ is nondecreasing (or nonincreasing) whenever* card $(E) < c$.

Further, due to the above-mentioned compatibility we obtain the following result from Theorem 9.1.5, which in turn may be compared with Theorem 9.3.3 of Saks.

16.5.3. THEOREM. *Suppose G-derivates are admissible and $f$ is G-derivable at nearly all the points where it is lower derivable. Given $a, b \in I$, $a < b$, let $E_-$ and $E_+$ denote the sets of points in $(a, b)$ where $f$ has a negative or positive G-derivative, respectively. Then if $f$ is $R^*LI$, then $f(a) - f(b) \le |f(E_-)|$; and if $f$ is $L^*UI$, then $f(b) - f(a) \le |f(E_+)|$.*

Next, using the same compatibility we obtain the following monotonicity theorem from the Third Monotonicity Theorem 9.1.7 in terms of lower $G$-regularity and lower $G$-singularity of the function. It may be recalled here that we call a function $f$ *lower G-regular* if $|\{x : Gf'(x) = -\infty\}| = 0$, and *lower G-singular* if $Gf' \ge 0$ at almost all the points where it is $G$-derivable.

16.5.4. THEOREM. *Suppose G-derivates are admissible and $f$ is either* (i) *$I^*LC$, or* (ii) *$LI_-$ and $I^*LC_+$. Then if $f$ is G-derivable at nearly all the points where it is lower derivable, and it is lower G-regular and lower G-singular, then $f$ is nondecreasing.*

Further, from the Fifth Monotonicity Theorem 9.1.10 we obtain the following monotonicity result using the compatibility obtained in Theorem 16.2.2.

16.5.5. THEOREM. *Suppose G-derivates are admissible and $f$ is $I^*LSC$, nowhere decreasing, and it is G-derivable at the points where it is lower differentiable. If $E$ is the set of points where $f$ has a zero G-derivative, then $f$ is increasing whenever either* (i) *$E$ is empty, or* (ii) *$f$ is $WLC_-$ and $E$ is finite.*

Next, we obtain an extended version of the Goldowski-Tonelli theorem in terms of subregular generalized derivatives. This theorem does not follow from any of the two Goldowski-Tonelli theorems of §9.6, and it is deduced here from Theorem 16.5.1. We need here the following reduction lemma.

16.5.6. LEMMA (FOURTH REDUCTION). *Let P be any function-theoretic property. If every continuous function on an open interval with property P is nondecreasing, then so is every $I^*C$ function on an open interval which has property P.*

PROOF. It is clear from the Second Reduction Lemma 9.1.1 that in this monotonicity proposition, continuity can be reduced to $I^*LSC$ and $USC$ together. Further, on making $I^*LSC$ a part of $P$ and then applying this lemma to the functions $-f_-$, it follows that $USC$ can be reduced to $I^*USC$. □

16.5.7. THEOREM (THIRD GOLDOWSKI-TONELLI). *Suppose f is $I^*C$ and G-derivates are subregular. Then if f is G-derivable* n.e. *and $Gf' \geq 0$* a.e., *then f is nondecreasing.*

PROOF. We can assume here without loss of generality that $G$-derivates are regular; for when they are subregular, they are stronger than some regular $G_1$-derivates, and so the given hypothesis in terms of $G$-derivative of $f$ would also hold for $G_1$-derivative. Further, on account of the above lemma, we can assume without loss of generality that $f$ is continuous. Then by Lemma 9.1.2 the set $N \equiv N(f)$ is perfect in $I$, and it is enough to show that $N = \varnothing$.

Suppose $N \neq \varnothing$, and set

$$A = \{x \in N : G\overline{D}f(x) \geq 0\}, \qquad B = \{x \in N : G\underline{D}f(x) = -\infty\}.$$

First, we claim that $A$ is dense in some portion $P$ of $N$. Since $Gf' \geq 0$ a.e., this holds clearly when $N^0 \neq \varnothing$. Hence suppose $N$ is nowhere dense, and let $(a, b)$ be any contiguous interval of $N$. Then since $f$ is nondecreasing on $[a, b]$, we have $D_-f(b) \geq 0$, and so it follows that $G\overline{D}f(b) \geq GD_-f(b) \geq D_-f(b) \geq 0$. Hence $A$ is dense in $N$.

Now, since $f$ is continuous, $G\underline{D}f$ is of $LB_1$ and $G\overline{D}f$ is of $UB_1$ (see Theorem 15.4.1). Hence $A$ and $B$ are $G_\delta$-sets. But since $A \cap B$ is countable by hypothesis, $A$ and $B$ thus cannot be simultaneously dense in any portion of $N$. Consequently, there is a subinterval $J$ of $I$, open relative to $I$, such that $Q \equiv N \cap J$ is a nonempty portion of $P$ which does not meet $B$. Now, if $x \in Q$, then $G\underline{D}f(x) > -\infty$, and if $x \in J \sim N$, then since $f$ is nondecreasing in some neighborhood of $x$, it is clear that $G\underline{D}f(x) \geq \underline{D}f(x) \geq 0$. Further, by hypothesis, $Gf' \geq 0$ a.e.. Hence it follows from Theorem 16.5.1 that $f$ is nondecreasing on $J$. Thus $N \cap J = \varnothing$, which is a contradiction. □

Next, in terms of subregular generalized derivates we deduce from Theorem 16.5.1 weaker versions of the Third and Fourth Derivate Theorems 2.4.1 and 2.4.4. What we mean by weaker versions is that these results are obtained here only when $E = X = I$.

16.5.8. THEOREM. *Suppose G-derivates are subregular and $f : I \to \mathbb{R}$ is IC. Then if $GD_+f > 0$ at a nonmeager set of points, then f is strongly increasing on* *some portion of $I$. Further, when $G$-derivates are totally subregular, the same holds provided that $f$ is either* (i) *$I^*LC$, or* (ii) *$LI_+$ and $I^*LC_-$.*

PROOF. To obtain the first part, there is no loss of generality in assuming $f$ to be continuous. Suppose $G$-derivates are subregular and $GD_+f > 0$ at a nonmeager set of points. Then there exist some regular $G_1$-derivates which are weaker than $G$-derivates. Now, since $G_1D_+f$ is of $LB_1$, it is by the Baire-Young theorem 1.4.1, $LSC$ r.e. Hence we can find a point $x \in I$ such that $G_1D_+f$ is $LSC$ at $x$ and $G_1D_+f(x) \geq GD_+f(x) > 0$. Let $0 < c < G_1D_+f(x)$. Then we can find a nondegenerate closed subinterval $J$ of $I$ such that $x \in J^0$ and $G_1D_+f > c$ on $J$. Hence it follows from axiom (D3) that $G_1D_+f_{-c} > 0$ on $J$, and so it follows from Theorem 16.5.1 that $f_{-c}$ is nondecreasing on $J$. Consequently, $f$ is strongly increasing on $J$.

When the $G$-derivates are totally subregular, the second part is obtained from Theorem 16.5.1 by a similar argument. □

Finally, we deduce from the above theorem the following weaker version of the Fourth Derivate Theorem 2.4.4.

16.5.9. THEOREM. *Suppose $G$-derivates are subregular and $f: I \to \mathbb{R}$ is $IC$. If $GD_+f$ and $GD^+f$ are finite at a nonmeager set of points, then $f$ is Lipschitz on some portion of $I$.*

*Moreover, if there is some $c \in \mathbb{R}_+$ such that $GD_+f > -c$ and $GD^+f < c$ at a nonmeager set of points, then $f$ is $c$-Lipschitz on some portion of $I$.*

PROOF. It is clearly enough to prove here the second part. Hence suppose the set $E = \{x : GD_+f(x) > -c, GD^+f(x) < c\}$ is nonmeager in $I$. Then there is clearly a subinterval $J$ of $I$ such that $E$ is nonmeager in every subinterval of $J$. Now it follows from axiom (D3), along with Theorem 16.5.8 on applying it to the restriction of $f_{+c}$ to $J$, that $f_{+c}$ is increasing on some subinterval $J_0$ of $J$. Similarly, on applying this theorem to the restriction of $-f_{-c}$ to $J_0$ it follows that $f_{-c}$ is decreasing on some subinterval $J_1$ of $J_0$. It is now clear that $f$ is $c$-Lipschitz on $J_1$. □

16.5.10. REMARK. We refer to the monographs of Bruckner [Br2] and Thomson [T1], as well as to Bruckner and Leonard [BL], for the known monotonicity results related to Theorems 16.5.1, 16.5.2, and 16.5.7 on some of the known generalized derivates and derivatives, as well as for their references. The known results in connection with Theorem 16.5.2 were obtained, however, only under the hypothesis of derivability in the considered sense everywhere.

It is natural to ask here whether the general results on strong monotonicity obtained in the First and Third Derivate Theorems 2.2.2 and 2.4.1 can be extended to admissible generalized derivates. Such results do not follow from these two theorems since admissible generalized derivates are weaker than ordinary derivates. Thus they require an independent analysis. We will return to this question later in [G20], where an abstract theory of derivates and derivatives is developed.

**§16.6 Relations between generalized and strong derivates.** In this section we obtain an extension of the Dini's theorems to admissible generalized derivates (see Theorem 16.6.1). Further, in Theorem 16.6.3, we obtain an extended version of Denjoy's Theorem 2.5.4 for subregular generalized derivates. This, in turn, leads to the identity of subregular generalized derivatives with ordinary and strong derivatives at a residual set of points.

The following theorem is similar to the extended version of Dini's theorems obtained in Theorem 14.3.1.

16.6.1. THEOREM. *Suppose G-derivates are admissible and* $x \in I$.

(a) *If* $f$ *is either* (i) $I^*USC$, (ii) $I^*LC$, *or* (iii) $LI_+$ *and* $I^*LC_-$, *then* $GD_+f(x) = \underline{D}_*f(x)$ *iff* $GD_+f$ *is LSC at* $x$.

(b) *If* $f$ *is either* (i) $I^*LSC$, (ii) $I^*UC$, *or* (iii) $UI_+$ *and* $I^*UC_-$, *then* $GD_+f(x) = \overline{D}_*f(x)$ *iff* $GD_+f$ *is USC at* $x$.

PROOF. The necessity parts of the results hold without any continuity hypothesis. Since $\underline{D}_*f \le \underline{D}f \le G\underline{D}f \le GD_+f$ and $GD_+f \le G\overline{D}f \le \overline{D}f \le \overline{D}_*f$ (see axiom (D2)), these parts follow clearly from Theorem 14.1.1.

To prove the sufficiency part of (a), suppose that one of its continuity hypotheses holds and that $GD_+f$ is $LSC$ at $x$, but $\underline{D}_*f(x) < GD_+f(x)$. Choose $c$ such that $\underline{D}_*f(x) < c < GD_+f(x)$. Then there is an open interval $U$ containing $x$ such that $GD_+f(y) > c$ for each $y \in U$. Then, by (D2), for each $y \in U$, $D^+f(y) \ge GD_+f(y) > c$, that is, $D^+f_{-c}(y) > 0$. Hence, in case (i) it follows from the Second Monotonicity Theorem 9.1.3 on applying it to $-(f_{-c})_-$, while in the other two cases it follows from the Fourth Monotonicity Theorem 9.1.8, that $f_{-c}$ is nondecreasing on $U$. Thus $\underline{D}_*f_{-c}(x) \ge 0$, or $\underline{D}_*f(x) \ge c$, which is a contradiction.

Next, to prove the sufficiency part of (b), suppose that one of its continuity hypotheses holds and that $GD_+f$ is $USC$ at $x$, but $GD_+f(x) < \overline{D}_*f(x)$. Choose $c$ such that $GD_+f(x) < c < \overline{D}_*f(x)$. Then there is this time an open interval $U$ containing $x$ such that $GD_+f(y) < c$ for each $y \in U$. Hence it follows from (D2) that $D_+f(y) < c$ for each $y \in U$. Thus $D_+f_{-c} < 0$ on $U$. Hence, in case (i) it follows from Theorem 9.1.3, and in the other two cases on applying Theorem 9.1.8 to $-f_{-c}$, that $f_{-c}$ is nonincreasing on $U$. Thus $\overline{D}_*f_{-c}(x) \le 0$, that is, $\overline{D}_*f(x) \le c$, which is a contradiction. □

On combining the continuity hypotheses of parts (a) and (b) of the above theorem appropriately (namely, the hypotheses (iii) of (a) and (b), and (ii) of (a) with (i) of (b), and vice versa), we obtain the following characterization of strong derivability which is similar to Corollary 14.3.2.

16.6.2. COROLLARY. *Suppose* $f$ *is* $I^*C_-$ *and either* (i) *right internal,* (ii) $I^*LC_+$, *or* (iii) $I^*UC_+$. *Then if G-derivates are admissible, then* $f$ *is strongly derivable at* $x \in I$ *iff* $GD_+f$ *is continuous at* $x$.

Next, in the case of subregular generalized derivates, we obtain the following version of Denjoy's Theorem 2.5.4 for generalized derivates.

16.6.3. THEOREM. *Suppose $f$ is IC. If G-derivates are subregular, then* $GD_-f = D_-f = \underline{D}_*f = D_+f = GD_+f$ *and* $GD^-f = D^-f = \overline{D}_*f = D^+f = GD^+f$ r.e. *in I.*

PROOF. There is clearly no loss of generality here in assuming $f$ to be continuous. It is enough to prove here the first set of relations, for the second set of relations follows from it with the help of axiom (D1).

Suppose the $G$-derivates are subregular. Then there exist regular $G_1$-derivates which are weaker than the $G$-derivates. Now, since $f$ is continuous, $G_1D_+f$ is of $LB_1$, and so it follows from the Baire-Young Theorem 1.4.1 that there is a residual set $E$ in $I$ such that $G_1D_+f$ is $LSC$ at each point of $E$. Now, for each $x \in E$ it follows from Theorem 16.6.1(a) that $G_1D_+f(x) = \underline{D}_*f(x)$. But since the $G$-derivates are stronger than $G_1$-derivates, we thus obtain the following with the help of (D2): $\underline{D}_*f(x) \leq D_+f(x) \leq GD_+f(x) \leq G_1D_+f(x) = \underline{D}_*f(x)$. This proves the second half of the first set of relations; the first half of this set of relations is proved similarly. □

Finally, from the above theorem we obtain the following equivalence of generalized, ordinary, and strong derivatives at a residual set of points; the last part of this corollary is obtained on the other hand from the first part with the help of Theorem 14.4.2.

16.6.4. COROLLARY. *Suppose $f$ is IC and the G-derivates are subregular. Then there is a residual set of points in I where the G-derivability of f is equivalent to its ordinary and strong derivabilities.*

*Further, if f is G-derivable* r.e. *in I, then there is a dense open set U in I such that f is derivable* a.e. *in U.*

## §16.7 Properties of generalized derivatives.

In this section we deduce from the properties of new derivatives various properties of generalized derivatives.

First, in Theorem 16.7.1, we obtain the Darboux property of admissible generalized derivatives. The next theorem deals with the Borel class of various derivability and knot sets relative to regular generalized derivates. Then, in Theorem 16.7.3, we obtain the Baire class of such derivatives. Finally, in Theorem 16.7.5, we establish the Denjoy property of subregular generalized derivatives.

These properties are known for some of the known generalized derivatives, but they are known only when the function under consideration is derivable in the sense in question everywhere. In the following theorems, except for the last one, this derivability hypothesis is reduced considerably.

First, we deduce the Darboux property of generalized derivatives from the Fifth Mean Value Theorem 16.4.1.

16.7.1. THEOREM. *Suppose $f$ is $I^*C$. If $G$-derivates are admissible and $f$ is $G$-derivable at the points where it is semidifferentiable, then $Gf'$ possesses the Darboux property.*

PROOF. Suppose $Gf'(x) < c < Gf'(y)$, where $x, y \in I$. We need to show here that there is a point $z$ between $x$ and $y$ where $f$ is $G$-derivable and $Gf'(z) = c$. Assume for the sake of definiteness that $x < y$. Then we have

$$D_+f(x) \le Gf'(x) < c < Gf'(y) \le D^-f(y).$$

Let $g$ denote the restriction of $f_{-c}$ to $J \equiv [x, y]$. Then $D_+g(x) < 0 < D^-g(y)$, and so $g$ is not monotone on $J$. Hence, according to Lemma 8.1.1, there is a nondegenerate subinterval $J_0$ of $J$ on which $g$ is continuous but not monotone. Consequently, as seen in the proof of Theorem 10.4.1, there exist two points $s, t$ in $J_0$ where $g(s) = g(t)$. It is now easy to see that $Qf(x, t) = c$, and so the existence of the desired point $z$ follows from Theorem 16.4.1. □

Next, in the following theorem we deal with the Borel class of various generalized derivability and knot sets.

Following the notations of §4.3, we will use $G\Delta(f)$ and $GK(f)$ to denote the $G$-derivability and $G$-knot sets of $f$. The symbols $G\Delta^*(f)$, $G\Delta_+(f)$, $GK_+(f)$, and so on, are defined similarly as in §4.3.

16.7.2. THEOREM. *Suppose $f$ is continuous and $G$-derivates are regular. Then the $G$-derivability sets $G\Delta_+(f)$, $G\Delta^*_+(f)$, $G\Delta(f)$, and $G\Delta^*(f)$ are $F_{\sigma\delta}$-sets; and the $G$-knot sets $GK_+(f)$, $GK(f)$, and $GK_b(f)$ are $G_\delta$-sets.*

PROOF. Each of the results follows easily from the regularity of the involved derivates by arguments similar to the ones used for Theorems 10.1.2 and 10.1.4. □

The following theorem deals with the Baire class of subregular generalized derivatives. Following the notations of §4.3, we will use $Gf'$ to denote the $G$-derivative of $f$; $Gf'_+$ and $Gf'_-$ denote similarly the right and left $G$-derivatives of $f$. Following our earlier convention, the Baire class of these $G$-derivatives are obtained in the following theorem relative to the sets of points where they exist.

16.7.3. THEOREM. *Suppose $f$ is continuous and $G$-derivates are subregular. Then $Gf'_+$ and $Gf'$ are of $B_1$. Further, when $G$-derivates are totally subregular, the same holds for every function $f$.*

PROOF. First, suppose $f$ is continuous and $G$-derivates are subregular. Then there exist regular $G_1$-derivates which are weaker than $G$-derivates. Let $A$ and $B$ denote the sets of points where $f$ is $G$- and $G_1$-derivable, respectively. Then since $G_1\underline{D}f$ is of $LB_1$ and $G_1\overline{D}f$ is of $UB_1$ (see Theorem 15.4.1), the $G_1$-derivative is simultaneously of $LB_1$ and $UB_1$ on $B$, and thus it is of $B_1$. Now since $G$-derivative is stronger than $G_1$-derivative, $Gf'$ is only a restriction of $G_1f'$ to $A$, and so is also of $B_1$. The last part is obtained by a similar argument. □

Now, on taking into account the Third and Fourth Status Theorems 15.4.2 and 15.4.4, we obtain the following result from the above theorem for various known generalized derivatives.

16.7.4. COROLLARY. *The preponderant, approximate, and ordinary derivatives of every function* $f \in \mathscr{F}(I)$ *are of* $B_1$. *Further, when* $f$ *is continuous, the derivatives of* $f$ *induced by the other generalized derivates considered in Theorem* 15.4.4 *are also of* $B_1$.

Finally, in the following theorem we obtain the Denjoy property of subregular generalized derivatives. This result does not follow from the Denjoy property of lower and weak derivatives obtained in §10.4, and it is obtained here by a somewhat different argument.

16.7.5. THEOREM. *Suppose* $f$ *is continuous and* $G$*-derivates are subregular. Then if* $f$ *is nonangular and* $G$*-derivable* n.e., *then* $Gf'$ *possesses the Denjoy property.*

*Consequently, the same holds when* $f$ *is* $G$*-derivable everywhere.*

PROOF. We can assume here without loss of generality that the $G$-derivates are regular; for when they are subregular, they are stronger than some regular $G_1$-derivates, and the $G$-derivative is thus stronger than $G_1$-derivative. Also, there is no loss of generality in assuming $I$ to be closed.

Given $a, b \in R$, $a < b$, suppose the set $E \equiv \{x : a < Gf'(x) < b\}$ is of measure zero. Then we need to show that $E = \varnothing$. Suppose $E \neq \varnothing$, and set

$$(1) \qquad A = \{x : G\underline{D}f(x) \le a\}, \qquad B = \{x : G\overline{D}f(x) \ge b\}.$$

First, we claim that each (connected) component of $A$, as well as of $B$, is closed. Let $K$ be any nondegenerate component of $A$, and let $c$ and $d$ denote the left and right endpoints, respectively, of $K$. Set $J = [c, d]$, and define $g(x) = f_{-a}(x)$, $x \in J$. Then since $G\underline{D}f \le a$ on $K$ and $f$ is $G$-derivable n.e., we have $Gf' \le a$ n.e. in $J$. Hence, by axiom (D2), $D_+f \le Gf' \le a$ n.e. in $J$, and so $D_+g = D_+f - a \le 0$ n.e. in $J$. Thus it follows from the Fourth Monotonicity Theorem 9.1.8 that $g$ is nonincreasing on $J$. Consequently, $D^+g(c) \le 0$ and $D^-g(d) \le 0$, or $D^+f(c) \le a$ and $D^-f(d) \le a$. Hence with the help of axiom (D2) we obtain $G\underline{D}f(c) \le GD^+f(c) \le D^+f(c) \le a$, and similarly $G\underline{D}f(d) \le D^-f(d) \le a$. Thus $c, d \in A$, and so $K$ is closed. A similar argument holds for the components of $B$.

Now, let $\mathscr{K}$ denote the family of all nondegenerate components of $A$ and $B$, and set $P = I \sim \cup\{K^0 : K \in \mathscr{K}\}$, where the interior is taken relative to $I$. Then since $E \ne \varnothing$, $P$ is a nonempty closed set. We claim that $P$ is further dense in itself. Suppose $P$ has some isolated point $x$. Then since the elements of $\mathscr{K}$ are closed, it is clear that $x \in I^0$. Hence there exist $y, z \in I$, $y < x < z$, such that either (i) $(y, x) \subset A$ and $(x, z) \subset B$ or (ii) vice versa. To be definite, suppose (i) holds. Then since $f$ is $G$-derivable n.e., it follows from Theorem 9.1.8 as before that $f_{-a}$ is nonincreasing on $[y, x]$ and $f_{-b}$ is nondecreasing on $[x, z]$. Thus $D^-f(x) \le a < b \le D_+f(x)$, which contradicts the nonangularity of $f$. A similar argument holds in case (ii). This proves that $P$ is perfect.

Now set $P_- = P \cap A$ and $P_+ = P \cap B$. Then it is enough to show that $P_-$ and $P_+$ are both dense in $P$. For, since $G\underline{D}f$ is of $LB_1$ and $G\overline{D}f$ is of $UB_1$ (see Theorem 15.4.1), $A$ and $B$ are $G_\delta$-sets, and so then $P_-$ and $P_+$ are both residual in $P$. Hence, $P_0 = P_- \cap P_+$ also is residual in $P$, and so it is uncountable. But since $f$ is, by (1), not $G$-derivable at any point of $P_0$, this contradicts the hypothesis.

We will give the proof for the denseness of $P_-$ in $P$; the denseness of $P_+$ is proved by a similar argument.

First, we claim that $P \subset \overline{A}$. Suppose there is a point $x \in P \sim \overline{A}$. Then there is a subinterval $J$ of $I$, open in $I$, such that $x \in J$ and $J \cap A = \varnothing$. Then $G\underline{D}f > a$ on $J$, and so $G\underline{D}f \ge b$ a.e. in $J$. Hence it follows from Theorem 9.1.8 as before that $G\overline{D}f \ge b$ at each point of $J$. Hence $x \in B^0$, which contradicts the fact that $x \in P$. This proves that $P \subset \overline{A}$.

Next, to see that $P_-$ is dense in $P$, let $x \in P$ and $U$ be any subinterval of $I$, open in $I$, which includes $x$. Then we need to show that $U \cap P_- \ne \varnothing$. But since $P \subset \overline{A}$ and $x$ is a limit point of $P$, there exists a component $K$ of $A$ which is included in $U$, and since $K$ is closed, each endpoint of $K$ is in $U \cap P_-$. □

16.7.6. Remark. We refer once again to Bruckner and Leonard [BL, pp. 33–36], Bruckner [Br2, pp. 156–190], and Thomson [T1], as well as to their references, for the known results on various known generalized derivatives.

In connection with Corollary 16.7.4, it should be pointed out here that in the case of preponderant derivative this result was obtained originally by Denjoy [D4] for continuous functions; the result on approximate derivative was obtained by Preiss [Pr1] in the general case; but both of these results were obtained under the hypothesis of everywhere derivability. In the case of ordinary derivative this result was obtained in the general form presented here in [G11, p. 322].

In connection with the notion of path derivates defined in §15.3, it should be pointed out here that Bruckner, O'Malley, and Thomson [BOT] have obtained some of the results of this chapter for path derivates and derivatives under various intersection conditions on the paths and other hypotheses. However, as recognized by them [BOT, p. 103], it is not clear which generalized derivatives other than approximate and preponderant derivatives satisfy these conditions. Also, the choice of path is dependent on the function under consideration [BOT, p. 100]. The hypothesis of subcontinuity of paths as defined in §15.4

may provide some insight into this problem, for the various results do hold in the case of subcontinuous paths.

Finally, it may be pointed out as mentioned earlier that the following results of this chapter hold also in the case when $I$ is a closed interval: Theorems 15.1.6, Lemma 15.4.3, Theorem 16.5.7, Theorem 16.6.3, Corollary 16.6.4, and Theorems 16.7.1 to 16.7.5. In many of these cases this fact follows directly on applying the result obtained to $I^0$, and in some cases it follows from the proofs of the results.

**§16.8 New generalized derivates and derivatives.** To obtain the various results of this chapter for any new generalized derivates, or the derivative induced by them, it should be clear by now that instead of obtaining each of these results independently all over again, it is enough to obtain the status of admissibility and regularity of these derivates. For these properties of the derivates determine entirely which of the results of this chapter hold for these derivates and the derivative induced by them.

To illustrate this phenomenon, we present in this section several new generalized derivates not considered earlier, and we investigate their admissibility and regularity. We do not use titles for the theorems of this section, but each of the five theorems obtained in this section will be referred simply as a Status Theorem.

First, let $s$ be any given selection and let $G$-limits be any of the generalized limits considered in Theorem 15.3.3 except for (a), (g), (j), and (k). Then, as defined in §15.2B, the derivates induced by $G$-limits and the selective quotient $sQ$ are called *Gs-derivates*. We will call all such derivates simply *generalized selective derivates*. For example, if $K$ is some given congruence, then

$$KsD_+f(x) = K\text{-}\liminf_{h\to 0+} sQf(x, x+h) = \liminf_{h\in K;\, h\to 0+} Qf(x, s_{[x,x+h]});$$

and if $s'$ is some other given selection, then

$$s'sD_+f(x) = s'\text{-}\liminf_{h\to 0+} sQf(x, x+h) = \liminf_{h\to 0+} Qf(x, s_{[x, s'_{[x,x+h]}]}).$$

In the following theorem we deal with the status of admissibility of all the generalized selective derivates defined above.

16.8.1. THEOREM. *The following generalized selective derivates are $s^*$-admissible*:

(a) *approximate selective derivates,*

(b) *sequentially congruent selective derivates,*

(c) *congruent selective derivates,*

(d) *qualitative selective derivates,*

(e) *negligent selective derivates,*

(f) *selective selective derivates, and*

(g) *path selective derivates.*

*Further,* (h) *preponderant selective derivates are admissible.*

PROOF. Let $s$ be any given selection. Then the selective quotient $sQ$ is $s^*$-admissible by the First Status Theorem 15.3.1. Hence, the result follows in each case from the First Induction Theorem 15.2.5 and the status of admissibility of the involved generalized limits that follows from the Second Status Theorem 15.3.3. □

Next, let $G$-limits be as before any of the generalized limits considered in Theorem 15.3.3 except for (a), (g), (j), and (k). Then the derivates induced by $G$-limits and the Cesàro quotient $CQ$ are called *GC-derivates*. We will call all such derivates simply *generalized Cesàro derivates*. For example, if $K$ is some given congruence, then

$$KCD_+f(x) = K\text{-}\liminf_{h\to 0+} CQf(x, x+h) = \liminf_{h\in K;\, h\to 0+} Qf_x^a(x, x+h);$$

and if $s$ is some given selection, then

$$sCD_+f(x) = s\text{-}\liminf_{h\to 0+} CQf(x, x+h) = \liminf_{h\to 0+} Qf_x^a(x, s_{[x,x+h]}).$$

In the following theorem we deal with the status of admissibility of all the generalized Cesàro derivates defined above.

16.8.2. THEOREM. *The following generalized Cesàro derivates are s-admissible:*

(a) *approximate Cesàro derivates,*
(b) *sequentially congruent Cesàro derivates,*
(c) *congruent Cesàro derivates,*
(d) *qualitative Cesàro derivates,*
(e) *negligent Cesàro derivates,*
(f) *selective Cesàro derivates, and*
(g) *path Cesàro derivates.*

*Further,* (h) *preponderant Cesàro derivates are admissible.*

PROOF. Since the Cesàro quotient $CQ$ is $s$-admissible by the First Status Theorem 15.3.1, in each of these cases the result follows directly from the First Induction Theorem 15.2.5 and the Second Status Theorem 15.3.3. □

Next, we investigate the status of regularity of all the new generalized derivates considered in the above two theorems. First, in the following theorem we obtain total regularity of preponderant selective and preponderant Cesàro derivates as in the Third Status Theorem 15.4.2.

16.8.3. THEOREM. *The* (a) *preponderant selective derivates and* (b) *preponderant Cesàro derivates are totally regular.*

*Consequently, the* (c) *selective derivates,* (d) *Cesàro derivates,* (e) *approximate selective derivates, and* (f) *approximate Cesàro derivates are totally subregular.*

PROOF. Let $s$ be any given selection. In cases (a) and (b) the result is obtained by an argument similar to the one used for the Third Status Theorem 15.4.2 on replacing $Qf$ by $sQf$ and $CQf$, respectively. The second part follows on the other hand from the first since the derivates in (c) and (e) are stronger than the ones in (a), and the derivates in (d) and (f) are stronger than the ones in (b). □

As a consequence of the above theorem we obtain the following result, as seen earlier, similar to Corollary 16.7.4. It should be noted here that in cases (a) and (d) this result was obtained in the last section only for continuous functions (see Corollary 16.7.4).

16.8.4. COROLLARY. *For every function* $f : I \to \mathbb{R}$, *the following derivatives of* $f$ *are of* $B_1$:

(a) *selective derivative,*

(b) *approximate selective derivative,*

(c) *preponderant selective derivative,*

(d) *Cesàro derivative,*

(e) *approximate Cesàro derivative,*

(f) *preponderant Cesàro derivative.*

Next, in the following theorem we deal with regularity of the remaining generalized selective derivates considered in Theorem 16.8.1.

16.8.5. THEOREM. *Relative to a continuous selection, the following generalized selective derivates are regular:*

(a) *sequentially congruent selective derivates,*

(b) *congruent selective derivates,*

(c) *qualitative selective derivates,*

(d) *negligent selective derivates, and*

(e) *continuously selective selective derivates.*

*Further,* (f) *subcontinuous path selective derivates are subregular.*

PROOF. The admissibility of each of these derivates follows directly from Theorem 16.8.1. Suppose $s$ is a continuous selection and $f : I \to \mathbb{R}$ is continuous. Then it is clear that the selective quotient $sQf(x, x+h) = Qf(x, s_{[x, x+h]})$ is continuous in $x$ for each $h$. Hence in each of the cases (a), (b), (e), and (f), the result is obtained by an argument similar to the one used for the Fourth Status Theorem 15.4.4 in the corresponding case on replacing $Qf$ by $sQf$. Further, (c) is a particular case of (d); and in case (d) it is easy to see that since $f$ is continuous, the negligent $s$-derivates of $f$ coincide with its $s$-derivates, which in turn are regular by Theorem 15.4.4. □

Finally, in the following theorem we deal with regularity of the remaining generalized Cesàro derivates considered in Theorem 16.8.2.

16.8.6. THEOREM. *The following generalized Cesàro derivates are regular:*
(a) *sequentially congruent Cesàro derivates,*
(b) *congruent Cesàro derivates,*
(c) *qualitative Cesàro derivates,*
(d) *negligent Cesàro derivates, and*
(e) *continuously selective Cesàro derivates.*
*Further,* (f) *subcontinuous path Cesàro derivates are subregular.*

PROOF. The admissibility of each of these derivates follows directly from Theorem 16.8.2. Suppose $f: I \to \mathbb{R}$ is continuous. Then it is clear that the Cesàro quotient $CQf(x, x + h)$ is continuous in $x$ for each $h$. Hence in each of the cases (a), (b), (e), and (f), the result is obtained by an argument similar to the one used for Theorem 15.4.4 in the corresponding case on replacing $Qf$ by $CQf$. Further, (c) is as before a particular case of (d); and in case (d) it is easy to see that the negligent Cesàro derivates of $f$ coincide with its Cesàro derivates, which in turn are regular by Theorem 15.4.4. □

# CHAPTER 17

# A Unified Theory of Generalized Symmetric Derivatives

In this chapter we first develop an axiomatic model similar to the one of Chapter 15 for generalized symmetric derivates. Then in terms of this model we present a unified theory of generalized symmetric derivatives similar to the one of the last chapter. The results of this chapter are organized roughly in the same order as in the last chapter.

First, in §17.1, we define various admissibilities and regularities of generalized symmetric derivates, and we investigate these properties for various known generalized symmetric derivates. Then, in §17.2, we obtain a conditional form of compatibility of generalized symmetric derivatives with new derivatives in terms of nonangularity in the corresponding sense.

Next, in §17.3, we deal with the Calculus of generalized symmetric derivatives. In §17.4 we obtain two mean value theorems and l'Hôpital's rule in terms of generalized symmetric derivatives. Then, in §17.5, we obtain some monotonicity theorems in terms of generalized symmetric derivates and derivatives, and we obtain the Goldowski-Tonelli theorem for such derivatives.

Next, in §17.6, we deal with the relations between generalized symmetric derivates and strong derivates, and in §17.7 we obtain the Darboux and Denjoy properties, and the Baire class, of generalized symmetric derivatives. Finally, in §17.8, we present several new generalized symmetric derivates to demonstrate the applicability of the theory developed in this chapter to new generalized symmetric derivates and derivatives.

Several of the results on generalized symmetric derivatives are found here to hold only for functions that are nonangular in the corresponding generalized sense. This, however, seems to be a natural condition, for the mean value theorem holds in terms of ordinary symmetric derivative only for nonangular functions. Also, this condition holds in fact for typical continuous functions relative to all known generalized derivates (see Theorem 19.2.1).

Most of the results that follow from this unified theory in the case of various known generalized symmetric derivates and derivatives are indeed new. We include here, in particular, the results that follow on ordinary and approximate symmetric derivatives which are new.

We assume throughout this chapter, unless stated otherwise, that $G$-limits are "admissible" generalized limits, and that $f: I \to \mathbb{R}$, where $I$ is "open", and $x \in I$.

## §17.1 Admissible and regular generalized symmetric derivates.

In this section we introduce axioms similar to the ones considered in Chapter 15 to define various admissibilities and regularity of generalized symmetric derivates. Also, we introduce here axioms to define various admissibilities of generalized symmetric quotients. A new axiom is added here to determine the relations between generalized symmetric derivates and the corresponding generalized unilateral derivates.

Further, we define various generalized symmetric derivates similar to the ones considered in §15.3, and we investigate their status of admissibility and regularity (see Status Theorems 17.1.7, 17.1.10, and 17.1.11). Moreover, in Theorems 17.1.8 and 17.1.9, we compare various generalized symmetric derivates with ordinary symmetric derivates and the corresponding generalized bilateral derivates.

We begin with the definition of ordinary symmetric derivates. Given $f \in \mathcal{F}(I)$, $x \in I$, and $h > 0$, suppose $x \pm h \in I$. Then we define the *ordinary symmetric quotient* $Q_s f(x, h)$ as follows:

$$Q_s f(x, h) = \frac{f(x+h) - f(x-h)}{2h}.$$

Now the *lower* and *upper symmetric derivates* of $f$ at $x$ are defined to be the lower and upper limits of $Q_s f(x, h)$ as $h \to 0$ from the right. We will denote these derivates by $\underline{D}_s f(x)$ and $\overline{D}_s f(x)$, respectively; and when they are equal, their common value is called the *symmetric derivative* of $f$ at $x$, which we will denote by $f'_s(x)$.

The various generalized symmetric derivates are defined as before in terms of generalized limits and generalized symmetric quotients. We first introduce here axioms similar to the ones considered in §15.2 to define various admissibilities of generalized symmetric derivates.

Given any generalized symmetric $G$-derivates, we will use $G\underline{D}_s$ and $G\overline{D}_s$ to denote these derivates. We will call these derivates *admissible* if the following three axioms hold for each $f \in \mathcal{F}(I)$ at each point of $I$:

(S1) $$G\overline{D}_s f = -G\underline{D}_s(-f);$$

(S2) $$\underline{D}_s f \le G\underline{D}_s f \le G\overline{D}_s f;$$

and for each $c \in \mathbb{R}$ and $x \in I$ we obtain

$$(S3) \qquad G\underline{D}_s(f_{+c})(x) = G\underline{D}_s f(x) + c.$$

Now, when the derivates $G\underline{D}_s f(x)$ and $G\overline{D}_s f(x)$ are equal, $f$ will be said to be *symmetrically G-derivable* at $x$, and the common value of these two derivates will be called the *symmetric G-derivative* of $f$ at $x$ and denoted by $Gf_s'(x)$. Also, when $Gf_s'(x)$ is finite, $f$ will be called (as before) *symmetrically G-differentiable* at $x$. Further, we will sometimes use $GM_s f(x)$ to denote the *symmetric G-median* of $f$ at $x$, which is defined to be the interval $[G\underline{D}_s f(x), G\overline{D}_s f(x)]$.

Next, given any admissible symmetric $G$-derivates, we will call them *s-admissible* if the following two axioms hold for every $f, g \in \mathscr{F}(I)$, $x \in I$, and $c \in \mathbb{R}$:

$$(S4) \qquad G\underline{D}_s f(x) + G\underline{D}_s g(x) \le G\underline{D}_s(f+g)(x);$$

$$(S5) \qquad G\underline{D}_s(cf)(x) = c\, G\underline{D}_s f(x) \quad \text{if} \quad c > 0.$$

We begin with two results similar to Lemma 15.2.3 and Theorem 15.2.4 in the case of $s$-admissible generalized symmetric derivates. The following lemma is obtained from the axioms (S1), (S4), and (S5) by arguments similar to the ones used for Lemma 15.2.3.

17.1.1. LEMMA. *Suppose symmetric G-derivates are s-admissible, $f, g \in \mathscr{F}(I)$, $x \in I$, and $c \in \mathbb{R}$. Then:*

(a) $G\underline{D}_s(f+g)(x) \le G\underline{D}_s f(x) + G\overline{D}_s g(x) \le G\overline{D}_s(f+g)(x) \le G\overline{D}_s f(x) + G\overline{D}_s g(x)$.

(b) *If $c > 0$, then $G\overline{D}_s(cf)(x) = c\, G\overline{D}_s f(x)$, and if $c < 0$, then $G\underline{D}_s(cf)(x) = c\, G\overline{D}_s f(x)$ and $G\overline{D}_s(cf)(x) = c\, G\underline{D}_s f(x)$. Further, each of these equations, and the one in axiom (S5), holds also when $c = 0$.*

Now, when $f$ is symmetrically $G$-differentiable at $x$, we obtain the following theorem needed later by combining part (a) of the above lemma with axiom (S4).

17.1.2. THEOREM. *Suppose symmetric G-derivates are s-admissible, $f, g \in \mathscr{F}(I)$, and let $h = f + g$. Then if $f$ is symmetrically G-differentiable at $x \in I$, then*

$$G\underline{D}_s h(x) = Gf_s'(x) + G\underline{D}_s g(x), \qquad G\overline{D}_s h(x) = Gf_s'(x) + G\overline{D}_s g(x).$$

Now, since axiom (S3) is not used in proving the above theorem, it is clear from this theorem that this axiom becomes redundant for the $s$-admissibility of symmetric $G$-derivates.

**A. Admissibility of generalized symmetric quotients.** To deal with some of the generalized symmetric derivates, we need to consider (as before) generalizations of ordinary symmetric quotient $Q_s f(x, h)$. By a *generalized symmetric quotient* $GQ_s$ we will mean an operator which assigns a real number $GQ_s f(x, h)$ to each $f \in \mathscr{F}(I)$, $x \in I$, and $h > 0$ provided that $x \pm h \in I$.

Given such a symmetric $G$-quotient, we will call it *admissible* if the following three axioms hold for each $f \in \mathscr{F}(I)$, $x \in I$, $h > 0$, and $c \in \mathbb{R}$ provided that $x \pm h \in I$:

(T1) $$GQ_s(-f)(x, h) = -GQ_s f(x, h);$$

(T2) $$\liminf_{h\to 0+} Q_s f(x, h) \le \liminf_{h\to 0+} GQ_s f(x, h);$$

(T3) $$GQ_s(f_{+c})(x, h) = GQ_s f(x, h) + c.$$

Next, given any admissible symmetric $G$-quotient, we will call it *s-admissible* if the following two axioms hold for each $f, g \in \mathscr{F}(I)$, $x \in I$, $h > 0$, and $c \in \mathbb{R}$ provided that $x \pm h \in I$:

(T4) $$GQ_s(f + g)(x, h) = GQ_s f(x, h) + GQ_s g(x, h);$$

(T5) $$GQ_s(cf)(x, h) = c\, GQ_s f(x, h).$$

Clearly, in the case of $s$-admissibility, the axioms (T1) and (T3) become redundant since they are implied by (T5) and (T4), respectively.

Finally, given any $s$-admissible symmetric $G$-quotient, we will call it *$s^*$-admissible relative to* some generalized *G-limits* if to each $f \in \mathscr{F}(I)$, $x \in I$, and $h \ne 0$, there corresponds a function $f_G(x, h)$ such that the following three axioms hold for each $f, g \in \mathscr{F}(I)$, $x \in I$, and $h > 0$ provided that $x \pm h \in I$:

(T6) if $f$ is $G$-continuous at $x$, then $G\text{-}\lim_{h\to 0} f_G(x, h) = f(x)$;

(T7) $GQ_s(fg)(x, h) = f_G(x, h) GQ_s g(x, h) + g_G(x, -h) GQ_s f(x, h)$;

(T8) if $f(x) \ne 0$, then $GQ_s(1/f)(x, h)$
$$= -GQ_s f(x, h)/\{f_G(x, h) f_G(x, -h)\}.$$

Further, when the symmetric $G$-quotient is $s^*$-admissible relative to every admissible generalized limits, it will be called as before simply *$s^*$-admissible*.

Next, to compare generalized symmetric derivates with the corresponding generalized unilateral derivates, we introduce here another axiom on symmetric $G$-quotient which relates it with the corresponding (ordinary) $G$-quotient (see §15.3).

Given any $s$-admissible symmetric $G$-quotient, we will call it *$s_1$-admissible* if the following axiom holds for each $f \in \mathscr{F}(I)$, $x \in I$, and $h > 0$ provided that $x \pm h \in I$:

$$\text{(T9)} \qquad GQ_s f(x,h) = \tfrac{1}{2}\{GQf(x, x+h) + GQf(x, x-h)\}.$$

Further, when a symmetric $G$-quotient is simultaneously $s_1$-admissible and $s^*$-admissible, it will be called *$s_1^*$-admissible.*

**B. Induced generalized symmetric derivates.** We now define induced generalized symmetric derivates similar to induced generalized derivates defined in §15.2.B. Given any generalized $G$-limits and generalized symmetric $G$-quotient, we will call the following *symmetric $G$-derivates* the *symmetric derivates induced by the $G$-limits and symmetric $G$-quotient*, that is, for each $f \in \mathscr{F}(I)$ and $x \in I$,

$$\text{(1)} \qquad G\underline{D}_s f(x) = G\text{-}\liminf_{h \to 0+} GQ_s f(x,h),$$

$$G\overline{D}_s f(x) = G\text{-}\limsup_{h \to 0+} GQ_s f(x,h).$$

The same prefix $G$- is used here as before in $G$-limits and symmetric $G$-quotient for notational simplicity, but they need not be related. The symmetric derivates induced by $G$-limits and symmetric $G_1$-quotient, obtained on replacing $GQ_s$ by $G_1Q_s$ in (1), will be called on the other hand *symmetric $GG_1$-derivates.*

Further, the symmetric $G$-derivates induced by $G$-limits and the ordinary symmetric quotient $Q_s$ will be called (as before) the *symmetric derivates induced by $G$-limits*; and similarly the symmetric $G$-derivates induced by the ordinary limits and symmetric $G$-quotient will be called the *symmetric derivates induced by symmetric $G$-quotient*. All the examples of generalized symmetric derivates considered in this section are of one of these two kinds.

Next, we define any generalized symmetric $G$-derivates to be *$s_1$-admissible* if they are induced by some $G$-limits and symmetric $G$-quotient, both of which are $s_1$-admissible (see §15.5 for the definition of $s_1$-admissible limits). Further, we define the symmetric $G$-derivates to be *$s^*$-admissible* if they are induced by some $G$-limits and symmetric $G$-quotient such that the $G$-limits are $s^*$-admissible and the symmetric $G$-quotient is $s^*$-admissible relative to $G$-limits. Also, when the symmetric $G$-derivates are simultaneously $s_1$- and $s^*$-admissible, they will be said simply to be *$s_1^*$-admissible.*

Now we obtain the following induction theorem on the admissibility of induced symmetric derivates.

17.1.3. Theorem (Third Induction). *If $G$-limits and symmetric $G$-quotient are simultaneously admissible, or $s$-, $s_1$-, or $s^*$-admissible, then so are the symmetric $G$-derivates induced by them.*

Proof. First, suppose the $G$-limits and symmetric $G$-quotient are admissible. Then for the symmetric $G$-derivates induced by them, the axiom (S1) follows from the definition (1) and the axioms (T1) and (A1); (S2) follows similarly from (T2) and (A2); and (S3) is obtained in turn from (T3) and (A3). Hence the symmetric $G$-derivates are admissible.

Next, suppose the $G$-limits and symmetric $G$-quotient are $s$-admissible. Then for the symmetric $G$-derivates the axiom (S4) follows from (T4) and (A4); and (S5) is obtained from (T5) and (A5), which proves the symmetric $G$-derivates to be $s$-admissible.

As for $s_1$- and $s^*$-admissibilities, the result follows directly from the above definitions of $s_1$- and $s^*$-admissible $G$-derivates. □

Now, since the ordinary limits are $s_1^*$-admissible by the Fifth Status Theorem 15.5.2, we obtain the following result from the above theorem.

17.1.4. Corollary. *If some symmetric G-quotient is admissible, or* $s$-, $s_1$-, *or* $s^*$-*admissible, then so are the symmetric G-derivates induced by it.*

**C. Examples of generalized symmetric derivates.** We now consider several examples of generalized symmetric derivates. Most of these symmetric derivates are induced by various generalized limits considered in Theorem 15.5.2. But some of them are induced by generalized symmetric quotients as in §15.3.

(i) **Cesàro symmetric quotient:** When $f$ is $D_*$-integrable, its *Cesàro symmetric quotient* $CQ_sf(x,h)$ is defined once again in terms of the average function $f_x^a$ as follows:

$$(2)\quad CQ_sf(x,h) = Q_sf_x^a(x,h) = \frac{1}{4h^2}\int_0^{2h}\{f(x+t)-f(x-t)\}\,dt.$$

(ii) **Selective symmetric quotient:** We define selective symmetric quotient only relative to a symmetric selection $s$—that is, a selection such that

$$(3)\qquad s_{[x,x+h]} - x = x - s_{[x-h,x]}.$$

Then the *selective symmetric quotient* $sQ_s$ is defined in terms of the ordinary symmetric quotient as follows:

$$(4)\qquad sQ_sf(x,h) = Q_sf(x, s_{[x,x+h]} - x).$$

Thus it is clear from (3) that

$$(5)\qquad sQ_sf(x,h) = \frac{f(s_{[x,x+h]}) - f(s_{[x-h,x]})}{2(s_{[x,x+h]} - x)}.$$

We now deal in the following theorem with the status of admissibility of the various symmetric quotients considered here.

17.1.5. THEOREM (SIXTH STATUS). *The ordinary and selective symmetric quotients are $s_1^*$-admissible, and the Cesàro symmetric quotient is $s_1$-admissible.*

PROOF. Let $s$ be any symmetric selection. We will first prove the $s$-admissibility of the three symmetric quotients $Q_s$, $sQ_s$, and $CQ_s$.

It is easy to verify from the definitions of $Q_s$ and $sQ_s$ that axioms (T1) to (T5) hold for them. Hence these symmetric quotients are $s$-admissible.

Further, suppose $f, g \in \mathscr{F}(I)$ are $D_*$-integrable, $x \in I$, $h > 0$, and $x \pm h \in I$. Then since $CQ_s f(x, h)$ is by definition $Q_s f_x^a(x, h)$, it is easy to see that the axioms (T1) and (T3) hold for $CQ_s$. Also, (T4) and (T5) follow from the fact that $(f+g)_x^a = f_x^a + g_x^a$ and $(cf)_x^a = cf_x^a$. Next, to obtain (T2) for $CQ_s$, we need to prove simply that $\underline{D}_s f(x) \le C\underline{D}_s f(x)$. Suppose $\underline{D}_s f(x) > -\infty$, for there is nothing to prove otherwise. Let $-\infty < c < \underline{D}_s f(x)$. Then there is a $\delta > 0$ such that $Q_s f(x, t) > c$ for each $t \in (0, \delta)$. Hence if $0 < h < \delta$, then for each $t \in (0, h)$, $f(x+t) - f(x-t) > 2ct$, and so

$$CQ_s f(x,h) = \frac{1}{4h^2}\int_0^{2h}\{f(x+t) - f(x-t)\}\,dt \ge \frac{1}{4h^2}\int_0^{2h} 2ct\,dt = c.$$

This proves that $C\underline{D}_s f(x) \ge c$, from which the desired inequality follows. Hence $CQ_s$ also is $s$-admissible.

Next, to obtain $s_1$-admissibility of these symmetric quotients, we need to verify the axiom (T9) for each of them. Suppose $f \in \mathscr{F}(I)$, $x \in I$, $h > 0$, and $x \pm h \in I$. Then in the case of the ordinary symmetric quotient $Q_s$ we clearly have

$$\begin{aligned} Q_s f(x,h) &= \frac{f(x+h) - f(x-h)}{2h} \\ &= \frac{1}{2}\left\{\frac{f(x+h) - f(x)}{h} + \frac{f(x) - f(x-h)}{h}\right\} \\ &= \frac{1}{2}\{Qf(x, x+h) + Qf(x, x-h)\}. \end{aligned}$$

This proves the axiom (T9) for $Q_s$. Further, from this relation we obtain the following with the help of (3) and (4):

$$\begin{aligned} sQ_s f(x,h) &= Q_s f(x, s_{[x,x+h]} - x) \\ &= \tfrac{1}{2}\{Qf(x, s_{[x,x+h]}) + Qf(s_{[x-h,x]}, x)\} \\ &= \tfrac{1}{2}\{sQf(x, x+h) + sQf(x, x-h)\}. \end{aligned}$$

Also, in the case of $CQ_s$ we obtain similarly from (2) the following:

$$CQ_sf(x,h) = Q_sf_x^a(x,h) = \tfrac{1}{2}\{Qf_x^a(x,x+h) + Qf_x^a(x,x-h)\}$$
$$= \tfrac{1}{2}\{CQf(x,x+h) + CQf(x,x-h)\}.$$

Next, to prove the $s^*$-admissibility of $Q_s$ and $sQ_s$, let $G$-limits be any given admissible generalized limits. First, choose $f_G(x,h)$ to be $f(x+h)$, so that $g_G(x,-h) = g(x-h)$. Then if $f$ is $G$-continuous at $x$, then $G\text{-}\lim_{h\to 0} f_G(x,h) = G\text{-}\lim_{h\to 0} f(x+h) = f(x)$. Further, it is easy to verify that

$$(6)\qquad Q_s(fg)(x,h) = f(x+h)Q_sg(x,h) + g(x-h)Q_sf(x,h),$$

and if $f(x) \neq 0$, then

$$(7)\qquad Q_s(1/f)(x,h) = -Q_sf(x,h)/\{f(x+h)f(x-h)\}.$$

It is thus clear that axioms (T6) to (T8) hold for $Q_s$, and hence this quotient is $s^*$-admissible.

Further, to obtain this property of $sQ_s$, choose $f_G(x,h)$ to be $f(s_{[x,x+h]})$, so that $g_G(x,-h) = g(s_{[x,x-h]})$. Now, if $f$ is $G$-continuous at $x$, then since $s_{[x,x+h]} \to x$ as $h \to 0$, $G\text{-}\lim_{h\to 0} f(s_{[x,x+h]}) = f(x)$. Further, with the help of (4) we obtain from (6) the following:

$$\begin{aligned} sQ_s(fg)(x,h) &\\ &= Q_s(fg)(x, s_{[x,x+h]} - x)\\ &= f(s_{[x,x+h]})Q_sg(x, s_{[x,x+h]} - x) + g(s_{[x-h,x]})Q_sf(x, s_{[x,x+h]} - x)\\ &= f(s_{[x,x+h]})sQ_sg(x,h) + g(s_{[x,x-h]})sQ_sf(x,h); \end{aligned}$$

and if $f(x) \neq 0$, we obtain by a similar argument from (7) the following:

$$sQ_s(1/f)(x,h) = -sQ_sf(x,h)/\{f(s_{[x,x+h]})f(s_{[x,x-h]})\}.$$

This clearly establishes the $s^*$-admissibility of $sQ_s$. □

Now, with the help of the above status theorem, we obtain the following version of the Third Induction Theorem for generalized symmetric derivates that are induced either by some generalized limits or by some generalized symmetric quotient, as defined in part B.

17.1.6. THEOREM (FOURTH INDUCTION).

(a) *If some generalized G-limits are admissible, or s-, $s_1$-, or $s^*$-admissible, then so are the symmetric G-derivates induced by them.*

(b) *If some generalized symmetric G-quotient is admissible, or s-, $s_1$-, or $s^*$-admissible, then so are the symmetric G-derivates induced by it.*

PROOF. Since the ordinary symmetric quotient is $s_1^*$-admissible by the above theorem, part (a) follows clearly from the Third Induction Theorem 17.1.3. Part (b) has on the other hand been obtained in Corollary 17.1.4. □

Next, in the following theorem we obtain the status of admissibility of the various generalized symmetric derivates that we consider here.

As for the terminology used in cases (d), (f), (g), and (h) of the following theorem, to avoid repeated use of the term "symmetric" in their terminology, the entities $K, \mathcal{N}, s$, and $\mathcal{P}$ relative to which these symmetric derivates are defined will be assumed implicitly throughout this chapter to be "symmetric" as defined in §15.5.

17.1.7. THEOREM (SEVENTH STATUS). *The following symmetric derivates are $s_1^*$-admissible*:

(a) *ordinary symmetric derivates,*

(b) *approximate symmetric derivates,*

(c) *sequentially congruent symmetric derivates,*

(d) *congruent symmetric derivates,*

(e) *qualitative symmetric derivates,*

(f) *negligent symmetric derivates,*

(g) *selective symmetric derivates, and*

(h) *path symmetric derivates.*

*Further, the following symmetric derivates are $s_1$-admissible*:

(i) *Borel symmetric derivates, and*

(j) *Cesàro symmetric derivates.*

*Furthermore,* (k) *the preponderant symmetric derivates are admissible.*

PROOF. In each of the above cases except for (g) and (j), the generalized symmetric derivates are induced by the corresponding generalized limits, and so the result for these symmetric derivates follows from the Second and Fifth Status Theorems 15.3.3 and 15.5.2 with the help of part (a) of the Fourth Induction Theorem 17.1.6. Further, since the symmetric derivates in (g) and (j) are induced by $sQ_s$ and $CQ_s$, the results for these symmetric derivates follow from the Sixth Status Theorem 17.1.5 with the help of part (b) of Theorem 17.1.6. □

**D. Comparison of generalized symmetric derivates.** We next introduce some terminology to compare any generalized symmetric derivates with other such derivates and with generalized bilateral derivates (see axiom (D0) for their definition). Let $G$- and $G_1$-derivates be any two admissible generalized symmetric or bilateral derivates. Then we will call $G$-derivates *stronger than* $G_1$-derivates, or $G_1$-derivates *weaker than* $G$-derivates, if for each $f \in \mathcal{F}(I)$ we have $G\underline{D}f \le G_1\underline{D}f$ at every point of $I$. It is easy to see in such a case from axioms (D0), (D1), and (S1) that $G\overline{D}f \ge G_1\overline{D}f$ everywhere; and hence in this case the derivative induced by the $G$-derivates is stronger than the one induced by $G_1$-derivates.

First in the following theorem we compare admissible generalized symmetric derivates with ordinary symmetric derivates. This theorem follows directly from axiom (S2).

17.1.8. THEOREM. *Given any admissible symmetric G-derivates, these derivates are weaker than the ordinary symmetric derivates; hence if* $f \in \mathscr{F}(I)$ *is symmetrically derivable at* $x \in I$, *then it is also symmetrically G-derivable at* $x$ *and* $Gf'_s(x) = f'_s(x)$.

Next in the following theorem we compare $s_1$-admissible symmetric $G$-derivates with the corresponding bilateral $G$-derivates.

17.1.9. THEOREM. *Given any* $s_1$*-admissible symmetric G-derivates, these derivates are weaker than the bilateral G-derivates; hence if* $f \in \mathscr{F}(I)$ *is G-derivable at* $x \in I$, *then it is also symmetrically G-derivable at* $x$ *and* $Gf'_s(x) = Gf'(x)$.

PROOF. Since the symmetric $G$-derivates are $s_1$-admissible, they are induced by $G$-limits and symmetric $G$-quotient, both of which are $s_1$-admissible. Given $f \in \mathscr{F}(I)$ and $x \in I$, we need to show here that $G\underline{D}_s f(x) \geq G\underline{D} f(x)$. Let $h > 0$ such that $x \pm h \in I$. Then according to axiom (T9) we have

$$GQ_s f(x,h) = \tfrac{1}{2}\{GQf(x, x+h) + GQf(x-h, x)\}.$$

Now since the $G$-limits are $s_1$-admissible, we obtain the following with the help of axioms (A4), (A5), and (A8):

$$\begin{aligned}
G\underline{D}_s f(x) &\geq \tfrac{1}{2}\left\{G\text{-}\liminf_{h\to 0+} GQf(x, x+h) + G\text{-}\liminf_{h\to 0+} GQf(x-h, x)\right\} \\
&= \tfrac{1}{2}\left\{GD_+ f(x) + G\text{-}\liminf_{h\to 0-} GQf(x, x+h)\right\} \\
&= \tfrac{1}{2}\{GD_+ f(x) + GD_- f(x)\} \geq G\underline{D} f(x).
\end{aligned}$$

The last inequality follows here from axiom (D0), for if the sum on its left side is indeterminate, then one of the derivates $GD_+ f(x)$ and $GD_- f(x)$ is $-\infty$, and so it follows from (D0) that $G\underline{D} f(x) = -\infty$.

The last part is of course a direct consequence of the first. □

As we see later in Example 17.1.12, the $s_1$-admissibility is indeed essential for the validity of the last theorem.

**E. Regularity of generalized symmetric derivates.** We finally define the notions of regular and subregular generalized symmetric derivates similar to the ones introduced in §15.4.

Given any admissible symmetric $G$-derivates, we will call them *regular*, as before, if $G\underline{D}_s f$ is of $LB_1$ for every continuous function $f$, and *totally regular* if this holds for every function $f$. This of course implies that $G\overline{D}_s f$ is of $UB_1$, for $G\overline{D}_s f = -G\underline{D}_s(-f)$ by axiom (S1).

Further, we will call any admissible symmetric $G$-derivates *subregular* (or *totally subregular*) if there exist some regular (or totally regular) symmetric $G_1$-derivates which are weaker than the symmetric $G$-derivates.

We first obtain total regularity of preponderant symmetric derivates in the following theorem, which is similar to the Third Status Theorem 15.4.2.

17.1.10. THEOREM (EIGHTH STATUS). *The preponderant symmetric derivates are totally regular. Consequently, the ordinary and approximate symmetric derivates are totally subregular.*

PROOF. We omit the proof here since this result is obtained by an argument similar to the one used for Theorem 15.4.2 on replacing $PD_+ f$ by $P\underline{D}_s f$ and $Qf(x, y)$ by $Qf(2x - y, y)$. □

Next, in the following theorem we consider the regularity of other generalized symmetric derivates.

17.1.11. THEOREM (NINTH STATUS). *The following symmetric derivates are regular:*

(a) *ordinary symmetric derivates,*

(b) *sequentially congruent symmetric derivates,*

(c) *congruent symmetric derivates,*

(d) *Borel symmetric derivates,*

(e) *Cesàro symmetric derivates,*

(f) *qualitative symmetric derivates,*

(g) *negligent symmetric derivates, and*

(h) *continuously selective symmetric derivates.*

*Further,* (i) *subcontinuous path symmetric derivates are subregular.*

PROOF. We omit the proof once again since in each of these cases the result is obtained by an argument similar to the one used in the proof of the Fourth Status Theorem 15.4.4 in the corresponding case, using Theorem 17.1.7 in the place of Theorem 15.3.3, and on replacing $D_+ f$ by $\underline{D}_s f$, $Qf(x, x+h)$ by $Q_s f(x, h)$, and $Qf(x, x+t)$ by $Q_s f(x, t)$, and similarly on replacing $CQf(x, x+h)$ by $CQ_s f(x, h)$ and $sQf(x, x+h)$ by $sQ_s f(x, h)$. Moreover, in case (g) it is again clear that for a continuous function $f$ the derivates in (g) coincide with the ones in (a) everywhere. □

Finally, to see the necessity of $s_1$-admissibility in Theorem 17.1.9, we include here an example to show that this theorem does not hold for preponderant derivative which is not $s$-admissible, as seen earlier in Example 15.3.4.

17.1.12. EXAMPLE. Decompose $(0, 1)$ into three disjoint sets $A$, $B$, and $C$, each of which has right density $1/3$ at 0, and set $E = A \cup (-B) \cup \{0\}$. Now, define $f(x) = 0$ if $x \in E$, and define $f(x) = x$ otherwise. Then it is clear that $Qf(0, h) = 0$ or 1 according to whether $h \in A \cup (-B)$ or $R \sim E$, respectively. Hence $Pf'(0) = 1$. However, it is further easy to see that $Q_s f(0, h) = 1/2$ or 1 according to whether $h \in A \cup B$ or $C$, respectively, so that $Pf'_s(0) = 1/2$. Thus the $P$-derivative does not coincide in general with the symmetric $P$-derivative.

17.1.13. REMARK. We include here some comments regarding the historical background of various symmetric derivatives. The notion of ordinary symmetric derivative was introduced by de la Vallée Poussin [V]. The Borel symmetric derivative was introduced in turn by Khintchine [Kh, p. 220]. The Cesàro symmetric derivative was considered on the other hand by Burkill [Bur3], and the approximate symmetric derivative has been considered by Mukhopadhyay [Mu]. We are not aware if any of the other notions of generalized symmetric derivates or derivatives considered in the Seventh Status Theorem 17.1.7 has been considered earlier in the literature. Most of the results obtained on ordinary symmetric derivates in the remaining sections of this chapter are, however, also new.

## §17.2 Conditional compatibility of generalized symmetric derivatives with new derivatives.

To apply the results of the theory of new derivatives to generalized symmetric derivatives, we obtain in this section a conditional form of compatibility of these derivatives with new derivatives. This compatibility is obtained in terms of generalized nonangularity relative to the notion under consideration.

Given any admissible generalized $G$-limits, we first define various $G$-angularities and $G$-nonangularities analogous to ordinary angularities and nonangularities (see §4.2).

We will call $f$ *lower* or *upper G-angular* at $x$ if $GD^- f(x) < GD_+ f(x)$ or $GD^+ f(x) < GD_- f(x)$, respectively; and if $f$ is not lower or upper $G$-angular at $x$, then it will be said to be *lower* or *upper G-nonangular*, respectively, at $x$. Also, $f$ is called *G-angular* at $x$ if it is either lower or upper $G$-angular at $x$; and if $f$ is not $G$-angular at $x$, then it will be said to be *G-nonangular* at $x$.

In the following theorem, which is obtained directly from the axiom (D2), we compare the $G$-angularity with ordinary angularity.

17.2.1. THEOREM. *If $f$ is lower angular at $x$, then it is also lower G-angular at $x$. Consequently, if $f$ is lower G-nonangular at $x$, then it is lower nonangular there; and if $f$ is G-nonangular at $x$, then it is nonangular there.*

Next, when symmetric $G$-derivates are $s_1$-admissible, we obtain a conditional form of compatibility of the symmetric $G$-derivative with weak derivative at the points of $G$-nonangularity. It is easy to see that these two derivatives are not compatible in general; consider, for example, the function $f(x) = |x|$, $x \in \mathbb{R}$; it is clear in this case that $f'_s(0) = 0$, but $Wf'(0) = Lf'(0) = [-1, 1]$.

First, in the following lemma we investigate the relation between the symmetric $G$-median and $G$-derivates.

17.2.2. LEMMA. *Suppose symmetric G-derivates are $s_1$-admissible. Then $GM_sf(x)$ meets the convex hulls of $\{GD_+f(x), GD^-f(x)\}$ and $\{GD_-f(x), GD^+f(x)\}$.*

PROOF. Since the symmetric $G$-derivates are $s_1$-admissible, they are induced by $G$-limits and symmetric $G$-quotient, both of which are $s_1$-admissible. First, let $a$ and $b$ denote the minimum and maximum of $GD_+f(x)$ and $GD^-f(x)$. Then with the help of the axioms (T9), (S5), and (A8), and Lemma 15.1.3, we obtain

$$\begin{aligned} G\underline{D}_sf(x) &= G\text{-}\liminf_{h\to 0+} \tfrac{1}{2}\{GQf(x, x+h) + GQf(x-h, x)\} \\ &\le \tfrac{1}{2}\Big\{GD_+f(x) + G\text{-}\limsup_{h\to 0-} GQf(x, x+h)\Big\} \\ &= \tfrac{1}{2}\{GD_+f(x) + GD^-f(x)\} \le b. \end{aligned}$$

It is proved by a similar argument that $G\overline{D}_sf(x) \ge a$. Consequently, $GM_sf(x)$ meets $[a, b]$. The other part is obtained by a similar argument. □

Now from the above lemma we obtain directly the following relations between symmetric and unilateral $G$-derivates.

17.2.3. LEMMA. *Suppose symmetric G-derivates are $s_1$-admissible. Then if $f$ is lower G-nonangular at $x$, then*

$$G\underline{D}_sf(x) \le GD^-f(x) \quad \text{and} \quad G\overline{D}_sf(x) \ge GD_+f(x);$$

*and if $f$ is upper G-nonangular at $x$, then*

$$G\underline{D}_sf(x) \le GD^+f(x) \quad \text{and} \quad G\overline{D}_sf(x) \ge GD_-f(x).$$

We now obtain the following conditional compatibility of generalized symmetric derivative with lower and weak derivatives. Two derivatives are called here *compatible at a point x* if they are equal at $x$ whenever both of them exist there.

17.2.4. THEOREM (SECOND COMPATIBILITY). *Suppose symmetric G-derivates are $s_1$-admissible. Then the symmetric G-derivative of $f$ is compatible with lower derivative at the points where it is lower G-nonangular; and it is compatible with weak derivative at the points where it is G-nonangular.*

PROOF. Suppose $f$ is symmetrically $G$-derivable at $x$. If $f$ is lower derivable and lower $G$-nonangular at $x$, then it follows from the last lemma and axiom (D2) that

$$Gf'_s(x) \le GD^-f(x) \le D^-f(x) \le D_+f(x) \le GD_+f(x) \le Gf'_s(x),$$

which implies that $Lf'(x)$ is unique and equal to $Gf'_s(x)$.

Similarly, when $f$ is weakly derivable and $G$-nonangular at $x$, we obtain the following on combining the two parts of the last lemma with (D2):

$$Gf_s'(x) \leq \wedge GD^{\pm}f(x) \leq \wedge D^{\pm}f(x) \leq \vee D_{\pm}f(x) \leq \vee GD_{\pm}f(x) \leq Gf_s'(x). \quad \square$$

In the case of ordinary symmetric derivative the above theorem can be strengthened somewhat. Let any $G$-derivative be called *weakly compatible with weak derivative* if for every function $f: I \to \mathbb{R}$ and $x \in I$, if $f$ is simultaneously $G$-derivable and weakly derivable at $x$, then $Gf'(x) \in Wf'(x)$.

17.2.5. THEOREM. *The ordinary symmetric derivative is weakly compatible with weak derivative.*

PROOF. Suppose $f: I \to \mathbb{R}$ is simultaneously symmetrically and weakly derivable at $x \in I$. Then since the ordinary symmetric derivates are $s_1$-admissible, it follows from Lemma 17.2.2 that $f_s'(x)$ is an element of the convex hulls of $\{D_+f(x), D^-f(x)\}$ and $\{D_-f(x), D^+f(x)\}$. Hence the result holds clearly when $f$ is lower or upper derivable at $x$. When $f$ is not semiderivable at $x$, $f$ is indeed nonangular and unilaterally derivable at $x$, say right derivable. Hence then $D_-f(x) \leq f_+'(x) \leq D^-f(x)$, and so it follows from Lemma 17.2.2 that $f_s'(x) = f_+'(x) = Wf'(x)$. $\square$

We include here an example to illustrate the necessity of the hypothesis in Theorem 17.2.4 and to show that Theorem 17.2.5 does not hold for generalized symmetric derivatives in general.

17.2.6. EXAMPLE. Define $f(x) = x$ when $x$ is positive and irrational, and define $f(x) = -x$ otherwise. Then it is easy to see that $f$ has an approximate symmetric derivative, as well as a qualitative symmetric derivative, equal to zero at $x = 0$, but it has on the other hand a unique lower derivative equal to $-1$ at that point. Thus Theorem 17.2.4 does not hold on replacing the $G$-nonangularity by ordinary nonangularity, and Theorem 17.2.5 does not hold for approximate and qualitative symmetric derivatives.

17.2.7. REMARK. It should be pointed out here that the points of $G$-angularity do not always form a countable set like the points of ordinary angularity.

This fact has been known for some time in the case of approximate angularity—that is, angularity relative to approximate derivates. Given a function $f$, let $A(f)$ denote the set of points where $f$ is approximately lower angular. It was proved by Ward [W1] that for every $F_\sigma$-set $E$ of measure zero there is a continuous function $f$ for which $E \subset A(f)$. Thus the set $A(f)$ can be uncountable; this fact was also proved later by Zajíček [Za1]. However, $A(f)$ is always of measure zero which follows from Theorem 1 of Ward [W1]; and in the case when $f$ is continuous, this set is meager by Theorems 15.4.2 and 16.6.3. Ward proved further that for every set $E$ of measure zero there is an approximately continuous function $f$ for which $E \subset A(f)$.

17.2.8. PROBLEM. It will be worthwhile to investigate hypotheses under which the points of $G$-angularity of a function form a countable set. Under this hypothesis, many of the results obtained in §§17.5 and 17.7 on generalized symmetric derivative would assume a simpler form as in the case of ordinary symmetric derivative.

It will be seen, however, in Chapter 19 (see Theorem 19.2.1) that a typical continuous function is $G$-nonangular in terms of all the generalized $G$-derivates considered in §15.3.

**§17.3 Calculus of generalized symmetric derivatives.** In this section we deal with the Calculus of generalized symmetric derivatives under various admissibility hypotheses.

First, in Theorem 17.3.1, we obtain the additivity and linearity of $s$-admissible generalized symmetric derivatives. Then, in Theorems 17.3.2 and 17.3.4, the product and quotient rules are obtained for $s^*$-admissible generalized symmetric derivatives. Further, in the cases of Borel and Cesàro symmetric derivatives which are not $s^*$-admissible, these rules are obtained under hypotheses similar to the ones used in §16.3B.

We assume throughout this section that $f,\ g \in \mathcal{F}(I)$, $x \in I$, and $c \in \mathbb{R}$.

We begin with the additivity and linearity of $s$-admissible generalized symmetric derivatives.

17.3.1. THEOREM. *Suppose symmetric $G$-derivates are $s$-admissible, $f$ is symmetrically $G$-derivable at $x$, and $g$ is symmetrically $G$-differentiable there. Then $f+g$ and $cf$ are symmetrically $G$-derivable at $x$ and*

$$G(f+g)'_s(x) = Gf'_s(x) + Gg'_s(x), \qquad G(cf)'_s(x) = c\,Gf'_s(x).$$

PROOF. The two results are obtained from Lemma 17.1.1 and Theorem 17.1.2 as in the case of Theorem 16.3.2. □

Next, in the following theorem we obtain the product rule for $s^*$-admissible generalized symmetric derivatives.

17.3.2. THEOREM. *Suppose symmetric $G$-derivates are $s^*$-admissible and $f$ and $g$ are $G$-continuous at $x$. Then if $f$ and $g$ are symmetrically $G$-differentiable at $x$, then so is $fg$ and*

$$(1) \qquad G(fg)'_s(x) = f(x)Gg'_s(x) + g(x)Gf'_s(x).$$

PROOF. The symmetric $G$-derivates are induced by hypothesis by $G$-limits and symmetric $G$-quotient where the $G$-limits are $s^*$-admissible and $GQ_s$ is $s^*$-admissible relative to $G$-limits. Now since $f$ and $g$ are $G$-continuous at $x$, there exist functions $f_G(x,h)$ and $g_G(x,h)$, $h \neq 0$, such that $G\text{-}\lim_{h\to 0} f_G(x,h) = f(x)$ and $G\text{-}\lim_{h\to 0} g_G(x,h) = g(x)$, and such that the following holds for $h > 0$ provided that $x+h \in I$:

$$GQ_s(fg)(x,h) = f_G(x,h)GQ_s g(x,h) + g_G(x,-h)GQ_s f(x,h).$$

Now since the $G$-limits are $s^*$-admissible, we obtain the following with the help of Theorems 15.1.6 and 15.1.9:

$$\begin{aligned} G(fg)'_s(x) &= G\text{-}\lim_{h\to 0+} f_G(x,h)GQ_s g(x,h) \\ &\quad + G\text{-}\lim_{h\to 0+} g_G(x,-h)GQ_s f(x,h) \\ &= f(x)Gg'_s(x) + g(x)Gf'_s(x). \quad \square \end{aligned}$$

17.3.3. LEMMA. *Suppose symmetric G-derivates are $s^*$-admissible, $f$ is G-continuous at $x$, and $f(x) \neq 0$. Then if $f$ is symmetrically G-differentiable at $x$, then so is $1/f$ and*

$$(2) \qquad G(1/f)'_s(x) = -\{1/f(x)\}^2 Gf'_s(x).$$

PROOF. This result is obtained from axiom (T8) and Theorem 15.1.9 by an argument similar to the one used for the last theorem. □

Now on combining Theorem 17.3.2 with this lemma we obtain the following quotient rule for $s^*$-admissible generalized symmetric derivatives.

17.3.4. THEOREM. *Suppose symmetric G-derivates are $s^*$-admissible, $f$ and $g$ are G-continuous at $x$, and $g(x) \neq 0$. Then if $f$ and $g$ are symmetrically G-differentiable at $x$, then so is $f/g$ and*

$$(3) \qquad G(f/g)'_s(x) = \{1/g(x)\}^2\{g(x)Gf'_s(x) - f(x)Gg'_s(x)\}.$$

Next, since the $B$-limits and symmetric $C$-quotient are defined in terms of integrals, the Borel and Cesàro symmetric derivates do not seem to be $s^*$-admissible as before, and so the product and quotient rules do not hold for Borel and Cesàro symmetric derivatives in general. In the rest of this section we obtain some hypotheses under which the product and quotient rules hold for Borel and Cesàro symmetric derivatives.

Given $f \in \mathscr{F}(I)$ and $x \in I$, we will call $f$ *symmetrically B-finite at* $x$, or simply *$B_s$-finite at* $x$, if the following limit is finite:

$$(4) \qquad \limsup_{h\to 0+} \limsup_{\delta\to 0+} \frac{1}{h}\int_\delta^h |Q_s f(x,t)|\,dt.$$

Similarly, we will call $f$ *symmetrically C-finite at* $x$, or simply *$C_s$-finite at* $x$, if the following limit is finite:

$$(5) \qquad \limsup_{h\to 0+} \frac{1}{4h^2}\int_0^{2h} |f(x+t) - f(x-t)|\,dt.$$

We will combine here, as before, the results on Borel and Cesàro symmetric derivatives due to their similar nature.

We begin with the following lemma on $B_s$- and $C_s$-finiteness similar to Lemma 16.3.7.

In analogy with the notion of $D$-finiteness, here we will call $f$ *symmetrically D-finite at* $x$, or simply *$D_s$-finite at* $x$, if the symmetric derivates of $f$ are finite at $x$. Similarly, we will call $f$ *$BD_s$-finite* or *$CD_s$-finite at* $x$ if its Borel or Cesàro symmetric derivates, respectively, are finite at $x$.

17.3.5. LEMMA. (a) *If $f$ is $D_s$-finite at $x$, then it is $B_s$- and $C_s$-finite at $x$.*
(b) *If $f$ is $B_s$- or $C_s$-finite at $x$, then it is $BD_s$- or $CD_s$-finite, respectively, at $x$.*

PROOF. To prove (a), suppose $f$ is $D_s$-finite at $x$. Then there is an $M \in \mathbb{R}_+$ and $\eta > 0$ such that $|Q_s f(x,t)| < M$ for $0 < t < \eta$. Hence if $0 < \delta < h < \eta$, then

$$\limsup_{\delta \to 0+} \frac{1}{h} \int_\delta^h |Q_s f(x,t)|\, dt \le \limsup_{\delta \to 0+} \frac{M}{h}(h - \delta) = M.$$

Hence the limit in (4) is bounded by $M$.

Further, if $0 < t < \eta$, then $|f(x+t) - f(x-t)| < 2Mt$. Hence,

$$\frac{1}{4h^2} \int_0^{2h} |f(x+t) - f(x-t)|\, dt \le \frac{1}{4h^2} \int_0^{2h} 2Mt\, dt = M,$$

which proves the limit in (5) to be bounded by $M$.

Next, to prove (b), first suppose $f$ is $B_s$-finite at $x$. Then the limit in (4) is bounded by some $M \in \mathbb{R}_+$. Hence

$$\begin{aligned} B\overline{D}_s f(x) &= \limsup_{h \to 0+} \limsup_{\delta \to 0+} \frac{1}{h} \int_\delta^h Q_s f(x,t)\, dt \\ &\le \limsup_{h \to 0+} \limsup_{\delta \to 0+} \frac{1}{h} \int_\delta^h |Q_s f(x,t)|\, dt \le M; \\ B\underline{D}_s f(x) &= \liminf_{h \to 0+} \liminf_{\delta \to 0+} \frac{1}{h} \int_\delta^h Q_s f(x,t)\, dt \\ &\ge -\limsup_{h \to 0+} \limsup_{\delta \to 0+} \frac{1}{h} \int_\delta^h |Q_s f(x,t)|\, dt \ge -M. \end{aligned}$$

The other part of (b) is proved by a similar argument. □

Next, we obtain the product rule for Borel and Cesàro symmetric derivatives in the following theorem.

17.3.6. THEOREM. *Suppose $G = B$ or $C$, and $f$ and $g$ are $G_s$-finite and continuous at $x$. Then if $f$ and $g$ are symmetrically $G$-differentiable at $x$, then so is $fg$ and the relation* (1) *holds.*

PROOF. First, suppose $f$ and $g$ are $B_s$-finite and symmetrically $B$-differentiable at $x$. It is easy to see here that

$$\begin{aligned} Q_s(fg)(x,t) &= f(x+t) Q_s g(x,t) + g(x-t) Q_s f(x,t) \\ &= f(x) Q_s g(x,t) + g(x) Q_s f(x,t) + S_1 + S_2, \end{aligned}$$

where

$$S_1 = \{f(x+t) - f(x)\}Q_s g(x,t), \qquad S_2 = \{g(x+t) - g(x)\}Q_s f(x,t).$$

Hence to obtain relation (1) in this case it is enough to show that

$$\lim_{h\to 0+} \lim_{\delta\to 0+} \frac{1}{h}\int_\delta^h S_i\,dt = 0 \quad \text{for} \quad i = 1,2.$$

However, using the $B_s$-finiteness and continuity of $f$ and $g$ at $x$, these limits are easily seen to be zero using an argument similar to the one used in the proof of Theorem 16.3.8.

Next, suppose $f$ and $g$ are $C_s$-finite and symmetrically $C$-differentiable at $x$. It is easy to see here that

$$\begin{aligned} f(x+t)g(x+t) &- f(x-t)g(x-t) \\ &= f(x)\{g(x+t) - g(x-t)\} \\ &\quad + g(x)\{f(x+t) - f(x-t)\} + T_1 + T_2 + T_3, \end{aligned}$$

where

$$\begin{aligned} T_1 &= \{f(x-t) - f(x)\}\{g(x+t) - g(x-t)\}, \\ T_2 &= \{g(x-t) - g(x)\}\{f(x+t) - f(x-t)\}, \\ T_3 &= \{f(x+t) - f(x-t)\}\{g(x+t) - g(x-t)\}. \end{aligned}$$

Hence we obtain

$$\begin{aligned} CQ_s f(x,h) &= Q_s(fg)_x^a(x,h) \\ &= \frac{1}{4h^2}\int_0^{2h}\{f(x+t)g(x+t) - f(x-t)g(x-t)\}\,dt \\ &= f(x)Q_s g_x^a(x,h) + g(x)Q_s f_x^a(x,h) + \sum_{i=1}^{3}\frac{1}{4h^2}\int_0^{2h} T_i\,dt. \end{aligned}$$

Thus to obtain relation (1) in this case it is enough to show that

$$\lim_{h\to 0+} \frac{1}{4h^2}\int_0^{2h} T_i\,dt = 0 \quad \text{for} \quad i = 1,2,3.$$

However, using the $C_s$-finiteness and continuity of $f$ and $g$ at $x$, each of these three limits is easily seen to be zero as in the proof of Theorem 16.3.8.

□

Next, to obtain the quotient rule we need the following lemma.

17.3.7. LEMMA. *Suppose $G = B$ or $C$, $f$ is continuous at $x$, and $f(x) \neq 0$. Then if $f$ is $G_s$-finite and symmetrically $G$-differentiable at $x$, then so is $1/f$ and the relation (2) holds.*

PROOF. We need here the following relation which is easy to verify:

$$(6)\quad \frac{1}{f(x+t)} - \frac{1}{f(x-t)}$$
$$= -\frac{f(x+t) - f(x-t)}{\{f(x)\}^2}$$
$$+ \frac{f(x+t)f(x-t) - \{f(x)\}^2}{f(x+t)f(x-t)\{f(x)\}^2}\{f(x+t) - f(x-t)\}.$$

First, suppose $f$ is $B_s$-finite and symmetrically $B$-differentiable at $x$. It is clear from (6) that

$$Q_s(1/f)(x,t) = -\{f(x)\}^{-2}Q_sf(x,t) + T,$$

where

$$T = \frac{f(x+t)f(x-t) - \{f(x)\}^2}{f(x+t)f(x-t)\{f(x)\}^2}Q_sf(x,t).$$

We claim that the integral $(1/h)\int_\delta^h |T|\,dt$ converges to zero as $\delta \to 0+$ and $h \to 0+$.

Since $f$ is $B_s$-finite at $x$, the limit in (4) is bounded by some $M \in \mathbb{R}_+$. Further, given $\varepsilon > 0$, since $f$ is continuous at $x$, $f(x+t)f(x-t)$ is continuous in $t$ at $t = 0$, and since $|f(x)| > 0$, we can choose an $\eta > 0$ such that if $0 < t < \eta$, then $|f(x+t)f(x-t)| > \{f(x)\}^2/2$ and

$$\left|f(x+t)f(x-t) - \{f(x)\}^2\right| < \varepsilon|f(x)|^4/(2M).$$

Hence if $0 < t < \eta$, then

$$|T| < \frac{\varepsilon|f(x)|^4 2}{2M|f(x)|^4}|Q_sf(x,t)| = \frac{\varepsilon}{M}|Q_sf(x,t)|.$$

Consequently, we obtain

$$\limsup_{h\to 0+}\ \limsup_{\delta\to 0+} \frac{1}{h}\int_\delta^h |T|\,dt \le \frac{\varepsilon}{M}\limsup_{h\to 0+}\ \limsup_{\delta\to 0+}\frac{1}{h}\int_\delta^h |Q_sf(x,t)|\,dt = \varepsilon.$$

It is now clear from above that $1/f$ is $B_s$-finite and symmetrically $B$-differentiable at $x$, and that relation (2) holds in this case for $G = B$.

Next, suppose $f$ is $C_s$-finite and symmetrically $C$-differentiable at $x$. We obtain here the following with the help of (6):

$$\begin{aligned} CQ_s(1/f)(x,h) &= Q_s(1/f)_x^a(x,h) \\ &= \frac{1}{4h^2}\int_0^{2h}\left\{\frac{1}{f(x+t)} - \frac{1}{f(x-t)}\right\}dt \\ &= -\frac{1}{\{f(x)\}^2}\frac{1}{4h^2}\int_0^{2h}\{f(x+t) - f(x-t)\}\,dt + I \\ &= -CQ_s f(x,h)/\{f(x)\}^2 + I, \end{aligned}$$

where

$$I = \frac{1}{4h^2}\int_0^{2h}\frac{f(x+t)f(x-t) - \{f(x)\}^2}{f(x+t)f(x-t)\{f(x)\}^2}\{f(x+t) - f(x-t)\}\,dt.$$

It is proved by an argument similar to above that the integral

$$\frac{1}{4h^2}\int_0^{2h}\frac{|f(x+t)f(x-t) - \{f(x)\}^2|}{|f(x+t)f(x-t)|\{f(x)\}^2}|f(x+t) - f(x-t)|\,dt$$

converges to zero as $h \to 0+$. Hence it is clear that $1/f$ is $C_s$-finite and symmetrically $C$-differentiable at $x$ and that (2) holds in this case for $G = C$.

□

Finally, on combining the above lemma with Theorem 17.3.6, we obtain the following quotient rule for Borel and Cesàro symmetric derivatives.

17.3.8. THEOREM. *Suppose $G = B$ or $C$, $f$ and $g$ are $G_s$-finite and continuous at $x$, and $g(x) \neq 0$. Then if $f$ and $g$ are symmetrically $G$-differentiable at $x$, then so is $f/g$ and the relation* (3) *holds.*

17.3.9. PROBLEM. We have not investigated here as before the chain rule for generalized symmetric derivatives, and so it remains to investigate hypotheses under which the chain rule holds for each admissible generalized symmetric derivative.

**§17.4 Mean value theorems and l'Hôpital's rule in terms of generalized symmetric derivatives.** In this section we obtain a mean value theorem along with Cauchy's mean value theorem in terms of generalized symmetric derivatives. Further, from the latter we deduce l'Hôpital's rule in terms of such derivatives.

We include here the results obtained from these three theorems in the particular case of ordinary symmetric derivative since they are not known. Also, we include here the mean value theorem in terms of approximate symmetric derivative that is obtained.

The following mean value theorem is obtained directly from the General Mean Value Theorem 8.1.2 with the help of the Second Compatibility Theorem 17.2.4.

17.4.1. THEOREM (SIXTH MEAN VALUE). *Let $I = [a, b]$. Suppose symmetric G-derivates are $s_1$-admissible, $f : I \to \mathbb{R}$ is $I^*C$, and it is symmetrically G-derivable and G-nonangular at the points where it is semidifferentiable. Then there is a point $x \in I^0$ such that f is symmetrically G-derivable at x and $Qf(a, b) = Gf_s'(x)$.*

In the case of ordinary symmetric derivative it is well known that the mean value theorem does not hold for it in general; consider $f(x) = |x|$, $x \in \mathbb{R}$. However, since the ordinary and approximate symmetric derivates are $s_1$-admissible by the Seventh Status Theorem 17.1.7, from the above theorem we obtain the following mean value theorem for ordinary and approximate symmetric derivatives.

17.4.2. COROLLARY. *Suppose $f : I \to \mathbb{R}$ is $I^*C$, where $I = [a, b]$.*

(a) *If f is symmetrically derivable and nonangular at the points where it is semidifferentiable, then there is a point $x \in I^0$ such that f is symmetrically derivable at x and $Qf(a, b) = f_s'(x)$.*

(b) *If f is approximately symmetrically derivable and A-nonangular at the points where it is semidifferentiable, then there is a point $x \in I^0$ such that f is approximately symmetrically derivable at x and $Qf(a, b) = Af_s'(x)$.*

Next, we deduce from the above theorem the following version of Cauchy's mean value theorem in terms of generalized symmetric derivative.

17.4.3. THEOREM (THIRD CAUCHY'S MEAN VALUE). *Let $f, g : I \to \mathbb{R}$, where $I = [a, b]$. Suppose* (i) *symmetric G-derivates are $s_1$-admissible,* (ii) *f is $I^*C$ and g is continuous on I,* (iii) *f is symmetrically G-derivable and G-nonangular on $I^0$, and* (iv) *g is G-differentiable on $I^0$ and $Gg'(x) \neq 0$ for $x \in I^0$. Then there is a point $x \in I^0$ such that*

(1) $$\frac{f(b) - f(a)}{g(b) - g(a)} = \frac{Gf_s'(x)}{Gg'(x)}.$$

PROOF. It is clear from the Fifth Mean Value Theorem 16.4.1 that $g(b) \neq g(a)$, so that the ratio on the left side of (1) is defined. Let $c$ denote this ratio, and define

$$h(x) = f(x) - f(a) - c\{g(x) - g(a)\}, \qquad x \in I.$$

Then it follows from Theorem 17.3.1 that $h$ is symmetrically $G$-derivable on $I^0$, and it follows from Theorem 15.2.4 that $h$ is $G$-nonangular. Further, since $h(a) = h(b) = 0$, it follows now from Theorem 17.4.1 that there is a point $x \in I^0$ such that $Gh'_s(x) = 0$. Consequently, by Theorem 17.3.1, $0 = Gh'_s(x) = Gf'_s(x) - cGg'(x)$, which proves (1). □

In the case of ordinary symmetric derivative we obtain the following version of Cauchy's mean value theorem from the above theorem as before.

17.4.4. Corollary. *Let $f$, $g : I \to \mathbb{R}$, where $I = [a, b]$. Suppose* (i) *$f$ is $I^*C$ and $g$ is continuous on $I$,* (ii) *$f$ is symmetrically derivable and nonangular on $I^0$, and* (iii) *$g$ is differentiable on $I^0$ and $g'(x) \neq 0$ for $x \in I^0$. Then there is a point $x \in I^0$ such that*

$$\frac{f(b) - f(a)}{g(b) - g(a)} = \frac{f'_s(x)}{g'(x)}.$$

Finally, from the last theorem we obtain the following version of l'Hôpital's rule in terms of generalized symmetric derivative.

17.4.5. Theorem (Third L'Hôpital's Rule). *Let $f, g : I \to \mathbb{R}$, where $I = (a, b)$, $-\infty \leq a < b < \infty$. Suppose* (i) *symmetric $G$-derivates are $s_1$-admissible,* (ii) *$f$ is $I^*C$ and $g$ is continuous on $I$,* (iii) *$f$ is symmetrically $G$-derivable and $G$-nonangular, and* (iv) *$g$ is $G$-differentiable and $Gg'(x) \neq 0$ for $x \in I$. Then if the following limit on the left side is of indeterminate form $0/0$ or $\infty/\infty$, then*

$$\lim_{x \to a+} \frac{f(x)}{g(x)} = \lim_{x \to a+} \frac{Gf'_s(x)}{Gg'(x)},$$

*provided that the limit on the right exists* (*finite or infinite*).

Proof. We omit the proof here since this result is obtained by an argument similar to the one used for the General l'Hôpital's Rule 8.2.3, using this time the above Cauchy's Mean Value Theorem 17.4.3 in the place of Theorem 8.2.1, and on replacing $Sf'$ by $Gf'_s$, $g'$ by $Gg'$, $\varepsilon$ by $=$, and $\subset$ by $\in$. □

In terms of ordinary symmetric derivative we thus obtain the following version of l'Hôpital's rule as before.

17.4.6. Corollary. *Let $f,g : I \to \mathbb{R}$, where $I = (a, b)$, $-\infty \leq a < b < \infty$. Suppose* (i) *$f$ is $I^*C$ and $g$ is continuous on $I$,* (ii) *$f$ is nonangular and symmetrically derivable, and* (iii) *$g$ is differentiable and $g'(x) \neq 0$ for $x \in I$. Then if the following limit on the left side is of indeterminate form $0/0$ or $\infty/\infty$, then*

$$\lim_{x \to a+} \frac{f(x)}{g(x)} = \lim_{x \to a+} \frac{f'_s(x)}{g'(x)}$$

*provided that the limit on the right exists* (*finite or infinite*).

In conclusion it may be pointed out here that as in Corollary 17.4.2, results similar to Corollaries 17.4.4 and 17.4.6 also follow from Theorems 17.4.3 and 17.4.5, respectively, in the case of approximate symmetric derivative.

**§17.5 Monotonicity and Goldowski-Tonelli theorems in terms of generalized symmetric derivatives.** In this section we obtain some monotonicity theorems in terms of generalized symmetric derivates and derivative. Also, a version of the Goldowski-Tonelli theorem is obtained in terms of such derivatives, and weaker versions of the Third and Fourth Derivate Theorems are obtained as in §16.5 in terms of generalized symmetric derivates.

First, in Theorems 17.5.1, we obtain a monotonicity theorem in terms of generalized symmetric derivates. Then in Theorems 17.5.3, 17.5.4, and 17.5.6 we deduce from the monotonicity results of Chapter 9 similar results in terms of generalized symmetric derivative. Then in Theorem 17.5.8 we obtain the Goldowski-Tonelli theorem in terms of generalized symmetric derivative. Similar monotonicity results are deduced from these theorems in terms of ordinary symmetric derivates and derivative.

Finally, in Theorems 17.5.10 and 17.5.11, we deduce from Theorem 17.5.1 weaker versions of the Third and Fourth Derivate Theorems in terms of generalized symmetric derivates.

It may be recalled here that we assume here that $f \in \mathscr{F}(I)$, where $I$ is assumed to be open unless stated otherwise.

The following monotonicity theorem in terms of generalized symmetric derivates is obtained directly from Theorem 16.5.1 with the help of Lemma 17.2.3.

17.5.1. THEOREM. *Suppose symmetric G-derivates are $s_1$-admissible, $I$ is closed, $f$ is upper G-nonangular* n.e., *and $f$ is either* (i) $I^*LC$, *or* (ii) $LI_+$ *and* $I^*LC_-$. *Then if* $G\underline{D}_s f \geq 0$ a.e. *and* $> -\infty$ n.e., *then $f$ is nondecreasing.*

Now since the ordinary and approximate symmetric derivates are $s_1$-admissible by the Seventh Status Theorem 17.1.7, and every function is nonangular in the ordinary sense n.e. (see Theorem 2.3.1), we obtain the following monotonicity results from the above theorem in terms of ordinary and approximate symmetric derivates.

17.5.2. COROLLARY. *Suppose $I$ is closed and $f$ is either* (i) $I^*LC$, *or* (ii) $LI_+$ *and* $I^*LC_-$. (a) *If* $\underline{D}_s f \geq 0$ a.e. *and* $> -\infty$ n.e., *then $f$ is nondecreasing.*

(b) *If $f$ is upper A-nonangular* n.e. *and* $A\underline{D}_s f \geq 0$ a.e. *and* $> -\infty$ n.e., *then $f$ is nondecreasing.*

Next, on account of the Second Compatibility Theorem 17.2.4, we obtain the following monotonicity result from the Second Monotonicity Theorem 9.1.3.

17.5.3. THEOREM. *Suppose symmetric G-derivates are $s_1$-admissible, f is $I^*LSC$, and f is symmetrically G-derivable and lower G-nonangular at the points where it is lower differentiable. If the set of points where f has a finite negative (or positive) symmetric G-derivative is of cardinality $< c$, then f is nondecreasing (or nonincreasing).*

Using the above-mentioned compatibility once again, we obtain the following result from Theorem 9.1.5 which may be compared with Theorem 9.3.3.

17.5.4. THEOREM. *Suppose symmetric G-derivates are $s_1$-admissible and f is symmetrically G-derivable and lower G-nonangular at nearly all the points where it is lower derivable. Given $a, b \in I$, $a < b$, let $E_-$ and $E_+$ denote the sets of points in $(a, b)$ where f has a negative or positive symmetric G-derivative, respectively. Then if f is $R^*LI$, then $f(a) - f(b) \leq |f(E_-)|$; and if f is $L^*UI$, then $f(b) - f(a) \leq |f(E_+)|$.*

This theorem yields in turn, as before, the following result on ordinary symmetric derivative.

17.5.5. COROLLARY. *Suppose f is symmetrically derivable at nearly all the points where it is lower derivable. Given $a, b \in I$, $a < b$, let $E_-$ and $E_+$ denote the sets of points in $(a, b)$ where f has a negative or positive symmetric derivative, respectively. Then if f is $R^*LI$, then $f(a) - f(b) \leq |f(E_-)|$; and if f is $L^*UI$, then $f(b) - f(a) \leq |f(E_+)|$.*

Let $f$ be called, as before, *lower symmetrically G-regular* or *lower symmetrically G-singular* if it is lower regular or lower singular, respectively, relative to the symmetric $G$-derivative (see §6.2 for definitions). Using these notions, the following monotonicity theorem holds in terms of generalized symmetric derivative similar to Theorem 16.5.4.

17.5.6. THEOREM. *Suppose symmetric G-derivates are $s_1$-admissible, f is symmetrically G-derivable and lower G-nonangular at nearly all the points where it is lower derivable, and f is either* (i) *$I^*LC$, or* (ii) *$LI_-$ and $I^*LC_+$. Then if f is lower symmetrically G-regular and lower symmetrically G-singular, then it is nondecreasing.*

PROOF. Under the present hypothesis it follows clearly from Theorem 17.2.4 that $Gf_s'$ is compatible with $Lf'$ at nearly all the points where the latter is defined. Hence the result follows from the Third Monotonicity Theorem 9.1.7. □

From this theorem we obtain as before the following monotonicity result in terms of ordinary symmetric derivative.

17.5.7. Corollary. *Suppose $f$ is either* (i) $I^*LC$, *or* (ii) $LI_-$ *and* $I^*LC_+$. *Then if $f$ is symmetrically derivable at nearly all the points where it is lower derivable, and it is lower symmetrically regular and lower symmetrically singular, then it is nondecreasing.*

It may be observed here that a result similar to the Fifth Monotonicity Theorem 9.1.10 can also be deduced from this theorem with the help of Theorem 17.2.4.

Next, in the following theorem we obtain a version of the Goldowski-Tonelli theorem in terms of generalized symmetric derivative similar to Theorem 16.5.7.

17.5.8. Theorem (Fourth Goldowski-Tonelli). *Suppose symmetric $G$-derivates are $s_1$-admissible and regular, and $f$ is $I^*C$ and upper $G$-nonangular. Then if $f$ is symmetrically $G$-derivable* n.e. *and* $Gf'_s \geq 0$ a.e., *then $f$ is nondecreasing.*

Proof. We omit the proof here since this result is obtained by an argument similar to the one given for Theorem 16.5.7 on replacing $G\underline{D}f$, $G\overline{D}f$, and $Gf'$ by $G\underline{D}_sf$, $G\overline{D}_sf$, and $Gf'_s$, respectively, and using Theorem 17.5.1 in the place of Theorem 16.5.1; for under the present hypothesis it follows from Lemma 17.2.3 that $G\overline{D}_sf \geq GD_-f$, and from Theorem 17.1.9 and axiom (D2) it follows that $G\underline{D}_sf \geq G\underline{D}f \geq \underline{D}f$. □

Now since the ordinary symmetric derivates are $s_1$-admissible and regular by Theorems 17.1.7 and 17.1.11, from the above theorem we obtain the following version of the Goldowski-Tonelli theorem in terms of ordinary symmetric derivative.

17.5.9. Corollary. *Suppose $f$ is $I^*C$ and upper nonangular. Then if $f$ is symmetrically derivable* n.e. *and* $f'_s \geq 0$ a.e., *then $f$ is nondecreasing.*

It will be seen later in Theorem 17.7.7 that Theorem 17.5.8 holds also for $s_1$-admissible subregular symmetric $G$-derivates provided that $f$ is symmetrically $G$-derivable everywhere.

Next, in the case of $s_1$-admissible regular generalized symmetric derivates, we obtain the following weaker version of the Third Derivate Theorem 2.4.1 from Theorem 17.5.1 and axiom (S3) by an argument similar to the one used for Theorem 16.5.8.

17.5.10. Theorem. *Suppose symmetric $G$-derivates are $s_1$-admissible and regular, and $f : I \to \mathbb{R}$ is $IC$ and upper $G$-nonangular* n.e. *Then if $G\underline{D}_sf > 0$ at a nonmeager set of points, then $f$ is strongly increasing on some portion of $I$.*

Further, from the above theorem we obtain the following weaker version of the Fourth Derivate Theorem 2.4.4 by an argument similar to the one used for Theorem 16.5.9.

17.5.11. THEOREM. *Suppose symmetric G-derivates are $s_1$-admissible and regular, and $f: I \to \mathbb{R}$ is IC and G-nonangular* n.e. *Then if $G\underline{D}_s f$ and $G\overline{D}_s f$ are finite at a nonmeager set of points, then f is Lipschitz on some portion of I.*

*Moreover, if there is some $c \in \mathbb{R}_+$ such that $G\underline{D}_s f > -c$ and $G\overline{D}_s f < c$ at a nonmeager set of points, then f is c-Lipschitz on some portion of I.*

Finally, in the case of ordinary symmetric derivates we obtain the following result as before from the last two theorems with the help of Theorem 2.3.1.

17.5.12. COROLLARY. *Suppose $f: I \to \mathbb{R}$ is IC.* (a) *If $\underline{D}_s f > 0$ at a nonmeager set of points, then f is strongly increasing on some portion of I.*

(b) *If there is some $c \in \mathbb{R}_+$ such that $\underline{D}_s f > -c$ and $\overline{D}_s f < c$ at a nonmeager set of points, then f is c-Lipschitz on some portion of I.*

17.5.13. REMARK. In connection with part (b) of Corollary 17.5.2, it may be noted here that Mukhopadhyay tried to prove the following proposition in [Mu]: If $f$ is approximately continuous and approximately symmetrically derivable everywhere, and $Af'_s \geq 0$ everywhere, then $f$ is nondecreasing. However, an error was found later in his proof as well as in the proofs given by some others, which unfortunately could not be salvaged. The following result was proved finally by Freiling and Rinne [FR] in this direction with the help of a symmetric density property: If $f:[a,b] \to \mathbb{R}$ is continuous and $A\underline{D}_s f \geq 0$ on $(a, b)$, then $f$ is nondecreasing. It will be interesting to find out in this connection how the hypothesis of upper $A$-nonangularity n.e. used in Corollary 17.5.2(b) has been overcome in the proof given by Freiling and Rinne for this theorem.

Further, it should be noted here that although Theorems 16.5.8 and 16.5.9 hold in the cases of preponderant, approximate, and subcontinuous path derivates (see Theorems 15.4.2 and 15.4.4), Theorems 17.5.10 and 17.5.11 do not seem to hold for the corresponding symmetric derivates; this is because the preponderant symmetric derivates are not $s$-admissible, and the approximate and subcontinuous path symmetric derivates do not seem to be regular in general.

## §17.6 Relations between generalized symmetric derivates and strong derivates.

In this section we obtain two theorems on the relations between generalized symmetric derivates and strong derivates similar to the theorems of Dini and Denjoy. Also, from these theorems we deduce similar results on ordinary and approximate symmetric derivates.

The following theorem in the direction of Dini's Theorem 14.3.1(a) is obtained with the help of Lemma 17.2.3 using arguments similar to the ones used for Theorem 16.6.1(a).

17.6.1. THEOREM. *Suppose symmetric G-derivates are $s_1$-admissible and f is either* (i) *$I^*USC$,* (ii) *$I^*LC$, or* (iii) *$LI_+$ and $I^*LC_-$. Then if f is upper G-nonangular* n.e., *then $G\underline{D}_s f(x) = \underline{D}_* f(x)$ iff $G\underline{D}_s f$ is LSC at x.*

It is easy to see that the above theorem holds also when $f$ is lower $G$-nonangular under the continuity hypotheses obtained on applying the above hypotheses to $-f_-$.

Now since the ordinary and approximate symmetric derivates are $s_1$-admissible by Theorem 17.1.7, and every function is nonangular in the ordinary sense n.e., from this theorem we obtain the following relation between the ordinary and approximate symmetric derivates and strong derivates.

17.6.2. COROLLARY. *Suppose $f$ is either* (i) $I^*USC$, (ii) $I^*LC$, *or* (iii) $LI_+$ *and* $I^*LC_-$. *Then*

(a) $\underline{D}_s f(x) = \underline{D}_* f(x)$ *iff* $\underline{D}_s f$ *is LSC at* $x$; *and*

(b) *if $f$ is upper A-nonangular* n.e., *then* $A\underline{D}_s f(x) = \underline{D}_* f(x)$ *iff* $A\underline{D}_s f$ *is LSC at* $x$.

Next, in the following theorem we obtain a result similar to the Denjoy's Theorem 2.5.4 for generalized symmetric derivates. This theorem is obtained from Theorem 17.6.1, and the result obtained on applying it to $-f$, by an argument similar to the one used for Theorem 16.6.3.

17.6.3. THEOREM. *Suppose symmetric G-derivates are $s_1$-admissible and subregular. Then if $f$ is IC and lower or upper G-nonangular* n.e., *then* $G\underline{D}_s f = \underline{D}f = \underline{D}_* f$ *and* $G\overline{D}_s f = \overline{D}f = \overline{D}_* f$ r.e. *in* $I$.

Now since the ordinary and approximate symmetric derivates are $s_1$-admissible and subregular by the Seventh and Eighth Status Theorems 17.1.7 and 17.1.10, from this theorem we obtain the following result on these derivates with the help of Theorem 2.3.1.

17.6.4. COROLLARY. *Suppose $f$ is IC. Then*

(a) $\underline{D}_s f = \underline{D}f = \underline{D}_* f$ *and* $\overline{D}_s f = \overline{D}f = \overline{D}_* f$ r.e. *in* $I$; *and*

(b) *if $f$ is lower or upper A-nonangular* n.e., *then* $A\underline{D}_s f = \underline{D}f = \underline{D}_* f$ *and* $A\overline{D}_s f = \overline{D}f = \overline{D}_* f$ r.e. *in* $I$.

**§17.7 Properties of generalized symmetric derivatives.** In this section we obtain the Darboux and Denjoy properties of generalized symmetric derivatives. Also, the Borel class of such derivability sets and the Baire class of these derivatives are obtained. The corresponding results in the particular cases of ordinary and approximate symmetric derivatives are deduced from the results obtained.

First, in the following theorem we deal with the Darboux property of generalized symmetric derivatives.

17.7.1. THEOREM. *Suppose symmetric G-derivates are $s_1$-admissible and $f$ is $I^*C$ and G-nonangular, and it is symmetrically G-derivable at the points where it is semidifferentiable. Then the symmetric G-derivative of $f$ possesses the Darboux property.*

Proof. This result is obtained by an argument similar to the one used for Theorem 16.7.1 on replacing $Gf'$ by $Gf'_s$, and using Theorem 17.4.1 in the place of Theorem 16.4.1; for under the present hypothesis it follows from Lemma 17.2.3 and axiom (D2) that $D_+f \leq GD_+f \leq G\overline{D}_sf$ and $G\underline{D}_sf \leq GD^-f \leq D^-f$. □

Now since the ordinary and approximate symmetric derivates are $s_1$-admissible by Theorem 17.1.7, we thus obtain the following result on the Darboux property of ordinary and approximate symmetric derivatives.

17.7.2. Corollary. *Suppose $f$ is $I^*C$.* (a) *If $f$ is nonangular and symmetrically derivable at the points where it is semidifferentiable, then its symmetric derivative possesses the Darboux property.*

(b) *If $f$ is A-nonangular and approximately symmetrically derivable at the points where it is semidifferentiable, then its approximate symmetric derivative possesses the Darboux property.*

Next, let $GS\Delta(f)$ and $GS\Delta^*(f)$ denote the two symmetric $G$-derivability sets of $f$ as defined earlier, and let $GK_s(f)$ denote the symmetric $G$-knot set of $f$. In the particular case of ordinary symmetric derivative, the reference to $G$ will be dropped from these notations.

The following result on the Borel class of these sets is obtained by usual arguments (see the proofs of Theorems 10.1.2 and 10.1.4).

17.7.3. Theorem. *Suppose $f$ is continuous and symmetric $G$-derivates are regular. Then the symmetric $G$-derivability sets $GS\Delta(f)$ and $GS\Delta^*(f)$ are $F_{\sigma\delta}$-sets, and the symmetric $G$-knot set $GK_s(f)$ is a $G_\sigma$-set.*

*Further, in the case when the symmetric $G$-derivates are totally regular, the same holds for every function $f$.*

Now, in the particular case of ordinary symmetric derivative we obtain the following result since such derivates are regular by Theorem 17.1.11.

17.7.4. Corollary. *Suppose $f$ is continuous. Then the symmetric derivability sets $S\Delta(f)$ and $S\Delta^*(f)$ are $F_{\sigma\delta}$-sets, and the symmetric knot set $K_s(f)$ is a $G_\delta$-set.*

The following theorem on the Baire class of generalized symmetric derivative is obtained by an argument similar to the one used for Theorem 16.7.3.

17.7.5. Theorem. *If symmetric $G$-derivates are subregular and $f$ is continuous, then $Gf'_s$ is of $B_1$; and when the symmetric $G$-derivates are totally subregular, the same holds for every function $f$.*

Now with the help of the Eighth Status Theorem 17.1.10 we obtain the following.

17.7.6. COROLLARY. *The preponderant, approximate, and ordinary symmetric derivatives of every function $f$ are of $B_1$.*

In the case of ordinary symmetric derivative, this result has been obtained by Filipczak [F] for continuous functions.

Next, with the help of Theorem 17.7.5, we obtain from the Fourth Goldowski-Tonelli Theorem 17.5.8 its following version in the case of $s_1$-admissible subregular symmetric $G$-derivates.

17.7.7. THEOREM. *Suppose symmetric $G$-derivates are $s_1$-admissible and subregular, and $f$ is $I^*C$ and upper $G$-nonangular. Then if $f$ is symmetrically $G$-derivable everywhere and $Gf'_s \geq 0$ a.e., then $f$ is nondecreasing.*

From this theorem we obtain the following version of the Goldowski-Tonelli theorem in terms of approximate symmetric derivative with the help of Theorems 17.1.7 and 17.1.10.

17.7.8. COROLLARY. *Suppose $f$ is $I^*C$, upper $A$-nonangular and approximately symmetrically derivable everywhere. Then if the approximate symmetric derivative of $f$ is $\geq 0$ a.e., then $f$ is nondecreasing.*

Finally, in the following theorem we obtain the Denjoy property of generalized symmetric derivative.

17.7.9. THEOREM. *Suppose symmetric $G$-derivates are $s_1$-admissible and regular, and $f$ is continuous and $G$-nonangular. Then if $f$ is symmetrically $G$-derivable n.e., then $Gf'_s$ possesses the Denjoy property.*

*Further, the same holds when the symmetric $G$-derivates are $s_1$-admissible and subregular provided that $f$ is symmetrically $G$-derivable everywhere.*

PROOF. We omit the proof of the first part here since it is obtained by an argument similar to the one used for Theorem 16.7.5 on replacing $G\underline{D}f$, $G\overline{D}f$, and $Gf'$ by $G\underline{D}_s f$, $G\overline{D}_s f$, and $Gf'_s$, respectively; for under the hypothesis of this part it follows from Lemma 17.2.3 and axiom (D2) that $D_{\pm} f \leq GD_{\pm} f \leq G\overline{D}_s f$ and $G\underline{D}_s f \leq GD^{\pm} f \leq D^{\pm} f$.

The second part follows on the other hand from the first, for under the hypothesis of the second part $Gf'_s$ is of $B_1$ by Theorem 17.7.5, and so the symmetric $G$-derivates of $f$ are regular. □

From the two parts of the above theorem we obtain as before the following result on ordinary and approximate symmetric derivatives with the help of Theorems 17.1.7, 17.1.10, and 17.1.11.

17.7.10. COROLLARY. *Suppose f is continuous.*

(a) *If f is nonangular and symmetrically derivable* n.e., *then the symmetric derivative of f possesses the Denjoy property.*

(b) *If f is A-nonangular and approximately symmetrically derivable everywhere, then the approximate symmetric derivative of f possesses the Denjoy property.*

**§17.8 New generalized symmetric derivates and derivatives.** To demonstrate the applicability of the theory presented in this chapter to new generalized symmetric derivates and derivatives, we present in this section several new generalized symmetric derivates, similar to the ones considered in §16.8, and we investigate the status of their admissibility and regularity. We do not use titles, as before, for the theorems of this section, but each of the five theorems obtained in this section will be referred simply as a Status Theorem.

First, let $s$ be any given symmetric selection and let $G$-limits be any of the generalized limits considered in Theorem 15.3.3 except for (a), (g), (j), and (k). Then, as defined in §17.1B, the symmetric derivates induced by $G$-limits and the selective symmetric quotient $sQ_s$ are called *symmetric Gs-derivates.* We will call all such derivates simply *generalized selective symmetric derivates.* For example, if $K$ is some given symmetric congruence, then

$$Ks\underline{D}_s f(x) = K\text{-}\liminf_{h\to 0+} sQ_s f(x,h) = \liminf_{h\in K;\, h\to 0+} Q_s f(x, s_{[x,x+h]} - x);$$

and if $s'$ is some other given symmetric selection, then

$$s's\underline{D}_s f(x) = s'\text{-}\liminf_{h\to 0+} sQ_s f(x,h) = \liminf_{h\to 0+} Q_s f(x, s_{[x, s'_{[x,x+h]}]} - x).$$

It may be recalled here that according to the convention of this chapter (see §17.1C), if $G$-limits are defined in terms of any of the entities $K, \mathcal{N}, s$, or $\mathcal{P}$ defined in §15.3, then this entity will be assumed implicitly to be "symmetric" as defined in §15.5. This symmetry is indeed needed for the $s_1$-admissibility of the corresponding generalized symmetric derivates.

In the following theorem we deal with the status of admissibility of all the generalized selective symmetric derivates defined above.

17.8.1. THEOREM. *The following generalized selective symmetric derivates are* $s_1^*$*-admissible*:

(a) *approximate selective symmetric derivates,*

(b) *sequentially congruent selective symmetric derivates,*

(c) *congruent selective symmetric derivates,*

(d) *qualitative selective symmetric derivates,*

(e) *negligent selective symmetric derivates,*

(f) *selective selective symmetric derivates, and*

(g) *path selective symmetric derivates.*

*Further,* (h) *preponderant selective symmetric derivates are admissible.*

PROOF. Let $s$ by any given symmetric selection. Then the selective symmetric quotient $sQ_s$ is $s_1^*$-admissible by the Sixth Status Theorem 17.1.5. Hence in each case the result follows from the Third Induction Theorem 17.1.3 with the help of the Fifth Status Theorem 15.5.2. □

Next, let $G$-limits be as before any of the generalized limits considered in Theorem 15.3.3 except for (a), (g), (j), and (k). Then the symmetric derivates induced by $G$-limits and the Cesàro symmetric quotient $CQ_s$ are called *symmetric GC-derivates*. We will call all such derivates simply *generalized Cesàro symmetric derivates*. For example, if $K$ is some given symmetric congruence, then

$$KC\underline{D}_s f(x) = K\text{-}\liminf_{h\to 0+} CQ_s f(x,h) = \liminf_{h\in K;\, h\to 0+} Q_s f_x^a(x,h);$$

and if $s$ is some given symmetric selection, then

$$sC\underline{D}_s f(x) = s\text{-}\liminf_{h\to 0+} CQ_s f(x,h) = \liminf_{h\to 0+} Q_s f_x^a(x, s_{[x,x+h]} - x).$$

In the following theorem we deal with admissibility of all the generalized Cesàro symmetric derivates defined above.

17.8.2. THEOREM. *The following generalized Cesàro symmetric derivates are* $s_1$*-admissible:*

(a) *approximate Cesàro symmetric derivates,*
(b) *sequentially congruent Cesàro symmetric derivates,*
(c) *congruent Cesàro symmetric derivates,*
(d) *qualitative Cesàro symmetric derivates,*
(e) *negligent Cesàro symmetric derivates,*
(f) *selective Cesàro symmetric derivates, and*
(g) *path Cesàro symmetric derivates.*

*Further,* (h) *preponderant Cesàro symmetric derivates are admissible.*

PROOF. Since the Cesàro symmetric quotient $CQ_s$ is $s_1$-admissible by the Sixth Status Theorem 17.1.5, in each of these cases the result follows directly from the Third Induction Theorem 17.1.3 and the Fifth Status Theorem 15.5.2 as above. □

Next, we investigate regularity of all the new generalized symmetric derivates considered in the above two theorems. First, in the following theorem we obtain total regularity of preponderant selective and preponderant Cesàro symmetric derivates as in Theorem 16.8.3.

17.8.3. THEOREM. *The* (a) *preponderant selective symmetric derivates and* (b) *preponderant Cesàro symmetric derivates are totally regular.*

*Consequently, the* (c) *selective symmetric derivates,* (d) *Cesàro symmetric derivates,* (e) *approximate selective symmetric derivates, and* (f) *approximate Cesàro symmetric derivates are totally subregular.*

PROOF. Let $s$ be any given symmetric selection. In cases (a) and (b) the result is obtained by an argument similar to the one used for the Third Status Theorem 15.4.2 on replacing $Qf(x, x+h)$ by $sQ_s(x, h)$ and $CQ_s(x, h)$, respectively.

The second part follows clearly from the first since the derivates in (c) and (e) are stronger than the ones in (a), and the derivates in (d) and (f) are stronger than the ones in (b). □

As a consequence of the above theorem we obtain the following result, as seen earlier, similar to Corollary 17.7.6. It should be noted here that in cases (a) and (d) this result follows from the Ninth Status Theorem 17.1.11 only in the case of continuous functions.

17.8.4. COROLLARY. *For every function $f : I \to \mathbb{R}$, the following derivatives of $f$ are of $B_1$:*

(a) *selective symmetric derivative,*

(b) *approximate selective symmetric derivative,*

(c) *preponderant selective symmetric derivative,*

(d) *Cesàro symmetric derivative,*

(e) *approximate Cesàro symmetric derivative, and*

(f) *preponderant Cesàro symmetric derivative.*

Next, in the following theorem we deal with regularity of the remaining generalized selective symmetric derivates considered in Theorem 17.8.1.

17.8.5. THEOREM. *Relative to a continuous symmetric selection, the following generalized selective symmetric derivates are regular:*

(a) *sequentially congruent selective symmetric derivates,*

(b) *congruent selective symmetric derivates,*

(c) *qualitative selective symmetric derivates,*

(d) *negligent selective symmetric derivates, and*

(e) *continuously selective selective symmetric derivates.*

*Further,* (f) *subcontinuous path selective symmetric derivates are subregular.*

PROOF. The admissibility of each of these derivates follows directly from Theorem 17.8.1. Suppose $s$ is a continuous symmetric selection and $f : I \to \mathbb{R}$ is continuous. Then it is clear that the selective symmetric quotient $sQ_s f(x, h) = Q_s f(x, s_{[x, x+h]} - x)$ is continuous in $x$ for each $h$. Hence in each of the cases (a), (b), (e), and (f), the result is obtained by an argument similar to the one used for the Fourth Status Theorem 15.4.4 in the corresponding case on replacing $Qf(x, x + h)$ by $sQ_s f(x, h)$. Further, (c) is a particular case of (d); and in case (d) it is easy to see that since $f$ is continuous, the negligent symmetric $s$-derivates of $f$ coincide with its symmetric $s$-derivates, which in turn are regular by Theorem 17.1.11. □

Finally, in the following theorem we deal with regularity of the remaining generalized Cesàro symmetric derivates considered in Theorem 17.8.2.

17.8.6. THEOREM. *The following generalized Cesàro symmetric derivates are regular*:

(a) *sequentially congruent Cesàro symmetric derivates,*

(b) *congruent Cesàro symmetric derivates,*

(c) *qualitative Cesàro symmetric derivates,*

(d) *negligent Cesàro symmetric derivates, and*

(e) *continuously selective Cesàro symmetric derivates.*

*Further*, (f) *subcontinuous path Cesàro symmetric derivates are subregular.*

PROOF. The admissibility of each of these derivates follows directly from Theorem 17.8.2. Suppose $f: I \to \mathbb{R}$ is continuous. Then it is clear that the Cesàro symmetric quotient $CQ_s f(x, h) = Q_s f_x^a(x, h)$ is continuous in $x$ for each $h$. Hence in each of the cases (a), (b), (e), and (f), the result is obtained by an argument similar to the one used for Theorem 15.4.4 in the corresponding case on replacing $Qf(x, x + h)$ by $CQ_s f(x, h)$. Further, (c) is as before a particular case of (d); and in case (d) it is easy to see that the negligent Cesàro symmetric derivates of $f$ coincide with its Cesàro symmetric derivates, which in turn are regular by Theorem 17.1.11. □

# CHAPTER 18

# A Unified Theory of Generalized New Derivatives

Utilizing the axiomatic model developed in Chapter 15, we present here a unified theory of generalized new derivatives obtained on combining the notions of new derivatives with those of various known generalized derivatives. Once again, we organize the results here in the same order as in Chapter 16.

First, in §18.1, we introduce the notions of generalized lower, upper, semi-, and weak derivatives which are obtained on replacing the ordinary limits by various admissible generalized limits. We refer to these derivatives as "generalized new derivatives"; they generalize on the one hand the new derivatives and on the other the corresponding generalized derivatives.

Next, in §18.2, we obtain two derivability theorems in terms of generalized new derivatives; and in §18.3 we deal with the Calculus of such derivatives. Then, in §18.4, some mean value theorems and Cauchy's mean value theorem are obtained for generalized lower and semi- derivatives, and l'Hôpital's rule is obtained in terms of these derivatives.

Next, in §18.5, we obtain some monotonicity theorems and the Goldowski-Tonelli theorem in terms of generalized new derivatives. Then, in §18.6, we deal with the relation between generalized new derivatives and strong derivative. Finally, in §18.7, we obtain the Darboux and Denjoy properties, and the Baire class, of generalized new derivatives.

We assume throughout this chapter, unless stated otherwise, that $G$-limits and $G$-derivates are *admissible* and that $f : I \to \mathbb{R}$ where $I$ is open, and $x \in I$.

**§18.1 Generalized lower, upper, semi-, and weak derivatives.** We first define in this section generalized versions of various new derivatives considered in §§4.2 and 4.3 corresponding to each notion of admissible generalized derivates (see §15.3).

Also, we include here some elementary results which show, in particular, that the generalized new derivatives generalize on the one hand the new derivatives and on the other the generalized derivatives.

Let $G$-derivates be any admissible generalized (unilateral) derivates. Then on using $G$-derivates in the place of ordinary derivates in §§4.2 and 4.3, we obtain the various notions of new $G$-derivatives. We present here some of the basic notions in this direction for the sake of clarity and to introduce the necessary nomenclature and notations.

Let us recall here, as mentioned in the introduction of this chapter, that we assume here throughout that $f \in \mathscr{F}(I)$ where $I$ is open, and $x \in I$.

Then we define $f$ to be *lower G-derivable* at $x$ if $GD^-f(x) \le GD_+f(x)$; and if so, then we call the interval $[GD^-f(x), GD_+f(x)]$ the *lower G-derivative* of $f$ at $x$ and denote it by $LGf'(x)$. Similarly, $f$ is called *upper G-derivable* at $x$ if $GD^+f(x) \le GD_-f(x)$; and if so, then $[GD^+f(x), GD_-f(x)]$ is called the *upper G-derivative* of $f$ at $x$ and denoted by $UGf'(x)$.

It is thus clear from axiom (D1) that $f$ is upper $G$-derivable at $x$ iff $-f$ is lower $G$-derivable at $x$; and if so, then $UGf'(x) = -LG(-f)'(x)$. Consequently, it is enough to consider once again the lower $G$-derivative.

Further, $f$ will be called *G-semiderivable* at $x$ if it is either lower or upper $G$-derivable at $x$; and if so, then its lower or upper $G$-derivative at $x$ will be called the *G-semiderivative* of $f$ at $x$ and denoted by $SGf'(x)$.

Next, to define the weak $G$-derivative, we first define the *lower* and *upper weak G-derivates* of $f$ similar to the ordinary weak derivates defined in §4.2 as follows:

$$G\underline{D}_w f = \vee GD_\pm f, \qquad G\overline{D}_w f = \wedge GD^\pm f.$$

Now we define $f$ to be *weakly G-derivable* at $x$ if $G\overline{D}_w f(x) \le G\underline{D}_w f(x)$; and if so, then the interval $[G\overline{D}_w f(x), G\underline{D}_w f(x)]$ will be called the *weak G-derivative* of $f$ at $x$ and denoted by $WGf'(x)$. It is easy to see here, as before, that $f$ is weakly $G$-derivable at $x$ iff it is either $G$-semiderivable or unilaterally $G$-derivable from some side at $x$.

Further, when the lower, upper, semi-, or weak $G$-derivative of $f$ at $x$ is a singleton, this derivative will be said to be *unique* and $f$ will be called *uniquely lower*, *upper*, *semi-*, or *weakly G-derivable*, respectively, at $x$.

Also, each element of the lower, upper, semi-, or weak $G$-derivative of $f$ at $x$ will be called, as before, a *lower*, *upper*, *semi-*, or *weak G-gradient*, respectively, of $f$ at $x$; and when $f$ has a finite lower, upper, semi-, or weak $G$-gradient at $x$, it will be said to be *lower*, *upper*, *semi-*, or *weakly G-differentiable*, respectively, at $x$.

Next, $f$ is called *lower G-knotted* at $x$ if $GD_+f(x) = -\infty$ and $GD^-f(x) = +\infty$; and *upper G-knotted* at $x$ if $GD_-f(x) = -\infty$ and $GD^+f(x) = +\infty$. Also, when $f$ is simultaneously lower and upper $G$-knotted at $x$, it will be said to be *G-biknotted* at $x$. We will use $G\underline{K}(f)$, $G\overline{K}(f)$, and $GK_b(f)$ to denote the *lower*, *upper*, and *bilateral G-knot sets* of $f$—that is, the sets of points where $f$ is lower or upper $G$-knotted or $G$-biknotted, respectively.

If $f$ is uniquely lower $G$-derivable at $x$ and it is upper $G$-knotted at $x$, then $f$ will be said as before to have a *knotted lower G-derivative* at $x$; the *knotted upper G-derivative* of $f$ at $x$ is defined similarly.

Further, $f$ is called *normally lower* (or *upper*) *G-derivable* at $x$ if either it is $G$-derivable at $x$ or it has a knotted lower (or upper) $G$-derivative at $x$. Finally, $f$ will be said to have a *knotted* (or *normal*) *G-semiderivative* at $x$ if it has a knotted (or normal) lower or upper $G$-derivative at $x$.

It is clear that the above new $G$-derivatives (except for the knotted ones) generalize the $G$-derivative. In the following theorem we compare the new $G$-derivatives with the corresponding ordinary new derivatives.

18.1.1. THEOREM. *If $f$ is lower derivable at $x$, then it is also lower $G$-derivable at $x$ and $Lf'(x) \subset LGf'(x)$. Similarly, if $f$ is weakly derivable at $x$, then it is also weakly $G$-derivable at $x$ and $Wf'(x) \subset WGf'(x)$.*

PROOF. First, suppose $f$ is lower derivable at $x$. Then with the help of axiom (D2) we obtain

$$GD^{-}f(x) \le D^{-}f(x) \le D_{+}f(x) \le GD_{+}f(x).$$

Hence it is clear that $f$ is lower $G$-derivable at $x$ and $Lf'(x) \subset LGf'(x)$. A similar argument holds for weak derivability. □

The following theorem is obtained on the other hand by an argument similar to the one used for Theorem 4.2.2.

18.1.2. THEOREM. *The function $f$ is $G$-derivable at $x$ iff it is lower and upper $G$-derivable there; and if so, then the lower and upper $G$-derivatives of $f$ at $x$ are unique and equal to $Gf'(x)$.*

In terms of the $G$-angularities defined in §17.2, it is clear as before that if $f$ is lower $G$-angular at $x$, then it is lower $G$-derivable there. Further, if $f$ is lower $G$-derivable and lower $G$-nonangular at $x$, then its lower $G$-derivative is unique at $x$. Similarly, if $f$ is $G$-nonangular and weakly $G$-derivable at $x$, then its weak $G$-derivative is unique at $x$.

Finally, we present some terminology and notations in connection with the above notions.

We will use the notations developed in §4.3 with a prefix $G$ to denote the various lower, upper, semi-, weak, and unilateral $G$-derivability sets and to denote the lower and upper $G$-knot sets of a function. For example, $G\underline{\Delta}(f)$ and $G\underline{\Delta}^{*}(f)$ will denote the sets of points where $f$ is lower $G$-derivable or lower $G$-differentiable, respectively, and $G\underline{\Delta}_{u}(f)$ will denote the set of points in $G\underline{\Delta}(f)$ where $LGf'$ is unique. Similarly, $G\underline{K}(f)$ and $G\overline{K}(f)$ will denote the sets of points where $f$ is lower or upper $G$-knotted, respectively.

Also, we will use $G\underline{A}(f)$ and $G\overline{A}(f)$ to denote the sets of points where $f$ is lower or upper $G$-angular, respectively. These two sets will be called the *lower* and *upper G-angularity sets*, respectively, of $f$, and the set $GA(f) = G\underline{A}(f) \cup G\overline{A}(f)$ will be called the *G-angularity set* of $f$. As was seen earlier in Remark 17.2.7, these $G$-angularity sets are not always countable.

As for the terminology in particular cases, consider for example the case when $G = A$. When $x \in A\underline{\Delta}(f)$, $A\underline{K}(f)$ or $A\underline{A}(f)$, $f$ will be said to be *approximately lower derivable*, *approximately lower knotted*, or *approximately lower angular*, respectively, at $x$. Similar terminology will be used for the points of $A\overline{\Delta}(f)$, $A\overline{K}(f)$, and $A\overline{A}(f)$. Also, $LAf'$, $UAf'$, and $WAf'$ will be called the *approximate lower*, *upper*, and *weak derivatives*, respectively, of $f$.

## §18.2 Derivability theorems for generalized new derivatives.

We first state in this section two derivability theorems for generalized new derivatives which follow from the earlier derivability theorems.

Let us recall here that $G$-derivates are assumed here to be any admissible generalized derivates, and $f \in \mathcal{F}(I)$.

The following result follows directly from the First Derivability Theorem 5.1.2 with the help of Theorem 18.1.1.

18.2.1. THEOREM (TENTH DERIVABILITY). *Suppose $f$ is LSC. Then $f$ is lower G-differentiable at a dense set of points in $I$. Moreover, if $f$ is lower nonangular, then it is lower G-differentiable at a c-dense set of points in $I$.*

On applying the above theorem to $f$ and $-f$, we obtain the following result on $G$-differentiability when $f$ has symmetrical $G$-derivates; a class of functions with this property will be considered later in Chapter 20.

18.2.2. COROLLARY. *Suppose $f$ is either LSC or USC. If $f$ has symmetrical G-derivates, then it is G-differentiable at a c-dense set of points in $I$.*

From the Third Derivability Theorem 6.2.2 we obtain on the other hand the following partial result on lower $G$-derivability; because the lower $G$-derivative of $f$ need not be unique or knotted at a point where it has a knotted lower derivative.

18.2.3. THEOREM (ELEVENTH DERIVABILITY). *Suppose $f$ is either $R^*LI$ of $L^*UI$. Then $f$ is lower G-derivable at a dense set of points in almost all of its level sets, as well as at a c-dense set of points in $I$.*

Further, with the help of Theorem 18.1.1 we obtain the following from the Fourth Derivability Theorem 6.3.5.

18.2.4. THEOREM. *Suppose $f$ is either $R^*LI$ or $L^*UI$.* (a) *If $f$ is L-regular, then it is lower G-differentiable at a dense set of points in almost all of its level sets, as well as at a metrically dense set of points in $I$.*

(b) *If $f$ is L-singular, then it has an infinite LG-gradient at a dense set of points in almost all of its level sets, as well as at a c-dense set of points in $I$ provided that $f$ is a nowhere constant Darboux function.*

18.2.5. PROBLEM. In connection with Theorem 18.2.3, it will be interesting to investigate if its lower $G$-derivability can be strengthened to normal lower $G$-derivability as in Theorem 6.2.2. As will be seen later in Theorem 19.5.4, in the case of some of the generalized derivates this does hold for a typical continuous function.

**§18.3 Calculus of generalized new derivatives.** In this section we deal with the Calculus of generalized new derivatives. Indeed, all the results obtained earlier in §§7.3 and 7.4 on ordinary new derivatives are found to hold for generalized new derivatives under suitable hypotheses.

First, we obtain the additivity and linearity of $s$-admissible generalized new derivatives in Theorems 18.3.3 to 18.3.8. Then in Part A of this section we obtain product and quotient rules for $s^*$-admissible generalized new derivatives (see Theorems 18.3.12 to 18.3.17). Finally, in Part B, these rules are obtained for Borel and Cesàro new derivatives which are not $s^*$-admissible.

We will assume throughout this section that $f, g \in \mathscr{F}(I)$, where $I$ is open, and $x \in I$. Further, we will use here the various terminology developed in §7.1 regarding similarity of semiderivatives, symmetry of unilateral derivatives, and compatibility of weak derivatives for such generalized new derivatives as well.

We begin with the following relation between generalized new derivatives of $f_{+c}$ and $f$ which holds for any admissible generalized derivates.

18.3.1. THEOREM. *Suppose G-derivates are admissible. Then for each $c \subset \mathbb{R}$, the function $f_{+c}$ is weakly G-derivable at $x$ iff $f$ is so; and if so, then the two weak G-derivatives are totally similar and $WG(f_{+c})'(x) = WGf'(x) + c$.*

PROOF. As mentioned earlier, the two relations of axiom (D3) on lower $G$-derivates hold also for upper $G$-derivates, as follows from axiom (D1). The result follows directly from these four relations on $G$-derivates at $x$. □

Next, to obtain the additivity and linearity of generalized new derivatives, we will assume up to Theorem 16.3.8 without mentioning that the $G$-derivates are $s$-admissible. We need here the following lemma on such derivates.

18.3.2. LEMMA. *Suppose G-derivates are s-admissible. Then if $f$ is right G-derivable at $x$, then each of the following equations holds provided that the* *sum on its right side is defined:*

$$GD_+(f+g)(x) = Gf'_+(x) + GD_+g(x),$$
$$GD^+(f+g)(x) = Gf'_+(x) + GD^+g(x).$$

*Further, similar relations hold for left G-derivates when $f$ is left G-derivable at $x$.*

PROOF. First, suppose $f$ is right $G$-derivable at $x$. Then the first equation follows from axiom (D4) and the first inequality of Lemma 15.2.3(a). The second equation follows on the other hand from the remaining two inequalities of that lemma. Similar arguments hold when $f$ is left $G$-derivable at $x$. □

Now, with the help of the above lemma, axioms (D4) and (D5), and Lemma 15.2.3, it can be verified that all the results of §7.3 remain valid for $s$-admissible generalized new derivatives. We state below some of the more important results in this direction. The next four results are obtained with the help of the results just mentioned by arguments similar to the ones used for Theorems 7.3.1, 7.3.2, and 7.3.7 and Lemma 7.3.9.

18.3.3. THEOREM. *Suppose $f$ is G-differentiable at $x$. Then $f+g$ is weakly G-derivable at $x$ iff $g$ is so; and if so, then the two weak G-derivatives are totally similar and*

$$WG(f+g)'(x) = Gf'(x) + WGg'(x).$$

18.3.4. THEOREM. *Suppose $f$ and $g$ are lower G-derivable at $x$.* (a) *If the following sum is defined, then $f+g$ is lower G-derivable at $x$ and*

$$LGf'(x) + LGg'(x) \subset LG(f+g)'(x). \tag{1}$$

(b) *If from each side either $f$ or $g$ is unilaterally G-differentiable at $x$, then $f+g$ is lower G-derivable at $x$ and* (1) *becomes an identity; and if $LGf'(x)$ and $LGg'(x)$ are simultaneously unique or normal, then so is $LG(f+g)'(x)$.*

18.3.5. THEOREM. *Suppose $f$ and $g$ have unique weak G-derivatives at $x$ of which the sum is defined, and also suppose that $f+g$ is weakly G-derivable at $x$. Then*

$$WGf'(x) + WGg'(x) \in WG(f+g)'(x). \tag{2}$$

*Further, if $f$ and $g$ have either* (i) *dissimilar G-semiderivatives at $x$, or* (ii) *asymmetric unilateral G-derivatives at $x$, then* (2) *becomes an identity.*

18.3.6. LEMMA. *Let $c \in \mathbb{R}$. If $f$ is weakly $G$-derivable at $x$, then so is $cf$; and if so, then*

$$WG(cf)'(x) = cWGf'(x),$$

*where $WG(cf)'(x)$ is totally $c$-compatible with $WGf'(x)$.*

Now, on combining the above lemma with Theorems 18.3.4 and 18.3.5 we obtain the following two theorems on the linearity of generalized semi- and weak derivatives.

18.3.7. THEOREM. *Let $h = af + bg$, where $a, b \in \mathbb{R}$ and $a \neq 0$. Suppose $f$ and $g$ have $ab$-compatible $G$-semiderivatives at $x$.* (a) *If the following sum is defined, then $h$ has a $G$-semiderivative at $x$, $a$-compatible with $SGf'(x)$, and*

$$SGh'(x) \supset aSGf'(x) + bSGg'(x). \tag{3}$$

(b) *If from each side either $f$ or $g$ is unilaterally $G$-differentiable at $x$, then $h$ has a $G$-semiderivative at $x$ as in* (a) *and* (3) *becomes an identity; and if $SGf'(x)$ and $SGg'(x)$ are simultaneously unique or normal, then so is $SGh'(x)$.*

18.3.8. THEOREM. *Let $h = af + bg$, where $a, b \in \mathbb{R}$. Suppose $f$ and $g$ have unique weak $G$-derivatives at $x$, the following sum is defined, and $h$ is weakly $G$-derivable at $x$. Then*

$$aWGf'(x) + bWGg'(x) \in WGh'(x). \tag{4}$$

*Further, if $f$ and $g$ have either* (i) *$-ab$-compatible $G$-semiderivatives at $x$, or* (ii) *asymmetric unilateral $G$-derivatives at $x$, then* (4) *becomes an identity.*

Next, we deal with the product and quotient rules for generalized new derivatives. We need to consider here two cases separately.

**A. Product and quotient rules for $s^*$-admissible generalized new derivatives.** We will assume throughout this part that the $G$-derivates are $s^*$-admissible, that is they are induced by some $G$-limits and a $G$-quotient, where the $G$-limits are $s^*$-admissible and the $G$-quotient is so relative to $G$-limits. Further, generalizing the notion of $D$-finiteness, we will call a function $f \in \mathcal{F}(I)$ *$GD$-finite at* a point $x \in I$ if its $G$-derivates are finite at $x$.

We need here some lemmas. We will adopt here our earlier convention that any given relation holds only if the sum involved in it is defined.

18.3.9. LEMMA. *Suppose $f$ is G-continuous at $x$. Then if $f(x) > 0$ and $g(x) \geq 0$, then*

$$\begin{aligned} &f(x)GD_+g(x) + g(x)GD_+f(x) \\ &\quad \leq GD_+(fg)(x) \leq f(x)GD_+g(x) + g(x)GD^+f(x) \\ &\quad \leq GD^+(fg)(x) \leq f(x)GD^+g(x) + g(x)GD^+f(x), \end{aligned}$$

*where $f$ and $g$ can be interchanged in the term after the second inequality. Further, the same holds in terms of left G-derivates.*

PROOF. Since $f$ is $G$-continuous at $x$, there is by hypothesis a function $f_G(x, h)$, $h \neq 0$, such that $G\text{-}\lim_{h \to 0} f_G(x, h) = f(x)$, and the following holds for $h \neq 0$ provided that $x + h \in I$:

$$(5) \quad GQ(fg)(x, x+h) = f_G(x, h)GQg(x, x+h) + g(x)GQf(x, x+h).$$

Now since $G$-limits are $s^*$-admissible, we thus obtain the following with the help of axioms (A4), (A5), and (A6):

$$\begin{aligned} GD_+(fg)(x) &\geq G\text{-}\liminf_{h \to 0+} f_G(x, h)GQg(x, x+h) \\ &\quad + G\text{-}\liminf_{h \to 0+} g(x)GQf(x, x+h) \\ &= f(x)GD_+g(x) + g(x)GD_+f(x), \end{aligned}$$

where the second term on the right is obviously zero when $g(x) = 0$. The other three inequalities are obtained similarly using the three inequalities of Lemma 15.1.3 in place of (A4). □

The following lemma is obtained by a similar argument.

18.3.10. LEMMA. *Suppose $f$ is G-continuous at $x$.* (a) *If $f$ is right G-derivable at $x$ and $f(x) > 0$, then*

$$GD_+(fg)(x) = f(x)GD_+g(x) + g(x)Gf'_+(x),$$
$$GD^+(fg)(x) = f(x)GD^+g(x) + g(x)Gf'_+(x).$$

(b) *If $g$ is right G-derivable at $x$ and $g(x) \geq 0$, then*

$$GD_+(fg)(x) = f(x)Gg'_+(x) + g(x)GD_+f(x),$$
$$GD^+(fg)(x) = f(x)Gg'_+(x) + g(x)GD^+f(x).$$

*Further, similar relations hold when $f$ or $g$ is left G-derivable at $x$.*

18.3.11. LEMMA. *Suppose $f$ is G-continuous at $x$ and $f(x) = g(x) = 0$. Then if $g$ is GD-finite at $x$, then $fg$ has a G-derivative at $x$ equal to zero.*

PROOF. Since $g(x) = 0$, in this case we have the following by (5): $GQ(fg)(x, x + h) = f_G(x + h)GQg(x, x + h)$, where $GQg(x, x + h)$ has finite $G$-limits as $h \to 0$. Hence the result follows on applying the last part of axiom (A7) to $f_G$ and $GQg$. □

Now, with the help of the above lemmas, axiom (A6), Lemma 15.1.8, and Theorem 15.1.9, it can be verified that all the results of §7.4 remain valid for $s^*$-admissible generalized new derivatives. We state below, as before, some of the more important results in this direction. The next four results are obtained with the help of the results just mentioned by arguments similar to the ones used for Theorems 7.4.1, 7.4.3, and 7.4.6 and Lemma 7.4.8.

18.3.12. THEOREM. *Suppose $f$ is G-continuous at $x$.* (a) *If $f$ is G-differentiable at $x$ and $f(x) \neq 0$, then $fg$ is weakly G-derivable at $x$ iff $g$ is so; and if so, then $WG(fg)'(x)$ is totally $f(x)$-compatible with $WGg'(x)$ and*

$$WG(fg)'(x) = f(x)WGg'(x) + g(x)Gf'(x).$$

(b) *If $g$ is weakly G-derivable at $x$, $g(x) = 0$ and either* (i) $f(x) \neq 0$ *or* (ii) *$g$ is GD-finite at $x$, then $fg$ has a weak G-derivative at $x$, totally $f(x)$-compatible with $WGg'(x)$, and*

$$WG(fg)'(x) = f(x)WGg'(x).$$

18.3.13. THEOREM. *Suppose $f$ is G-continuous at $x$, $f(x) \neq 0$, and $f$ and $g$ have $f(x)g(x)$-compatible G-semiderivatives at $x$.* (a) *If the following sum is defined, then $fg$ has a G-semiderivative at $x$, $f(x)$-compatible with $SGg'(x)$, and*

$$f(x)SGg'(x) + g(x)SGf'(x) \subset SG(fg)'(x). \tag{6}$$

(b) *If from each side either $f$ or $g$ is unilaterally G-differentiable at $x$, then $fg$ has a G-semiderivative at $x$ as in* (a) *and* (6) *becomes an identity; and if $SGf'(x)$ and $SGg'(x)$ are simultaneously unique or normal, then so is $SG(fg)'(x)$.*

*Moreover, the above holds also when $f(x) = 0$ provided that $f$ is G-differentiable at $x$ and $g$ is G-continuous at $x$.*

18.3.14. THEOREM. *Suppose $f$ is G-continuous at $x$, $f(x) \neq 0$, $f$ and $g$ have unique weak G-derivatives at $x$ such that the following sum is defined, and $fg$ is weakly G-derivable at $x$. Then*

$$f(x)WGg'(x) + g(x)WGf'(x) \in WG(fg)'(x). \tag{7}$$

*Further, if $f$ and $g$ have either* (i) $-f(x)g(x)$*-compatible G-semiderivatives at $x$, or* (ii) *asymmetric unilateral G-derivatives at $x$, then* (7) *becomes an identity.*

*Moreover, the same holds when $f(x) = 0$ provided that $f$ is GD-finite at $x$ and $g$ is G-continuous at $x$.*

18.3.15. LEMMA. *Suppose $f$ is G-continuous at $x$, and $f(x) \neq 0$. Then $1/f$ is weakly G-derivable at $x$ iff $f$ is so; and if so, then the two weak G-derivatives at $x$ are totally dissimilar and*

$$WG(1/f)'(x) = -\{1/f(x)\}^2 WGf'(x).$$

Finally, on combining the above lemma with Theorems 18.3.13 and 18.3.14 we obtain the following quotient rules for generalized semi- and weak derivatives.

18.3.16. THEOREM. *Suppose $g$ is G-continuous at $x$, $g(x) \neq 0$, and $f$ and $g$ have $-f(x)g(x)$-compatible G-semiderivatives at $x$.* (a) *If the following difference is defined, then $f/g$ has a G-semiderivative at $x$, $g(x)$-compatible with $SGf'(x)$, and*

$$(8) \qquad SG(f/g)'(x) \supset \{1/g(x)\}^2\{g(x)SGf'(x) - f(x)SGg'(x)\}.$$

(b) *If from each side either $f$ or $g$ is unilaterally G-differentiable at $x$, then $f/g$ has a G-semiderivative at $x$ as in* (a), *and* (8) *becomes an identity; and if $SGf'(x)$ and $SGg'(x)$ are simultaneously unique or normal, then so is $SG(f/g)'(x)$.*

18.3.17. THEOREM. *Suppose $g$ is G-continuous at $x$, $g(x) \neq 0$, $f$ and $g$ have unique weak G-derivatives at $x$ such that the following difference is defined, and $f/g$ is weakly G-derivable at $x$. Then*

$$(9) \qquad \{1/g(x)\}^2\{g(x)WGf'(x) - f(x)WGg'(x)\} \in WG(f/g)'(x).$$

*Further, if $f$ and $g$ have either* (i) $f(x)g(x)$*-compatible G-semiderivatives at $x$, or* (ii) *asymmetric unilateral G-derivatives at $x$, then* (9) *becomes an identity.*

**B. Product and quotient rules for Borel and Cesàro new derivatives.** Since the Borel and Cesàro derivates are not $s^*$-admissible, we obtain in this part the product and quotient rules for the new derivatives induced by these derivates under $B$- and $C$-finiteness defined in §16.3B. We need here some lemmas as before.

18.3.18. LEMMA. *Suppose $G = B$ or $C$, $f$ is $G$-finite at $x$, and $g$ is continuous at $x$. Then if $f(x) \geq 0$ and $g(x) \geq 0$, then*

$$\begin{aligned} &f(x)GD_+g(x) + g(x)GD_+f(x) \\ &\quad \leq GD_+(fg)(x) \leq f(x)GD_+g(x) + g(x)GD^+f(x) \\ &\quad \leq GD^+(fg)(x) \leq f(x)GD^+g(x) + g(x)GD^+f(x), \end{aligned}$$

*where $f$ and $g$ can be interchanged in the term after the second inequality. Further, the same holds in terms of left $G$-derivates.*

PROOF. First, suppose $f$ is $B$-finite at $x$. Then as seen in the proof of Theorem 16.3.8, we have

$$\begin{aligned} &\frac{1}{h}\int_\delta^h Q(fg)(x, x+t)\,dt \\ &\quad = \frac{f(x)}{h}\int_\delta^h Qg(x, x+t)\,dt + \frac{g(x)}{h}\int_\delta^h Qf(x, x+t)\,dt + I, \end{aligned}$$

where $I$ is an integral which converges to zero as $\delta \to 0+$ and $h \to 0+$. Each of the above inequalities for $G = B$ follows clearly from this relation.

Next, suppose $f$ is $C$-finite at $x$. Then as seen in the later part of the proof of Theorem 16.3.8, we have

$$Q(fg)_x^a(x, x+h) = f(x)Qg_x^a(x, x+h) + g(x)Qf_x^a(x, x+h) + J,$$

where $J$ is an integral which converges to zero as $h \to 0$. Each of the above inequalities for $G = C$ follows clearly from this relation.

Further, in the case of left $G$-derivates the result is obtained by a similar argument on replacing $x + t$ by $x - t$ and $x + h$ by $x - h$. □

The following lemma is obtained by a similar argument with the help of the proof of Theorem 16.3.8.

18.3.19. LEMMA. *Suppose $G = B$ or $C$, $f$ is $G$-finite at $x$, and $g$ is continuous at $x$.*

(a) *If $f$ is right $G$-derivable at $x$ and $f(x) \geq 0$, then*

$$\begin{aligned} GD_+(fg)(x) &= f(x)GD_+g(x) + g(x)Gf'_+(x), \\ GD^+(fg)(x) &= f(x)GD^+g(x) + g(x)Gf'_+(x). \end{aligned}$$

(b) *If $g$ is right $G$-derivable at $x$ and $g(x) \geq 0$, then*

$$GD_+(fg)(x) = f(x)Gg'_+(x) + g(x)GD_+f(x),$$
$$GD^+(fg)(x) = f(x)Gg'_+(x) + g(x)GD^+f(x).$$

*Further, similar relations hold when $f$ or $g$ is left $G$-derivable at $x$.*

We begin with the following product rule when one of the two functions is $B$- or $C$-differentiable at the point in question.

18.3.20. THEOREM. *Suppose $G = B$ or $C$, $f$ is $G-$finite at $x$, and $g$ is continuous at $x$.* (a) *If $f$ is $G$-differentiable at $x$, then $fg$ is weakly $G$-derivable at $x$ iff $g$ is so; and if so, then $WG(fg)'(x)$ is totally $f(x)$-compatible with $WGg'(x)$ and*

$$WG(fg)'(x) = f(x)WGg'(x) + g(x)Gf'(x).$$

(b) *If $g$ is weakly $G$-derivable at $x$ and $g(x) = 0$, then $fg$ has a weak $G$-derivative at $x$, totally $f(x)$-compatible with $WGg'(x)$, and*

$$WG(fg)'(x) = f(x)WGg'(x).$$

PROOF. First, suppose $f(x) \geq 0$. In this case it is easy to see that the part (a) follows directly from Lemma 18.3.19(a). Further, when $g(x) = 0$, it follows clearly from Lemma 18.3.18 that

$$GD_\pm(fg)(x) = f(x)GD_\pm g(x), \qquad GD^\pm(fg)(x) = f(x)GD^\pm g(x),$$

and the part (b) follows in this case from these relations.

Further, in the case when $f(x) < 0$, the two results are obtained on applying the above results to $-f$ and $g$ with the help of Lemma 18.3.6. □

Next, with the help of Lemmas 18.3.18 and 18.3.19 and Theorem 18.3.20, we obtain the following product rule for Borel and Cesàro semiderivatives by an argument similar to the one used for Theorem 7.4.3.

18.3.21. THEOREM. *Suppose $G = B$ or $C$, $f$ is $G$-finite at $x$, and $g$ is continuous at $x$. Further, suppose $f$ and $g$ have $f(x)g(x)$-compatible $G$-semiderivatives at $x$.* (a) *If the following sum is defined, then $fg$ has a $G$-semiderivative at $x$, $f(x)$-compatible with $SGg'(x)$, and*

$$f(x)SGg'(x) + g(x)SGf'(x) \subset SG(fg)'(x). \tag{10}$$

(b) *If from each side either $f$ or $g$ is unilaterally $G$-differentiable at $x$, then $fg$ has a $G$-semiderivative at $x$ as in* (a) *and* (10) *becomes an identity; and if*

*$SGf'(x)$ and $SGg'(x)$ are simultaneously unique or normal, then so is $SG(fg)'(x)$.*

Further, with the help of Lemmas 18.3.18 and 18.3.19 and Theorem 18.3.21, we obtain the following conditional product rule for Borel and Cesàro weak derivatives by an argument similar to the one used for Theorem 7.4.6.

18.3.22. THEOREM. *Suppose $G = B$ or $C$, $f$ is $G$-finite at $x$, and $g$ is continuous at $x$. Further, suppose $f$ and $g$ have unique weak $G$-derivatives at $x$ such that the following sum is defined, and also suppose that $fg$ is weakly $G$-derivable at $x$. Then*

$$f(x)WGg'(x) + g(x)WGf'(x) \in WG(fg)'(x). \tag{11}$$

*Further, if $f$ and $g$ have either* (i) *$-f(x)g(x)$-compatible $G$-semiderivatives at $x$, or* (ii) *asymmetric unilateral $G$-derivatives at $x$, then* (11) *becomes an identity.*

Next, to obtain the quotient rule we need the following lemma.

18.3.23. LEMMA. *Suppose $G = B$ or $C$, $f$ is $G$-finite and continuous at $x$, and $f(x) \neq 0$. Then $1/f$ also is $G$-finite at $x$; and $1/f$ is weakly $G$-derivable at $x$ iff $f$ is so; and if so, then the two weak $G$-derivatives at $x$ are totally dissimilar and*

$$WG(1/f)'(x) = -\{1/f(x)\}^2 WGf'(x).$$

PROOF. We need to prove here only the second part, for the $G$-finiteness of $1/f$ under the present hypothesis was obtained earlier in Lemma 16.3.10.

First, suppose $f$ is $B$-finite at $x$. Then as seen in the proof of Lemma 16.3.10, we have

$$Q(1/f)(x, x+t) = -(1/f(x))^2 Qf(x, x+t) + T,$$

where $(1/h)\int_\delta^h |T|\,dt$ converges to zero as $\delta \to 0+$ and $h \to 0+$. Hence if $f(x) > 0$, it follows clearly from this relation that

$$BD_\pm(1/f)(x) = -(1/f(x))^2 BD^\pm f(x), \tag{12}$$

$$BD^\pm(1/f)(x) = -(1/f(x))^2 BD_\pm f(x),$$

and it is easy to see that the result follows in this case from these equations. A similar argument holds when $f(x) < 0$.

Next, suppose $f$ is $C$-finite at $x$. Then as seen in the later part of the proof of Lemma 16.3.10, we have

$$Q(1/f)_x^a(x, x+h) = -\{1/f(x)\}^2 Qf_x^a(x, x+h) + I,$$

where $I$ converges to zero as $h \to 0$. Hence if $f(x) > 0$, from this relation we obtain relations similar to (12) in terms of $C$-derivates, from which the result follows clearly in this case. A similar argument holds again when $f(x) < 0$. □

Finally, on combining the above lemma with Theorems 18.3.21 and 18.3.22 we obtain the following quotient rules for Borel and Cesàro semi- and weak derivatives.

18.3.24. THEOREM. *Suppose $G = B$ or $C$, $g$ is $G$-finite at $x$, and $f$ and $g$ are continuous at $x$. Further, suppose $g(x) \neq 0$ and that $f$ and $g$ have $-f(x)g(x)$-compatible $G$-semiderivatives at $x$.* (a) *If the following difference is defined, then $f/g$ has a $G$-semiderivative at $x$, $g(x)$-compatible with $SGf'(x)$, and*

$$(13) \quad SG(f/g)'(x) \supset \{1/g(x)\}^2\{g(x)SGf'(x) - f(x)SGg'(x)\}.$$

(b) *If from each side either $f$ or $g$ is unilaterally $G$-differentiable at $x$, then $f/g$ has a $G$-semiderivative at $x$ as in* (a), *and* (13) *becomes an identity; and if $SGf'(x)$ and $SGg'(x)$ are simultaneously unique or normal, then so is $SG(f/g)'(x)$.*

18.3.25. THEOREM. *Suppose $G = B$ or $C$, $g$ is $G$-finite at $x$, and $f$ and $g$ are continuous at $x$. Further, suppose $g(x) \neq 0$, $f$ and $g$ have unique weak $G$-derivatives at $x$ such that the following difference is defined, and $f/g$ is weakly $G$-derivable at $x$. Then*

$$(14) \quad \{1/g(x)\}^2\{g(x)WGf'(x) - f(x)WGg'(x)\} \in WG(f/g)'(x).$$

*Further, if $f$ and $g$ have either* (i) *$f(x)g(x)$-compatible $G$-semiderivatives at $x$, or* (ii) *asymmetric unilateral $G$-derivatives at $x$, then* (14) *becomes an identity.*

**§18.4 Mean value theorems and l'Hôpital's rule in terms of generalized new derivatives.** In this section we obtain mean value theorems along with Cauchy's mean value theorem in terms of various generalized new derivatives. Further, we deduce from the latter the l'Hôpital's rule in terms of these derivatives.

First, we obtain in Theorems 18.4.1 and 18.4.3 two mean value theorems in terms of generalized semi- and lower derivatives. As mentioned earlier, we assume throughout this chapter that the $G$-derivates are admissible.

On taking into account Theorem 18.1.1, we obtain directly from the General Mean Value Theorem 8.1.2 the following mean value theorem in terms of generalized semiderivative.

18.4.1. THEOREM (SEVENTH MEAN VALUE). *Suppose* $f : I \to \mathbb{R}$ *is* $I^*C$, *where* $I = [a, b]$. *Then there is a point* $x \in I^0$ *such that* $f$ *is G-semidifferentiable at* $x$ *and* $Qf(a, b) \in SGf'(x)$.

From this theorem we obtain the following result in the case of functions with symmetrical $G$-derivates—for example, $G$-smooth functions (see Chapter 20).

18.4.2. COROLLARY. *Let* $I = [a, b]$, *and suppose* $f : I \to \mathbb{R}$ *is* $I^*C$ *and has symmetrical G-derivates. Then there is a point* $x \in I^0$ *such that* $f$ *is G-differentiable at* $x$ *and* $Qf(a, b) = Gf'(x)$.

Further, due to Theorem 18.1.1, we obtain directly from the Second Mean Value Theorem 8.1.4 the following mean value theorem in terms of generalized lower derivative.

18.4.3. THEOREM (EIGHTH MEAN VALUE). *Let* $I = [a, b]$, *and suppose* $f : I \to \mathbb{R}$ *is* $I^*C$ *and upper nonangular. Then there is a point* $x \in I^0$ *such that* $f$ *is lower G-differentiable at* $x$ *and* $Qf(a, b) \in LGf'(x)$. *Consequently, if* $f$ *is further lower G-nonangular, then* $Qf(a, b) = LGf'(x)$.

Next, we deduce from the above two mean value theorems the following version of Cauchy's mean value theorem in terms of generalized new derivatives which may be compared with the General Cauchy's Mean Value Theorem 8.2.1.

18.4.4. THEOREM (FOURTH CAUCHY'S MEAN VALUE). *Let* $f, g : I \to \mathbb{R}$ *where* $I = [a, b]$. *Suppose* (i) *G-derivates are s-admissible,* (ii) $f$ *is* $I^*C$ *and* $g$ *is continuous on* $I$, *and* (iii) $g$ *is G-differentiable on* $I^0$ *and* $Gg'(x) \neq 0$ *for* $x \in I^0$. *Then there is a point* $x \in I^0$ *such that* $f$ *is G-semidifferentiable at* $x$ *and*

$$\frac{f(b) - f(a)}{g(b) - g(a)} \in \frac{SGf'(x)}{Gg'(x)}. \tag{1}$$

*Further, when* $f$ *is upper G-nonangular, the same holds in terms of the lower G-derivative of* $f$; *and if* $f$ *is G-nonangular, then identity holds in place of* (1) *in terms of* $LGf'(x)$.

PROOF. It is clear from (iii) and Theorem 18.4.1 that $g(a) \neq g(b)$. Hence the ratio on the left side of (1) is defined. Let $c$ denote this ratio, and define

$$h(x) = f(x) - f(a) - c\{g(x) - g(a)\}, \qquad x \in I. \tag{2}$$

Then $h(a) = 0 = h(b)$, and since $h$ is clearly $I^*C$ on $I$, it follows from Theorem 18.4.1 that there is a point $x \in I^0$ such that $h$ is $G$-semidifferentiable at $x$ and $0 \in SGh'(x)$. Now, it follows from Theorem 18.3.2 and Lemma

18.3.5 that $f$ is $G$-semiderivable at $x$, and $0 \in SGh'(x) = SGf'(x) - cGg'(x)$. Consequently, $cGg'(x) \in SGf'(x)$, which proves (1).

Next, to prove the second part, suppose $f$ is upper $G$-nonangular. Then it follows from Theorem 15.2.4 that $h$ also is upper $G$-nonangular, and hence it is upper nonangular. Now the result is obtained by an argument similar to above from Theorem 18.4.3. The last part is of course an immediate consequence of this result. □

When $f$ has symmetrical $G$-derivates, the above theorem yields the following extension of the Second Cauchy's Mean Value Theorem 16.4.2 in terms of $G$-derivative.

18.4.5. COROLLARY. *Let $f, g : I \to \mathbb{R}$, where $I = [a, b]$. Suppose* (i) *G-derivates are s-admissible,* (ii) *$f$ is $I^*C$ and $g$ is continuous on $I$,* (iii) *$f$ has symmetrical G-derivates, and* (iv) *$g$ is G-differentiable on $I^0$ and $Gg'(x) \neq 0$ for $x \in I^0$. Then there is a point $x \in I^0$ such that $f$ is G-differentiable at $x$ and*

$$\frac{f(b) - f(a)}{g(b) - g(a)} = \frac{Gf'(x)}{Gg'(x)}.$$

Next, we deduce from the above Cauchy's mean value theorem the following version of l'Hôpital's rule in terms of lower and semi- $G$-derivatives. We state the result as before in terms of the right limit.

18.4.6. THEOREM (FOURTH L'HÔPITAL'S RULE). *Let $f, g : I \to \mathbb{R}$, where $I = (a, b)$, $-\infty \leq a < b < \infty$. Suppose* (i) *G-derivates are s-admissible,* (ii) *$f$ is $I^*C$ and $g$ is continuous on $I$, and* (iii) *$g$ is G-differentiable and $Gg'(x) \neq 0$ for $x \in I$. Let $E$ denote the set of points in $I$ where $f$ is G-semidifferentiable. Then if the following limit on the left side is of indeterminate form $0/0$ or $\infty/\infty$, then*

$$\lim_{x \to a+} \frac{f(x)}{g(x)} \in U\text{-}\lim_{x \in E\,;\, x \to a+} \frac{SGf'(x)}{Gg'(x)}, \tag{3}$$

*provided that the limit on the right exists (finite or infinite).*

*Further, when $f$ is upper G-nonangular, the same holds in terms of lower G-derivative of $f$ with $E = G\underline{\Delta}^*(f)$; and if $f$ is G-nonangular, then* (3) *holds for $LGf'$ in terms of ordinary limit.*

PROOF. We omit the proof here since this theorem is obtained by an argument similar to the one used for Theorem 8.2.3, using this time the above Cauchy's mean value theorem 18.4.4 in place of Theorem 8.2.1, and on replacing $Sf'$ by $SGf'$ and $g'$ by $Gg'$. □

When $f$ has symmetrical $G$-derivates, the above theorem yields the following extension of the Second l'Hôpital's Rule 16.4.3 in terms of ordinary $G$-derivative.

18.4.7. COROLLARY. *Let $f, g : I \to \mathbb{R}$, where $I = (a, b)$, $-\infty \le a < b < \infty$. Suppose* (i) *$f$ is $I^*C$ and has symmetrical G-derivates, and* (ii) *$g$ is continuous and G-differentiable and $Gg'(x) \neq 0$ for $x \in I$. Let $E$ denote the set of points in $I$ where $f$ is G-differentiable. Then if the following limit on the left side is of indeterminate form $0/0$ or $\infty/\infty$, then*

$$\lim_{x \to a+} \frac{f(x)}{g(x)} = \lim_{x \in E; x \to a+} \frac{Gf'(x)}{Gg'(x)},$$

*provided that the limit on the right exists (finite or infinite).*

**§18.5 Monotonicity and Goldowski-Tonelli theorems in terms of generalized new derivatives.** In this section we obtain some monotonicity theorems in terms of generalized new derivatives including Goldowski-Tonelli theorems for such derivatives.

The first four theorems deal in fact with monotonicity results in terms of generalized lower derivative, and in Theorem 18.5.5 we obtain the Goldowski-Tonelli theorem in terms of generalized weak derivative.

It may be recalled here that we assume here $G$-limits and $G$-derivates to be admissible, and $f : I \to \mathbb{R}$, where $I$ is an open subinterval of $\mathbb{R}$.

The following monotonicity theorem in terms of generalized lower derivative is obtained directly from the Second Monotonicity Theorem 9.1.3 with the help of Theorem 18.1.1.

18.5.1. THEOREM. *Suppose $f$ is $I^*LSC$, and let $E$ be the set of points where $f$ has a finite negative (or positive) lower G-gradient. Then $f$ is nondecreasing (or nonincreasing) whenever one of the following holds:* (i) *$E$ is empty,* (ii) *$f$ is $WC_-$ (or $WC_+$) and $E$ is finite, or* (iii) *$f$ is lower G-nonangular and* $\operatorname{card}(E) < c$.

The following result related to monotonicity is obtained similarly from Theorem 9.1.5.

18.5.2. THEOREM. *Let $a, b \in I$, $a < b$, and $E_-$ and $E_+$ be the sets of points in $(a, b)$ where $f$ has a negative or positive lower G-gradient, respectively. Then if $f$ is $R^*LI$, then $f(a) - f(b) \le |f(E_-)|$; and if $f$ is $L^*UI$, then $f(b) - f(a) \le |f(E_+)|$.*

Modifying the earlier definitions of lower regularity and lower singularity slightly, we will call $f$ *lower LG-regular* if it maps the points where it has a lower $G$-gradient $-\infty$ into a set of measure zero, and *lower LG-singular* if the set of points where it has a negative lower $G$-gradient is of measure zero. In terms of these two notions we obtain the following monotonicity theorem from the Third Monotonicity Theorem 9.1.7 which is similar to Theorem 16.5.4.

18.5.3. THEOREM. *Suppose f is either* (i) $I^*LC$, *or* (ii) $LI_-$ *and* $I^*LC_+$. *If f is lower LG-regular and lower LG-singular, then it is nondecreasing.*

Further, from the Fifth Monotonicity Theorem 9.1.10 we obtain the following result in terms of a single lower $G$-gradient.

18.5.4. THEOREM. *Suppose f is $I^*LSC$ and nowhere decreasing, and let E be the set of points where f has a zero lower G-gradient. Then f is increasing whenever either* (i) *E is empty, or* (ii) *f is $WLC_-$ and E is finite.*

Next, when the $G$-derivates are regular, we obtain the following version of the Goldowski-Tonelli theorem in terms of generalized weak derivative.

18.5.5. THEOREM (FIFTH GOLDOWSKI-TONELLI). *Suppose f is $I^*C$ and G-derivates are regular. Then if f is weakly G-derivable* n.e. *and* $WGf' \geq 0$ a.e., *then f is nondecreasing.*

*Consequently, the same holds in terms of $LGf'$ when f is lower G-derivable* n.e. *and* $LGf' \geq 0$ a.e.

PROOF. On account of the Fourth Reduction Lemma 16.5.6, we can assume here without loss of generality that $f$ is continuous. Then, by Lemma 9.1.2, the set $N \equiv N(f)$ is perfect in $I$, and it is enough to show that $N = \varnothing$. Suppose $N \neq \varnothing$, and set

$$A = \{x \in N : GD^-f(x) \wedge GD^+f(x) \geq 0\},$$
$$B = \{x \in N : GD_-f(x) \vee GD_+f(x) = -\infty\}.$$

First, we claim that $A$ is dense in some portion $P$ of $N$. Since $WGf' \geq 0$ a.e., this holds clearly when $N^0 \neq \varnothing$. Hence suppose $N$ is nowhere dense. Then for each contiguous interval $(a, b)$ of $N$, since $f$ is nondecreasing on $[a,b]$, $GD^+f(a) \geq D_+f(a) \geq 0$ and $GD^-f(x) \geq D_-f(b) \geq 0$. Thus each of the sets $A_- \equiv \{x \in N : GD^-f(x) \geq 0\}$ and $A_+ \equiv \{x \in N : GD^+f(x) \geq 0\}$ is dense in $N$. Further, since $GD^-f$ and $GD^+f$ are of $UB_1$, $A_-$ and $A_+$ are $G_\delta$-sets, and so they are residual in $N$. Hence $A = A_- \cap A_+$ also is a residual $G_\delta$-set in $N$, and so it is dense in $N$.

Now, since $B$ also is similarly a $G_\delta$-set, $A$ and $B$ cannot be simultaneously dense in any portion of $N$. Hence there is a subinterval $J$ of $I$, open relative to $I$, such that $Q \equiv N \cap J$ is a nonempty portion of $P$ which does not meet $B$. Now, if $x \in Q$, then $GD^+f(x) \geq GD_+f(x) > -\infty$, and if $x \in J \sim N$, then since $f$ is nondecreasing in some neighborhood of $x$, it follows clearly from axiom (D2) that $GD^+f(x) \geq 0$. Further, by hypothesis, $WGf' \geq 0$ a.e., so that $|I \sim A| = 0$. Hence it follows from Theorem 16.5.1 that $f$ is nondecreasing on $J$. Thus $N \cap J = \varnothing$, which is a contradiction. □

18.5.6. PROBLEM. In connection with the Fifth Goldowski-Tonelli Theorem 18.5.5, it will be worthwhile to investigate if this theorem remains valid when the $G$-derivates are subregular; for, if so, then this will prove the Goldowski-Tonelli theorem for approximate weak derivative as well (see Theorem 15.4.2).

**§18.6 Relations between generalized new derivatives and strong derivative.** In this section we obtain two theorems dealing with the relations between generalized new derivatives and strong derivative.

First, in the following theorem we obtain strong derivative in terms of $U$-limit of generalized new derivatives.

18.6.1. THEOREM. *Suppose G-derivates are admissible and $f: I \to \mathbb{R}$ is $I^*C$. Then f has a strong derivative $c \in \overline{\mathbb{R}}$ at $x \in I^0$ iff c is the U-limit of $SGf'$ at x relative to $G\Delta_s(f)$. Further, when f is upper nonangular, the same holds iff c is the U-limit of $LGf'$ at x relative to $G\underline{\Delta}(f)$.*

PROOF. The first part is obtained from the Seventh Mean Value Theorem 18.4.1 by an argument similar to the one used for Theorem 8.3.1. The second part is obtained similarly from the Eighth Mean Value Theorem 18.4.3. □

The following theorem on the identity of generalized new derivatives with strong derivative is obtained, on the other hand, directly from Theorem 16.6.3.

18.6.2. THEOREM. *Suppose G-derivates are subregular and $f: I \to \mathbb{R}$ is IC. Then there is a residual set E in I such that at each $x \in E$, f is lower, or weakly, G-derivable at x iff it is strongly derivable there; and if so, then the two derivatives are equal at x.*

**§18.7 Properties of generalized new derivatives.** In this section we obtain the properties of generalized new derivatives similar to the properties of new derivatives obtained earlier.

First, in Theorem 18.7.1, we obtain the Darboux property of generalized lower and semi- derivatives. Then, in the case of regular $G$-derivates, we obtain in Theorem 18.7.2 the Borel class of various new $G$-derivability sets, and we obtain in Theorem 18.7.3 the Baire class of new $G$-derivatives. Finally, in Theorem 18.7.4, the Denjoy property of lower $G$-derivative is obtained in the same case.

Let us recall here that we assume here, unless stated otherwise, the $G$-derivates to be admissible, and $f: I \to \mathbb{R}$ where $I$ is open.

First, we deduce from the Mean Value Theorems 18.4.1 and 18.4.3 the Darboux property of generalized new derivatives.

18.7.1. THEOREM. *Suppose f is $I^*C$ and G-derivates are admissible. Then $SGf'$ and $WGf'$ possess the Darboux property.*

*Moreover, if f is further upper G-nonangular, then $LGf'$ also possesses the Darboux property.*

PROOF. We omit the proof here since this theorem is obtained by arguments similar to the ones used for Theorem 10.4.1 on replacing ordinary new derivatives by new $G$-derivatives and replacing ordinary derivates by $G$-derivates, and using (D2) and Theorems 18.3.1, 18.4.1, and 18.4.3 in place of Theorems 7.3.1, 8.1.2, and 8.1.4, respectively. □

Next, we will investigate the Borel class of various generalized new derivability sets, angularity sets, and knot sets. We will drop here as before the reference to $f$ in the following notations for the sake of brevity.

We will call here, as in §4.3, the sets $G\underline{\Delta}$, $G\underline{\Delta}_u$, $G\underline{\Delta}_k$, $G\underline{\Delta}_n$, $G\underline{\Delta}^*$, $G\underline{\Delta}^*_u$, $G\underline{\Delta}^*_k$, and $G\underline{\Delta}^*_n$ the *lower G-derivability sets* of $f$. The sets $G\overline{\Delta}$, $G\overline{\Delta}_u$, $G\overline{\Delta}_k$, $G\overline{\Delta}_n$, $G\overline{\Delta}^*$, $G\overline{\Delta}^*_u$, $G\overline{\Delta}^*_k$, and $G\overline{\Delta}^*_n$ will be called similarly the *upper G-derivability sets* of $f$; and the sets $G\Delta_s$, $G\Delta_{su}$, $G\Delta_{sk}$, $G\Delta_{sn}$, $G\Delta^*_s$, $G\Delta^*_{su}$, $G\Delta^*_{sk}$, and $G\Delta^*_{sn}$ will be called the *semi-G-derivability sets* of $f$. Further, we will call the sets $G\Delta_w$, $G\Delta_{wu}$, $G\Delta^*_w$, and $G\Delta^*_{wu}$ the *weak G-derivability sets* of $f$, and the sets $G\Delta_+$, $G\Delta_-$, $G\Delta_u$, $G\Delta^*_+$, $G\Delta^*_-$, and $G\Delta^*_u$ will be called the *unilateral G-derivability sets* of $f$.

18.7.2. THEOREM. *Suppose $f$ is continuous and G-derivates are regular. Then*

(a) *the G-angularity sets $G\underline{A}$, $G\overline{A}$, and $GA$ of $f$ are $F_\sigma$-sets;*

(b) *the G-knot sets $G\underline{K}$, $G\overline{K}$, $GK$, and $GK_b$ of $f$ are $G_\delta$-sets; and*

(c) *all the lower, upper, semi-, weak, and unilateral G-derivability sets of $f$ are $F_{\sigma\delta}$-sets.*

*Further, when the G-derivates are totally regular, the same holds for every function $f$.*

PROOF. First, suppose $f$ is continuous and the $G$-derivates are regular. Then the lower $G$-derivates $GD_+f$ and $GD_-f$ are of $LB_1$, and the upper $G$-derivates $GD^+f$ and $GD^-f$ are of $UB_1$.

**Re** (a): Let $Q$ denote the set of all rational numbers. Then it is easy to see that

$$G\underline{A}(f) = \bigcup_{r,s\in Q\,:\,r<s} [\{x : GD^-f(x) < r\} \cap \{x : GD_+f(x) > s\}].$$

Hence it follows clearly from the Baire classes of $GD^-f$ and $GD_+f$ that $G\underline{A}(f)$ is an $F_\sigma$-set. Now on applying this result to $-f$ it follows that $G\overline{A}(f)$ is an $F_\sigma$-set; and hence $GA(f) = G\underline{A}(f) \cup G\overline{A}(f)$ is also an $F_\sigma$-set.

**Re** (b): Since $G\underline{K}(f) = \{x : GD_+f(x) = -\infty\} \cap \{x : GD^-f(x) = +\infty\}$, it is clear once again from the Baire classes of $G$-derivates of $f$ that $G\underline{K}(f)$ is a $G_\sigma$-set. Now on applying this result to $-f$ it follows that $G\overline{K}(f)$ is a $G_\delta$-set; and hence $GK(f) = G\underline{K}(f) \cup G\overline{K}(f)$ and $GK_b(f) = G\underline{K}(f) \cap G\overline{K}(f)$ are also $G_\delta$-sets.

**Re** (c): It is easy to see here that

$$G\underline{\Delta}(f) = I \sim \bigcup_{r,s\in Q\,:\,r<s} [\{x : GD_+f(x) \le r\} \cap \{x : GD^-f(x) \ge s\}].$$

Hence it follows clearly from the Baire classes of $G$-derivates of $f$ that $G\underline{\Delta}(f)$ is an $F_{\sigma\delta}$-set. Now, it follows clearly from (a) that $G\underline{\Delta}_u(f) = G\underline{\Delta}(f) \sim G\underline{A}(f)$ is an $F_{\sigma\delta}$-set, and it follows from (b) that $G\underline{\Delta}_k(f) = G\underline{\Delta}(f) \cap G\overline{K}(f)$ is also an $F_{\sigma\delta}$-set. Further, we obtain with the help of Theorem 16.7.2 that $G\underline{\Delta}_n(f) = G\underline{\Delta}_k(f) \cup G\Delta(f)$ is an $F_{\sigma\delta}$-set.

Next, it is easy to see that

$$G\underline{\Delta}^*(f) = G\underline{\Delta}(f) \sim [\{x : GD_+f(x) = -\infty\} \cup \{x : GD^-f(x) = +\infty\}].$$

Hence it follows clearly from the Baire classes of $G$-derivates of $f$ that $G\underline{\Delta}^*(f)$ is an $F_{\sigma\delta}$-set. Now it is clear that $G\underline{\Delta}^*_u(f) = G\underline{\Delta}_u(f) \cap G\underline{\Delta}^*(f)$ is an $F_{\sigma\delta}$-set; and similar arguments hold for $G\underline{\Delta}^*_k(f)$ and $G\underline{\Delta}^*_n(f)$.

Thus it has been proved that all the lower $G$-derivability sets of $f$ are $F_{\sigma\delta}$-sets. On applying these results to $-f$ it follows that all the upper $G$-derivability sets of $f$ are $F_{\sigma\delta}$-sets. Now since $G\Delta_s(f) = G\underline{\Delta}(f) \cup G\overline{\Delta}(f)$, it follows from the above results that $G\Delta_s(f)$ is an $F_{\sigma\delta}$-set. Similar arguments hold for the other semi-$G$-derivability sets of $f$.

Next, it is clear that $G\underline{D}_w f = \vee GD_{\pm} f$ is of $LB_1$, and $G\overline{D}_w f$ is similarly of $UB_1$. Hence the results on various weak $G$-derivability sets of $f$ follow as above on replacing $GD_+f$ and $GD^-f$ by $G\underline{D}_w f$ and $G\overline{D}_w f$, respectively. We omit here the proofs for the various unilateral $G$-derivability sets of $f$, for the results on these sets are obtained by arguments which are very similar to above.

Finally, when the $G$-derivates are totally regular, the above arguments remain valid for every function $f$. □

Next, in the following theorem we deal with the Baire class of generalized new derivatives.

18.7.3. THEOREM. *Suppose f is continuous and G-derivates are regular. Then the lower and weak G-derivatives of f are of $LB_1$, and thus they are of $B_1$ on $G\underline{\Delta}_u(f)$ and $G\Delta_{wu}(f)$, respectively.*

*Further, when the G-derivates are totally regular, the same holds for every function f.*

PROOF. First, let $\varphi$ denote the lower $G$-derivative of $f$ which is defined on $G\underline{\Delta}(f)$. Then $\varphi_* = GD^-f$ is of $UB_1$ and $\varphi^* = GD_+f$ is of $LB_1$. Hence it follows from Theorem 1.5.1 that $\varphi$ is of $LB_1$. A similar argument holds for the weak $G$-derivative.

The last part follows again as in the last theorem. □

Finally, in the following theorem we obtain the Denjoy property of generalized lower derivative by a suitable modification in the proof of Theorem 16.7.5. The derivative $LGf'$ will be said here to possess the *Denjoy property* if for each $a, b \in \mathbb{R}$, $a < b$, the set $\{x : LGf'(x) \cap (a, b) \neq \varnothing\}$ has a positive measure if it is not empty.

18.7.4. THEOREM. *Suppose f is continuous and G-derivates are regular. Then if f is nonangular and lower G-derivable* n.e. *then* $LGf'$ *possesses the Denjoy property.*

PROOF. We can clearly assume here without loss of generality that $I$ is closed. Given $a, b \in \mathbb{R}$, $a < b$, suppose the set $E \equiv \{x : LGf'(x) \cap (a, b) \neq \varnothing\}$ is of measure zero. Then we need to show that $E = \varnothing$. Suppose $E \neq \varnothing$, and set

$$(1) \qquad A = \{x : GD_+f(x) \le a\}, \qquad B = \{x : GD^-f(x) \ge b\}.$$

First, we claim that each component of $A$ or $B$ includes its left or right endpoint, respectively. Let $C$ be any nondegenerate component of $A$, and let $c$ and $d$ denote the left and right endpoints, respectively, of $C$. Define $g(x) = f_{-a}(x)$, $c \le x \le d$. Now, since at each point of $A$, $D_+f \le GD_+f \le a$, we have $D_+g \le 0$ n.e. in $[c, d]$. Hence it follows from the Fourth Monotonicity Theorem 9.1.8 that $g$ is nonincreasing on $[c, d]$. Consequently, $D^+g(c) \le 0$, and so by (D2), $GD_+f(c) \le D^+f(c) \le a$. Thus $c \in A$, which proves that $C$ is closed from the left. It is proved by a similar argument that the components of $B$ are closed from the right.

Now, let $\mathscr{C}$ denote the family of all nondegenerate components of $A$ and $B$, and set $P = I \sim \bigcup\{C^0 : C \in \mathscr{C}\}$, where the interior is taken relative to $I$. Then since $E \neq \varnothing$, $P$ is a nonempty closed set. We claim that $P$ is further dense in itself. Suppose $P$ has some isolated point $x$. Then it is clear that there exist $y, z \in I$, $y < x < z$, such that either (i) $(y, x) \subset A$ and $(x, z) \subset B$, or (ii) vice versa. To be definite, suppose (i) holds. Then since $f$ is lower $G$-derivable n.e., it follows from Theorem 9.1.8 as before that $f_{-a}$ is nonincreasing on $[y, x]$ and $f_{-b}$ is nondecreasing on $[x, z]$. Thus $D^-f(x) \le a < b \le D_+f(x)$, which contradicts the nonangularity of $f$. A similar argument holds in case (ii). This proves that $P$ is perfect.

Now, set $P_- = P \cap A$ and $P_+ = P \cap B$. Then it is enough to show that $P_-$ and $P_+$ are both dense in $P$. For, since $GD_+f$ and $GD^-f$ are by hypothesis of $LB_1$ and $UB_1$, respectively, $A$ and $B$ are $G_\delta$-sets, and so $P_-$ and $P_+$ are then both residual in $P$. Consequently, $P_0 \equiv P_- \cap P_+$ is also residual in $P$, and thus it is uncountable. But for each $x \in P_0$ it is clear from (1) that $f$ is not lower $G$-derivable at $x$, which contradicts the hypothesis.

We will give the proof here for the denseness of $P_-$ in $P$; the denseness of $P_+$ is proved by a similar argument.

First, we show that $P \subset \overline{A}$. If it is not so, then there exists a point $x \in P \sim \overline{A}$. Hence there is a subinterval $J$ of $I$, open in $I$, such that $x \in J$ and $J \cap A = \varnothing$. Then $GD_+f > a$ on $J$, and it is $\ge b$ a.e. on $J$. Hence it follows from Theorem 9.1.8 as before that $f_{-b}$ is nondecreasing on $J$, and so

$GD^{-}f \geq D_{-}f \geq b$ at each point of $J$. Hence $x \in B^0$, which contradicts the fact that $x \in P$. This proves that $P \subset \bar{A}$.

Next, to see that $P_-$ is dense in $P$, let $x \in P$ and $U$ be any subinterval of $I$, open in $I$, which contains $x$. Then we need to show that $U \cap P_- \neq \varnothing$. Since $P \subset \bar{A}$ and $x$ is a limit point of $P$, there clearly exists a component $C$ of $A$ such that $\bar{C} \subset U$, and then the left endpoint of $C$ is in $U \cap P_-$. □

18.7.5. PROBLEM. We state here two problems in connection with Theorem 18.7.4 which remain unresolved. Firstly, it will be worthwhile to investigate if this theorem remains valid when the $G$-derivates are subregular. For, if so, then this will prove the Denjoy property of lower approximate derivative (see Theorem 15.4.2).

Secondly, it will be interesting to investigate if Theorem 18.7.4 holds also for generalized weak derivative. The above proof does not extend to this derivative.

# PART VI

# UNIFIED THEORIES OF SOME OTHER ASPECTS OF GENERALIZED DERIVATES AND DERIVATIVES

**Introduction to Part VI.** In this final part we present three independent unified theories dealing with generalized differentiation. In the first theory we deal with properties of generalized derivates of typical continuous functions, including generalized nonangularity, generalized nonderivability, and the nature of generalized new derivatives of such functions. In the second theory we deal with generalized differentiability of generalized smooth functions, and the third unified theory is devoted to generalized Perron integrals. The axiomatic models developed in Chapters 15 and 17 of Part V are utilized in these three theories.

First, in Chapter 19, we present a unified theory of generalized derivates of typical continuous functions. It may be recalled here that we use $\mathbb{C}$ to denote the Banach space of all continuous real-valued functions on the interval $I \equiv [0, 1]$ under the uniform norm. To unify the proofs for various generalized derivates, we define here the notions of $d$-connected and $d$-subconnected generalized derivates on $\mathbb{C}$ and prove that the various generalized derivates considered in Chapter 15 are at least $d$-subconnected on $\mathbb{C}$. Then for every $d$-subconnected generalized $G$-derivates on $\mathbb{C}$, we obtain the following properties of a typical function $f$ in $\mathbb{C}$: (i) $f$ is $G$-nonangular; (ii) $f$ has at least one $G$-derivate infinite from each side at each point of $I$, and so $f$ is $G$-nondifferentiable from each side; and (iii) $f$ is $G$-biknotted a.e. in $I$. Further, in this case we deduce with the help of a result of Part IV on strongly nowhere monotone functions that (iv) each $r \in \mathbb{R}$ is a knotted lower and upper $G$-derivative of $f$ at two dense sets of points, and so $f$ is knottedly (or normally) lower and upper $G$-differentiable at two $c$-dense sets of points.

Further, for several of the generalized $G$-derivates of Chapter 15, we prove here for a typical function $f$ in $\mathbb{C}$ that (v) $f$ is $G$-knotted (i.e., the bilateral lower and upper $G$-derivates of $f$ are everywhere $-\infty$ and $+\infty$, respectively), and so $f$ is $G$-nonderivable; and that (vi) $f$ has knotted lower (and upper) $G$-derivatives $+\infty$ and $-\infty$ at two $c$-dense sets of points.

In terms of various generalized symmetric $G$-derivates considered in Chapter 17, we prove here similarly that (vii) a typical function $f$ in $\mathbb{C}$ is symmetrically $G$-nondifferentiable; and we prove for several of these derivates that (viii) a typical function $f$ in $\mathbb{C}$ is symmetrically $G$-nonderivable.

The results of Chapter 19 include in particular the deep results of Mazurkiewicz, Banach, Jarník, and Saks on ordinary unilateral derivates of typical functions in $\mathbb{C}$, and similarly they include a result of Evans on ordinary symmetric derivates.

Next, in Chapter 20, we present a unified theory of generalized smooth functions to obtain classes of continuous functions which are differentiable in any generalized sense considered in Chapter 15 at least at some points. Relative to any $s_1$-admissible generalized limits defined in Chapter 15, we define here the notions of generalized lower and upper smooth functions and of smooth functions. Further, we introduce here a notion of admissibility of generalized smoothness which is found to hold in many cases. Then we deduce from the various derivability theorems of Part II the existence of points where an admissible lower or upper $G$-smooth function is $G$-derivable or $G$-differentiable. Also, for an admissible generalized smooth function we obtain here the mean value and Darboux properties of its generalized derivative without any derivability hypothesis. The results of this chapter generalize in a sense the results obtained in Chapter 12 on ordinary smooth functions.

Finally, in Chapter 21, we present a unified axiomatic theory of generalized Perron integrals in terms of another axiomatic model of generalized derivates which applies simultaneously to generalized derivatives of Chapter 16, generalized symmetric derivatives of Chapter 17, and generalized new derivatives of Chapter 18. We deal here with the basic problem of reconstruction of a primitive function from its various kinds of generalized derivatives, and we extend the various known results on the ordinary Perron integral to generalized Perron integrals. These results were used earlier in Chapter 11 to obtain corresponding results on new Perron integrals.

# *CHAPTER 19*

# A Unified Theory of Generalized Derivates of Typical Continuous Functions

In this chapter we present a unified theory of generalized derivates of typical continuous functions. We obtain here, in particular, the generalized nonangularity and generalized nonderivability of typical continuous functions in terms of various generalized derivates and generalized symmetric derivates, and we investigate the nature of generalized lower and upper derivatives of such functions relative to various generalized derivates.

A function is called here nonderivable (or nondifferentiable) in any generalized sense if it is not derivable (or differentiable) in that sense at any point. Further, a property of functions is said to hold for typical continuous functions if it holds for a residual set of functions in $\mathbb{C}$.

First, to unify the proofs for various generalized derivates, we define in §19.1 the notions of $d$-connected and $d$-subconnected generalized derivates on $\mathbb{C}$, and we obtain these properties for various known generalized derivates. Then, in §19.2, we obtain several results on the nature of such generalized derivates of typical continuous functions. This establishes, in particular, the nonangularity and nondifferentiability of typical continuous functions in terms of various generalized derivates.

In §19.3, we define similarly the notions of $s$-connected and $s$-subconnected generalized symmetric derivates on $\mathbb{C}$, and we obtain symmetric nondifferentiability of typical continuous functions in terms of various generalized symmetric derivates.

Next, in §19.4, we obtain nonderivability of typical continuous functions in terms of several generalized derivates and generalized symmetric derivates. Finally, in §19.5, we investigate the nature of generalized lower and upper derivatives of typical continuous functions.

**§19.1 Connected generalized derivates on $\mathbb{C}$.** To obtain some of the results on various generalized derivates in a unified manner, we first define in this section $d$-connected and $d$-subconnected generalized derivates on $\mathbb{C}$, and we investigate these properties for various known generalized derivates.

Throughout this chapter we will use $I$ to denote the interval $[0,1]$. Further, in terms of our earlier notations, $I^+$ and $I^-$ will denote the semiclosed intervals $[0,1)$ and $(0,1]$, respectively. We need here some notations to define the desired notions.

Let $G$-derivates be any admissible generalized derivates. Then for each $n \in \mathbb{N}$, we define $G\text{-}A_n$ to be the set of functions $f \in \mathbb{C}$ for which there is a point $x \in I^+$ such that $GD_+f(x) > -n$ and $GD^+f(x) < n$; and we define $G\text{-}B_n$ to be the set of functions $f \in \mathbb{C}$ for which $GD^+f(0) < n$. Further, for each real number $r$, we define $G\text{-}C_r$ to be the set of functions $f \in \mathbb{C}$ for which there is a point $x \in I^0$ such that $GD^-f(x) < r < GD_+f(x)$. Finally, for each $f \in \mathbb{C}$ and $n \in \mathbb{N}$, we define

$$G\text{-}E_{f,n} = \{x \in I^+ : GD^+f(x) < n\}.$$

Now, in terms of the above sets, we define any admissible $G$-derivates to be *d-connected on* $\mathbb{C}$ if the following three conditions hold:

(i) For each $n \in \mathbb{N}$, $G\text{-}A_n$ and $G\text{-}B_n$ are $F_\sigma$-sets (in $\mathbb{C}$).

(ii) For each $r \in \mathbb{R}$, $G\text{-}C_r$ is an $F_\sigma$-set.

(iii) For each $f \in \mathbb{C}$ and $n \in \mathbb{N}$, $G\text{-}E_{f,n}$ is the union of a sequence of sets $\{E_{f,n,i}\}$ such that the following property holds for each $E_{f,n,i}$:

(P) If $x \in E_{f_k,n,i}$ for a sequence of functions $\{f_k\}$ in $\mathbb{C}$ which converges uniformly to $f$, then $x \in E_{f,n,i}$.

Further, we will call any admissible $G$-derivates *d-subconnected on* $\mathbb{C}$ if there exist some $d$-connected $G_1$-derivates which are weaker than the $G$-derivates.

To include the results on ordinary derivates, we will consider throughout this chapter the ordinary derivates to be a trivial case of admissible generalized derivates. Further, it is not necessary to consider here Marcus' qualitative derivates and Császar's negligent derivates, for in the case of a continuous function these derivates coincide with ordinary derivates (see the proof of Theorem 15.4.4(g)).

We introduce here further some convenient notations of intervals which will be used frequently in the following proof and in the next section. For each integer $i > 1$, we will use $J_i$, $K_i$, and $U_i$ to denote the following intervals:

$$(1) \qquad J_i = [0, 1 - 1/i], \qquad K_i = [1/i, 1 - 1/i], \qquad U_i = (0, 1/i).$$

19.1.1. THEOREM (TENTH STATUS). *The following derivates are d-connected on* $\mathbb{C}$:

(a) *ordinary derivates,*

(b) *preponderant derivates,*

(c) *sequentially congruent derivates,*

(d) *congruent derivates,*
(e) *Borel derivates,*
(f) *Cesàro derivates, and*
(g) *continuously selective derivates.*

*Moreover, the following derivates are d-subconnected on* $\mathbb{C}$:
(h) *approximate derivates, and*
(i) *subcontinuous path derivates.*

PROOF. Firstly, all the derivates considered here are admissible according to the Second Status Theorem 15.3.3. We begin here with the derivates in (d), for the derivates in (a) and (c) are only particular cases of these derivates.

**Re** (d): *Congruent derivates.* Let $K$ be any subset of $\mathbb{R} \sim \{0\}$ such that $0 \in K^b$. To obtain the result for $K$-derivates, for each $n \in \mathbb{N}$ and $i > 1$, we define $A_{n,i}$ to be the set of functions $f \in \mathbb{C}$ for which there is a point $x \in J_i$ such that $|Qf(x, x+h)| \le n - 1/i$ for $h \in K \cap U_i$, and we define $B_{n,i}$ to be the set of functions $f \in \mathbb{C}$ such that $Qf(0, h) \le n - 1/i$ for $h \in K \cap U_i$. Further, for each $r \in \mathbb{R}$ and $i > 2$, define $C_{r,i}$ to be the set of functions $f \in \mathbb{C}$ for which there is a point $x \in K_i$ such that

$$Qf(x, x+h) \ge r + 1/i \quad \text{for } h \in K \cap U_i$$

and

$$Qf(x, x+h) \le r - 1/i \quad \text{for } h \in K \cap (-U_i).$$

Finally, for each $f \in \mathbb{C}$, $n \in \mathbb{N}$, and $i > 1$, define

$$E_{f,n,i} = \{x \in J_i : Qf(x, x+h) \le n - 1/i \quad \text{for } h \in K \cap U_i\}.$$

Then it is easy to see that $K\text{-}A_n = \bigcup_{i>1} A_{n,i}$, $K\text{-}B_n = \bigcup_{i>1} B_{n,i}$, $K\text{-}C_r = \bigcup_{i>2} C_{r,i}$, and $K\text{-}E_{f,n} = \bigcup_{i>1} E_{f,n,i}$.

Now, to see that $A_{n,i}$ is closed in $\mathbb{C}$, suppose $f \in \mathbb{C}$ is the uniform limit of a sequence of functions $\{f_k\}$ in $A_{n,i}$. Then for each $k$ there is a point $x_k \in J_i$ such that

$$(2) \qquad |Qf_k(x_k, x_k + h)| \le n - 1/i \quad \text{for } h \in K \cap U_i.$$

Now since $J_i$ is compact, on replacing $\{f_k\}$ by a subsequence if necessary, we can assume that $\{x_k\}$ converges to some point $x \in J_i$. Then for each $h \in K \cap U_i$, it is clear that

$$(3) \qquad Qf_k(x_k, x_k + h) \to Qf(x, x+h) \quad \text{as } k \to \infty.$$

Hence it follows from (2) that $f \in A_{n,i}$. This proves that $A_{n,i}$ is closed.

The sets $B_{n,i}$ and $C_{r,i}$ are proved by a similar argument to be closed in $\mathbb{C}$. Hence the sets $K\text{-}A_n$, $K\text{-}B_n$, and $K\text{-}C_r$ are $F_\sigma$-sets. Property (P) of the sets $E_{f,n,i}$ follows on the other hand from (3) on choosing $x_k$ to be $x$ for each $k$.

**Re** (b): *Preponderant derivates*. Let $\{r_j\}$ be any increasing sequence of positive numbers which converges to $1/2$. To obtain the result for $P$-derivates, for each $n, j \in \mathbb{N}$, and $i > 1$, we define $A_{n,j,i}$ to be the set of functions $f \in \mathbb{C}$ for which there is a point $x \in J_i$ such that the following relations hold for each $h \in U_i$:

$$|\{t \in (0,h) : Qf(x, x+t) > n - 1/j\}| \leq r_j h, \tag{4}$$

$$|\{t \in (0,h) : Qf(x, x+t) < -n + 1/j\}| \leq r_j h; \tag{5}$$

and we define $B_{n,j,i}$ to be the set of functions $f \in \mathbb{C}$ such that for each $h \in U_i$,

$$|\{t \in (0,h) : Qf(0,t) > n - 1/j\}| \leq r_j h.$$

Further, for each $r \in \mathbb{R}$, $k \in \mathbb{N}$, and $i > 2$, define $C_{r,j,i}$ to be the set of functions $f \in \mathbb{C}$ for which there is a point $x \in K_i$ such that the following relations hold for each $h \in U_i$:

$$|\{t \in (0,h) : Qf(x, x+t) < r + 1/j\}| \leq r_j h,$$

$$|\{t \in (-h,0) : Qf(x, x+t) > r - 1/j\}| \leq r_j h.$$

Finally, for each $f \in \mathbb{C}$, $n, j \in \mathbb{N}$, and $i > 1$, define

$$E_{f,n,j,i} = \{x \in J_i : |\{t \in (0,h) : Qf(x, x+t) > n - 1/j\}| \leq r_j h \quad \text{for } h \in U_i\}.$$

Then it is easy to verify that $P\text{-}A_n = \bigcup_j \bigcup_{i>1} A_{n,j,i}$, $P\text{-}B_n = \bigcup_j \bigcup_{i>1} B_{n,j,i}$, $P\text{-}C_r = \bigcup_j \bigcup_{i>2} C_{r,j,i}$, and $P\text{-}E_{f,n} = \bigcup_j \bigcup_{i>1} E_{f,n,j,i}$.

Now, to see that $A_{n,j,i}$ is closed in $\mathbb{C}$, suppose $f \in \mathbb{C}$ is the uniform limit of a sequence of functions $\{f_k\}$ in $A_{n,j,i}$. Then for each $k$ there is a point $x_k \in J_i$ such that the following relations hold for each $h \in U_i$:

$$|\{t \in (0,h) : Qf_k(x_k, x_k+t) > n - 1/j\}| \leq r_j h, \tag{6}$$

$$|\{t \in (0,h) : Qf_k(x_k, x_k+t) < -n + 1/j\}| \leq r_j h. \tag{7}$$

We can assume here as before that $\{x_k\}$ converges to some point $x \in J_i$. Now, given $h \in U_i$, let $E$ denote the set defined within chain brackets in (4), and let $E_k$ denote for each $k$ the set defined within chain brackets in (6). Then since $Qf_k(x_k, x_k+t)$ converges to $Qf(x, x+t)$ for each $t \in (0,h)$, it is clear that

$$E \subset \liminf_{k\to\infty} E_k.$$

Hence it follows from the Fatou's lemma and (6) that

$$|E| \leq \liminf_{k\to\infty} |E_k| \leq r_j h.$$

This proves the inequality (4) for $f$. The inequality (5) is obtained similarly from (7). Consequently, $f \in A_{n,j,i}$, which proves that $A_{n,j,i}$ is closed.

The sets $B_{n,j,i}$ and $C_{r,j,i}$ are proved similarly to be closed in $\mathbb{C}$. Hence the sets $P$-$A_n$, $P$-$B_n$, and $P$-$C_r$ are $F_\sigma$-sets. Finally, to obtain property (P) of the sets $E_{f,n,j,i}$, suppose $x \in E_{f_k,n,j,i}$ for a sequence of functions $\{f_k\}$ which converges uniformly to $f$. Then it follows by an argument similar to above that $x \in E_{f,n,j,i}$.

**Re** (e): *Borel derivates.* Let us recall here that

$$BD_+ f(x) = \liminf_{h\to 0+} \liminf_{\delta\to 0+} \frac{1}{h} \int_\delta^h Qf(x, x+t)\, dt.$$

Hence to obtain the result for $B$-derivates, for each $n \in \mathbb{N}$ and $i > 1$, we define $A_{n,i}$ to be the set of functions $f \in \mathbb{C}$ for which there is a point $x \in J_i$ such that for each $h \in U_i$ we obtain

$$\limsup_{\delta\to 0+} \frac{1}{h} \left| \int_\delta^h Qf(x, x+t)\, dt \right| \leq n - 1/i;$$

and we define $B_{n,i}$ to be the set of functions $f \in \mathbb{C}$ such that for each $h \in U_i$ we obtain

$$\limsup_{\delta\to 0+} \frac{1}{h} \int_\delta^h Qf(0, t)\, dt \leq n - 1/i.$$

Further, for each $r \in \mathbb{R}$ and $i > 1$, define $C_{r,i}$ to be the set of functions $f \in \mathbb{C}$ for which there is a point $x \in K_i$ such that

$$\liminf_{\delta\to 0+} \frac{1}{h} \int_\delta^h Qf(x, x+t)\, dt \geq r + 1/i \quad \text{for } h \in U_i$$

and

$$\limsup_{\delta\to 0-} \frac{1}{h} \int_\delta^h Qf(x, x+t)\, dt \leq r - 1/i \quad \text{for } h \in -U_i.$$

Finally, for each $f \in \mathbb{C}$, $n \in \mathbb{N}$, and $i > 1$, define

$$E_{f,n,i} = \left\{ x \in J_i : \limsup_{\delta\to 0+} \frac{1}{h} \int_\delta^h Qf(x, x+t)\, dt \leq n - 1/i \quad \text{for } h \in U_i \right\}.$$

Then it is easy to see that $B\text{-}A_n = \bigcup_{i>1} A_{n,i}$, $B\text{-}B_n = \bigcup_{i>1} B_{n,i}$, $B\text{-}C_r = \bigcup_{i>2} C_{r,i}$, and $B\text{-}E_{f,n} = \bigcup_{i>1} E_{f,n,i}$.

Now, to see that $A_{n,i}$ is closed in $\mathbb{C}$, suppose $f \in \mathbb{C}$ is the uniform limit of a sequence of functions $\{f_k\}$ in $A_{n,i}$. Then for each $k$ there is a point $x_k \in J_i$ such that for each $h \in U_i$,

$$\limsup_{\delta\to 0+} \frac{1}{h}\left|\int_\delta^h Qf_k(x_k, x_k + t)\,dt\right| \le n - 1/i.$$

Replacing $\{f_k\}$ by a subsequence if necessary, we can assume here that $\{x_k\}$ converges to some point $x \in J_i$ and that $\| f_k - f\| \le 1$ for each $k$. Let $M = \| f\|$. Then for each $k$, $\|f_k\| \le M + 1$, and hence if $\delta \le t \le h$, then

$$|Qf_k(x_k, x_k + t)| \le 2(M+1)/t \le 2(M+1)/\delta.$$

Now since $\{Qf_k(x_k, x_k + t)\}$ converges pointwise to $Qf(x, x+t)$ for $\delta \le t \le h$ as $k \to \infty$, it follows from Lebesgue's convergence theorem that

$$(8) \qquad \int_\delta^h Qf_k(x_k, x_k + t)\,dt \to \int_\delta^h Qf(x, x+t)\,dt \quad \text{as } k \to \infty.$$

It is thus clear that $f \in A_{n,i}$, and so $A_{n,i}$ is closed.

The sets $B_{n,i}$ and $C_{r,i}$ are proved similarly to be closed in $\mathbb{C}$. Hence the sets $B\text{-}A_n$, $B\text{-}B_n$, and $B\text{-}C_r$ are $F_\sigma$-sets. Further, property (P) of the sets $E_{f,n,i}$ is obtained from (8) on replacing $x_k$ by $x$ for each $k$.

**Re** (f): *Cesàro derivates.* Let us first observe here, as may be verified easily,

$$CD_+f(x) = \liminf_{h\to 0+} \frac{2}{h}\left\{\frac{1}{h}\int_0^h f(x+t)\,dt - f(x)\right\}.$$

Hence to obtain the result for $C$-derivates, for each $n \in \mathbb{N}$ and $i > 1$, we define $A_{n,i}$ to be the set of functions $f \in \mathbb{C}$ for which there is a point $x \in J_i$ such that for each $h \in U_i$,

$$(9) \qquad \frac{2}{h}\left|\frac{1}{h}\int_0^h f(x+t)\,dt - f(x)\right| \le n - 1/i,$$

and

$$B_{n,i} = \left\{f \in \mathbb{C} : \frac{2}{h}\int_0^h f(t)\,dt - f(0) \le n - 1/i \quad \text{for } h \in U_i\right\}.$$

Further, for each $r \in \mathbb{R}$ and $i > 1$, define $C_{r,i}$ to be the set of functions

$f \in \mathbb{C}$ for which there is a point $x \in K_i$ such that

$$\frac{2}{h}\left\{\frac{1}{h}\int_0^h f(x+t)\,dt - f(x)\right\} \geq r + 1/i \quad \text{for } h \in U_i$$

and

$$\frac{2}{h}\left\{\frac{1}{h}\int_0^h f(x+t)\,dt - f(x)\right\} \leq r - 1/i \quad \text{for } h \in -U_i.$$

Finally, for each $f \in \mathbb{C}$, $n \in \mathbb{N}$, and $i > 1$, define

$$(10)\quad E_{f,n,i} = \left\{x \in J_i : \frac{2}{h}\left\{\frac{1}{h}\int_0^h f(x+t)\,dt - f(x)\right\} \leq n - 1/i \quad \text{for } h \in U_i\right\}.$$

Then it is clear that $C\text{-}A_n = \bigcup_{i>1} A_{n,i}$, $C\text{-}B_n = \bigcup_{i>1} B_{n,i}$, $C\text{-}C_r = \bigcup_{i>2} C_{r,i}$, and $C\text{-}E_{f,n} = \bigcup_{i>1} E_{f,n,i}$.

Now, to see that $A_{n,i}$ is closed in $\mathbb{C}$, suppose $f \in \mathbb{C}$ is the uniform limit of a sequence of functions $\{f_k\}$ in $A_{n,i}$. Then for each $k$ there is a point $x_k \in J_i$ such that for each $h \in U_i$,

$$(11)\quad \frac{2}{h}\left|\frac{1}{h}\int_0^h f_k(x_k+t)\,dt - f_k(x_k)\right| \leq n - 1/i.$$

We can assume here as before that $\{x_k\}$ converges to some point $x \in J_i$. Then $f_k(x_k) \to f(x)$ as $k \to \infty$, and since $\{f_k\}$ converges uniformly to $f$, it is further easy to see that

$$(12)\quad \int_0^h f_k(x_k+t)\,dt \to \int_0^h f(x+t)\,dt \quad \text{as } k \to \infty.$$

Hence it follows from (11) that $f \in A_{n,i}$. Thus $A_{n,i}$ is closed.

The sets $B_{n,i}$ and $C_{r,i}$ are proved similarly to be closed in $\mathbb{C}$. Hence the sets $C\text{-}A_n$, $C\text{-}B_n$, and $C\text{-}C_r$ are $F_\sigma$-sets. Property (P) of the sets $E_{f,n,i}$ also follows easily from (12).

**Re** (g): *Continuously selective derivates.* Let $s$ be any continuous selection. To obtain the result for $s$-derivates, for each $n \in \mathbb{N}$ and $i > 1$, define $A_{n,i}$ to be the set of functions $f \in \mathbb{C}$ for which there is a point $x \in J_i$ such that for each $h \in U_i$,

$$|Qf(x, s_{[x,x+h]})| \leq n - 1/i,$$

and

$$B_{n,i} = \{f \in \mathbb{C} : Qf(0, s_{[0,h]}) \leq n - 1/i \quad \text{for } h \in U_i\}.$$

Further, for each $r \in \mathbb{R}$ and $i > 2$, define $C_{r,i}$ to be the set of functions $f \in \mathbb{C}$ for which there is a point $x \in K_i$ such that

$$Qf(x, s_{[x,x+h]}) \geq r + 1/i \quad \text{for } h \in U_i$$

and

$$Qf(x, s_{[x,x+h]}) \leq r - 1/i \quad \text{for } h \in -U_i.$$

Finally, for each $f \in \mathbb{C}$, $n \in \mathbb{N}$ and $i > 1$, define

$$E_{f,n,i} = \{x \in J_i : Qf(x, s_{[x,x+h]}) \leq n - 1/i \quad \text{for } h \in U_i\}.$$

Then it is easy to see that $s\text{-}A_n = \bigcup_{i>1} A_{n,i}$, $s\text{-}B_n = \bigcup_{i>1} B_{n,i}$, $s\text{-}C_r = \bigcup_{i>2} C_{r,i}$, and $s\text{-}E_{f,n} = \bigcup_{i>1} E_{f,n,i}$.

Now, the fact that the sets $A_{n,i}$, $B_{n,i}$, and $C_{r,i}$ are closed in $\mathbb{C}$ follows easily from the fact that $Qf(x, s_{[x,x+h]})$ is continuous in $x$ for each $h$. Hence the sets $s\text{-}A_n$, $s\text{-}B_n$, and $s\text{-}C_r$ are $F_\sigma$-sets. Property (P) of the sets $E_{f,n,i}$ is, on the other hand, obvious in this case.

Finally, the result for the derivates in (h) and (i) follow directly from above since these derivates are stronger than the derivates in (b) and (g), respectively. □

## §19.2 Generalized derivates of typical continuous functions.

In this section we obtain some results with the help of Theorem 19.1.1 on the nature of various generalized derivates of typical continuous functions. These results will be used in the next two sections to obtain further results on these derivates.

We need here some nomenclature. Given any admissible $G$-derivates and $f \in \mathbb{C}$, we will call $f$ *right GD-infinite at* a point $x \in I^+$ if at least one right $G$-derivate of $f$ is infinite at $x$—that is, if either $GD_+f(x) = -\infty$ or $GD^+f(x) = +\infty$. Similarly, $f$ will be called *left GD-infinite at* $x \in I^-$ if at least one left $G$-derivate of $f$ is infinite at $x$.

Further, $f$ will be called simply *left* or *right GD-infinite* if it is so at each point of $I^-$ or $I^+$, respectively; and when $f$ is simultaneously left and right $GD$-infinite, it will be said to be *bilaterally GD-infinite*.

Finally, we will call $f$ *G-nonderivable* if it is not $G$-derivable at any point of $I$; and $f$ will be called *G-nondifferentiable from left* or *right* if it is not left or right $G$-differentiable at any point of $I^-$ or $I^+$, respectively. Also, in the case of ordinary derivates the prefix "$G$-" will be dropped as usual.

Clearly, if $f$ is right $GD$-infinite, then it is $G$-nondifferentiable from the right. But if $f$ is right $G$-knotted, then it is $G$-nonderivable from the right.

We will further use some notations in the following proof and the next section. Given $f \in \mathbb{C}$ and $r \in \mathbb{R}_+$, we will use $\overline{B}(f, r)$ to denote the closed ball in $\mathbb{C}$ with center $f$ and radius $r$. Further, we will use $f^\wedge$ to denote the following function in $\mathbb{C}$:

$$f^\wedge(x) = f(1-x), \qquad x \in I.$$

Also, for each set $E \subset \mathbb{C}$, we will use $-E$ and $E^\wedge$ to denote the following sets of functions:

$$-E = \{-f : f \in E\}, \qquad E^\wedge = \{f^\wedge : f \in E\}.$$

A set $E \subset \mathbb{C}$ is called on the other hand a *boundary set* [K1, p. 66] if its interior is empty, or if its complement $\mathbb{C} \sim E$ is dense in $\mathbb{C}$. It is clear that a boundary $F_\sigma$-set in $\mathbb{C}$ is meager in $\mathbb{C}$ (see, e.g., [K1, p. 81]).

In the following theorem we obtain some results on the generalized derivates of typical continuous functions which hold for all $d$-subconnected derivates on $\mathbb{C}$. In the particular case of ordinary derivates, part (a) of this theorem has been obtained by Banach [B5] and Mazurkiewicz [Maz3], and parts (c) and (d) were obtained by Jarník [J1].

19.2.1. THEOREM. *Suppose G-derivates are d-subconnected on $\mathbb{C}$. Then the following holds for a typical function f in $\mathbb{C}$:*

(a) *$f$ is bilaterally GD-infinite,*

(b) *$f$ is G-knotted at* 0 *and* 1,

(c) *$f$ is G-nonangular, and*

(d) *$f$ is G-biknotted a.e.*

*Consequently, the above holds for each of the derivates* (a) *to* (i) *considered in Theorem* 19.1.1.

PROOF. We can assume here without loss of generality that the $G$-derivates are $d$-connected on $\mathbb{C}$. For, there exist by the hypothesis some $G_1$-derivates which are $d$-connected on $\mathbb{C}$ and are weaker than $G$-derivates. Then, given $f \in \mathbb{C}$, it is easy to see that if (a), (b), (c), or (d) holds for $G_1$-derivates, then it also holds for $G$-derivates.

Let $A_n$, $B_n$, $C_r$, and $E_{f,n}$ denote the sets $G\text{-}A_n$, $G\text{-}B_n$, $G\text{-}C_r$, and $G\text{-}E_{f,n}$ defined in the last section. Then since $G$-derivates are $d$-connected on $\mathbb{C}$, the sets $A_n$, $B_n$, and $C_r$ are $F_\sigma$-sets.

**Re** (a): We will first prove here that the set $A$ of right $GD$-infinite functions is residual in $\mathbb{C}$. It is clear that $A = \mathbb{C} \sim \bigcup_n A_n$. Now since $A_n$'s are $F_\sigma$-sets, it is thus enough to show that $A_n$ is a boundary set for each $n$.

Given $n$ and an open set $U$ in $\mathbb{C}$, choose a polynomial $u \in U$ and $r > 0$ such that the ball $\bar{B}(u, r) \subset U$. Let $c = \|u'\|$, and choose an integer $q > (c+n)/(2r)$. Set $s = 1/(2q)$, so that $2qs = 1$. Now define

$$f(x) = \begin{cases} 0 & \text{for } x = 2ks, \quad k = 0, 1, \dots, q, \\ r & \text{for } x = (2k+1)s, \quad k = 0, 1, \dots, q-1, \end{cases}$$

and let $f$ be defined linearly on the intervals $[(k-1)s, ks]$, $k = 1, 2, \dots, 2q$.

Next, let $g = u + f$. Then since $\|f\| = r$, it is clear that $g \in U$. Further, for each $x \in I^+$, $f$ is right differentiable at $x$ and $|f'_+(x)| = r/s = 2rq > c + n$. Hence $g$ also is right differentiable at $x$ and $g'_+(x) = u'(x) + f'_+(x)$, and so it is clear that $|g'_+(x)| > n$. Further, it follows from Theorem 15.2.2 that $g$ is right $G$-derivable at $x$ and $|Gg'_+(x)| = |g'_+(x)| > n$. Since this holds for each $x \in I^+$, $g \notin A_n$, which proves that $A_n$ is a boundary set.

The set $A$ is thus residual in $\mathbb{C}$. Further, it is clear that each $f \in A^\wedge$ is left $GD$-infinite, and $A \cap A^\wedge$ is again residual in $\mathbb{C}$.

**Re** (b): Given $n \in \mathbb{N}$, it is easy to see that the above defined function $g \notin B_n$. Hence the set $B_n$ also is a boundary set, and thus it is meager in $\mathbb{C}$. The set $B = \mathbb{C} \sim \bigcup_n B_n$ is consequently residual in $\mathbb{C}$ where for each $f \in B$, $GD^+f(0) = \infty$. The set $E = B \cap (-B)$ is thus a residual set in $\mathbb{C}$ whose elements are $G$-knotted at 0. Further, it is clear that each $f \in E^\wedge$ is $G$-knotted at 1, and $E \cap E^\wedge$ is again residual in $\mathbb{C}$.

**Re** (c): Let $F$ and $P$ denote the sets of functions in $\mathbb{C}$ which are lower $G$-nonangular or $G$-nonangular, respectively. Then if $Q$ denotes the set of rational numbers, it is clear that $F = \mathbb{C} \sim \bigcup\{C_r : r \in Q\}$. Hence $F$ is a $G_\delta$-set; and since $P = F \cap (-F)$, $P$ also is a $G_\delta$-set.

Further, since $G$-derivates are admissible, it follows from Theorem 15.2.2 that each polynomial in $\mathbb{C}$ is $G$-derivable, and thus it is $G$-nonangular. The set $P$ is thus a dense $G_\delta$-set in $\mathbb{C}$, and so is residual in $\mathbb{C}$.

**Re** (d): To prove this part, it is enough to show that the set $S$ of functions $f$ in $\mathbb{C}$ for which $GD^+f = \infty$ a.e. is residual in $\mathbb{C}$. For, then $T = S \cap (-S)$ is also residual in $\mathbb{C}$, and each $f \in T$ is right $G$-knotted a.e. Consequently, $T \cap T^\wedge$ is a residual set in $\mathbb{C}$ whose elements are $G$-biknotted a.e.

Now let $E_f$ denote for each $f \in \mathbb{C}$ the set of points $x \in I^+$ such that $GD^+f(x) < \infty$, and set

$$H = \{f \in \mathbb{C} : |E_f| > 0\}.$$

Then it is enough to show that $H$ is meager in $\mathbb{C}$.

For each $f \in \mathbb{C}$ and $n \in \mathbb{N}$, define $E_{f,n} = \{x \in I^+ : GD^+f(x) < n\}$. Then $E_f = \bigcup_n E_{f,n}$. Further, for each $f$ and $n$, there is by hypothesis a sequence of sets $\{E_{f,n,i}\}$ such that $E_{f,n} = \bigcup_i E_{f,n,i}$, where the sets $E_{f,n,i}$ have property (P). Thus

$$E_f = \bigcup_n \bigcup_i E_{f,n,i}. \tag{1}$$

Now for each $n, i, j \in \mathbb{N}$, set

$$H_{n,i,j} = \{f \in \mathbb{C} : |E_{f,n,i}| \geq 1/j\}.$$

Then it is clear from (1) that

$$H = \bigcup_n \bigcup_i \bigcup_j H_{n,i,j}. \tag{2}$$

We claim that the sets $H_{n,i,j}$ are closed in $\mathbb{C}$. Given $n, i, j$, suppose $f \in \mathbb{C}$ is the uniform limit of a sequence of functions $\{f_k\}$ in $H_{n,i,j}$. Then if $x \in E_{f_k,n,i}$ for infinitely many $k$'s, then it follows from property (P) of these sets that $x \in E_{f,n,i}$. Thus $E_{f,n,i} \supset \limsup_{k\to\infty} E_{f_k,n,i}$, and hence

$$|E_{f,n,i}| \geq \limsup_{k\to\infty} |E_{f_k,n,i}| \geq 1/j.$$

Consequently, $f \in H_{n,i,j}$, which proves that $H_{n,i,j}$ is closed. Hence to prove $H$ to be meager, it is enough by (2) to show that $H_{n,i,j}$ is a boundary set for each $n, i, j$.

Given $n, i, j$ and an open set $U$ in $\mathbb{C}$, choose a polynomial $u \in U$ and $r > 0$ such that $\overline{B}(u, r) \subset U$. Let $c = \|u'\|$, and choose an integer $q$ such that

$$q \geq \max\left\{n, \frac{(c+n)j}{r(j+1)}\right\}. \tag{3}$$

Further, let

$$s = \frac{j}{q(j+1)} \quad \text{and} \quad t = \frac{1}{q(j+1)}. \tag{4}$$

Then $s + t = 1/q$, so that $q(s+t) = 1$.

Now define

$$f(x) = \begin{cases} 0 & \text{for } x = k(s+t), & k = 0, 1, \dots, q, \\ r & \text{for } x = (k+1)s + kt, & k = 0, 1, \dots, q-1, \end{cases}$$

and let $f$ be defined linearly on the intervals

$$I_k = [k(s+t), (k+1)s + kt], \qquad k = 0, 1, \dots, q-1,$$
$$J_k = [(k+1)s + kt, (k+1)(s+t)], \qquad k = 0, 1, \dots, q-1.$$

Clearly, $|I_k| = s$ and $|J_k| = t$ for each $k$.

Next, let $g = u + f$. Then since $\|f\| = r$, $g \in U$. Further, if $x \in I_k^0$ for some $k$, then it is clear that $f$ is derivable at $x$ and $f'(x) = r/s$. Hence $g$ also is derivable at $x$ and, by (3) and (4), we obtain

$$g'(x) = u'(x) + f'(x) \geq -c + \frac{r}{s} = -c + \frac{rq(j+1)}{j} \geq n.$$

Hence, by Theorem 15.2.2, $g$ is $G$-derivable at $x$ and $Gg'(x) = g'(x) \geq n$. Consequently, with the help of (4) we obtain

$$|E_{g,n}| \leq \sum_{i=0}^{k-1} |J_k| = qt = \frac{1}{j+1} < \frac{1}{j}.$$

Thus $g \notin H_{n,i,j}$, which proves that $H_{n,i,j}$ is a boundary set. Consequently, $H$ is meager in $\mathbb{C}$. □

We include here a direct consequence of part (a) of the above theorem. The other parts of this theorem will be used later in §§19.4 and 19.5.

19.2.2. COROLLARY. *Let G-derivates by any of the derivates* (a) *to* (i) *considered in Theorem* 19.1.1. *Then a typical function f in* $\mathbb{C}$ *is G-nondifferentiable from each side.*

In connection with this corollary it may be observed here, as seen later in §19.5, that a typical continuous function is not $G$-nonderivable from any one side.

**§19.3 Generalized symmetric derivates of typical continuous functions.** In this section we obtain results similar to the ones of the last section for various generalized symmetric derivates.

To obtain a unified proof for various generalized symmetric derivates, we first introduce as before a notion of their connectedness on $\mathbb{C}$. The ordinary symmetric derivates will be considered here, as before, to be a trivial case of admissible generalized symmetric derivates.

Given any admissible generalized symmetric $G$-derivates, for each $n \in \mathbb{N}$, we define $G\text{-}S_n$ to be the set of functions $f \in \mathbb{C}$ for which there is a point $x \in I^0$ such that $G\underline{D}_s f(x) > -n$ and $G\overline{D}_s f(x) < n$. Further, for each $f \in \mathbb{C}$ and $n \in \mathbb{N}$, we define

$$G\text{-}T_{f,n} = \left\{x \in I^0 : G\overline{D}_s f(x) < n\right\}.$$

Now we define any admissible symmetric $G$-derivates to be *s-connected on* $\mathbb{C}$ if the following two conditions hold:

(i) For each $n \in \mathbb{N}$, $G\text{-}S_n$ is an $F_\sigma$-set (in $\mathbb{C}$).

(ii) For each $f \in \mathbb{C}$ and $n \in \mathbb{N}$, $G\text{-}T_{f,n}$ is the union of a sequence of sets $\{T_{f,n,i}\}$, each of which has property (P).

Further we will call, as before, any admissible symmetric $G$-derivates *s-subconnected on* $\mathbb{C}$ if there exist some $s$-connected symmetric $G_1$-derivates which are weaker than the symmetric $G$-derivates.

19.3.1. THEOREM (ELEVENTH STATUS). *The following symmetric derivates are s-connected on* $\mathbb{C}$:

(a) *ordinary symmetric derivates,*

(b) *preponderant symmetric derivates,*

(c) *sequentially congruent symmetric derivates,*

(d) *congruent symmetric derivates,*

(e) *Borel symmetric derivates,*

(f) *Cesàro symmetric derivates, and*

(g) *continuously selective symmetric derivates.*

*Moreover, the following symmetric derivates are s-subconnected on* $\mathbb{C}$:

(h) *approximate symmetric derivates, and*

(i) *subcontinuous symmetric path derivates.*

PROOF. The admissibility of each of these symmetric derivates follows directly from the Seventh Status Theorem 17.1.7. We omit the rest of the proof here since in each of these cases the result is obtained by arguments similar to the ones used in the proof of Theorem 19.1.1 for the corresponding derivates, with the following modifications: Replace (i) $J_i$ by $K_i$, (ii) $G\text{-}A_n$ by $G\text{-}S_n$, (iii) $G\text{-}E_{f,n}$ by $G\text{-}T_{f,n}$, (iv) $Qf(x, x+h)$ and $Qf(x, x+t)$ by $Q_s f(x, h)$ and $Q_s f(x, t)$, respectively, and (v) $Qf(x, s_{[x,x+h]})$ by $Q_s f(x, s_{[x,x+h]} - x)$; and (vi) in the case of Cesàro symmetric derivates, replace (9) and (10) of §19.1 by the following:

$$\text{(9}'\text{)} \qquad \frac{1}{h^2}\left|\int_0^h \{f(x+t) - f(x-t)\}\,dt\right| \le n - 1/i,$$

$$\text{(10}'\text{)} \qquad E_{f,n,i} = \left\{x \in J_i : \frac{1}{h^2}\left|\int_0^h \{f(x+t) - f(x-t)\}\,dt\right| \le n - 1/i \quad \text{for } h \in U_i\right\}. \quad \square$$

We will further use here the following terminology similar to the one introduced in the last section. Given any admissible generalized symmetric $G$-derivates and $f \in \mathbb{C}$, we will call $f$ *symmetrically GD-infinite* at a point $x \in I^0$ if at least one symmetric $G$-derivate of $f$ is infinite at $x$—that is, if either $G\underline{D}_s f(x) = -\infty$ or $G\overline{D}_s f(x) = \infty$. Similarly, $f$ will be called *symmetrically G-knotted* at $x$ if $G\underline{D}_s f(x) = -\infty$ and $G\overline{D}_s f(x) = \infty$. Further, $f$ will be called simply symmetrically $GD$-infinite or symmetrically $G$-knotted if it is so at each point of $I^0$.

Finally, we will call $f$ *symmetrically G-nonderivable* or *symmetrically G-nondifferentiable* if it is not symmetrically $G$-derivable or symmetrically $G$-differentiable, respectively, at any point of $I^0$. Also, in the case of ordinary symmetric derivates the prefix "$G$-" will be dropped as usual.

Clearly, if $f$ is symmetrically $GD$-infinite or symmetrically $G$-knotted, then it is symmetrically $G$-nondifferentiable or symmetrically $G$-nonderivable, respectively.

Now in the following theorem we obtain results similar to parts (a) and (d) of Theorem 19.2.1 for $s$-subconnected generalized symmetric derivates. The other two parts of that theorem are not applicable in the present case.

19.3.2. THEOREM. *Suppose symmetric G-derivates are s-subconnected on* $\mathbb{C}$. *Then the following holds for a typical function f in* $\mathbb{C}$:

(a) *f is symmetrically GD-infinite, and*

(b) *f is symmetrically G-knotted* a.e.

*Consequently, the above holds for each of the symmetric derivates* (a) *to* (i) *considered in Theorem* 19.3.1.

PROOF. We can assume here without loss of generality, as seen earlier in the proof of Theorem 19.2.1, that the given symmetric $G$-derivates are $s$-connected on $\mathbb{C}$. Then the sets $G\text{-}S_n$ are $F_\sigma$-sets.

**Re** (a): Let $S$ denote the set of functions in $\mathbb{C}$ which are symmetrically $GD$-infinite. Then it is clear that $S = \mathbb{C} \sim \bigcup_n G\text{-}S_n$. Hence, to prove $S$ to be residual in $\mathbb{C}$, it is enough to show that $G\text{-}S_n$ is a boundary set for each $n$.

Given $n$ and an open set $U$ in $\mathbb{C}$, choose a polynomial $u \in U$ and $r > 0$ such that the ball $\overline{B}(u, r) \subset U$. Let $c = \|u'\|$, and choose an integer $q > 4(c + n)/(3r)$. Set $s = 1/(3q)$, so that $3qs = 1$. Now define

$$f(x) = \begin{cases} 0 & \text{for } x = 3ks, \quad k = 0, 1, \ldots, q, \\ r & \text{for } x = 3ks + s, \quad k = 0, 1, \ldots, q - 1, \end{cases}$$

and let $f$ be defined linearly on the intervals $I_k = [3ks, 3ks + s]$ and $J_k = [3ks + s, 3(k + 1)s]$, $k = 0, 1, \ldots, q - 1$. Then it is clear that $|I_k| = s$ and $|J_k| = 2s$ for each $k$.

Next, let $g = u + f$. Then since $\|f\| = r$, $g \in U$. We first claim that for each $x \in I^0$, $f$ is symmetrically derivable at $x$ and $|f_s'(x)| \geq r/(4s)$. If $x \in I_k$ or $J_k$ for some $k$, then it is clear that $f'(x) = r/s$ or $-r/(2s)$, respectively. Further, when $x = 3ks$ or $3ks + s$ for some $k$, then it is easy to see that $f_s'(x) = \frac{1}{2}\{f_+'(x) + f_-'(x)\} = \frac{1}{2}\{r/s + (-r/(2s)\} = r/(4s)$, which proves the claim.

Thus for each $x \in I^0$, $|f_s'(x)| \geq r/(4s) = 3qr/4 > c + n$. Hence $g$ also is symmetrically derivable at $x$ and $g_s'(x) = u'(x) + f_s'(x)$, from which it is clear that $|g_s'(x)| > n$. Hence by Theorem 17.1.8, $g$ is also symmetrically $G$-derivable at $x$ and $|Gg_s'(x)| > n$. Thus $g \notin G\text{-}S_n$, which proves that $G\text{-}S_n$ is a boundary set.

**Re** (b): We omit the proof of this part since it follows by an argument similar to the one used for Theorem 19.2.1(d) on replacing $GD^+f$ by $G\overline{D}_s f$ and $E_{f,n}$ by $G\text{-}T_{f,n}$. □

We include here as before the following direct consequence of part (a) of the above theorem.

19.3.3. COROLLARY. *Let symmetric G-derivates be any of the symmetric derivates* (a) *to* (i) *considered in Theorem* 19.3.1. *Then a typical function f is* $\mathbb{C}$ *is symmetrically G-nondifferentiable.*

**§19.4 Generalized nonderivability and generalized symmetric nonderivability of typical continuous functions.** In this section we obtain nonderivability and symmetric nonderivability of typical continuous functions in terms of some of the notions of generalized derivatives.

Given any generalized $G$-derivates, it may be recalled here that a function $f \in \mathbb{C}$ is called *G-knotted* at a point $x \in I$ if $G\underline{D}f(x) = -\infty$ and $G\overline{D}f(x) = +\infty$, and $f$ is called simply $G$-knotted if it is so at each point of $I$. Clearly, if $f$ is $G$-knotted, then it is $G$-nonderivable.

We begin with some of the symmetric derivates in the following theorem. Parts (a) and (b) of this theorem are known. Part (a) on ordinary symmetric derivates was obtained originally by Kostyrko [Ko] with the help of a symmetrically nonderivable function constructed by L. Filipczak [F.L1], and part (b) on approximate symmetric derivates was obtained later by Evans [Ev]. We strengthen these two results here to preponderant symmetric derivates, and the proofs presented here are quite different.

19.4.1. THEOREM. *Let symmetric G-derivates be any of the following symmetric derivates:*

(a) *ordinary symmetric derivates,*

(b) *approximate symmetric derivates,*

(c) *preponderant symmetric derivates,*

(d) *sequentially congruent symmetric derivates,*

(e) *congruent symmetric derivates.*

*Then a typical function $f$ in $\mathbb{C}$ is symmetrically G-knotted, and hence symmetrically G-nonderivable.*

PROOF. It is enough to prove the result here for the derivates in parts (c) and (e). For the derivates in (a) and (b) are stronger than the ones in (c), and the derivates in (d) constitute a particular case of (e).

First, let symmetric $G$-derivates denote any of the derivates in (c) and (e). Now let $G\text{-}S_n$ denote for each $n \in \mathbb{N}$ the set of functions $f \in \mathbb{C}$ for which there is a point $x \in I^0$ such that $G\overline{D}_s f(x) < n$. Then it is enough to show that $G\text{-}S_n$ is meager in $\mathbb{C}$ for each $n$. For, then the set $S = \mathbb{C} \sim \bigcup_n G\text{-}S_n$ is a residual set in $\mathbb{C}$ such that for each $f \in S$, $G\overline{D}_s f(x) = \infty$ for $x \in I^0$; and consequently $S \cap (-S)$ is a residual set of functions in $\mathbb{C}$ whose elements are symmetrically $G$-knotted. We will use here the notations $K_i$ and $U_i$ as defined earlier in (1) of §19.1, but $J_i$ will be used differently.

Now, to deal with (c), let $\{r_j\}$ be an increasing sequence of positive real numbers which converges to $1/2$. Then for each $n, j \in \mathbb{N}$ and $i > 2$, we define $P\text{-}S_{n,j,i}$ to be the set of functions $f \in \mathbb{C}$ for which there is a point $x \in K_i$ such that for each $h \in U_i$,

$$\left|\{t \in (0,h) : Q_s f(x,t) > n - 1/j\}\right| \le r_j h.$$

Next, to deal with (e), let $K$ be a given subset of $\mathbb{R}_+$ such that $0 \in K^+$. Then for each $n \in \mathbb{N}$ and $i > 2$, we define $K\text{-}S_{n,i}$ to be the set of functions $f \in \mathbb{C}$ for which there is a point $x \in K_i$ such that

$$Q_s f(x,h) \leq n - 1/i \quad \text{for } h \in K \cap U_i.$$

It is now easy to verify that $P\text{-}S_n = \bigcup_j \bigcup_{i>2} P\text{-}S_{n,j,i}$ and $K\text{-}S_n = \bigcup_{i>2} K\text{-}S_{n,i}$. Further, using arguments similar to the ones used in the proof of Theorem 19.1.1 (see cases (b) and (d)), the sets $P\text{-}S_{n,j,i}$ and $K\text{-}S_{n,i}$ are proved to be closed in $\mathbb{C}$. Hence it is enough to prove that these sets are boundary sets in $\mathbb{C}$, for this will prove that they are meager in $\mathbb{C}$.

**Re** (c): *Or the sets* $P\text{-}S_{n,j,i}$. Given $n, j \in \mathbb{N}$ and $i > 2$, let $U$ be a given open set in $\mathbb{C}$. Then to prove the set $P\text{-}S_{n,j,i}$ to be a boundary set, we need to find a function $g \in U$ which is not in $P\text{-}S_{n,j,i}$—that is, such that for each $x \in K_i$ there is an $h \in U_i$ for which the following holds:

$$(1) \qquad |\{t \in (0,h) : Q_s g(x,t) > n - 1/j\}| > r_j h.$$

Choose a polynomial $u \in U$ and $r > 0$ such that $\overline{B}(u,r) \subset U$. Let $c = \|u'\|$. Further choose a real number $a$ such that $1/2 < a < 1 - r_j$, and let $m$ be the smallest integer such that $m > 3/\{2(2a-1)\}$. Then it is clear that $m \geq 2$ and that $b \equiv 2a - 1 - 3/(2m) > 0$.

Now set

$$(2) \qquad M = \max\left\{i, \frac{2(c+n)}{rb}\right\}, \qquad s_0 = \frac{1}{(m+1)M + m},$$

and let $s$ be any element of the interval $S \equiv (0, s_0)$. Then $\{(m+1)M + m\}s < 1$, and so $(m+1)s < (m+1)Ms < 1 - ms$. Let $q$ be the largest positive integer such that $q \leq (1 - ms)/\{(m+1)s\}$. Then it is clear that $d \equiv \{(m+1)q + m\}s \leq 1$ and that $1 - ms < (q+1)(m+1)s$. Hence it is clear that

$$1 - d = 1 - \{(m+1)q + m\}s < (m+1)s.$$

Further, since $M \geq i$ and $s < s_0$, it follows from (2) that $(m+1)si < \{(m+1)M + m\}s < 1$. Thus we obtain

$$(3) \qquad 1 - d < (m+1)s < 1/i.$$

Also, it is easy to see from the two relations of (2) that $ms < (m+1)s_0 < 1/M \leq rb/\{2(c+n)\}$, and so we obtain

$$(4) \qquad \frac{rb}{2ms} > c + n.$$

We now define the following saw-like function $f$ in terms of the parameters $m$, $s$, and $q$. Let

$$f(x) = \begin{cases} 0 & \text{for } x = (m+1)ks, \qquad k = 0,1,\dots,q, \\ r & \text{for } x = \{(m+1)k+m\}s, \quad k = 0,1,\dots,q, \quad \text{and for } x = 1; \end{cases}$$

and let $f$ be defined linearly on the intervals

$$I_k = [(m+1)ks, \{(m+1)k+m\}s], \qquad k = 0,1,\dots,q,$$
$$J_k = [\{(m+1)k+m\}s, (m+1)(k+1)s], \qquad k = 0,1,\dots,q-1,$$

and on $J_q = [d, 1]$.

Next, define $g = u + f$. Then since $\|f\| = r$, it is clear that $g \in U$. We claim that $g \notin P\text{-}S_{n,j,i}$. First note that $|I_k| = ms$ for each $k$, $|J_k| = s$ for $k \le q-1$, and, by (3), $|J_q| = 1 - d < 1/i$. Thus $f$ is a periodic function on $[0, (m+1)qs]$ with period $(m+1)s$. Further, it is clear from (3) that $K_i$ does not meet $I_0$, $J_0$, and $J_q$. Hence each point of $K_i$ is included in either $(I_k)^0$ for some $k > 0$ or in $J_k$ for some $k < q$.

First, suppose $x \in (I_k)^0$ for some $k > 0$. Then due to the linearity of $f$ on $I_k$, $f$ is derivable at $x$ and $f'(x) = r/(ms)$. But since $a < 1$, it is clear that $b < 1$, and so it follows from (4) that $f'(x) > (c+n)/b > c+n$. Further, since $\|u'\| = c$, it follows from Dini's Theorem 14.3.1 that $|Q_s u(y,z)| \le c$ for each $y, z \in I$. Hence due to the linearity of $f$ on $I_k$, on choosing $h$ small enough such that $x \pm h \in I_k$, for each $t \in (0,h)$ we obtain

$$Q_s g(x,t) = Q_s u(x,t) + Q_s f(x,t) > -c + c + n = n,$$

which proves (1) for $x$.

Next, for the sake of simplicity we consider the interval $J_0$. The result on the intervals $J_k$ $(k = 1,2,\dots,q-1)$ follows from it by the periodicity of $f$. Hence suppose $x \subset J_0$, and let $h = ms$. Then by (3), $h < 1/i$; that is, $h \in U_i$. Now set $V \equiv V_s = (ah, h) = (ams, ms)$. Then since $a < 1 - r_j$, $|V| = (1-a)h > r_j h$. Thus, to prove (1) for $x$, it is enough to show that for each $t \in V$, $Q_s g(x,t) > n$, or, as seen above, that $Q_s f(x,t) > c + n$. Hence let $ams < t < ms$ and note that $ms \le x \le (m+1)s$. We need to consider here two cases separately.

**Case 1:** $ms \le x \le (m + \frac{1}{2})s$. Since $t \in V$ and $am > 1$, in this case we have $(m+1)s < (a+1)ms < x + t < (2m + \frac{1}{2})s$. Hence it follows clearly from the definition (or the graph) of $f$ that

$$f(x+t) = \frac{r}{ms}\{x + t - (m+1)s\}$$
$$> \frac{r}{ms}\{(a+1)ms - (m+1)s\} = r(a - 1/m).$$

Further, it is easy to see as before that $0 < x - t < \{(1 - a)m + \frac{1}{2}\}s < ms$, and so we obtain

$$f(x-t) = \frac{r}{ms}(x-t) < \frac{r}{ms}\left\{(1-a)m + \frac{1}{2}\right\}s = r\{(1-a) + 1/(2m)\}.$$

Hence in this case with the help of (3) we obtain

$$Q_s f(x,t) = \frac{f(x+t) - f(x-t)}{2t} > \frac{r(a - 1/m) - r\{(1-a) + 1/(2m)\}}{2t}$$
$$= \frac{r}{2t}\left(2a - 1 - \frac{3}{2m}\right) = \frac{rb}{2t} > \frac{rb}{2ms} > c + n.$$

**Case 2:** $(m + \frac{1}{2})s \le x \le (m+1)s$. In this case it is clear that $(m+1)s < \{(a+1)m + \frac{1}{2}\}s < x + t < (2m+1)s$, and so

$$f(x+t) = \frac{r}{ms}\{x + t - (m+1)s\}$$
$$> \frac{r}{ms}\left[\left\{(a+1)m + \frac{1}{2}\right\}s - (m+1)s\right] = r(a - 1/(2m)).$$

Further, again since $am > 1$, it is easy to see that $\frac{1}{2}s < x - t < \{(1-a)m + 1\}s < ms$, and so we obtain

$$f(x-t) = \frac{r}{ms}(x-t) < \frac{r}{ms}\{(1-a)m + 1\}s = r(1 - a + 1/m).$$

Hence in this case we obtain as before

$$Q_s f(x,t) > \frac{r(a - 1/(2m)) - r(1 - a + 1/m)}{2t}$$
$$> \frac{r}{2ms}\left(2a - 1 - \frac{3}{2m}\right) > c + n.$$

This proves that $g \notin P\text{-}S_{n,j,i}$, and hence $P\text{-}S_{n,j,i}$ is a boundary set.

**Re** (e): *Or the sets* $K\text{-}S_{n,i}$. Given $K$, $n \in \mathbb{N}$ and $i > 2$, let $U$ be a given open set in $\mathbb{C}$. Then to prove the set $K\text{-}S_{n,i}$ to be a boundary set, we need to find here a function $g \in U$ such that for each $x \in K_i$ there is an $h \in K \cap U_i$ for which $Q_s g(x,h) > n - 1/i$. We claim that this holds for the above constructed function $g$ with an appropriate choice of the parameter $s$.

Choose $a = 3/4$. Then since $3/\{2(2a-1)\} = 3$, $m = 4$. Hence $V_s = (ams, ms) = (3s, 4s)$. Now since zero is a right limit point of $K$, it is clear that on choosing $s$ small enough we can find an $h \in K \cap V_s$. Then it is clear from the above proof of (c) that for this choice of $s$, $g \notin K\text{-}S_{n,i}$. This, in turn, proves that $K\text{-}S_{n,i}$ is a boundary set. □

Next, in the following theorem we obtain the nonderivability of a typical continuous function in terms of the derivates that correspond to the symmetric derivates considered in the last theorem. This theorem also is known for the derivates in (a) and (b). In the case of ordinary derivates, this theorem was first observed by Saks [S5] without proof, and its proof was soon published by Jarník [J1]. Part (b) was also obtained subsequently by Jarník [J2].

19.4.2. THEOREM. *Let G-derivates be any of the following derivates:*
(a) *ordinary derivates,*
(b) *approximate derivates,*
(c) *preponderant derivates,*
(d) *sequentially congruent derivates,*
(e) *congruent derivates.*
*Then a typical function f in $\mathbb{C}$ is G-knotted, and hence G-nonderivable.*

PROOF. Let $G$-derivates denote any of the derivates under consideration. Then since the bilateral $G$-derivates are stronger than the corresponding symmetric $G$-derivates, it follows from Theorem 19.4.1 that a typical function $f$ in $\mathbb{C}$ is $G$-knotted at the interior points of $I$. The result follows now on combining this fact with part (b) of Theorem 19.2.1. □

19.4.3. REMARK. In connection with L. Filipczak's example of a continuous symmetrically nonderivable function mentioned above, it is interesting to note here that later she constructed a much more difficult example of a continuous function which is neither symmetrically derivable nor unilaterally derivable at any point (see [F.L2]).

19.4.4. PROBLEM. It remains to obtain Theorem 19.4.1 for the following symmetric derivates: (i) Borel symmetric derivates, (ii) Cesàro symmetric derivates, and (iii) continuously selective symmetric derivates. The result on (iii) would imply this theorem for (iv) subcontinuous symmetric path derivates as well, and then Theorem 19.4.2 would follow from these results as before for (v) Borel derivates, (vi) Cesàro derivates, (vii) continuously selective derivates, and (viii) subcontinuous path derivatives.

It is only natural to expect that these nonderivabilities can be handled by methods similar to the ones used in the proof of Theorem 19.4.1. However, due to some intrinsic difficulties that arise in these three cases, we have not been able to settle these questions yet.

**§19.5 Generalized knotted semiderivability and unilateral derivability of typical continuous functions.** In this section we establish the existence of finite and infinite generalized knotted (and so normal) lower and upper derivatives of typical continuous functions at some points. A theorem of Saks [S5] on the existence of infinite unilateral derivative for such functions follows from one of these results.

Further, in Theorem 19.5.4 we sum up the results obtained in this chapter on various generalized derivates of typical continuous functions, and in Theorem 19.5.5 we state the various results obtained in terms of ordinary derivates of typical continuous functions.

We first obtain in the following two theorems some general hypotheses under which any continuous function possesses finite or infinite generalized knotted lower and upper derivatives.

19.5.1. THEOREM. *Given any admissible G-derivates, suppose* $f \in \mathbb{C}$ *is G-nonangular and bilaterally GD-infinite. Then every real number is a knotted lower and upper G-derivative of f at two dense sets of points, and thus f is knottedly lower and upper G-differentiable at two c-dense sets of points.*

*Consequently, for each of the derivates* (a) *to* (i) *considered in Theorem* 19.1.1, *the above holds for a typical function f in* $\mathbb{C}$.

PROOF. First, suppose $f$ is $G$-nonangular and bilaterally $GD$-infinite. Then $f$ is clearly strongly nowhere monotone, and so it follows from Theorem 13.3.3 that each $r \in \mathbb{R}$ is a lower gradient of $f$ at a dense set of points $E$ in $I$. Now, given $x \in E$, $r \in GLf'(x)$ by Theorem 18.1.1; and since $f$ is $G$-nonangular, it is clear that $GLf'(x)$ is unique and equal to $r$. Now, since $f$ is bilaterally $GD$-infinite, it is clear that the derivative $GLf'(x)$ is knotted. The $c$-denseness of the knotted lower $G$-differentiability set of $f$ is now obvious. Further, on applying the result proved to $-f$, the result on knotted upper $G$-derivative of $f$ is obtained.

The last part follows now with the help of Theorem 19.2.1. □

19.5.2. THEOREM. *Given any admissible G-derivates, suppose* $f \in \mathbb{C}$ *is G-nonangular, G-knotted, and G-biknotted* a.e. *Then f has an infinite knotted lower* (*or upper*) *G-derivative at a dense set of points in almost all of its level sets; and it has knotted lower* (*or upper*) *G-derivatives* $+\infty$ *and* $-\infty$ *at two c-dense sets of points.*

*Consequently, for each of the derivates* (a) *to* (e) *considered in Theorem* 19.4.2, *the above holds for a typical function f in* $\mathbb{C}$.

PROOF. First, suppose the given hypothesis holds for $f$. We will prove the result here for the lower $G$-derivative; in the case of the upper $G$-derivative the result is obtained on applying this result to $-f$.

Now, since $G$-derivates are admissible, $f$ is clearly also knotted and biknotted a.e. in the ordinary sense. Hence it is clear that $f$ is $S$-singular, and so it follows from part (b) of the Fourth Derivability Theorem 6.3.5 that $f$ has an infinite normal lower derivative at a dense set of points in almost all of its level sets. First, suppose $f$ has a normal lower derivative $+\infty$ at a point $x$. Then $f$ has a right derivative $+\infty$ at $x$. Hence $Gf'_+(x) = +\infty$. Further, since $f$ is $G$-nonangular, $GD^-f(x) \geq Gf_+(x) = +\infty$; and since $f$ is $G$-knotted, it is clear that $GD_-f(x) = -\infty$. This proves that $f$ has a knotted lower $G$-derivative $+\infty$ at $x$. A similar argument holds when $f$ has a normal lower derivative $-\infty$ at $x$.

The first result of the first part is thus established. Its second result is obtained by a similar argument from Theorem 13.5.4(b). Now the second part follows from the first with the help of Theorems 19.2.1 and 19.4.2. □

In the case of ordinary derivates, the following result of Saks [S5] on the existence of infinite unilateral derivatives for a typical continuous function is implicit in the above theorem. This result was obtained by Saks using the sophisticated theory of analytic sets.

19.5.3. COROLLARY (SAKS). *A typical function f in* $\mathbb{C}$ *has right derivatives* $+\infty$ *and* $-\infty$ *at two c-dense sets of points, and similarly it has left derivatives* $+\infty$ *and* $-\infty$ *at two c-dense sets of points.*

In the following theorem we sum up the results obtained so far on various generalized derivates of typical continuous functions to put them in perspective.

19.5.4. THEOREM. *Let G-derivates denote any of the derivates considered in Theorem* 19.1.1. *Then the following properties hold for a typical function f in* $\mathbb{C}$:

(a) *f is G-nonangular and bilaterally GD-infinite*;

(b) *f is G-biknotted* a.e. *and* r.e. *in I*; *and*

(c) *every real number is a knotted lower and upper G-derivative of f at two dense sets of points in I.*

*Further, in the case of G-derivates considered in Theorem* 19.4.2, *the following properties also hold for f*:

(d) *f is G-knotted*; *and*

(e) *f has knotted lower (and upper) G-derivatives* $+\infty$ *and* $-\infty$ *at two c-dense sets of points in I.*

PROOF. Part (a) and the first part of (b) re "a.e." follow directly from Theorem 19.2.1. Further, since $f$ is thus strongly nowhere monotone, it follows from Theorem 13.3.1 that $f$ is biknotted r.e. in $I$. Now since the $G$-derivates are subregular by one of the Status Theorems 15.4.2 and 15.4.4, it follows from Theorem 16.6.3 that $f$ is also $G$-biknotted r.e. in $I$. Further, part (c) follows directly from Theorem 19.5.1. Parts (d) and (e) were established on the other hand in Theorems 19.4.2 and 19.5.2. □

Next, in the following theorem we present the various properties of ordinary derivates of typical continuous functions that follow from the above theorems. Another result on such derivates is obtained here with the help of Theorem 13.3.2.

19.5.5. THEOREM. *The following differential properties hold for a typical continuous function f in terms of its ordinary derivates*:

(a) *f is nonangular, knotted, and bilaterally D-infinite*;

(b) *f is biknotted* a.e. *and* r.e. *in I*;

(c) *every real number is a knotted lower and upper derivative of f at two dense sets of points*;

(d) *$f$ has knotted lower (and upper) derivatives $+\infty$ and $-\infty$ at two c-dense sets of points; and*

(e) *for each $r \in \mathbb{R}$ the following four sets are c-dense in $I$: the sets $E_{r,1}$ and $E_{r,2}$ of points where $f$ is upper knotted and* (i) $D_+f = r \le D^-f$ *or* (ii) $D_+f \le r = D^-f$, *respectively; and the sets $E_{r,3}$ and $E_{r,4}$ of points where $f$ is lower knotted and* (iii) $D^+f = r \ge D_-f$ *or* (iv) $D^+f \ge r = D_-f$, *respectively.*

*Moreover, the sets of points in parts* (b) *to* (e) *cover together all the points of* $(0, 1)$.

PROOF. Parts (a) to (d) have already been established in Theorem 19.5.4. Part (e) is obtained on the other hand easily from Theorem 13.3.2(b) since $f$ is knotted and bilaterally $D$-infinite by (a). The last part also can be seen easily to hold due to (a). □

It is interesting to note here the similarity between the above result and the results of §13.6 on the properties of almost all Brownian paths.

In connection with part (e) of the above theorem, it was pointed out in [G10, p. 539] that exactly one derivate of $f$ remains undetermined at the points of the sets $E_{r,i}$ ($i = 1$ to 4) for each $r \in \mathbb{R}$. This problem seems to have been resolved completely by the following unpublished theorem of D. Preiss which has been quoted by Zajiček [Za2, p. 119]. This delicate theorem is claimed to have been obtained using the so-called Banach-Mazur game method.

THEOREM P (PREISS, UNPUBLISHED). *Let $D^+$, $D_+$, $D^-$, $D_-$ be any extended real numbers for which* $\max(|D^+|, |D_+|) = \max(|D^-|, |D_-|) = \infty$ *and* $[D_-, D^-] \cup [D_+, D^+] = [-\infty, \infty]$. *Then for a typical function $f$ in $\mathbb{C}$ the following relations hold at a c-dense set of points*:

$$D^+f = D^+,\ D_+f = D_+,\ D^-f = D^-,\ D_-f = D_-.$$

If this theorem is valid, then with its help it is easy to see that Theorem 19.5.5 can be strengthened to its following version, which in turn provides a complete description of the ordinary derivates of a typical continuous function.

19.5.5′. THEOREM (PROVISIONAL). *The following differential properties hold for a typical function $f$ in $\mathbb{C}$ in terms of its ordinary derivates*:

(a) *$f$ is nonangular, knotted, and bilaterally $D$-infinite*;

(b) *$f$ is biknotted* a.e. *and* r.e.; *and*

(c) *for each $r, s \in \overline{\mathbb{R}}$, $r \le s$, the sets of points where* (i) *$f$ is upper knotted and $D_+f = r$ and $D^-f = s$, or where* (ii) *$f$ is lower knotted and $D_-f = r$ and $D^+f = s$, are c-dense.*

*Consequently, every $r \in \overline{\mathbb{R}}$ is a knotted lower and upper derivative of $f$ at two c-dense sets of points.*

A complete description of the level structure of typical continuous functions was obtained earlier in [BG]. In the following theorem we add some results on the differential structure of almost all the level sets of such functions.

19.5.6. THEOREM. *For a typical continuous function $f$, the following holds for almost all of its level sets $S$:*

(a) *$S$ is perfect,*

(b) *$f$ is biknotted at every bilateral limit point of $S$, and*

(c) *$f$ has infinite knotted lower and upper derivatives at two dense sets of points in $S$.*

PROOF. Since such a function $f$ is nonderivable by the above theorem, part (a) follows from Corollary 6.4.3. Part (b) follows on the other hand from the Second Level Structure Theorem 6.4.1, and (c) follows from Theorem 13.5.1(b) on applying it to $f$ and $-f$ since $f$ is biknotted a.e. by the last theorem. □

We conclude with some comments regarding the origins of Theorem 19.5.5.

19.5.7. REMARK. In the case of ordinary derivates the results of Theorem 19.5.5 have been obtained earlier at the interior points of the interval $I = [0, 1]$. The origins of this theorem go back to Banach [B5] and Mazurkiewicz [Maz3], who first proved the nondifferentiability of a typical function $f$ in $\mathbb{C}$ by proving it to be bilaterally $D$-infinite. Next, it was observed by Saks [S5] that $f$ can also be proved to be nonderivable by similar arguments and obtained Corollary 19.5.3 using analytic sets.

The above investigations were pursued further by Jarník in [J1], where he proved $f$ to be nonangular, knotted at the interior points of $I$, and biknotted a.e. Finally, in [G10], $f$ was proved to be biknotted r.e., and parts (c), (d), and (e) of Theorem 19.9.5 were obtained with the help of some results of [G4] and [G8]. The knottedness of $f$ at the endpoints 0 and 1 is, however, obtained here for the first time.

Finally, it may be observed here that to extend part (e) of Theorem 19.5.4 to any generalized $G$-derivates considered in Theorem 19.1.1 but not in Theorem 19.4.2, it is necessary to obtain $G$-nonderivability of typical functions in $\mathbb{C}$. To extend part (e) of Theorem 19.5.5 to any generalized $G$-derivates, it is necessary on the other hand to extend the Morse's Theorem 3.5.5 to $G$-derivates.

# CHAPTER 20

# A Unified Theory of Generalized Smooth Functions

In contrast with the results of the last chapter, we present in this chapter a unified theory of generalized smooth functions to obtain large classes of continuous functions which are differentiable in any generalized sense at least at some points.

First, in §20.1, we define lower and upper smooth functions, as well as smooth functions, relative to each $s_1$-admissible generalized limits. Then we define admissibility of generalized smoothness and investigate which of the generalized smoothnesses are admissible.

Next, in §20.2, we deduce from the various derivability theorems of Part II the existence of points where an admissible generalized lower or upper smooth function is generalized derivable or differentiable in the corresponding sense.

Finally, in §20.3, we obtain the mean value theorem for admissible generalized smooth functions in terms of generalized derivatives without any derivability hypothesis. Also, the Darboux property of generalized derivatives of such functions is obtained relative to the sets of points where they exist.

**§20.1 Generalized smooth functions and admissibility of generalized smoothness.** In this section we first define generalized lower and upper smooth functions, as well as smooth functions, relative to each $s_1$-admissible generalized limits considered in §15.5. Then we define admissibility of generalized smoothness in terms of a new axiom, and we investigate which of the generalized smoothnesses are admissible.

To define generalized smoothness in various cases, we first define here some smoothness operators, called *S-operators*. Given $f \in \mathscr{F}(I)$, $x \in I$, and $h > 0$, suppose $x \pm h \in I$. Then we define the *ordinary S-operator* $Sf(x, h)$ in terms of $\Delta^2 f(x,h)$ as follows:

$$Sf(x,h) = \frac{1}{h}\Delta^2 f(x,h) = \frac{f(x+h)+f(x-h)-2f(x)}{h}.$$

To define Cesàro and selective smoothnesses, however, we need different $S$-operators. When $f$ is $D_*$-integrable, the *Cesàro S-operator* of $f$ is defined in terms of the average function $f_x^a$ as follows:

$$\begin{aligned} CSf(x,h) = Sf_x^a(x,h) &= \frac{1}{h}\{f_x^a(x+h)+f_x^a(x-h)-2f_x^a(x)\} \\ &= \frac{1}{h}\left[\frac{1}{2h}\int_0^{2h}\{f(x+t)+f(x-t)\}\,dt - 2f(x)\right]. \end{aligned}$$

Further, when $s$ is a symmetric selection, the *selective S-operator* of $f$ relative to $s$ is defined as follows:

$$sSf(x,h) = \frac{f(s_{[x,x+h]})+f(s_{[x-h,x]})-2f(x)}{s_{[x,x+h]}-x}.$$

Now, first let $G$-limits denote any of the unilateral limits considered in the Fifth Status Theorem 15.5.2 except Cesàro and selective limits. Then we define $f$ to be *lower* or *upper G-smooth* at $x \in I$ if

(1) $\displaystyle G\text{-}\liminf_{h\to 0+} Sf(x,h) \ge 0$ or $\displaystyle G\text{-}\limsup_{h\to 0+} Sf(x,h) \le 0$, respectively.

In the case of selective smoothness, however, we use $sSf(x,h)$ in (1) in the place of $Sf(x,h)$. Thus $f$ is called *lower* or *upper s-smooth* at $x$ if

$$s\text{-}\liminf_{h\to 0+} sSf(x,h) \ge 0 \quad \text{or} \quad s\text{-}\limsup_{h\to 0+} sSf(x,h) \le 0, \quad \text{respectively}.$$

Further, the Cesàro smoothness is defined in terms of ordinary limits using $CSf(x,h)$. Thus we define $f$ to be *lower* or *upper C-smooth* at $x$ if

$$\liminf_{h\to 0+} CSf(x,h) \ge 0 \quad \text{or} \quad \limsup_{h\to 0+} CSf(x,h) \le 0, \quad \text{respectively}.$$

Once again, to obtain a unified treatment of all these smoothnesses, we will sometimes use $GSf(x,h)$ to denote the ordinary $S$-operator $Sf(x,h)$ when $G$ differs from $C$ and $s$; and in the case of $C$-smoothness we will use as before $G$-limits to denote the ordinary limits. Thus the following holds in all the cases: namely, $f$ is lower or upper $G$-smooth at $x$ if

$$G\text{-}\liminf_{h\to 0+} GSf(x,h) \ge 0 \quad \text{or} \quad G\text{-}\limsup_{h\to 0+} GSf(x,h) \le 0, \quad \text{respectively}.$$

Further, in each of these cases, $f$ will be called *G-smooth* at $x$ if it is simultaneously lower and upper $G$-smooth at $x$—that is, if $GSf(x,h)$ has $G$-limit equal to zero as $h \to 0$ from the right. Also, $f$ will be called as usual lower or upper $G$-smooth, or $G$-smooth, if it is so at each point of $I$.

Next, to obtain points of $G$-differentiability of $G$-smooth functions, we introduce an axiom relating the $GS$-operator with the $G$-quotient.

Given any $s_1$-admissible $G$-limits, we will call the $G$-smoothness *admissible* if (i) the corresponding $G$-derivates are admissible and if (ii) the following axiom holds for each $f \in \mathscr{F}(I)$, $x \in I$, and $h > 0$ provided that $x \pm h \in I$:

$$\text{(U1)} \qquad GSf(x,h) = GQf(x, x+h) - GQf(x-h, x).$$

In the following theorem we establish the admissibility of various generalized smoothnesses.

20.1.1. THEOREM (TWELFTH STATUS). *The following smoothnesses are admissible*:

(a) *ordinary smoothness,*
(b) *approximate smoothness,*
(c) *sequentially congruent smoothness,*
(d) *symmetrically congruent smoothness,*
(e) *Borel smoothness,*
(f) *Cesàro smoothness,*
(g) *qualitative smoothness,*
(h) *symmetrically selective smoothness, and*
(i) *symmetric path smoothness.*

PROOF. If $G$-smoothness denotes any of these smoothnesses, the $s_1$-admissibility of the corresponding $G$-limits follows from the Fifth Status Theorem 15.5.2, and the admissibility of the $G$-derivates follows from the Second Status Theorem 15.3.3. Thus we need to verify only the axiom (U1) in each of these cases.

As for the ordinary $S$-operator, we clearly have

$$\text{(2)} \qquad \begin{aligned} Sf(x,h) &= (1/h)[\{f(x+h) - f(x)\} - \{f(x) - f(x-h)\}] \\ &= Qf(x, x+h) - Qf(x-h, x). \end{aligned}$$

Hence (U1) holds in all the cases except for (f) and (h).

Further, in case (f), we obtain the following from (2):

$$\begin{aligned} CSf(x,h) &= Sf_x^a(x,h) = Qf_x^a(x, x+h) - Qf_x^a(x-h, x) \\ &= CQf(x, x+h) - CQf(x-h, x); \end{aligned}$$

and when $s$ is a symmetric selection, it is easy to see that

$$sSf(x,h) = \frac{\{f(s_{[x,x+h]}) - f(x)\} - \{f(x) - f(s_{[x-h,x]})\}}{s_{[x,x+h]} - x}$$
$$= sQf(x, x+h) - sQf(x-h, x). \quad \square$$

## §20.2 Generalized derivability of generalized smooth functions.

In this section we prove the existence of points where an admissible generalized lower or upper smooth function is generalized derivable or differentiable in the corresponding sense.

We will assume throughout the present and the next section that $G$-smoothness is admissible, $I$ is an open subinterval of $\mathbb{R}$, $f \in \mathcal{F}(I)$, and $x \in I$.

We begin with the following theorem similar to Theorem 12.6.1, which establishes the relationship between $G$-smoothness and $G$-differentiability.

20.2.1. THEOREM. *If $f$ is lower $G$-differentiable at $x$, then it is lower $G$-smooth at $x$. Consequently, if $f$ is $G$-differentiable at $x$, then it is $G$-smooth at $x$.*

PROOF. First, suppose $f$ is lower $G$-differentiable at $x$. Then there exists $c \in \mathbb{R}$ such that $GD^-f(x) \le c \le GD_+f(x)$. Hence, with the help of the axioms (U1), (A1), (A4), (A5), and (A8), we obtain

$$\begin{aligned}
G\text{-}\liminf_{h\to 0+} GSf(x,h) &= G\text{-}\liminf_{h\to 0+}\{GQf(x,x+h) - GQf(x-h,x)\} \\
&\ge G\text{-}\liminf_{h\to 0+} GQf(x,x+h) - G\text{-}\limsup_{h\to 0+} GQf(x-h,x) \\
&= GD_+f(x) - G\text{-}\limsup_{h\to 0-} GQf(x,x+h) \\
&= GD_+f(x) - GD^-f(x) \ge c - c = 0.
\end{aligned}$$

This proves that $f$ is lower $G$-smooth at $x$.

The second part follows, on the other hand, on combining the first with the result obtained on applying it to $-f$. $\square$

In the following theorem we obtain some results on lower $G$-smoothness similar to Lemmas 12.3.3 and 12.6.2.

20.2.2. THEOREM. *Suppose $f$ is lower $G$-smooth at $x$. Then*

(a) *$GD_-f(x) \le GD_+f(x)$ and $GD^-f(x) \le GD^+f(x)$;*

(b) *$f$ is upper $G$-nonangular (and hence upper nonangular) at $x$;*

(c) *if $f$ is upper $G$-derivable at $x$, then it is $G$-derivable at $x$; and if $f$ is upper derivable at $x$, then further $Uf'(x) = Gf'(x)$.*

PROOF. Since $f$ is lower $G$-smooth at $x$, we obtain from the axioms (U1) and (A8), with the help of Lemma 15.1.3,

$$0 \le G\text{-}\liminf_{h \to 0+} GSf(x, h) = G\text{-}\liminf_{h \to 0+} \{GQf(x, x + h) - GQf(x - h, x)\}$$
$$\le \min\{GD_+f(x) - GD_-f(x), GD^+f(x) - GD^-f(x)\},$$

where the last inequality holds for the differences that are defined. Now, when $GD_-f(x)$ is finite, it follows clearly from above that $GD_-f(x) \le GD_+f(x)$; and this inequality holds trivially when $GD_-f(x) = -\infty$ or $GD_+f(x) = +\infty$. Otherwise, $GD_+f(x) - GD_-f(x) = -\infty$, which is clearly not possible by above. This proves the first inequality of (a); its second inequality is obtained similarly from the second difference in the above inequality.

Now it follows from (a) and axiom (D2) that $GD_-f(x) \le GD_+f(x) \le GD^+f(x)$, and hence $f$ is upper $G$-nonangular at $x$. The other part of (b) within brackets follows now easily from (D2).

Further, when $f$ is upper $G$-derivable at $x$, it follows from (a) that

$$GD^-f(x) \le GD^+f(x) \le GD_-f(x) \le GD_+f(x),$$

from which it is clear that $f$ is $G$-derivable at $x$. Finally, if $f$ is upper derivable at $x$, then with the help of (D2) we obtain the following from (a):

$$GD^-f(x) \le GD^+f(x) \le D^+f(x) < D\ f(x) \le GD_-f(x) \le GD_+f(x);$$

hence $Uf'(x)$ is unique and equal to $Gf'(x)$. □

Now on combining part (a) of the above theorem with the result obtained on applying it to $-f$, we obtain the following.

20.2.3. COROLLARY. *If $f$ is $G$-smooth at $x$, then its $G$-derivates are symmetrical at $x$.*

Next, to deal with the existence of points of $G$-derivability for lower $G$-smooth functions, we obtain the following theorem, which is somewhat similar to Theorem 12.4.2.

20.2.4. THEOREM (TWELFTH DERIVABILITY). (a) *If $f$ is either $L^*LI$ or $R^*UI$, and it is lower $G$-smooth* n.e., *then $f$ is $G$-derivable at a dense set of points in almost all of its level sets, and at a c-dense set of points in $I$.*

(b) *If $f$ is a weakly Darboux, $G$-smooth function and it has property $(S_2)$, then it is everywhere $G$-derivable in almost all of its level sets.*

PROOF. Part (a) follows directly from the Third Derivability Theorem 6.2.2 (on applying it to $-f$) with the help of Theorem 20.2.2(c); and part (b) is obtained similarly from the Fifth Derivability Theorem 6.5.2. □

Next we deduce from the First Derivability Theorem the following theorem on $G$-differentiability which is similar to Theorem 12.6.3. In the particular case of approximate smoothness, the following theorem has been obtained by Neugebauer [Ne3, Lemma 4] in the case of continuous approximately smooth functions.

20.2.5. THEOREM (THIRTEENTH DERIVABILITY). *If $f$ is USC and lower $G$-smooth, then it is $G$-differentiable at a c-dense set of points in $I$.*

PROOF. Since $f$ is lower $G$-smooth, it is upper nonangular by Theorem 20.2.2(b). Hence it follows from the First Derivability Theorem 5.1.2, on applying it to $-f$, that $f$ is uniquely upper differentiable at a $c$-dense set of points $E$ in $I$. Consequently, it follows from the second part of Theorem 20.2.2(c) that $f$ is $G$-differentiable at each point of $E$. □

Finally, with the help of Theorem 20.2.2(c) we obtain the following theorem from the Second Derivability Theorem 5.3.1, which in turn may be compared with Theorem 12.4.4.

20.2.6. THEOREM. *Suppose $f$ is $GU$-$BV_*$. Then $f$ is $G$-differentiable at almost all the points where it is lower $G$-smooth; and if $f$ is lower $G$-smooth* n.e. *and $N$ is the set of points where $f$ is not $G$-derivable, then $|f(N)| = \Lambda\{Gf(N)\} = 0$.*

20.2.7. PROBLEM. It should be noted here that a result analogous to Theorem 20.2.4 was obtained earlier in Theorem 12.4.2 in the case of lower quasismooth functions. Using $G$-limits in the place of ordinary limits, it is clear that the notions of *lower* and *upper* *$G$-quasismooth functions*, as well as *$G$-quasismooth functions*, can be defined as in §12.1. Then it is easy to see that a result analogous to Lemma 12.4.1 holds for lower $G$-quasismooth functions in terms of $G$-derivates and $G$-derivability.

However, Theorem 12.4.2 does not seem to extend to lower $G$-quasismooth functions in general. For, this theorem was deduced in fact from the Third Derivability Theorem 6.2.2, which in turn was deduced from the Normal Level Structure Theorem 6.1.2, and these two theorems do not seem to extend to normal $G$-semiderivability in general.

It will therefore be interesting to find a suitable hypothesis under which Theorems 6.1.2 and 6.2.2 hold also for $G$-derivates; because under this hypothesis Theorem 20.2.4 will also hold for lower $G$-quasismooth functions.

## §20.3 Mean value and Darboux properties of generalized derivatives of generalized smooth functions.

In this section we obtain the mean value and Darboux properties of the ordinary lower derivative of an admissible generalized lower $G$-smooth function $f$, and we also obtain the same properties of the generalized derivative of $f$ in the case when $f$ is $G$-smooth.

We assume here as mentioned before that $G$-smoothness is admissible, and $f \in \mathscr{F}(I)$, where $I$ is open.

First, in the following theorem we obtain a generalized version of the Fourth Mean Value Theorem 12.7.1.

20.3.1. Theorem (Ninth Mean Value). *Suppose $f$ is $I^*C$. Then if $f$ is lower $G$-smooth, then $Lf'$ possesses the mean value property. Further, if $f$ is $G$-smooth, then $Gf'$ also possesses the mean value property.*

Proof. First, suppose $f$ is lower $G$-smooth. Then $f$ is upper nonangular by Theorem 20.2.2(b), and hence the result follows from the Second Mean Value Theorem 8.1.4.

Next, suppose $f$ is $G$-smooth. Then it follows from Theorem 20.2.2(c) that at each point where $f$ is lower derivable, it is also $G$-derivable and the two derivatives are equal. Hence the result follows from the first part. □

In the following theorem we obtain on the other hand the Darboux property of lower and generalized derivatives under generalized smoothness. This theorem may in turn be compared with Theorem 12.7.2.

20.3.2. Theorem. *Suppose $f$ is $I^*C$. Then if $f$ is lower $G$-smooth, then $Lf'$ possesses the Darboux property. Further, if $f$ is $G$-smooth, then $Gf'$ also possesses the Darboux property.*

Proof. First, suppose $f$ is lower $G$-smooth. Then since $f$ is upper nonangular by Theorem 20.2.2(b), the Darboux property of $Lf'$ follows directly from the second part of Theorem 10.4.1.

Next, suppose $f$ is $G$-smooth. Then it follows from the axiom (D2) and Theorem 20.2.2(c) that $f$ is $G$-derivable at each point where it is semiderivable. Hence the Darboux property of $Gf'$ is obtained from Theorem 16.7.1. □

# CHAPTER 21

# A Unified Axiomatic Theory of Generalized Perron Integrals

In this final chapter we present an axiomatic theory of generalized Perron integrals which applies simultaneously to various generalized derivatives of Chapter 16, generalized symmetric derivatives of Chapter 17, and generalized new derivatives of Chapter 18. Indeed, we define here generalized Perron integrals, generalized symmetric Perron integrals, and generalized lower, upper, and weak Perron integrals to resolve the problem of reconstruction of a primitive function from its various kinds of generalized derivatives considered in these three chapters.

First, in §21.1, we introduce another axiomatic model of generalized derivates needed for this purpose and define their various forms of $H$-admissibilities. Then, in §21.2, we define various $P$-admissibilities and $P$-regularity of generalized Perron integrals. Examples of generalized Perron integrals dealing with various generalized derivatives considered in Chapters 16–18 are considered in these two sections which meet these admissibility and regularity criteria. Also, these generalized Perron integrals are compared among themselves and with the ordinary Perron integral.

Next, in §21.3, we establish various basic properties of generalized Perron integrals induced by $P$-admissible generalized derivates. Then, in §21.4, we obtain various properties of indefinite generalized Perron integrals under suitable admissibility and regularity hypotheses. A comparison of generalized Perron integrals with Lebesgue integral is made, and two convergence theorems of Lebesgue are extended to generalized Perron integrals.

Finally, in §21.5, several examples of generalized Perron integrals are considered to which the theory developed in this chapter applies, and we indicate here some of the unresolved problems that arise naturally from the developments of this chapter.

**§21.1. $H$-admissible generalized derivates and generalized Perron integrals.** In this section we present a new axiomatic model of generalized $H$-derivates and define their various notions of $H$-admissibilities in terms of a series of axioms. Also, corresponding to $H$-derivates we define here a generalized Perron integral which we call $HP$-integral.

Further, in terms of the generalized $G$-derivates considered in Chapter 15, we define here generalized right, left, lower, upper, and weak $GP$-integrals, and in terms of the generalized symmetric $G$-derivates of Chapter 17 we define generalized symmetric $GP$-integral. Then we investigate the nature of $H$-admissibility of all of these generalized Perron integrals.

We use here the prefix "$H$-" in $H$-derivates merely to avoid confusion with the generalized $G$-derivates considered in Chapters 15–17.

Let $H\underline{D}$ and $H\overline{D}$ be any generalized lower and upper derivates, which are not necessarily unilateral or bilateral derivates. Also, for any $f \in \mathscr{F}(I)$ and $x \in I$, we do not assume here that $H\underline{D}f(x) \leq H\overline{D}f(x)$. However, when $H\overline{D}f(x) \leq H\underline{D}f(x)$, we will call $f$ *H-derivable* at $x$ and the interval $[H\overline{D}f(x), H\underline{D}f(x)]$ will be called the *H-derivative* of $f$ at $x$ and denoted by $Hf'(x)$. In the case when $Hf'(x)$ is a singleton, $f$ will be called, as before, *uniquely H-derivable* at $x$. Also, when $Hf'(x)$ includes some finite element, $f$ will be called *H-differentiable* at $x$.

We will call these $H$-derivates *partially H-admissible* if the following three axioms hold for every $f, g \in \mathscr{F}(I)$, $x \in I$, and $c \in \mathbb{R}$:

(P1) $\underline{D}f(x) \leq H\underline{D}f(x) \leq \overline{D}f(x), \qquad \underline{D}f(x) \leq H\overline{D}f(x) \leq \overline{D}f(x).$

(P2) If $f$ is differentiable at $x$ in the ordinary sense, then

$$H\underline{D}(f+g)(x) = f'(x) + H\underline{D}g(x), \qquad H\overline{D}(f+g)(x) = f'(x) + H\overline{D}g(x).$$

(P3) If $c > 0$, then $H\underline{D}(cf)(x) = cH\underline{D}f(x)$ and $H\overline{D}(cf)(x) = cH\overline{D}f(x)$.

Further, given any partially $H$-admissible $H$-derivates, we will call them

(a) *weakly H-admissible* if the following axiom holds for every $f \in \mathscr{F}(I)$, $x \in I$:

(P4) $$H\overline{D}f(x) = -H\underline{D}(-f)(x);$$

(b) *semi-H-admissible* if for every $f, g \in \mathscr{F}(I)$, $x \in I$, we obtain

(P5) $$H\underline{D}f(x) + H\underline{D}g(x) \leq H\underline{D}(f+g)(x),$$

$$H\overline{D}(f+g)(x) \leq H\overline{D}f(x) + H\overline{D}g(x).$$

Finally, we will call $H$-derivates *H-admissible* if they are simultaneously weakly and semi- $H$-admissible—that is, if all the above axioms (P1) to (P5) hold.

It should be mentioned here that the prefix "$H$-" in various $H$-admissibilities is independent of the choice of "$H$" in $H$-derivates. The reasons behind the choice of the other terminology used here will be apparent later from Theorem 21.1.4. It should be noted here that the weak and semi- $H$-admissibilities of $H$-derivates are not comparable in general, but each of them implies their partial $H$-admissibility. Further, in the case of $H$-admissible derivates, it is clear that the second relations of the axioms (P1), (P2), (P3), and (P5) become redundant; for they follow from their first relations with the help of the axiom (P4).

We include here some lemmas on $H$-admissible generalized derivates which will be needed in the subsequent sections.

The following lemma follows directly from the axiom (P1).

21.1.1. LEMMA. *Suppose $H$-derivates are partially $H$-admissible, $f \in \mathscr{F}(I)$ and $x \in I$. Then if $f$ is derivable in the ordinary sense at $x$, then it is uniquely $H$-derivable at $x$ and $Hf'(x) = f'(x)$.*

21.1.2. LEMMA. *Suppose $H$-derivates are weakly $H$-admissible, $f \in \mathscr{F}(I)$ and $x \in I$. Then if $c < 0$, then $H\underline{D}(cf)(x) = cH\overline{D}f(x)$ and $H\overline{D}(cf)(x) = cH\underline{D}f(x)$. Further, these two relations and the relations in axiom* (P3) *hold also when $c = 0$.*

PROOF. When $c < 0$, the two desired relations follow clearly from axiom (P3) with the help of (P4). Further, when $c = 0$, $cf \equiv 0$, and so it follows from Lemma 21.1.1 that $H(cf)'(x) = (cf)'(x) = 0$. The right sides of the desired relations are zero in this case by definition even when they are indeterminate. □

21.1.3. LEMMA. *Suppose $H$-derivates are $H$-admissible, $f, g \in \mathscr{F}(I)$ and $x \in I$. Then*

(a) $H\underline{D}(f+g)(x) \le H\underline{D}f(x) + H\overline{D}g(x) \le H\overline{D}(f+g)(x)$;

(b) *if $f$ is uniquely $H$-derivable at $x$, then*

$$H\underline{D}(f+g)(x) = Hf'(x) + H\underline{D}g(x),$$

$$H\overline{D}(f+g)(x) = Hf'(x) + H\overline{D}g(x);$$

(c) $H\underline{D}f(x) \le H\overline{D}f(x)$.

PROOF. On account of axiom (P5) we have

$$H\underline{D}(f+g)(x) + H\underline{D}(-g)(x) \le H\underline{D}f(x).$$

Hence with the help of axiom (P4) we obtain

$$H\underline{D}(f+g)(x) - H\overline{D}g(x) \le H\underline{D}f(x).$$

The first inequality of (a) follows now by an argument similar to the one used in the proof of Lemma 15.1.3. The second inequality of (a) follows on the other hand from the first with the help of axiom (P4).

Part (b) follows now on combining the inequalities of (a) with those of the axiom (P5) and inverting the roles of $f$ and $g$.

To obtain part (c), let $c \in \mathbb{R}$ and define $g(x) = c$ for $x \in I$. Then according to Lemma 21.1.1 we have $Hg'(x) = 0$. Hence with the help of parts (b) and (a) we obtain

$$H\underline{D}f(x) = H\underline{D}(f+g)(x) \leq H\underline{D}g(x) + H\overline{D}f(x) = H\overline{D}f(x). \qquad \square$$

In the case of $H$-admissible $H$-derivates it is clear from part (b) of the last lemma that axiom (P2) becomes redundant. Further, in this case it follows from part (c) of this lemma that the $H$-derivative of $f$ is unique at every point where it exists.

Next, we define the Perron integral induced by $H$-derivates.

Given any arbitrary generalized $H$-derivates, let $I = [a, b]$, and let $L$ be any given linear space of real-valued functions on $I$ that is closed under uniform convergence. Let $f : I \to \overline{\mathbb{R}}$ and $U, V \in L$. Then we will call $U$ an *H-major function* of $f$ and call $V$ an *H-minor function* of $f$ if the following conditions hold:

| | |
|---|---|
| (a) $U(a) = 0$, | (a′) $V(a) = 0$, |
| (b) $H\underline{D}U \geq f$ a.e., | (b′) $H\overline{D}V \leq f$ a.e., |
| (c) $H\underline{D}U > -\infty$ n.e., | (c′) $H\overline{D}V < +\infty$ n.e. |

We will call $f$ *HP-integrable* if it has at least one $H$-major function and one $H$-minor function and $\inf U(b) = \sup V(b)$, where inf and sup are taken over all $H$-major and $H$-minor functions, respectively, of $f$. Further, when it is so, this common bound will be called the *HP-integral* of $f$ and denoted by $(HP)\int_a^b f(x)\,dx$.

We will call this $HP$-integral the *Perron integral induced by H-derivates*. Also, to emphasize the role of $L$ in the above definition, we will call the above integral sometimes the *HP-integral relative to L*.

To include a direct generalization of the classical definition of Perron integral in our consideration, we will call $U$ and $V$ *usual H-major* and *usual H-minor functions* of $f$ when the above conditions (b), (b′), (c), and (c′) hold at every point of $I$; and the $HP$-integral obtained in this case will in turn be called the *usual HP-integral*.

Now, given an $HP$-integral induced by $H$-derivates, we will call this integral *H-admissible* if the $H$-derivates are so. Similarly, the $HP$-integral will be called *partially*, *weakly*, or *semi- H-admissible* if the $H$-derivates are so.

As for the choice of $L$, unless stated otherwise we will assume it to be $\mathscr{C}$, the space of all continuous functions on $I$. Other linear spaces will be used for $L$ only in the case of usual $HP$-integrals. We will use for this purpose $\mathscr{F}$ to denote the space of all real-valued functions on $I$, and we will use $\mathscr{M}$ to denote the space of measurable functions in $\mathscr{F}$.

Next, we deal with examples of generalized Perron integrals which meet various criteria of $H$-admissibilities. First, let $G$-derivates be any admissible generalized (unilateral) derivates as defined in §15.2. Then the Perron integral induced by the bilateral $G$-derivates $G\underline{D}$ and $G\overline{D}$ defined by the axiom (D0) will be called the *bilateral GP-integral*, or, simply, *GP-integral*, and it will be denoted by $(GP)\int_a^b f(x)dx$. Further, the Perron integrals induced by the pairs of derivates $(GD_+, GD^+)$ and$(GD_-, GD^-)$ will be called the *right* and *left GP-integrals*, respectively; and the ones induced by the pairs of derivates $(GD_+, GD^-)$, $(GD_-, GD^+)$, and $(G\underline{D}_w, G\overline{D}_w)$ will be called the *lower*, *upper*, and *weak GP-integrals*, respectively, or, simply, $LGP$-, $UGP$- and $WGP$-integrals, respectively. Further, we will use $(LGP)\int_a^b f(x)\,dx$, $(UGP)\int_a^b f(x)\,dx$, and $(WGP)\int_a^b f(x)\,dx$ to denote the $LGP$-, $UGP$- and $WGP$-integrals, respectively, of $f$.

In the following theorem we deal with the status of $H$-admissibility of all of these forms of $GP$-integrals.

21.1.4. THEOREM (THIRTEENTH STATUS). *Suppose any generalized G-derivates are s-admissible. Then*

(a) *the GP-integral and right and left GP-integrals are H-admissible;*

(b) *the lower and upper GP-integrals are semi-H-admissible; and*

(c) *the weak GP-integral is weakly H-admissible.*

PROOF. It is easy to see from axioms (D1) to (D5) of §15.2, Lemmas 15.2.1 and 15.2.3, and Theorem 15.2.4 that the axioms (P1) to (P5) hold for the pairs of derivates$(GD_+, GD^+)$ and $(GD_-, GD^-)$. Hence it is easy to see that these axioms hold also for the bilateral derivates $G\underline{D} = \wedge GD_\pm$ and $G\overline{D} = \vee GD^\pm$. Hence the integrals in (a) are $H$-admissible.

Further, when $H\underline{D} = G\underline{D}_w$ and $H\overline{D} = G\overline{D}_w$, it is easy to verify that axioms (P1) to (P4) hold for these $H$-derivates. Hence the weak $GP$-integral is weakly $H$-admissible.

Finally, for the pairs of derivates$(GD_+, GD^-)$ and $(GD_-, GD^+)$, it is clear that axioms (P1), (P2), (P3), and (P5) hold. Hence the lower and upper $GP$-integrals are semi-$H$-admissible. □

Next, consider any admissible generalized symmetric $G$-derivates defined in §17.1. We will call the Perron integral induced by these derivates the *symmetric GP-integral*. In the case of ordinary symmetric derivates this integral will be called in turn the *ordinary symmetric P-integral*.

21.1.5. THEOREM (FOURTEENTH STATUS). *If any generalized symmetric G-derivates are s-admissible, then the symmetric GP-integral induced by them is H-admissible.*

PROOF. It is easy to verify from axioms (S1) to (S5), Lemma 17.1.1, and Theorem 17.1.2 that axioms (P1) to (P5) hold for the symmetric $G$-derivates $G\underline{D}_s$ and $G\overline{D}_s$. □

## §21.2. *P*-admissible and *P*-regular generalized Perron integrals.

In this section we define various $P$-admissibilities of generalized Perron integrals, and in part A of this section we define $P$-regularity of these integrals. Also, we obtain these properties for various generalized $GP$-integrals considered in the last section.

Further, in part B, we define comparability and compatibility of generalized Perron integrals, and we obtain these relations for some generalized Perron integrals.

Let $HP$-integral be any Perron integral induced by generalized $H$-derivates. We will call this $HP$-integral *$P_0$-admissible* if the following two axioms hold for every $H$-major and $H$-minor functions $U$ and $V$ of any function $f$:

(P6) $\qquad H\overline{D}(U - V)(x) \geq H\underline{D}U(x) - H\overline{D}V(x)$ for n.e. $x$;

(P7) $\qquad U - V$ is nondecreasing.

Further, when the $HP$-integral is $P_0$-admissible and $H$-admissible, we will call it simply *P-admissible*. Similarly, when the $HP$-integral is $P_0$-admissible and partially, weakly, or semi- $H$-admissible, we will call it *partially*, *weakly* or *semi*- $P$-admissible, respectively.

It should be noted here, as before, that the weak and semi- $P$-admissibilities of the $HP$-integral are not comparable in general, but each of them implies its partial $P$-admissibility. Further, the $HP$-integral is $P$-admissible iff it is simultaneously weakly and semi- $P$-admissible.

To obtain $P_0$-admissibility of some of the generalized $P$-integrals considered in the last section, we will need some restrictions on the choice of major and minor functions used in their definitions.

Let us recall here that we use $\mathscr{C}$ here to denote the linear space of all continuous functions on $I$. Given any admissible generalized $G$-derivates, let $G\underline{N}$ and $G\overline{N}$ denote the sets of functions in $\mathscr{C}$ which are lower or upper $G$-nonangular, respectively, nearly everywhere, and let $GN$ be the largest linear space of functions in $\mathscr{C}$ which are $G$-nonangular nearly everywhere. In the case of ordinary derivates it should be noted here that $G\underline{N}$, $G\overline{N}$, and $GN$ coincide with $\mathscr{C}$ due to Theorem 2.3.1.

In the definitions of lower and upper $GP$-integrals we will assume that the major functions are chosen from $G\underline{N}$ and $G\overline{N}$, respectively, and the minor functions from $\mathscr{C}$. Further, in the definitions of weak and symmetric $GP$-integrals we will assume that the major and minor functions are chosen only from $GN$. In the definitions of $GP$-integral and right and left $GP$-integrals the major and minor functions will be chosen on the other hand arbitrarily from $\mathscr{C}$.

In the case of $H$-admissible $H$-derivates, as we see in the following lemma, axiom (P6) becomes redundant and axiom (P7) can be weakened as follows.

21.2.1. LEMMA. *Suppose H-derivates are H-admissible. Then the HP-integral is $P_0$-admissible provided that the following holds for every $F \in L$:*

(P7.1) *If* $H\underline{D}f \geq 0$ a.e. *and* $H\underline{D}f > -\infty$ n.e., *then F is nondecreasing.*

*Further, the usual HP-integral is $P_0$-admissible provided that the following holds for every $F \in L$:*

(P7.2) *If* $H\underline{D}f \geq 0$ *everywhere, the F is nondecreasing.*

PROOF. Suppose $U$ and $V$ are any $H$-major and $H$-minor functions of some function $f$. Let $W = U - V$. Then for each $x \in I$, by the axioms (P4) and (P5) we have

$$(1) \quad H\underline{D}W(x) \geq H\underline{D}U(x) + H\underline{D}(-V)(x) = H\underline{D}U(x) - H\overline{D}V(x).$$

Hence it follows from Lemma 21.1.3(c) that axiom (P6) holds for the $HP$-integral. Further, it follows from (1) and the conditions (c) and (c′) that $H\underline{D}W > -\infty$ n.e. Let $A$ denote the set of points $x \in I$ for which (i) $H\underline{D}U(x) \geq f(x) \geq H\overline{D}V(x)$, (ii) $H\underline{D}U(x) > -\infty$, and (iii) $H\overline{D}V(x) < +\infty$. Then it is clear that $|I \sim A| = 0$. Let $x \in A$. Then when $f(x)$ is finite, it follows from (1) and (i) that $H\underline{D}W(x) \geq f(x) - f(x) = 0$. Further, when $f(x)$ is infinite, it is easy to see from (1), (ii), and (iii) that $H\underline{D}W(x) = +\infty$. Thus $H\underline{D}W \geq 0$ a.e., and so in case (P7.1) holds, then $W$ is nondecreasing. Thus axiom (P7) also holds for the $HP$-integral, and so this integral is $P_0$-admissible.

The second part is obtained by a similar argument. □

In the following lemma we obtain another weakened form of axiom (P7) when axiom (P6) holds.

21.2.2. LEMMA. *Suppose axiom* (P6) *holds for an HP-integral. Then this integral is $P_0$-admissible provided that the following holds for every $F \in L$:*

(P7.3) *If* $H\overline{D}F \geq 0$ a.e. *and* $H\overline{D}F > -\infty$ n.e., *then F is nondecreasing.*

PROOF. Suppose $U$ and $V$ are any $H$-major and $H$-minor functions of some function $f$. Let $W = U - V$. Then according to the axiom (P6),

$$(2) \qquad H\overline{D}W(x) \geq H\underline{D}U(x) - H\overline{D}V(x) \quad \text{for n.e. } x.$$

Using arguments similar to the ones used for the above lemma, it follows in this case from (2) and the conditions (b), (c), (b′), and (c′) that $H\overline{D}W > -\infty$ n.e. and $H\overline{D}W \geq 0$ a.e. Hence if (P7.3) holds, then $W$ is nondecreasing, and so (P7) holds. □

Next, in the following theorem we deal with the status of $P$-admissibility of various generalized $GP$-integrals considered in Theorem 21.1.4.

21.2.3. THEOREM (FIFTEENTH STATUS). *Suppose any generalized G-derivates are s-admissible. Then*

(a) *the GP-integral and right and left GP-integrals are P-admissible;*

(b) *the lower and upper GP-integrals are semi-P-admissible; and*

(c) *the weak GP-integral is weakly P-admissible.*

*Further, the ordinary usual Perron integral is P-admissible relative to* $\mathcal{F}$, *and the approximate and Cesàro usual Perron integrals are P-admissible relative to* $\mathcal{M}$.

PROOF. First, consider any $s$-admissible generalized $G$-derivates. Then for each of the $GP$-integrals in (a), (b), and (c), the required form of $H$-admissibility has already been established in the Thirteenth Status Theorem 21.1.4. Hence we need to obtain here only their $P_0$-admissibility. Note that $L = \mathcal{C}$ in this case.

In the case of the $GP$-integrals in (a), since they are $H$-admissible, the result follows directly from Lemma 21.2.1 with the help of the Fourth Monotonicity Theorem 9.1.8. In the cases of $GP$-integrals in (b) and (c) it is enough to obtain on the other hand the axiom (P6), for then the result follows from Lemma 21.2.2 and Theorem 9.1.8.

First, consider the lower $GP$-integral. Suppose $U$ and $V$ are any lower $G$-major and lower $G$-minor functions of some function $f$. Then $U$ is, by the hypothesis, lower $G$-nonangular n.e. Let $A$ denote the set of points where $U$ is lower $G$-nonangular, $GD_+U > -\infty$ and $GD^-V < +\infty$. Then $I \sim A$ is countable by the conditions (c) and (c′). Set $W = U - V$. Then at each point of $A$ it follows from Lemma 15.2.3(a) and the axiom (D1) that

$$GD^-W \geq GD^-U + GD_-(-V) = GD^-U - GD^-V \geq GD_+U - GD^-V.$$

This proves axiom (P6) for the lower $GP$-integral. In the case of the upper $GP$-integral the result is obtained by a similar argument.

Next, to deal with the weak $GP$-integral, suppose $U$ and $V$ are any weak $G$-major and weak $G$-minor functions of some function $f$. Let $W = U - V$, and set $u = \vee GD_{\pm}U$ and $v = \wedge GD^{\pm}V$. Then we need to show that each of the derivates $GD^+W$ and $GD^-W$ is $\geq u - v$ n.e.. We give the proof here for $GD^+W$; a similar argument holds for the other derivate.

Let $E$ denote the set of points where $U$, $V$, and $W$ are $G$-nonangular, $u > -\infty$, and $v < +\infty$. Then since $U$, $V$, and $W$ are by the hypothesis $G$-nonangular n.e., it is clear from the conditions (c) and (c′) that $I \sim E$ is countable. We claim that $GD^+W \geq u - v$ at each point of $E$. It is clear that $u - v$ is defined at each point of $E$. Let $x \in E$. We need to consider here four cases separately.

**Case 1:** $u(x) = GD_+U(x)$, $v(x) = GD^+V(x)$. In this case we obtain the following from Lemma 15.2.3(a) and axiom (D1):

$$GD^+W(x) \geq GD^+U(x) - GD^+V(x) \geq GD_+U(x) - v(x) = u(x) - v(x).$$

**Case 2:** $u(x) = GD_-U(x), v(x) = GD^-V(x)$. Since $W$ is $G$-nonangular at $x$, in this case we obtain the following from the axioms (D4) and (D1):

$$GD^+W(x) \geq GD_-W(x) \geq GD_-U(x) - GD_-V(x) \geq u(x) - v(x).$$

**Case 3:** $u(x) = GD_+U(x)$, $v(x) = GD^-V(x)$. Since $V$ is $G$-nonangular at $x$, we obtain in this case as before

$$GD^+W(x) \geq GD_+W(x) \geq GD_+U(x) - GD_+V(x) \geq u(x) - v(x).$$

**Case 4:** $u(x) = GD_-U(x)$, $v(x) = GD^+V(x)$. Since $U$ is $G$-nonangular at $x$, in this case we obtain

$$GD^+W(x) \geq GD^+U(x) - GD^+V(x) \geq GD_-U(x) - v(x) = u(x) - v(x).$$

This proves the result for the weak $GP$-integral.

Next, the ordinary, approximate, and Cesàro bilateral derivates are $H$-admissible by the Second Status Theorem 15.3.3. In the case of ordinary usual $P$-integral the result follows from the second part of Lemma 21.2.1 since (P7.2) is known to hold for ordinary bilateral derivates with $L = \mathscr{F}$ (see, e.g., [N1, p. 266]). In the cases of approximate and Cesàro bilateral derivates the result follows similarly from (P7.2) which has been obtained for these derivates by Sunouchi and Utagawa [SU] with $L = \mathscr{M}$. □

Next, in the following theorem we deal with $P$-admissibility of generalized symmetric $P$-integrals.

21.2.4. Theorem (Sixteenth Status). *If any generalized symmetric $G$-derivates are $s_1$-admissible, then the symmetric $GP$-integral induced by them is $P$-admissible.*

Proof. The $H$-admissibility of the symmetric $GP$-integral has already been obtained in Theorem 21.1.5. To obtain its $P_0$-admissibility, suppose $U$ and $V$ are symmetric $G$-major and $G$-minor functions of some function $f$. Let $W = U - V$. Then since $U$ and $V$ are chosen in this case from $GN$, $W$ is continuous and $G$-nonangular n.e. Further, it follows from the axioms (S4) and (S1) that

$$G\underline{D}_sW(x) \geq G\underline{D}_sU(x) + G\underline{D}_s(-V)(x) = G\underline{D}_s\, U(x) - G\overline{D}_sV(x)$$

for each $x$. Hence it follows from axiom (S2) that axiom (P6) holds for the symmetric $GP$-integral. Further, it follows from conditions (b), (b′), (c), and (c′) in the present case that $G\underline{D}_s\, W \geq 0$ a.e. and $> -\infty$ n.e. It follows now from Theorem 17.5.1 that $W$ is nondecreasing. □

**A. *P*-regular generalized derivates and Perron integrals.** We next define a form of regularity of $H$-derivates which will also be needed for some results on $HP$-integrals.

We will call $H$-derivates *P-regular relative to L* if the following two axioms hold for every function $F \in L$:

$$\text{(P8)} \qquad H\underline{D}f \le H\overline{D}F \text{ a.e.}$$

(P9) If $F$ is $H$-derivable a.e., then $HF'$ is measurable.

Axiom (P8) holds trivially in the case of $H$-admissible $H$-derivates, which is clear from Lemma 21.1.3(c). If $H$-derivates are, on the other hand, $P$-regular relative to $L$, then for each $F \in L$ the $H$-derivative of $F$ is unique at almost all the points where it exists.

Now, given any $HP$-integral induced by $H$-derivates relative to $L$, we will call the $HP$-integral *P-regular* if the $H$-derivates are $P$-regular relative to $L$. In the next two theorems we deal with $P$-regularity of various $GP$-integrals considered in Theorems 21.1.4 and 21.1.5.

21.2.5. THEOREM (SEVENTEENTH STATUS). *If any generalized G-derivates are subregular or totally subregular, then the GP-integral and right and left GP-integrals are P-regular relative to $\mathscr{C}$ or $\mathscr{F}$, respectively.*

*Moreover, if G-derivates are regular and every function in $\mathscr{C}$ is G-nonangular* a.e., *then the lower, upper, and weak GP-integrals are P-regular relative to $\mathscr{C}$.*

PROOF. First, suppose $G$-derivates are subregular. Then they are admissible by hypothesis, and so axiom (P8) holds for the $GP$- and right and left $GP$-integrals on account of axioms (D0) and (D2). Further, for each of these integrals, axiom (P9) follows in the present case from Theorem 16.7.3 relative to $\mathscr{C}$. A similar argument holds when $G$-derivates are totally subregular.

Next, suppose $G$-derivates are regular and that every function in $\mathscr{C}$ is $G$-nonangular a.e. Then it is clear that axiom (P8) holds for lower, upper, and weak $GP$-integrals relative to $\mathscr{C}$. Further, for each of these integrals, axiom (P9) follows from Theorem 18.7.3 relative to $\mathscr{C}$. □

21.2.6. THEOREM (EIGHTEENTH STATUS). *If any generalized symmetric G-derivates are subregular or totally subregular, then the symmetric GP-integral induced by them is P-regular relative to $\mathscr{C}$ or $\mathscr{F}$, respectively.*

PROOF. First, suppose symmetric $G$-derivates are subregular. Then they are admissible by hypothesis; and so for the symmetric $GP$-integral, the axiom (P8) follows from axiom (S2). Further, axiom (P9) follows for this integral from Theorem 17.7.5 relative to $\mathscr{C}$. A similar argument holds in the other case. □

**B. Comparison of generalized Perron integrals.** We next introduce some terminology to compare generalized Perron integrals. Let $HP$- and $H_1P$- be any two generalized Perron integrals. We will call these two integrals *compatible* if for each function $f: I \to \overline{\mathbb{R}}$ which is simultaneously $HP$- and $H_1P$-integrable, these two integrals of $f$ are equal.

Further, we will call $HP$-integral *stronger than* $H_1P$-integral, or $H_1P$-integral *weaker than* $HP$-integral, if every function $f$ which is $HP$-integrable is also $H_1P$-integrable and if, for such a function $f$,

$$(H_1P)\int_a^b f(x)\,dx = (HP)\int_a^b f(x)\,dx.$$

Also, when the $HP$-integral is simultaneously stronger and weaker than the $H_1P$-integral, we will call these two integrals *equivalent.*

Next, to compare generalized derivates in the present setting, let $H$- and $H_1$-derivates be any two partially $H$-admissible generalized derivates. Then we will call $H$-derivates *stronger than* $H_1$-derivates, or $H_1$-derivates *weaker than* $H$-derivates, if for each $f \in \mathcal{F}(I)$ *we have* $H\underline{D}f \leq H_1\underline{D}f$ and $H\overline{D}f \geq H_1\overline{D}f$ at each point of $I$.

It should be noted here that if $H$-derivates are stronger than $H_1$-derivates, then $H$-derivability is stronger than $H_1$-derivability; indeed, if any function $f$ is $H$-derivable at a point $x$, then it is also $H_1$-derivable at $x$ and $Hf'(x) \subset H_1 f'(x)$.

We include here some results dealing with these comparisons.

21.2.7. THEOREM. *Suppose $HP$- and $H_1P$-integrals are partially $P$-admissible. Then if $H$-derivates are stronger than $H_1$-derivates, then the $HP$-integral is stronger than the $H_1P$-integral.*

PROOF. Suppose $H$-derivates are stronger than $H_1$-derivates, and suppose that $f: I \to \overline{\mathbb{R}}$ is $HP$-integrable. Let $\mathcal{U}$ and $\mathcal{V}$ denote the classes of $H$-major and $H$-minor functions of $f$, and let $\mathcal{U}_1$ and $\mathcal{V}_1$ denote the classes of $H_1$-major and $H_1$-minor functions of $f$. Then it is clear that $\mathcal{U} \subset \mathcal{U}_1$ and $\mathcal{V} \subset \mathcal{V}_1$. Hence we have

$$\inf_{U \in \mathcal{U}_1} U(b) \leq \inf_{U \in \mathcal{U}} U(b) = \sup_{V \in \mathcal{V}} V(b) \leq \sup_{V \in \mathcal{V}_1} V(b). \tag{3}$$

Further, for each $U \in \mathcal{U}_1$ and $V \in \mathcal{V}_1$ it follows from axiom (P7) that $U - V$ is nondecreasing, so that $U(b) \geq V(b)$. Hence the bounds in (3) are identical, and so $f$ is $H_1P$-integrable and the $H_1P$-integral of $f$ coincides with its $HP$-integral. □

Now, on account of the axiom (P1) we obtain the following result from the above theorem comparing the $HP$-integral with the ordinary Perron integral.

21.2.8. Corollary. *Suppose an HP-integral is partially P-admissible. Then this integral is weaker than the ordinary Perron integral. Consequently, if a function $f: I \to \overline{\mathbb{R}}$ is Perron integrable in the ordinary sense, then it is also HP-integrable and*

$$(HP)\int_a^b f(x)\,dx = (P)\int_a^b f(x)\,dx.$$

The following consequence of the above theorem follows on the other hand from the definitions of the various *GP*-integrals considered in the last section.

21.2.9. Corollary. *Suppose any generalized G-derivates are s-admissible. Then the right, left, lower, and upper GP-integrals are weaker than the (bilateral) GP-integral, but each of them is stronger than the weak GP-integral.*

In the following theorem we deal with compatibility of lower and upper *GP*-integrals.

21.2.10. Theorem. *Suppose any generalized G-derivates are s-admissible. Then the lower and upper GP-integrals induced by them are compatible.*

Proof. Suppose $f: I \to \overline{\mathbb{R}}$ is simultaneously lower and upper *GP*-integrable. Let $\underline{\mathscr{U}}$ and $\underline{\mathscr{V}}$ denote the classes of lower *G*-major and lower *G*-minor functions of $f$, and let $\overline{\mathscr{U}}$ and $\overline{\mathscr{V}}$ denote the classes of upper *G*-major and upper *G*-minor functions of $f$. Then if $U \in \overline{\mathscr{U}}$ and $V \in \underline{\mathscr{V}}$, then the conditions (b) and (c) hold for the derivate $GD_-U$ and (b′) and (c′) hold for $GD^-V$. Now, since the left *GP*-integral is $P_0$-admissible by Theorem 21.2.3(a), it follows from axiom (P7) that $U - V$ is nondecreasing. Hence $U(b) \geq V(b)$. When $U \in \underline{\mathscr{U}}$ and $V \in \overline{\mathscr{V}}$, it is proved similarly that $U - V$ is nondecreasing, and so again $U(b) \geq V(b)$. Hence we obtain

$$\inf_{U \in \underline{\mathscr{U}}} U(b) = \sup_{V \in \underline{\mathscr{V}}} V(b) \leq \inf_{U \in \overline{\mathscr{U}}} U(b) = \sup_{V \in \overline{\mathscr{V}}} V(b) \leq \inf_{U \in \underline{\mathscr{U}}} U(b).$$

Thus it is clear that the lower and upper *GP*-integrals of $f$ are equal. □

21.2.11. Remark. It should be noted here that in the particular case of *P*-admissible generalized usual Perron integrals, an axiomatic theory has been developed earlier by Kubota [Ku]. The axioms used by him are somewhat different but are equivalent to what is called here *P*-admissibility of usual *HP*-integrals. Thus axiom (P9) was not included in his theory, but its consequence obtained later in Corollary 21.4.5 was used inadvertently in his proofs of Theorems 21.4.7 and 21.4.9 (see Theorems 3.8 and 3.9 of [Ku]). Also, in that particular case, many of the results of the next three sections were not covered in [Ku].

**§21.3. Elementary properties of generalized Perron integrals.** In this section we obtain some of the basic properties of generalized Perron integrals under various admissibility hypotheses.

To ascertain the new hypotheses in the generalized setting, we include here complete proofs of the results obtained in the present and the next section. Many of these proofs are essentially adaptations of the classical proofs in the case of ordinary usual Perron integral (see, e.g., Natanson [N2, Chapter XVI] and McShane [Mc2, Chapter VIII]).

We begin with the following theorem which shows that the generalized *HP*-integral solves completely the problem of reconstructing a primitive function $F$ from its $H$-derivative.

21.3.1. THEOREM (RECONSTRUCTION). *Suppose an HP-integral is partially P-admissible and* $F \in L$ *is H-differentiable* n.e. *Then if* $f(x) \in HF'(x)$ *for* n.e. $x$, *then* $f$ *is HP-integrable and*

$$F(b) - F(a) = (HP)\int_a^b f(x)\,dx.$$

PROOF. Let $G(x) = F(x) - F(a)$, $x \in I$. Then according to axiom (P2), $G$ and $F$ have the same $H$-derivates everywhere. Hence there is a set $A$ in $I$ such that $I \sim A$ is countable and for each $x \in A, G$ is $H$-differentiable at $x$ and $f(x) \in HG'(x)$. There is thus a function $g : A \to \mathbb{R}$ such that for each $x \in A$, $H\overline{D}G(x) \le f(x)$ [or $g(x)$] $\le H\underline{D}G(x)$. Hence it is clear that $G$ is simultaneously an $H$-major and $H$-minor function of $f$. Now the result follows clearly on taking axiom (P7) into account. □

The following characterization of *HP*-integrability follows directly from its definition.

21.3.2. LEMMA. *A function* $f : I \to \overline{\mathbb{R}}$ *is HP-integrable iff for every* $\varepsilon > 0$ *there exist H-major and H-minor functions* $U$ *and* $V$ *of* $f$ *such that* $U(b) - V(b) < \varepsilon$.

In the following theorem we deal with finiteness of *HP*-integrable functions.

21.3.3. THEOREM. *Suppose an HP-integral is partially P-admissible. Then every HP-integrable function is finite* a.e.

PROOF. Suppose $f$ is *HP*-integrable. Choose any $H$-major and $H$-minor functions $U$ and $V$ of $f$. Then according to axiom (P7), $W \equiv U - V$ is nondecreasing. Further, according to axiom (P6), there is a set $E \subset I$ such that $I \sim E$ is countable and for each $x \in E$,

$$H\overline{D}W(x) \ge H\underline{D}U(x) - H\overline{D}V(x). \tag{1}$$

Let $A$ denote the set of points in $E$ at which $H\overline{D}V \leq f \leq H\underline{D}U$, $H\underline{D}U > -\infty$ and $H\overline{D}V < +\infty$. Then according to the conditions (b), (c), (b′), and (c′), $|A| = b - a$.

Now let $B$ denote the set of points in $A$ where $f$ is infinite. Given $x \in B$, first suppose $f(x) = +\infty$. Then $H\underline{D}U(x) = +\infty$, and since $H\overline{D}V(x) < +\infty$, it follows from (1) that $H\overline{D}W(x) = +\infty$. The same is easily seen to hold when $f(x) = -\infty$. Hence at each point $x \in B$, it follows from axiom (P1) that $\overline{D}W(x) = +\infty$. But since $W$ is nondecreasing, this implies that $|B| = 0$. Consequently, $f$ is finite a.e. □

21.3.4. THEOREM. *Suppose an HP-integral is partially P-admissible and $a < c < b$. Then if $f$ is HP-integrable on $[a, b]$, then it is also so on $[a, c]$ and $[c, b]$ and*

$$(HP)\int_a^b f(x)\,dx = (HP)\int_a^c f(x)\,dx + (HP)\int_c^b f(x)\,dx. \tag{2}$$

*Conversely, if $f$ is HP-integrable on $[a, c]$ and $[c, b]$, then it is also so on $[a, b]$.*

PROOF. First, suppose $f$ is $HP$-integrable on $[a, b]$. Then, given $\varepsilon > 0$, we can choose by Lemma 21.3.2, $H$-major and $H$-minor functions $U$ and $V$ of $f$ such that $U(b) - V(b) < \varepsilon$. Clearly, $U$ and $V$ are also $H$-major and $H$-minor functions of $f$ on $[a, c]$, and since $U - V$ is nondecreasing by axiom (P7), $U(c) - V(c) < \varepsilon$. Hence it follows from Lemma 21.3.2 that $f$ is $HP$-integrable on $[a, c]$, and hence

$$V(c) \leq (HP)\int_a^c f(x)\,dx \leq U(c). \tag{3}$$

Now define $U^*(x) = U(x) - U(c)$, $V^*(x) = V(x) - V(c)$ for $c \leq x \leq b$. Then according to axiom (P2), $H\underline{D}U^*(x) = H\underline{D}U(x)$ and $H\overline{D}V^*(x) = H\overline{D}V(x)$ for $c \leq x \leq b$. Hence $U^*$ and $V^*$ are clearly $H$-major and $H$-minor functions of $f$ on $[c, b]$. Further, since $U(c) - V(c) \geq 0$, we obtain $U^*(b) - V^*(b) \leq U(b) - V(b) < \varepsilon$. Hence $f$ is also $HP$-integrable on $[c, b]$, and

$$V(b) - V(c) = V^*(b) \leq (HP)\int_c^b f(x)\,dx \leq U^*(b) = U(b) - U(c). \tag{4}$$

Now on combining (3) and (4) we obtain

$$V(b) \leq (HP)\int_a^c f(x)\,dx + (HP)\int_c^b f(x)\,dx \leq U(b).$$

The relation (2) clearly follows from this relation.

Next, to obtain the converse, suppose $f$ is $HP$-integrable on $[a, c]$ and $[c, b]$. Then, given $\varepsilon > 0$, we can choose $H$-major and $H$-minor functions $U_1$ and $V_1$ of $f$ on $[a, c]$ such that $U_1(c) - V_1(c) < \varepsilon$, and $H$-major and $H$-minor functions $U_2$ and $V_2$ of $f$ on $[c, b]$ such that $U_2(b) - V_2(b) < \varepsilon$. Now define $U = U_1$ and $V = V_1$ on $[a, c]$, and for each $x \in [c, b]$, define $U(x) = U_1(c) + U_2(x)$ and $V(x) = V_1(c) + V_2(x)$. Then it is clear that $U$ and $V$ are $H$-major and $H$-minor functions of $f$ on $[a, b]$, and since $U(b) - V(b) < 2\varepsilon$, it follows from Lemma 21.3.2 that $f$ is $HP$-integrable on $[a, b]$. □

21.3.5. THEOREM. *If $f$ and $g$ are HP-integrable and $f \leq g$ a.e., then*

$$(HP)\int_a^b f(x)\,dx \leq (HP)\int_a^b g(x)\,dx.$$

PROOF. Suppose the hypothesis holds, and let $\mathscr{V}_f$ and $\mathscr{V}_g$ denote the classes of $H$-minor functions of $f$ and $g$, respectively. Then since $f \leq g$ a.e., it is clear that $\mathscr{V}_f \subset \mathscr{V}_g$. Hence we obtain

$$\int_a^b f(x)\,dx = \sup_{V \in \mathscr{V}_f} V(b) \leq \sup_{V \in \mathscr{V}_g} V(b) = \int_a^b g(x)\,dx. \qquad \square$$

21.3.6. THEOREM. *If $f$ is HP-integrable and $g = f$ a.e., then $g$ also is HP-integrable and*

$$(HP)\int_a^b g(x)\,dx = (HP)\int_a^b f(x)\,dx. \tag{5}$$

*Further, the same holds for the usual HP-integral provided that H-derivates are semi-H-admissible.*

PROOF. The result in the first part for the general $HP$-integral is obvious. For it is clear from the general conditions (a) to (c) and (a′) to (c′) that a function is an $H$-major or $H$-minor function of $f$ iff it is so for $g$.

To prove the second part, suppose $H$-derivates are semi-$H$-admissible and $f$ is $HP$-integrable in the usual sense. Let $E = \{x : g(x) \neq f(x)\}$. Then since $|E| = 0$, there exists a continuous nondecreasing function $h$ on $I$ such that $h'(x) = +\infty$ for each $x \in E$ (see [N1, p. 214] for an example). On replacing $h$ by the function $\{h - h(a)\}/\{h(b) - h(a)\}$ if necessary, we can assume here that $h(a) = 0$ and $h(b) = 1$.

Now, given $\varepsilon > 0$, there exist by Lemma 21.3.2 usual $H$-major and $H$-minor functions $U$ and $V$ of $f$ such that $U(b) - V(b) < \varepsilon$. Then $U(a) = V(a) = 0$, and for each $x \in I$ we have $H\underline{D}U(x) \geq f(x) \geq H\overline{D}V(x)$, $H\underline{D}U(x) > -\infty$ and $H\overline{D}V(x) < +\infty$. Now define

$$U^*(x) = U(x) + \varepsilon h(x), \qquad V^*(x) = V(x) - \varepsilon h(x), \qquad x \in I. \tag{6}$$

Then since $h$ is nondecreasing, for each $x \in I$ we obtain the following with the help of axioms (P5) and (P1):

$$\begin{aligned}(7)\qquad H\underline{D}U^*(x) &\geq H\underline{D}U(x) + H\underline{D}(\varepsilon h)(x)\\ &\geq H\underline{D}U(x) + \underline{D}(\varepsilon h)(x) \geq H\underline{D}U(x) > -\infty.\end{aligned}$$

Further, when $x \in I \sim E$, we obtain from (7), $H\underline{D}U^*(x) \geq H\underline{D}U(x) \geq f(x) = g(x)$, and if $x \in E$, then since$(\varepsilon h)'(x) = \varepsilon h'(x) = +\infty$, it is again clear from (7) that $H\underline{D}U^*(x) = +\infty \geq g(x)$. This proves that $U^*$ is a usual $H$-major function of $g$. It is proved by a similar argument that $V^*$ is a usual $H$-minor function of $g$. Further, since

$$U^*(b) - V^*(b) = U(b) - V(b) + 2\varepsilon h(b) < 3\varepsilon,$$

this proves according to Lemma 21.3.2 that $g$ is $HP$-integrable in the usual sense. The relation (5) follows on the other hand from the following inequalities:

$$V(b) - \varepsilon = V^*(b) \leq (HP)\int_a^b g(x)\,dx \leq U^*(b) = U(b) + \varepsilon. \qquad \square$$

Next, in the following theorem we deal with scalar products of $HP$-integrable functions.

21.3.7. THEOREM. *If $H$-derivates are partially $H$-admissible and $f$ is $HP$-integrable, then for each $c \geq 0$, $cf$ also is $HP$-integrable and*

$$(8)\qquad (HP)\int_a^b cf(x)\,dx = c(HP)\int_a^b f(x)\,dx.$$

*Further, the same holds for $c < 0$ provided that $H$-derivates are weakly $H$-admissible.*

PROOF. Because the result holds trivially for $c = 0$, suppose $c > 0$ and that $f$ is $HP$-integrable. Then, given $\varepsilon > 0$, there exist $H$-major and $H$-minor functions $U$ and $V$ of $f$ such that $U(b) - V(b) < \varepsilon/c$.

First, suppose $H$-derivates are partially $H$-admissible. Then according to axiom (P3), $H\underline{D}(cU)(x) = cH\underline{D}U(x)$ and $H\overline{D}(cU)(x) = cH\overline{D}U(x)$ for each $x \in I$; and the same holds on replacing $U$ by $V$ or $f$. Hence it is easy to see that $cU$ and $cV$ are $H$-major and $H$-minor functions of $cf$. Now since $cU(b) - cV(b) < \varepsilon$, this proves that $cf$ is $HP$-integrable. Further, the relation (8) follows from the fact that

$$cV(b) \leq (HP)\int_a^b cf(x)\,dx \leq cU(b).$$

Next, suppose $H$-derivates are weakly $H$-admissible. Then it follows from Lemma 21.1.2 that $H\underline{D}(-cU)(x) = -cH\overline{D}U(x)$ and $H\overline{D}(-cU)(x) = -cH\underline{D}U(x)$ for each $x$; and the same holds on replacing $U$ by $V$ or $f$. It is now easy to see in this case that $-cV$ and $-cU$ are $H$-major and $H$-minor functions of $-cf$. Further, since $-cV(b) + cU(b) < \varepsilon$, this proves that $-cf$ is $HP$-integrable. Also, it is seen as above that the relation (8) holds on replacing $c$ by $-c$. □

Now, on combining the second part of the above theorem with Theorem 21.3.5, we obtain the following result since $-|f| \le f \le |f|$ everywhere.

21.3.8. COROLLARY. *Suppose an HP-integral is weakly H-admissible. Then if f and $|f|$ are HP-integrable, then*

$$\left|(HP)\int_a^b f(x)\,dx\right| \le (HP)\int_a^b |f(x)|\,dx.$$

Finally, in the following theorem we obtain additivity of $HP$-integrals that are induced by semi-$H$-admissible generalized derivates.

21.3.9. THEOREM. *Suppose H-derivates are semi-H-admissible. Then if f and g are HP-integrable and $f + g$ is defined* a.e., *then $f + g$ is HP-integrable and*

$$(9)\qquad (HP)\int_a^b (f+g)(x)\,dx = (HP)\int_a^b f(x)\,dx + (HP)\int_a^b g(x)\,dx.$$

PROOF. Let $h = f + g$. Given $\varepsilon > 0$, choose $H$-major and $H$-minor functions $U_1$ and $V_1$ of $f$, and $H$-major and $H$-minor functions $U_2$ and $V_2$ of $g$, such that $U_i(b) - V_i(b) < \varepsilon/2$ for $i = 1, 2$. Now define $U = U_1 + U_2$ and $V = V_1 + V_2$. Then according to axiom (P5), $H\underline{D}U(x) \ge H\underline{D}U_1(x) + H\underline{D}U_2(x)$ and $H\overline{D}V(x) \le H\overline{D}V_1(x) + H\overline{D}V_2(x)$ for each $x$. Hence it is easy to see that $U$ and $V$ are $H$-major and $H$-minor functions of $h$. Further, it is clear that $U(b) - V(b) < \varepsilon$. This proves that $h$ is $HP$-integrable. Relation (9) follows on the other hand from the fact that

$$\begin{aligned} V(b) &= V_1(b) + V_2(b) \\ &\le (HP)\int_a^b f(x)\,dx + (HP)\int_a^b g(x)\,dx \le U(b). \quad \square \end{aligned}$$

**§21.4. Properties of indefinite generalized Perron integrals.** In this section we obtain various properties of indefinite generalized Perron integrals under appropriate hypotheses.

Suppose an *HP*-integral is partially *P*-admissible and that $f: I \to \overline{\mathbb{R}}$ is *HP*-integrable. Then according to Theorem 21.3.4, $f$ is *HP*-integrable on $[a, x]$ for $a < x \le b$. Define further

$$(HP)\int_a^a f(t)\,dt = 0.$$

We will now call the function

$$F(x) = (HP)\int_a^x f(t)\,dt, \qquad a \le x \le b,$$

the *indefinite HP-integral* of $f$, or the *indefinite Perron integral* of $f$ *induced by H-derivates.*

21.4.1. Lemma. *Suppose an HP-integral is partially P-admissible and F is the indefinite HP-integral of some function f. Then if U and V are H-major and H-minor functions of f, then $U - F$ and $F - V$ are nondecreasing on I, and $U(x) \ge F(x) \ge V(x)$ for each $x \in I$.*

Proof. Suppose $a \le c < d \le b$. Define $U^*(x) = U(x) - U(c)$ for $c \le x \le d$. Then it is clear from axiom (P2) that $U^*$ is an $H$-major function of $f$ on $[c, d]$. Hence we obtain with the help of Theorem 21.3.4,

$$F(d) - F(c) = (HP)\int_c^d f(t)\,dt \le U(d) - U(c).$$

Thus $U(c) - F(c) \le U(d) - F(d)$, which proves $U - F$ to be nondecreasing. Now, given $x \in I$, we thus have $U(x) - F(x) \ge U(a) - F(a) = 0$, and so $U(x) \ge F(x)$. The results on $V$ are obtained by a similar argument. □

21.4.2. Theorem. *Suppose an HP-integral is partially P-admissible. Then the indefinite HP-integral of every HP-integrable function is an element of L.*

Proof. Suppose $F$ is the indefinite *HP*-integral of some function $f$. Given $\varepsilon > 0$, choose $H$-major and $H$-minor functions $U$ and $V$ of $f$ such that $U(b) - V(b) < \varepsilon$. Now since $V(b) \le F(b)$, and $U - F$ is nondecreasing by the above lemma, for each $x \in I$ we have

$$0 = U(a) - F(a) \le U(x) - F(x) \le U(b) - F(b) \le U(b) - V(b) < \varepsilon.$$

On choosing $\varepsilon = 1/n$ for each positive integer $n$, we thus obtain a sequence of functions $\{U_n\}$ in $L$ which converges uniformly to $F$. Consequently, $F \in L$. □

In the following theorem we deal with extension of the classical fundamental theorem of Calculus to generalized Perron integrals.

21.4.3. THEOREM (FUNDAMENTAL). *Suppose an HP-integral is partially P-admissible. Then if F is the indefinite HP-integral of f, then for almost every point x of I, F is H-derivable at x and* $f(x) \in HF'(x)$.

*Moreover, if the H-derivates are H-admissible or P-regular, then* $f(x) = HF'(x)$ *for* a.e. $x$.

PROOF. We will first prove that $H\underline{D}f \geq f$ a.e. Given $\varepsilon > 0$, choose an $H$-major function $U$ of $f$ such that $U(b) - F(b) < \varepsilon^2$. Then the function $W = U - F$ is nondecreasing by Lemma 21.4.1. Hence $W$ is differentiable a.e. and

$$(1) \qquad (L)\int_a^b W'(x)\,dx \leq W(b) - W(a) = U(b) - F(b) < \varepsilon^2.$$

Let $E$ denote the set of points $x$ in $I$ where $W$ is differentiable, $H\underline{D}U(x) \geq f(x)$, and $H\underline{D}U(x) > -\infty$. Then it is clear from conditions (b) and (c) that $|E| = b - a$. Now set

$$(2) \quad A_\varepsilon = \{x : H\underline{D}f(x) < f(x) - \varepsilon\}, \qquad B_\varepsilon = \{x \in E : W'(x) > \varepsilon\}.$$

Suppose $x \in E \cap A_\varepsilon$. Then it is clear that $H\underline{D}F(x) < H\underline{D}U(x) - \varepsilon$. Further, since $H\underline{D}F(x) < +\infty$ and $H\underline{D}U(x) > -\infty$, it follows clearly from axiom (P2) that $W'(x) = H\underline{D}U(x) - H\underline{D}F(x) > \varepsilon$, that is, $x \in B_\varepsilon$. Thus $E \cap A_\varepsilon \subset B_\varepsilon$, and so $|A_\varepsilon| \leq |B_\varepsilon|$. Further, with the help of (1) we obtain

$$\varepsilon|B_\varepsilon| \leq (L)\int_{B_\varepsilon} W'(x)\,dx \leq (L)\int_a^b W'(x)\,dx < \varepsilon^2.$$

Hence $|B_\varepsilon| < \varepsilon$, and so $|A_\varepsilon| < \varepsilon$.

Now, since

$$\{x : H\underline{D}F(x) < f(x)\} = \bigcup_n \{x : H\underline{D}F(x) < f(x) - \varepsilon 2^{-n}\},$$

on replacing $\varepsilon$ in $A_\varepsilon$ by $\varepsilon 2^{-n}$ for each $n$, we obtain

$$|\{x :\ H\underline{D}F(x) < f(x)\}| \leq \sum_n |A_{\varepsilon 2^{-n}}| < \sum_n \varepsilon 2^{-n} = \varepsilon.$$

This proves that $H\underline{D}F \geq f$ a.e.

Using an $H$-minor function $V$ of $f$, we can prove similarly that $H\overline{D}F \leq f$ a.e. Hence at almost every point $x$ of $I$, $H\overline{D}F(x) \leq f(x) \leq H\underline{D}F(x)$, that is, $F$ is $H$-derivable at $x$ and $f(x) \in HF'(x)$.

The second part follows now from Lemma 21.1.3(c) in the case when the $H$-derivates are $H$-admissible, and in the other case it follows from axiom (P8). □

21.4.4. COROLLARY. *Suppose an HP-integral is partially P-admissible. Then if $f$ is HP-integrable, then each of the H-major and H-minor functions of $f$ is H-derivable* a.e.

Let $F$ be the indefinite $HP$-integral of $f$ and let $U$ and $V$ be any $H$-major and $H$-minor functions of $f$. Then since $U - F$ is nondecreasing by Lemma 21.4.1, it is differentiable a.e. Hence it follows from the above theorem with the help of axiom (P2) that $U = (U - F) + F$ is $H$-derivable a.e. A similar argument holds for $V$.

On combining the last part of the above theorem with axiom (P9), we further obtain the following result.

21.4.5. COROLLARY. *Suppose an HP-integral is partially P-admissible and P-regular. Then every HP-integrable function is measurable.*

Next, in the following theorem we compare generalized $HP$-integral with Lebesgue integral.

21.4.6. THEOREM. *Suppose an HP-integral is partially P-admissible and P-regular. Then if $f$ is HP-integrable and non-negative* a.e., *then it is Lebesgue integrable and*

$$(HP)\int_a^b f(x)\,dx = (L)\int_a^b f(x)\,dx. \tag{3}$$

PROOF. Suppose $f$ is $HP$-integrable and non-negative a.e. Then according to Corollary 21.4.5, $f$ is measurable. Choose any $H$-major function $U$ of $f$, and let $V(x) = 0$ for $x \in I$. Then according to Lemma 21.1.1, $HV'(x) = 0$ for each $x$, and so $V$ is an $H$-minor function of $f$. Hence, according to axiom (P7), $U = U - V$ is nondecreasing. Consequently, $U$ is differentiable a.e. and $U'$ is summable. But since $U$ is an $H$-major function of $f$, it follows from axiom (P1) that $U'(x) = H\underline{D}U(x) \geq f(x) \geq 0$ for a.e. $x$. Now since $f$ is measurable, this implies the Lebesgue integrability of $f$, and relation (3) follows now from Corollary 21.2.8. □

It is clear from the above proof that Theorem 21.4.6 holds also without the hypothesis of $P$-regularity in the case when $f$ is given to be measurable. The following result follows from this version of the above theorem with the help of Corollary 21.2.8.

21.4.7. COROLLARY. *Suppose an HP-integral is partially P-admissible. Then if $f$ is measurable and $|f|$ is HP-integrable, then $f$ is Lebesgue integrable and* (3) *holds.*

From the above theorem we further obtain the following with the help of Theorem 21.3.9.

21.4.8. COROLLARY. *Suppose an HP-integral is semi-P-admissible and P-regular. Then if $f$ is HP-integrable and there is a Lebesgue integrable function $g$ such that $g \leq f$* a.e., *then $f$ is Lebesgue integrable.*

The function $f - g$ is *HP*-integrable by Theorem 21.3.9, and so it is Lebesgue integrable by the above theorem; and hence $f = g + (f - g)$ is also Lebesgue integrable.

Next, in the following theorem we obtain an extension of the monotone convergence theorem on Lebesgue integral to *HP*-integral.

21.4.9. THEOREM. *Suppose an HP-integral is semi-P-admissible and P-regular, and that $\{f_n\}$ is a nondecreasing sequence of HP-integrable functions whose HP-integrals are bounded from above. Then if $-f_1$ is also HP-integrable, then so is the pointwise limit $f$ of $\{f_n\}$ and*

$$(4)\qquad (HP)\int_a^b f(x)\,dx = \lim_n\,(HP)\int_a^b f_n(x)\,dx.$$

PROOF. Given $n > 1$, the function $f_n - f_1$ is non-negative and *HP*-integrable by Theorem 21.3.9, and so it is Lebesgue integrable by Theorem 21.4.6. Hence, by the monotone convergence theorem on Lebesgue integral (see Halmos [Hal, p. 112]), we have

$$(5)\qquad (L)\int_a^b (f - f_1)(x)\,dx = \lim_n\,(L)\int_a^b (f_n - f_1)(x)\,dx.$$

Further, according to Corollary 21.2.8 we have

$$(6)\quad (L)\int_a^b (f_n - f_1)(x)\,dx = (HP)\int_a^b (f_n - f_1)(x)\,dx, \qquad n = 2, 3, \ldots$$

Now since the sequence of integrals $(HP)\int_a^b f_n(x)\,dx$, $n = 1, 2, \ldots$ is bounded from above, it follows from Theorem 21.3.9 that the sequence of integrals in (6) is also bounded from above. Hence, by (5),

$$0 \leq (L)\int_a^b (f - f_1)(x)\,dx < \infty.$$

Thus $f - f_1$ is Lebesgue integrable. Hence $f - f_1$ is *HP*-integrable by Corollary 21.2.8, and so is in turn $f$ by Theorem 21.3.9. Relation (4) follows now from (5) with the help of Theorem 21.3.9. □

Finally, in the following theorem we obtain an extension of Lebesgue's dominated convergence theorem to *HP*-integral.

21.4.10. THEOREM. *Suppose an HP-integral is semi-P-admissible and P-regular. Let* $\{f_n\}$ *be a sequence of HP-integrable functions which converges in measure to* $f$. *Then if* $-f_1$ *is also HP-integrable and there is a Lebesgue integrable function* $g$ *such that for each* $n$, $|f_n - f_1| \le g$ a.e. *then* $f$ *is HP-integrable and* (4) *holds.*

PROOF. Clearly, the sequence $\{f_n - f_1\}$ converges in measure to $f - f_1$. Now, for each $n > 1$, since $f_n - f_1 \ge -g$ a.e., it follows from Theorem 21.3.9 and Corollary 21.4.8 that $f_n - f_1$ is Lebesgue integrable. Hence it follows from Lebesgue's dominated convergence theorem (see Halmos [Hal, p. 110]) that $f - f_1$ is Lebesgue integrable and

$$(L)\int_a^b (f - f_1)(x)\,dx = \lim_n (L)\int_a^b (f_n - f_1)(x)\,dx.$$

Now the result follows clearly with the help of Corollary 21.2.8 and Theorem 21.3.9. □

**§21.5. Applications of the axiomatic theory and open problems.** In this final section we present several examples of generalized Perron integrals to which the theory developed in this chapter applies. Also, we present here some problems that arise naturally from this theory but remain unresolved.

We begin in the following theorem with examples of bilateral and right and left generalized Perron integrals. We do not include here qualitative and negligent derivates since they coincide with ordinary derivates in the case of continuous functions. Also, we do not include here sequentially congruent derivates since they constitute a particular case of congruent derivates.

21.5.1. THEOREM. *Let G-derivates be any of the following derivates:*

(a) *ordinary derivates,*
(b) *congruent derivates,*
(c) *Borel derivates,*
(d) *Cesàro derivates,*
(e) *continuously selective derivates,*
(f) *congruent Cesào derivates,*
(g) *congruent selective derivates,*
(h) *continuously selective Cesàro derivates,*
(i) *continuously selective selective derivates,*
(j) *approximate derivates,*
(k) *approximate Cesàro derivates,*
(l) *approximate selective derivates,*
(m) *subcontinuous path derivates,*

(n) *subcontinuous path Cesàro derivates,*
(o) *subcontinuous path selective derivates.*

*Then all the results of* §§21.3 *and* 21.4 *hold for the bilateral, right and left GP-integrals relative to* $\mathscr{C}$.

PROOF. It is clear from the Status Theorems 15.3.3, 16.8.1, and 16.8.2 that each of the derivates in (a) to (o) are $s$-admissible. Further, according to the Status Theorems 15.4.2, 15.4.4, 16.8.3, 16.8.5, and 16.8.6, each of them are subregular. Hence if $G$-derivates denote any of these derivates, then it follows from the Fifteenth and Sixteenth Status Theorems 21.2.3 and 21.2.5 that the bilateral, right, and left $GP$-integrals are $P$-admissible and $P$-regular relative to $\mathscr{C}$. Consequently, all the results of §§21.3 and 21.4 hold for these $GP$-integrals relative to $\mathscr{C}$. □

Next, in the following theorem we deal with examples of lower, upper, and weak generalized Perron integrals. We assume here that $L = \mathscr{C}$.

21.5.2. THEOREM. *Let G-derivates be any of the derivates* (a) *to* (o) *considered in Theorem* 21.5.1. *Then except for the second part of Theorem* 21.3.7 *and Corollary* 21.3.8, *all other results of* §21.3 *hold for lower and upper GP-integrals; and except for the last part of Theorem* 21.3.6 *and Theorem* 21.3.9, *all other results of* §21.3 *hold for weak GP-integral. Also, Lemma* 21.4.1, *Theorem* 21.4.2, *the first part of Theorem* 21.4.3, *and Corollaries* 21.4.4 *and* 21.4.7 *hold for each of these new GP-integrals.*

*Next, suppose G-derivates denote any of the derivates* (a) *to* (i) *of Theorem* 21.5.1 *and that every function in* $\mathscr{C}$ *is G-nonangular* a.e. *Then all the results of* §21.4 *hold for lower and upper GP-integrals; and the first seven results of* §21.4 *up to Corollary* 21.4.7 *hold for weak GP-integral.*

PROOF. First, let $G$-derivates denote any of the derivates (a) to (o) of Theorem 21.5.1. Then the $G$-derivates are $s$-admissible as seen earlier. Hence it follows from the Fifteenth Status Theorem 21.2.3 that the lower and upper $GP$-integrals are semi-$P$-admissible, and that the weak $GP$-integral is weakly $P$-admissible. Hence it is easy to verify that the various assertions of the first part hold for lower, upper, and weak $GP$-integrals.

Next, suppose $G$-derivates denote any of the derivates (a) to (i) of Theorem 21.5.1 and that every function in $\mathscr{C}$ is $G$-nonangular a.e. Then it follows from the Status Theorems 15.4.4, 16.8.5, and 16.8.6 that the $G$-derivates are regular. Hence it follows from the Seventeenth Status Theorem 21.2.5 that the lower, upper, and weak $GP$-integrals are $P$-regular (relative to $\mathscr{C}$). Now it is easy to verify that the various assertions of the second part hold for lower, upper, and weak $GP$-integrals. □

In the following theorem we deal with examples of usual bilateral generalized Perron integrals.

21.5.3. THEOREM. *Let G-derivates be any of the following derivates:*

(a) *ordinary derivates,*
(b) *approximate derivates,*
(c) *Cesàro derivates,*
(d) *continuously selective derivates,*
(e) *approximate Cesàro derivates,*
(f) *approximate selective derivates.*

*Then if axiom* (P7.2) *holds for G-derivates relative to a linear space L, then all the results of* §§21.3 *and* 21.4 *hold for the usual GP-integral relative to L.*

PROOF. Let $G$-derivates denote any of the derivates (a) to (f), and suppose that axiom (P7.2) holds for $G$-derivates relative to $L$. Then since the $G$-derivates are $s$-admissible as seen above, the usual $GP$-integral is $H$-admissible by the Thirteenth Status Theorem 21.1.4(a). Further, according to Lemma 21.2.1, the usual $GP$-integral is $P_0$-admissible relative to $L$, and so it is $P$-admissible relative to $L$. Also, since the $G$-derivates are totally subregular according to Theorem 15.4.2 or 16.8.3, it follows from the Seventeenth Status Theorem 21.2.5 that the usual $GP$-integral is $P$-regular relative to $\mathscr{F}$. Hence it is clear that all the results of §§21.3 and 21.4 hold for the usual $GP$-integral relative to $L$. □

We include here some consequences of the above theorem on usual Perron integrals that are known.

21.5.4. COROLLARY. *The results of* §§21.3 *and* 21.4 *hold for the following usual Perron integrals:*

(a) *the ordinary Perron integral induced by ordinary bilateral derivates relative to* $\mathscr{F}$;
(b) *the approximately continuous Perron integral induced by approximate bilateral derivates relative to* $\mathscr{M}$; *and*
(c) *the Cesàro–Perron integral induced by Cesàro bilateral derivates relative to the linear space of* $D_*$*-integrable functions.*

For, in case (a) it is well known that (P7.2) holds in terms of ordinary bilateral derivates (see, e.g., Natanson [N1, p. 266]); and in cases (b) and (c), (P7.2) has been obtained by Sunouchi and Utagawa [SU].

Here the usual Perron integral in (a) is the classical Perron integral, and the Perron integrals in (b) and (c) have been defined by Sunouchi and Utagawa in [SU]. The integral in (b) is known to be more general than Burkill's $AP$-integral [Bur1].

In the following theorem we deal on the other hand with examples of generalized symmetric Perron integrals.

21.5.5. THEOREM. *Let symmetric G-derivates be any of the following symmetric derivates:*

(a) *ordinary symmetric derivates,*
(b) *congruent symmetric derivates,*

(c) *Borel symmetric derivates,*
(d) *Cesàro symmetric derivates,*
(e) *continuously selective symmetric derivates,*
(f) *congruent Cesàro symmetric derivates,*
(g) *congruent selective symmetric derivates,*
(h) *continuously selective Cesàro symmetric derivates,*
(i) *continuously selective selective symmetric derivates,*
(j) *approximate symmetric derivates,*
(k) *approximate Cesàro symmetric derivates,*
(l) *approximate selective symmetric derivates,*
(m) *subcontinuous path symmetric derivates,*
(n) *subcontinuous path Cesàro symmetric derivates,*
(o) *subcontinuous path selective symmetric derivates.*

*Then all the results of* §21.3 *and* 21.4 *hold for the symmetric GP-integral induced by these derivates relative to* $\mathscr{C}$.

PROOF. It is clear from the Status Theorems 17.1.7, 17.8.1, and 17.8.2 that each of these symmetric derivates are $s_1$-admissible. Further, according to the Status Theorems 17.1.10, 17.1.11, 17.8.3, 17.8.5, and 17.8.6, each of them are subregular. Hence the symmetric Perron integral induced by each of these symmetric derivates is, according to the Sixteenth and Eighteenth Status Theorems 21.2.4 and 21.2.6, $P$-admissible and $P$-regular relative to $\mathscr{C}$. Consequently, all the results of §§21.3 and 21.4 hold for these generalized symmetric Perron integrals relative to $\mathscr{C}$. □

Next, we indicate some problems that arise naturally from the developments of the present chapter in the case of various generalized Perron integrals, but remain unresolved.

21.5.6. PROBLEM. (a) It will be worthwhile to investigate if the restrictions on the choice of major and minor functions in the definitions of lower, upper, weak, and symmetric generalized Perron integrals can be relaxed or eliminated. In the case of ordinary derivates, there is of course no such restriction since every function is nonangular in the ordinary sense n.e.

(b) It will also be worthwhile to investigate for which of the generalized $G$-derivates considered in Theorem 21.5.2, every function in $\mathscr{C}$ is $G$-nonangular a.e., because in the case of such $G$-derivates, according to that theorem all the results of §21.4 hold for lower and upper $GP$-integrals, and the first seven results of that section hold for weak $GP$-integral.

Apart from the ordinary derivates, this is known to hold also for approximate derivates as proved by Ward [W1]. However, it is not clear whether approximate derivates are regular, or whether the axiom (P9) holds for the approximate lower, upper, and weak Perron integrals.

(c) It will be interesting to investigate for which of the generalized Perron integrals considered in this section the integration by parts formula and the substitution rule hold.

(d) Finally, it will be interesting to investigate if for any of the generalized Perron integrals considered in this section a descriptive definition can be obtained similar to the ones obtained in §§11.6 and 11.8.

21.5.7. REMARK. We conclude this section with some comments on the historical background of generalized Perron integrals. After the discovery of ordinary (usual) Perron integral by Perron [Pe] and Bauer [Bau], several generalizations of this integral have appeared in the literature over the period of time. First, Burkill introduced the notions of approximate and Cesàro Perron integrals and obtained some of their properties (see [Bur1] and [Bur2]). See the survey article of Bullen [Bu1] for further developments on these Perron integrals. The symmetric Borel Perron integral was introduced in turn by Marcinkiewicz and Zygmund [MZ] in connection with some deeper problems of trigonometric series. Further, in the same connection, the symmetric Cesàro Perron integral was considered by Burkill [Bur3].

In connection with the last two references, it should be noted here that the theory of generalized symmetric Perron integrals presented in this chapter applies to symmetric Borel and Cesàro Perron integrals in full generality only when, relative to Borel and Cesàro derivates, every continuous function is nonangular n.e. (see §21.2). It will therefore be interesting to investigate if this holds for Borel and Cesàro derivates.

# References


[A1] Alexandroff, P., *Über die Äquivalenz des Perronschen und des Denjoyschen Integralbegriffes*, Math. Zeitschr. **20** (1924), 213–222.

[A2] Alexandroff, P., *L'intégration au sense de M. Denjoy considérée comme recherche des fonctions primitivès*, Rec. Math. Soc. Math. Moscou **31** (1924), 465–476.

[AP] Anderson, J. M., and Pitt, L. D., *Probabilistic behaviour of functions in the Zygmund spaces* $\Lambda^*$ *and* $\lambda^*$, Proc. London Math. Soc. (3) **59** (1989), 558–592.

[AF] Aubin, J. P. and Frankowska, H. *Set-Valued Analysis*, Birkhauser, Boston, 1990.

[Ba] Baire, R., *Sur les fonctions de variables réelles*, Ann. Mat. Pura Appl. (3) **3** (1899), 1–122.

[B1] Banach, S., *Sur les ensembles de points où la dérivée est infinie*, C. R. Acad. Sci. Paris **173** (1921), 457–459.

[B2] Banach, S., *Sur les fonctions dérivées des fonctions mesurables*, Fund. Math. **3** (1922), 128–132.

[B3] Banach, S., *Sur les lignes rectifiables et les surfaces dont l'aire est finie*, Fund. Math. **7** (1925), 225–236.

[B4] Banach, S., *Sur une classe de fonctions continues*, Fund. Math. **8** (1926), 166–173.

[B5] Banach, S., *Über die Baire'sche Kategorie gewisser Funktionenmengen*, Studia Math. **3** (1931), 174–179.

[BS] Banach, S., and Saks, S., *Sur les fonctions absolument continues des fonctions absolument continues*, Fund. Math. **11** (1928), 113–116.

[BM] Bary, N., and Menchoff, D., *Sur l'intégrale de Lebesgue-Stieltjes et les fonctions absolument continues des fonctions absolument continues*, Ann. Mat. Pura Appl. (4) **5** (1928), 19–54.

[Bau] Bauer, H., *Der Perronsche Integralbegriff und seine Beziehung zum Lebesgueschen*, Monatshefte Math. Phys. **26** (1915), 153–198.

[Be] Besicovitch, A. S., *Discussion der stetigen Funktionen im Zussammenhang mit der Frage über ihre Differenzierbarkeit*, Bull. Acad. Sci. URSS **19** (1925), 97–122, 527–540.

[BMa] Boboc, N., and Marcus, S., *Sur la détermination d'une fonction par les valeurs prises sur un certain ensemble*, Ann. Sci. Ecole Norm. Sup. (3) **76** (1959), 151–159.

[Bo1] Borel, E., *Leçons sur la théorie des fonctions*, Paris, 1898.

[Bo2] Borel, E., *Modèles arithmétiques et analytiques de l'irréversibilité apparente*, C. R. Acad. Sci. Paris **154** (1912), 1148–1150.

[Boi] Bois-Reymond, P. du, *Versuch einer Klassification der willkurlichen Funktionen reeler Argumente*, J. für Math. **79** (1875), 21–37.

[BLe] Borwein, J. M., and Lewis, A. S., *Convex Analysis and Nonlinear Optimization*, Wiley/CMS, to appear.

[BP] Borwein, J. M., and Preiss, D., *A smooth variational principle with applications to subdifferentiability and to differentiability of convex functions*, Trans. Amer. Math. Soc. **303** (1987), 517–527.

[Bou] Bouligand, G. *Introduction à la géométrie infinitésimale directe*, Paris, 1932.

[Br1] Bruckner, A. M., *An affirmative answer to a problem of Zahorski, and some consequences*, Michigan Math. J. **13** (1966), 15–26.

[Br2] Bruckner, A. M., *Differentiation of Real Functions*, 2nd ed., CRM Monograph Series, Vol. 5, American Mathematical Society, Providence, 1994.

[BG] Bruckner, A. M., and Garg, K. M., *The level structure of a residual set of continuous functions*, Trans. Amer. Math. Soc. **232** (1977), 307–321.

[BGo] Bruckner, A. M., and Goffman, C., *The boundary behavior of real functions in the upper half plane*, Rev. Roumaine Math. Pures Appl. **11** (1966), 507–518.

[BL] Bruckner, A. M., and Leonard, J. L., *Derivatives*, Amer. Math. Monthly **73** (1966) no. 4, part II, 24–56.

[BOT] Bruckner, A. M., O'Malley, R. J., and Thomson, B. S., *Path derivatives: a unified view of certain generalized derivatives*, Trans. Amer. Math. Soc. **283** (1984), 97–125.

[Bru] Brudno, A., *Continuity and differentiability* (in Russian), Mat. Sb. **13** (55) (1943), 119–134.

[Bu1] Bullen, P. S., *Non Absolute Integrals: A Survey, Real Analysis Exchange*, **5** (1979/80), 195–259.

[Bu2] Bullen, P. S., *A simple proof of integration by parts for the Perron integral*, Can. Math. Bull. **28** (2) (1985), 195–199.

[Bur1] Burkill, J. C., *The approximately continuous Perron integral*, Math. Zeitschr. **34** (1931), 270–278.

[Bur2] Burkill, J. C., *The Cesàro-Perron integral*, Proc. London Math. Soc. (2) **34** (1932), 314–322.

[Bur3] Burkill, J. C., *Integrals and trigonometric series*, Proc. London Math. Soc. (3) **1** (1951), 46–57.

[Ca] Carathéodory, C., Vorlesungen über reelle Funktionen, 3rd ed., Chelsea, New York, 1968.

[Car] Cartan, H., *Differential Calculus*, Hermann, Paris, 1971.

[Ce] Cellérier, C., *Note sur les principes fondamentaux de l'analyse*, Bull. Sci. Math. (2) **14** (1890), 142–160.

[C1] Choquet, G., *Application des propriétés descriptives de la fonction contingent à la théorie des fonctions de variable réelle et à la géometrie différentielle des variétés cartésiennes*, J. Math. Pures Appl. (9) **26** (1947), 115–226.

[C2] Choquet, G., *Convergences*, Ann. Univ. Grenoble **23** (1948), 57–112.

[C3] Choquet, G., *Topology*, Academic Press, New York, 1966.

[Cl] Clarke, F. H., *Optimization and Nonsmooth Analysis*, Wiley-Interscience, New York, 1983.

[CEL] Crandall, M. G., Evans, L. C., and Lions, P. L., *Some properties of viscosity solutions of Hamilton-Jacobi equations*, Trans. Amer. Math. Soc. **282** (1984), 487–502.

[CL] Crandall, M. G., and Lions, P. L., *Viscosity solutions of Hamilton–Jacobi equations*, Trans. Amer. Math. Soc. **277** (1983), 1–42.

[Cs] Császár, A., *Sur une généralization de la notion de dérivée*, Acta Sci. Math. (Szeged) **16** (1955), 137–159.

[Da] Darboux, G., *Mémoire sur la théorie des fonctions discontinues*, Ann. Sci. Ecole Norm. Sup. (2) **4** (1875), 57–112.

[D1] Denjoy, A., *Mémoire sur les nombres dérivés des fonctions continues*, J. Math. Pures Appl. (7) **1** (1915), 105–240.

[D2] Denjoy, A., *Sur les fonctions dérivées sommables*, Bull. Soc. Math. France **34** (1915), 161–248.

[D3] Denjoy, A., *Sur une propriété des fonctions dérivées*, Enseignement Math. **18** (1916), 320–328.

[D4] Denjoy, A., *Mémoire sur la totalisation des nombres dérivés non-sommables*, Ann. École Norm. Sup. **33** (1916), 127–222; **34** (1917), 181–238.

[D5] Denjoy, A., *Leçons sur le calcul des coefficients d'une série trigonométrique*. II$^{e}$ partie, Métrique et topologie d'ensembles parfaits et de fonctions, Gauthier-Villars, Paris 1941.

[DGZ] Deville, R., Godefroy, G., and Zizler, V., *Smoothness and Renormings in Banach Spaces*, Pitman Monographs and Surveys in Pure and Applied Mathematics 64, 1993.

[Di] Dini, U., *Fondamenti per la teorica della funzioni di variabili reali*, Pisa, 1878.

[DEK] Dvoretzky, A., Erdős, P., and Kakutani, S., *Nonincrease everywhere of the Brownian motion process*, in Proceeding of the Fourth Berkeley Symposium on Mathematics, Statistics, and Probability, 1960, Vol. II, University of California Press, Berkeley, 1961, 103–116.

[E] Ellis, H.W., *Darboux properties and applications to non-absolutely convergent integrals*, Can. J. Math. **3** (1951), 471–485.

[ES] Esser, M., and Shisha, O., *A modified differentiation*, Amer. Math. Monthly **71** (1964), 904–906.

[Ev] Evans, M. J., *On continuous functions and the approximate symmetric derivatives*, Colloq. Math. **31** (1974), 129–136.

[F] Filipczak, F. M., *Sur la structure de l'ensemble des points où une fonction continue n'admet pas de dérivée symmétrique*, Dissertationes Math. Vol. **130** (1976), 49 pp.

[F.L1] Filipczak, L., *Exemple d'une fonction continue privée de dérivée symmétrique partout*, Colloq. Math. **20** (1969), 249–253.

[F.L2] Filipczak, L., *On continuous function without unilateral and symmetrical derivatives*, Acta Univ. Lodz., Zeszyty Nauk. Uniw. Lodz., Nauk. Matem. —Przyrodnicze, Ser. II, Zeszyt **34** (1980), 161–223.

[Fla] Flatto, L., *Advanced Calculus*, Williams & Wilkins, Baltimore, 1976.

[Fr] Freedman, D., *Brownian Motion and Diffusion*, Springer-Verlag, New York/Berlin, 1983.

[FR] Freiling, C., and Rinne, D., *A symmetric density property: monotonicity and the approximate symmetric derivative*, Proc. Amer. Math. Soc. **104** (1988), 1098–1102.

[Ga] Gál, I. S., On the fundamental theorems of the calculus, Trans. Amer. Math. Soc. **86** (1957), 309–320.

[G1] Garg, K. M., *An analogue of Denjoy's theorem*, Ganita **12** (1961), 9–14.

[G2] Garg, K. M., *On nowhere monotone functions. I. Derivates at a residual set*, Ann. Univ. Sci. Budapest. Eötvös Sect. Math. **5** (1962), 173–177.

[G3] Garg, K. M., *On nowhere monotone functions. II. Derivates at sets of power c and at sets of positive measure*, Rev. Math. Pures Appl. (Bucharest) **7** (1962) 663–671.

[G4] Garg, K. M., *On nonwhere monotone functions. III. Functions of first and second species*, Rev. Math. Pures Appl. (Bucharest) **8** (1963), 83–90.

[G5] Garg, K. M., *Applications of Denjoy analogue. II. Local structure of level sets and Dini derivates*, Acta Math. Acad. Sci. Hungar. **14** (1963), 183–186.

[G6] Garg, K. M., *Applications of Denjoy analogue. III. Distribution of various typical level sets*, Acta Math. Acad. Sci. Hungar. **14** (1963), 187–195.

[G7] Garg, K. M., *Applications of Denjoy analogue. I. Sufficient conditions for a function to be monotone*, Ann. Polon. Math. **15** (1964), 159–165.

[G8] Garg, K. M., *On asymmetrical derivates of non-differentiable functions*, Can. J. Math. **20** (1968), 135–143.

[G9] Garg, K. M., *On singular functions*, Rev. Roumaine Math. Pures Appl. **14** (1969), 1441–1452.

[G10] Garg, K. M., *On a residual set of continuous functions*, Czechoslovak Math. J. **20** (95) (1970), 537–543.

[G11] Garg, K. M., *On bilateral derivates and the derivative*, Trans. Amer. Math. Soc. **210** (1975), 295–329.

[G12] Garg, K. M., *On a new definition of derivative*, Bull. Amer. Math. Soc. **82** (1976) no. 5, 768–770.

[G13] Garg, K. M., *A new notion of derivative*, Real Analysis Exchange **7** (1981/82), 65–84.

[G14] Garg, K. M., *A general nonseparable theory of functions and multifunctions*, Real Analysis Exchange **9** (1983/84), 317–335.

[G15] Garg, K. M., *Construction of absolutely continuous and singular functions that are nowhere of monotonic type*, Contemp. Math. **42** (1985), 61–79.

[G16] Garg, K. M., *Relativization of some aspects of the theory of functions of bounded variation*, Dissertationes Math. **320** (1992), 123 pp.

[G17] Garg, K. M., *Some unified axiomatic theories of generalized derivatives*, Real Analysis Exchange **23** (1) (1997/98), 59–76.

[G18] Garg, K. M., *A unified theory of bilateral derivates*, to appear.

[G19] Garg, K. M., *A unified nonseparable theory of functions and multifunctions*, to appear.

[G20] Garg, K. M., *An abstract theory of derivates and derivative*, to appear.

[GO] Gelbaum, B. R., and Olmsted, J. M. H., Counterexamples in Analysis, Holden-Day, San Francisco, 1964.

[GMT] De Giorgi, E., Marino, A., and Tosques, M., *Problemi di evoluzione in spazi metrici e curve di massima pendenza*, Rend. Classe Sci. Fis. Mat. Nat. Acad. Naz. Lincei **68** (1980), 180–187.

[Go] Goldowsky, G., *Note sur les dérivées exactes*, Mat. Sb. **35** (1928), 35–36.

[Ha1] Hahn, H., *Über halbstetige und unstetige Funktionen*, Wiener Akad. Ber. **126** (1917), 91–110.

[Ha2] Hahn, H., *Theorie der reellen Funktionen*, Vol. I, Berlin, 1921.

[Hak] Hake, H., *Ueber de la Vallée Poussins Ober. und Unterfunktionen einfacher Integrale und die Integraldefinition von Perron*, Math. Ann. **83** (1921), 119–142.

[Hal] Halmos, P. R., *Measure Theory*, Van Nostrand, New York, 1950.

[Har] Hardy, G. H., *Weierstrass's non-differentiable function*, Trans. Amer. Math. Soc. **17** (1916), 301–325.

[Hau1] Hausdorff, F., *Grundzuge der Mengenlehre* Verlag von Viet, Leipzig, 1914.

[Hau2] Hausdorff, F., *Uber halbstetige Funktionen und deren Verallgemeincrung*, Math. Zeit. **5** (1919), 292–309.

[Hau3] Hausdorff, F., *Set Theory*, Chelsea, New York, 1957.

[HS] Hewitt, E., and Stromberg, K., *Real and Abstract Analysis*, Springer-Verlag, New York, 1965.

[H1] Hobson, E. W., *The Theory of Functions of a Real Variable and the Theory of Fourier's Series*. Vol. I, Dover, New York, 1957.

[H2] Hobson, E. W., *The Theory of Functions of a Real Variable and the Theory of Fourier's Series*. Vol. II, Dover, New York, 1957.

[H3] Hobson, E. W., et al., *Squaring the Circle, and Other Monographs*, Chelsea, New York, 1953.

[I1] Ioffe, A. D., *Calculus of Dini subdifferentials of functions and contingent coderivatives of set valued maps*, Nonlinear Analysis: Theory, Methods & Applications **8** (1984), 517–539.

[I2] Ioffe, A. D., *Approximate subdifferentials and applications I: The finite dimensional theory*, Trans. Amer. Math. Soc. **281** (1984), 390–416.

[I-T] Ionescu Tulcea, C. T., *Sur l'intégration des nombres dérivés*, C. R. Acad. Sci. Paris **225** (1947), 558–560.

[Io1] Iosifescu, M., *Propriétés différentielles des fonctions réelles d'une variable réelle, jouissant de la propriété de Darboux*, C. R. Acad. Sci. Paris **248** (1959), 1918–1919.

[Io2] Iosifescu, M., *Über eine Erweiterung eines Satzes von S. Stoilow*, Rev. Math. Pures Appl. **4** (1959), 725–729.

[J1] Jarník, V., *Über die Differenzierbarkeit stetiger Funktionen*, Fund. Math. **21** (1933), 48–58.

[J2] Jarník, V., *Sur la dérivabilité des fonctions continues*, Spisy Přírodov, Fakulty Univ. Karlovy, No. 129 (1934), 3–9.

[Ke] Kempisty, S., *Sur les fonctions quasicontinues*, Fund. Math. **19** (1932), 184–197.

[Kh] Khintchine, A., *Recherches sur la structure des fonctions mesurables*, Fund. Math. **9** (1927), 212–279.

[Kni] Knight, F. B., *Essentials of Brownian Motion and Diffusion*, Mathematical Surveys, No. 18, American Mathematical Soc., Providence, 1981.

[Kn] Knopp, K., *Ein einfaches Verfahren zur Bildung stetiger nirgends differenzierbarer Funktionen*, Math. Zeitschr. 2 (1918), 1–26.

[KV1] Kolmogoroff, A., and Verčenko, J., *Ueber Unstetigkeitspunkte von Funktionen zweier Veränderlichen*, C. R. Acad. Sci. URSS **1** (1934), 1–3, 105–107.

[KV2] Kolmogoroff, A., and Verčenko, J., *Weitere Untersuchungen über Unstetigkeitspunkte von Funktionen zweier Veränderlichen*, C. R. Acad. Sci. URSS **4** (1934), 361–364.

[Ko] Kostyrko, P., *On the symmetric derivative*, Colloq. Math. **25** (1972), 265–267.

[Ku] Kubota, Y., *A generalized derivative and integrals of the Perron type*, Proc. Japan Acad. **41** (1965), 443–448.

[K1] Kuratowski, K., *Topology*, Vol. I, Academic Press, New York; PWN, Warsaw, 1966.

[K2] Kuratowski, K., *Topology*, Vol. II, Academic Press, New York; PWN, Warsaw, 1968.

[K3] Kuratowski, K., *On set-valued B-measurable mappings and a theorem of Hausdorff*, Theory of Sets and Topology, Berlin, 1972, pp. 355–362.

[K4] Kuratowski, K., *A general approach to the theory of set-valued mappings*, General Topology and Its Relation to Modern Analysis and Algebra, III, Academia, Prague 1972, pp. 271–280.

[L] Lebesgue, H., *Leçons sur l'intégration et la recherche des fonctions primitives*, 2me éd., Gauthier-Villars, Paris, 1928.

[Le] Leonard, J. L., *Some conditions implying the monotonicity of a real function*, Rev. Roumaine Math. Pures Appl. **17** (1972), 757–780.

[Lev] Levy, B., *Ancora alcune osservazioni sulle funzioni derivata*, Atti. Accad. Naz. Lincei Rend. Cl. Sci. Fis. Mat. Natur. (5) **15** (1906), 358–368.

[Li] Lions, P. L., *Generalized Solutions of Hamilton–Jacobi Equations*, Pitman, Boston, 1982.

[Lo] Looman, H., *Ueber die Perronsche Integraldefinition*, Math. Ann. **93** (1925), 153–156.

[Lu1] Lusin, N., *Sur les propriétés des fonctions mesurables*, C. R. Acad. Sci. Paris **154** (1912), 1688–1690.

[Lu2] Lusin, N., *Intégrale et série trigonometrique*, Moscow, 1915 (Russian).

[Man1] Manna, N. C., *On some properties of nowhere monotone semi-continuous functions*, Bull. Math. Soc. Sci. Math. R.S. Roumanie **13** (61) (1969), 65–72.

[Man2] Manna, N. C., *Dini derivatives of semi-continuous functions. II*, Rev. Roumaine Math. Pures Appl. **15** (1970), 75–80.

[Man3] Manna, N. C., *On a generalization of smooth and symmetric functions*, Colloq. Math. **22** (1971), 291–299.

[MM] Manna, N. C., and Mukhopadhyay, S. N., *Dini derivatives of semicontinuous functions. I*, Acta Sci. Math. (Szeged) **30** (1969), 93–97.

[Ma] Marchaud, A., *Sur une condition de quasi-rectifiabilité*, Fund. Math. **20** (1933), 105–116.

[Mar] Marcinkiewicz, J., *Sur les séries de Fourier*, Fund. Math. **27** (1936), 38–69.

[MZ] Marcinkiewicz, J., and Zygmund, A., On the differentiability of functions and summability of trigonometrical series, Fund. Math. **26** (1936), 1–43.

[M1] Marcus, S., *An analysis of real-valued functions based on the notion of category in Baire's sense* (Romanian), Studii Circ. Mat. **7** (1956), 251–272.

[M2] Marcus, S., *Sur un théorème de M. S. Stoilow, concernant les fonctions continues d'une variable reele*, Rev. Math. Pures Appl. **2** (1957), 409–412.

[M3] Marcus, S., *Sur un théorème de M. A. Marchaud et sur les fonctions dérivables presque partout*, C. R. Acad. Sci. Paris **244** (1957), 2345–2347.

[M4] Marcus, S., *Sur les ensembles stationnaires des fonctions dérivées—finie ou infinies*, Com. Acad. R.P. Romîne **12** (1962), 399–402. (Romanian).

[Maz1] Mazurkiewicz, S., *Sur la relation entre l'existence de la dérivée seconde generalisée et la continuité de la fonction*, Prace Mat. Fiz. **30** (1919), 225–242. (Polish).

[Maz2] Mazurkiewicz, S., *Sur la dérivée première généralisée*, Fund. Math. **11** (1928), 145–147.

[Maz3] Mazurkiewicz, S., *Sur les fonctions non dérivables*, Studia Math. **3** (1931), 92–94.

[Mc1] McShane, E. J., *On Perron integration*, Bull. Amer. Math. Soc. **48** (1942), 718–726.

[Mc2] McShane, E. J., *Integration*, Princeton University Press, Princeton, 1944.

[Mi] Minakshisundaram, S., *On the roots of a continuous non-differentiable function*, J. Indian Math. Soc. **4** (1940), 31–33.

[MSA] Mirică, S., Staicu, V., and Angelescu, N., *Fréchet semi-differentials of real-valued functions*, Instituto Mathematico 'Ulisse Dini', Viale Morgani, 67-A-50134 Firenze, 1987, 47 pp.

[Mis1] Mišík, L., *Halbborelsche Funktionen und extreme Ableitungen*, Math. Slov. **27** (1977), 409–421.

[Mis2] Mišík, L., *Extreme essential derivatives of Borel and Lebesgue measurable functions*, Math. Slov. **29** (1979), 25–38.

[Mo1] Morse, A. P., *A continuous function with no unilateral derivatives*, Trans. Amer. Math. Soc. **44** (1938), 496–507.

[Mo2] Morse, A. P., *Dini derivatives of continuous functions*, Proc. Amer. Math. Soc. **5** (1954), 126–130.

[Mu] Mukhopadhyay, S. N., *On approximate Schwarz differentiability*, Monatsh. Math. **70** (1966), 454–460.

[N1] Natanson, I. P., *Theory of Functions of a Real Variable*, Vol. I, Ungar, New York, 1955.

[N2] Natanson, I. P., *Theory of Functions of a Real Variable*, Vol. II, Ungar, New York, 1960.

[Ne1] Neugebauer, C. J., *A theorem on derivates*, Acta Sci. Math. (Szeged) **23** (1962), 79–81.

[Ne2] Neugebauer, C. J., *Symmetric, continuous, and smooth functions*, Duke Math. J. **31** (1964), 23–31.

[Ne3] Neugebauer, C. J., *Smoothness and differentiability in $L_p$*, Studia Math. **25** (1964), 81–91.

[O-M] O'Malley, R. J., *Selective derivates*, Acta. Math. Acad. Sci. Hung. **29** (1977), 77–97.

[O] Ostrowski, A., *Differential and Integral Calculus*, Scott, Foresman, Glenview, Illinois, 1968.

[P] Peano, G., *Sur la définition de la dérivée*, Mathesis (2) **2** (1892), 12–14.

[Pe] Perron, O., *Ueber den Integralbegriff*, S. B. Heidelberg. Akad. Wiss. **16** (1914).

[Pr1] Preiss, D., *Approximate derivatives and Baire classes*, Czechoslovak Math. J. **21** (96) (1971), 373–382.

[Pr2] Preiss, D., *Differentiability of Lipschitz functions on Banach spaces*, J. Functional Analysis **91** (1990), 312–345.

[R] Rademacher, H., *Eineindeutige Abbildungen und Messbarkeit*, Monatschefte Math. Phys. **27** (1916), 183–291.

[Ri1] Ridder, J., *Ueber Definitionen von Perron-Integralen. I*, Indag. Math. **9** (1947), 227–235.

[Ri2] Ridder, J., *Ueber Definitionen von Perron-Integralen. II*, Indag. Math. **9** (1947), 280–289.

[Rie] Riemann, B., *Über die Darstellbarkeit einer Funktion durch eine trigonometrische Reihe*, Ges. Werke. 2 Aufl. Leipzig (1892), 227–271.

[Ro1] Rockafellar, R. T., *Convex Analysis*, Princeton University Press, Princeton, 1970.

[Ro2] Rockafellar, R. T., *The Theory of Subgradients and Its Applications to Problems of Optimization: Convex and Nonconvex Functions*, Helderman Verlag, Berlin, 1981.

[RW] Rockafellar, R. T., and Wets, R. J. B., *Variational Analysis*, Springer-Verlag, New York, 1998.

[Roy] Royden, H. L., *Real Analysis*, 2nd ed., Macmillan, New York, 1968.

[Ru] Rudin, W., *Real and Complex Analysis*, McGraw-Hill, New York, 1966.

[S1] Saks, S., *Sur les nombres dérivés des fonctions*, Fund. Math. **5** (1924), 98–104.

[S2] Saks, S., *Sur un théorème de M. Lusin*, Fund. Math. **6** (1924), 111–116.

[S3] Saks, S., *Sur l'intégrale de M. Denjoy*, Fund. Math. **15** (1930), 242–262.

[S4] Saks, S., *Sur certaines classes de fonctions continues*, Fund. Math. **17** (1931), 124–151.

[S5] Saks, S., *On the functions of Besicovitch in the space of continuous functions*, Fund. Math. **19** (1932), 211–219.

[S6] Saks, S., *Théorie de l'intégrale*, Monografie Mat., Vol. 2, PWN, Warsaw, 1933.

[S7] Saks, S., *Sur quelque propriétés métriques d'ensembles*, Fund. Math. **26** (1936), 234–240.

[S8] Saks, S., *Theory of the Integral*, 2nd ed., Monografie Mat., Vol. 7, PWN, Warsaw, 1937.

[Sa1] Sargent, W. L. C., *The Borel derivates of a function*, Proc. London Math. Soc. (2) **38** (1934), 180–196.

[Sa2] Sargent, W. L. C., *A descriptive definition of Cesáro–Perron integrals*, Proc. London Math. Soc. **47** (1941), 212–247.

[Se] Sen, H. K., Darboux's property and its applications, Proc. Benares Math. Soc. (N.S.) **2** (1940), 17–23.

[Si1] Sierpiński, W., *Sur l'ensemble des points angulaires d'une courbe $y = f(x)$*, Bull. Acad. Sci. Cracovie, Sér. A (1912), 850–855.

[Si2] Sierpiński, W., *Un théorème sur les fonctions dérivées*, Bull. Acad. Sci. Cracovie, Sér. A (1914), 205–217.

[Si3] Sierpiński, W., *Sur les fonctions dérivées des fonctions discontinues*, Fund. Math. **3** (1922), 123–127.

[Sin1] Sindalovskii, G. H., *Continuity and differentiability with respect to congruent sets*, Soviet Math. Dokl. **1** (1961), 1217–1218.

[Sin2] Sindalovskii, G. H., *Differentiability with respect to congruent sets*, Izv. Akad. Nauk SSSR Ser. Mat. **29** (1965), 11–40.

[Sing1] Singh, A. N., *The Theory and Construction of Non-differentiable Functions*, Lucknow University, Lucknow, 1935.

[Sing2] Singh, A. N., *On functions without one-sided derivatives II*, Proc. Benares Math. Soc. (N.S.) **4** (1942), 95–108.

[SZ] Stein, E. M., and Zygmund, A., On the differentiability of functions, Studia Math. **23** (1964), 247–283.

[St1] Stoilow, S., *Sur les transformations continues d'une variable*, C. R. Acad. Sci. Paris **179** (1924), 807–810.

[St2] Stoilow, S., *Sur l'ensemble où une fonction continue a une valeur constante*, C. R. Acad. Sci. Paris **179** (1924), 1585–1586.

[St3] Stoilow, S., *Sur l'inversion des fonctions continues*, Bull. Soc. Math. France **53** (1925), 135–148.

[SU] Sunouchi, G., and Utagawa, M., *The generalized Perron integrals*, Tohoku Math. J. **1** (1949), 95–99.

[Sw] Świątkowski, T., *On the conditions of monotonicity of functions*, Fund. Math. **59** (1966), 189–201.

[Swo] Swokowski, E. W., *Calculus with Analytic Geometry*, 2nd ed., Prindle, Weber & Schmidt, Boston, 1979.

[T1] Thomson, B. S., *Real Functions*, Lecture Notes in Mathematics #1170, Springer-Verlag, 1985.

[T2] Thomson, B. S., *Symmetric Properties of Real Functions*, Marcel Dekker, New York, 1994.

[To] Tolstoff, G., *Sur la dérivée approximative exacte*, Mat. Sb. **4** (46) (1938), 499–504.

[Ton] Tonelli, L., *Sulle derivate esatte*, Mem. Accad. Sci. Ist. Bologna (8) **8** (1930/31), 13–15.

[V] Vallée Poussin, Ch. J. de la, *Intégrales de Lebesgue. Fonctions d'ensemble.* Classes de Baire, Paris, 1916.

[Vi] Vitali, G., *Sulle funzioni continue*, Fund. Math. **8** (1926), 175–188.

[W] Ward, A. J., *On the points where* $AD^+ < AD_-$, J. London Math. Soc. **8** (1933), 293–299.

[Wi] Willard, S., *General Topology*, Addison-Wesley, Reading, Massachusettes, 1970.

[Y.G1] Young, G. C., *A note on derivates and differential coefficients*, Acta Math. **37** (1914) 141–154.

[Y.G2] Young, G. C., *On infinite derivates*, Quart. J. Math. Oxford Ser. **47** (1916), 127–175.

[Y.G3] Young, G. C., *On the derivates of a function*, Proc. London Math. Soc. (2) **15** (1916), 360–384.

[Y1] Young, W. H., *On the infinite derivates of a function of a single real variable*, Ark. Mat. Astr. Fys. **1** (1903), 201–204.

[Y2] Young, W. H., *On the distinction of right and left at points of discontinuity*, Quart. J. Math. Oxford Ser. **39** (1908), 67–83.

[Y3] Young, W. H., *Oscillating successions of continuous functions*, Proc. London Math. Soc. (2) **6** (1908), 298–320.

[Y4] Young, W. H., *On the derivates of non-differentiable functions*, Messenger of Math. **38** (1908), 65–69.

[YY] Young, W. H., and Young, G. C., *On derivates and the theorem of the mean*, Quart. J. Math. Oxford Ser. **40** (1909), 1–26, 144–145.

[Z1] Zahorski, Z., *Punktmengen, in welchen eine stetige Funktion nicht differenzierbar ist*, Mat. Sb. **9** (51) (1941), 487–510 (Russian).

[Z2] Zahorski, Z., *Sur l'ensemble des points de non-dérivabilité d'une fonction continue*, Bull. Soc. Math. France **74** (1946), 147–178.

[Z3] Zahorski, Z., *O zbiorze punktów nieróżniczkowalności funkcji dowolnej*, Rocznika Polskiego Towarzystwa Matematycznego **21** (1947), 23–26.

[Z4] Zahorski, Z., *Sur la première dérivée*, Trans. Amer. Math. Soc. **69** (1950), 1–54.

[Za1] Zajiček, L., *On the intersection of the sets of right and left internal approximate derivatives*, Czech. Math. J. **23** (98) (1973), 521–526.

[Za2] Zajiček, L., *The differentiability structure of typical functions in* $C[0, 1]$, Real Analysis Exchange **13** (1987-88), pp. 119, 103–106, 93.

[Zy1] Zygmund, A., *Smooth functions*, Duke Math. J. **12** (1945), 47–76.

[Zy2] Zygmund, A., *Trigonometric Series*, Vol. I, Cambridge Univ. Press, New York and London, 1968.

[Zy3] Zygmund, A., *Trigonometric Series*, Vol. II, Cambridge Univ. Press, New York and London, 1968.

# Index of Symbols

## General:

## Entities related to $f$:



## Weaker forms of continuity and internal properties:

## Properties of functions:

## Derivates and derivatives:



## Medians of functions:



## Derivability sets:

## Generalized derivability sets:



## Integrals:

# Subject Index

## CANADIAN MATHEMATICAL SOCIETY SERIES OF MONOGRAPHS AND ADVANCED TEXTS

*Monographies et Études de la Société Mathématique du Canada*

**EDITORS/RÉDACTEURS: Jonathan M. Borwein and Peter B. Borwein**

| Author | Title | Vol. No. |
|---|---|---|
| Frank H. Clarke | ***Optimization and Nonsmooth Analysis** | 1 |
| Erwin Klein and Anthony C. Thompson | ***Theory of Correspondences: Including Applications to Mathematical Economics** | 2 |
| I. Gohberg, P. Lancaster, and L. Rodman | **Invariant Subspaces of Matrices with Applications** | 3 |
| Jonathan M. Borwein and Peter B. Borwein | †**Pi and the AGM—A Study in Analytic Number Theory and Computational Complexity** | 4 |
| Subhashis Nag | **The Complex Analytic Theory of Teichmüller Spaces** | 5 |
| Manfred Kracht and Erwin Kreyszig | ***Methods of Complex Analysis in Partial Differential Equations with Applications** | 6 |
| Ernest J. Kani and Robert A. Smith | ***The Collected Papers of Hans Arnold Heilbronn** | 7 |
| Victor P. Snaith | ***Topological Methods in Galois Representation Theory** | 8 |
| Kalathoor Varadarajan | **The Finiteness Obstruction of CTC Wall** | 9 |
| John H. Berglund, Hugo D. Jünghenn, and Paul Milne | ***Analysis of Semigroups: Function Spaces, Compactifications, Representations** | 10 |
| Robert V. Moody and Arturo Pianzola | **Lie Algebras with Triangular Decompositions** | 11 |
| F. A. Sherk, P. McMullen, A. Thompson, and A. Weiss | **Kaleidoscopes: Selected Writings of H. S. M. Coxeter** | 12 |
| Alf van der Poorten | **Notes on Fermat's Last Theorem** | 13 |
| Peter A. Fillmore | **A User's Guide to Operator Algebras** | 14 |
| Peter Lancaster and Kęstutis Šalkauskas | **Transform Methods in Applied Mathematics: An Introduction** | 15 |
| Ole A. Nielsen | **An Introduction to Integration and Measure Theory** | 16 |

*Indicates an out-of-print title. *Indique un titre épuisé.*

†This title is now available in paperback. *Une version à couverture souple est maintenant disponible.*

Canadian Mathematical Society Series *(Continued)*
*Collection d'ouvrages de la Société Mathématique du Canada (suite)*

| Author | Title | Vol. No. |
|---|---|---|
| Wieslaw Krawcewicz and Jianhong Wu | **Theory of Degrees with Application to Bifurcations and Differential Equations** | 17 |
| Marián J. Fabian | **Gâteaux Differentiability of Convex Functions and Topology: Weak Asplund Spaces** | 18 |
| Glen D. Anderson, Mavina K. Vamanamurthy, and Matti K. Vuorinen | **Conformal Invariants, Inequalities, and Quasiconformal Maps** | 19 |
| Ivan Singer | **Abstract Convex Analysis** | 20 |
| Bruce C. Berndt, Ronald J. Evans, and Kenneth S. Williams | **Gauss and Jacobi Sums** | 21 |
| Hugh C. Williams | **Édouard Lucas and Primality Testing** | 22 |
| Mario Petrich | **Completely Regular Semigroups** | 23 |
| Krishna M. Garg | **Theory of Differentiation: A Unified Theory of Differentiation Via New Derivate Theorems and New Derivatives** | 24 |

Printed and bound by CPI Group (UK) Ltd, Croydon, CR0 4YY
16/09/2026
14963441-0008